Proceedings of the 2nd International Conference of the UISPP Commission on Flint Mining in Pre- and Protohistoric Times (Madrid, 14-17 October 2009)

Edited by

Marta Capote
Susana Consuegra
Pedro Díaz-del-Río
Xavier Terradas

BAR International Series 2260
2011

Published in 2016 by
BAR Publishing, Oxford

BAR International Series 2260

Proceedings of the 2nd International Conference of the UISPP Commission on Flint Mining in Pre- and Protohistoric Times (Madrid, 14-17 October 2009)

ISBN 978 1 4073 0831 9

COVER IMAGE *Last mining event at Casa Montero, Madrid (c. 5200 cal BC). Illustration by Juan Álvarez-Cebrián*

BAR Publishing is the trading name of British Archaeological Reports (Oxford) Ltd. British Archaeological Reports was first incorporated in 1974 to publish the BAR Series, International and British. In 1992 Hadrian Books Ltd became part of the BAR group. This volume was originally published by Archaeopress in conjunction with British Archaeological Reports (Oxford) Ltd / Hadrian Books Ltd, the Series principal publisher, in 2011. This present volume is published by BAR Publishing, 2016.

Printed in England

BAR
PUBLISHING

BAR titles are available from:

BAR Publishing
122 Banbury Rd, Oxford, OX2 7BP, UK
EMAIL info@barpublishing.com
PHONE +44 (0)1865 310431
FAX +44 (0)1865 316916
www.barpublishing.com

Photograph (July 2008) courtesy of Dr. W. Schins, Chairman of the Limburg Section of the Dutch Geological Society, shows Werner to the left and Sjeuf to the right.

In memory of P.J. (Sjeuf) Felder (28.02.1928 - 15.05.2009) and Werner M. Felder (27.09.1930 - 15.12.2008)

Contents

Contributors

Kamil Adamczak
The Institute of Archaeology and Ethnology
Polish Academy of Sciences
Al. Solidarności 105
00-140 Warsaw (Poland)
kamil.adamczak@gazeta.pl

Manuel Alcaraz
Área de Prehistoria,
Universidad de Alcalá de Henares
C/ Colegios 2
28801 Alcalá de Henares, Madrid (Spain)

Marco António Andrade
MA in Prehistory and Archaeology
UNIARQ – Centro de Arqueologia da Universidade de Lisboa (Ancient Peasant Societies Workgroup)
Faculdade de Letras, Alameda da Universidade
P-1600-214 Lisboa (Portugal)
folha-de-acacia@iol.pt

Thomas Bachnetzer
Institut für Archäologien.
Universität Innsbruck
Langer Weg 11
A-6020 Innsbruck (Austria)

Javier Baena Preysler
Dep. Prehistoria y Arqueología,
Universidad Autónoma de Madrid
Campus de Cantoblanco.
28049 Madrid (Spain)
javier.Baena@uam.es

Sergio Bárez
Arquex S.L.
Sector Embarcaciones, 10. Local 2
28760 Tres Cantos, Madrid (Spain)
http://www.arquex.es

Ran Barkai
Institute of Archaeology,
Tel-Aviv University
69978, Tel-Aviv (Israel)
barkaran@post.tau.ac.il

Alfonso Benito-Calvo
Centro Nacional de Investigación sobre la Evolución Humana (CENIEH)
Paseo Sierra de Atapuerca s/n
09002, Burgos (Spain)
alfonso.benito@cenieh.es

Andrzej Boguszewski
17, route de Toulouse
65200 Bagnères de Bigorre (France)
andrzej.boguszewski@inrap.fr

Marta Capote
G.I. Prehistoria Social y Económica
Instituto de Historia, Centro de Ciencias Humanas y Sociales, CSIC
C/ Albasanz 26-28.
28037 Madrid (Spain)
marta.capote@cchs.csic.es

François Charraud
Doctorant, Université de Nice-Sophia Antiplois, UMR 6130-CEPAM
Boulevard de l'Europe.
14540 Bourguébus (France)
fcharraud@gmail.com

Ignacio Clemente
Consejo Superior de Investigaciones Científicas
Departamento de Arqueología y Antropología, Inst. «Milá y Fontanals»
c/ Egipcíaques, 15
08001 Barcelona (Spain)
ignacio@imf.csic.es

Susana Consuegra
G.I. Prehistoria Social y Económica
Instituto de Historia, Centro de Ciencias Humanas y Sociales, CSIC
C/ Albasanz 26-28.
28037 Madrid (Spain)
susana.consuegra@cchs.csic.es

Felipe Cuartero
Dep. Prehistoria y Arqueología
Universidad Autónoma de Madrid
Campus de Cantoblanco
28049 Madrid (Spain)

Valérie Deloze
Inrap – National Institute of Preventive Archaeological Research
Centre de Recherches Archéologiques du Mans
20 rue Hippolyte-Foucault
72200 Le Mans (France)

Pedro Díaz-del-Río
G.I. Prehistoria Social y Económica
Instituto de Historia, Centro de Ciencias Humanas y Sociales, CSIC
C/ Albasanz 26-28.
28037 Madrid (Spain)
pedro.diazdelrio@cchs.csic.es

Salvador Domínguez-Bella
Departamento de Ciencias de la Tierra. Facultad de Ciencias. Universidad de Cádiz.
Campus Rio San Pedro. 11510. Puerto Real, Cádiz (Spain)
salvador.dominguez@uca.es

Ana Escobar
Dep. Prehistoria y Arqueología
Universidad Autónoma de Madrid
Campus de Cantoblanco
28049 Madrid (Spain)

David Field

English Heritage
Firefly Avenue, Swindon, Wiltshire SN2 2GZ (United Kingdom)
david.field@english-heritage.org.uk

Attilio Galiberti

Department of Archaeology and Arts History, Prehistory Section
University of Siena
Via Roma 56
53100 Siena (Italy)
galiberti@unisi.it

Erik Gallouin

Inrap – National Institute of Preventive Archaeological Research
Centre de Recherches Archéologiques de Bourguébus
4, Boulevard de l'Europe
14540 Bourguébus (France)

Andrés Gascue

Dpto. de Arqueología, Facultad de Humanidades,
Universidad de la República
Magallanes 1577, Montevideo 11200 (Uruguay)
andresgascue@gmail.com

Emmanuel Ghesquière

Inrap – National Institute of Preventive Archaeological Research
Centre de Recherches Archéologiques de Bourguébus
4, Boulevard de l'Europe
14540 Bourguébus (France)
emmanuel.ghesquiere@inrap.fr

David Giazzon

Inrap Basse-Normandie
Boulevard de l'Europe, 14540 Bourguébus (France)
david.giazzon@inrap.fr

Sébastien Giazzon

Inrap Basse-Normandie
Boulevard de l'Europe, 14540 Bourguébus (France)
sebastien.giazzon@inrap.fr

Juan F. Gibaja

Consejo Superior de Investigaciones Científicas
Departamento de Arqueología y Antropología,
Inst. «Milá y Fontanals»
c/ Egipcíaques, 15
08001 Barcelona (Spain)
jfgibaja@gmail.com

François Giligny

Université de Paris 1-Panthéon-Sorbonne,
Protohistoire européenne-Arscan, Maison de l'Archéologie et de l'Ethnologie,
21 allée de l'Université, 92023 Nanterre cedex (France)
francois.giligny@univ-paris1.fr

Victor S. Gonçalves

Grupo de estudos sobre as antigas sociedades camponesas.
Centro de Arqueologia da Universidade de Lisboa (UNIARQ)
Faculdade de Letras, Alameda da Universidade
P-1600-214 Lisboa (Portugal)
vsg@campus.ul.pt

Avi Gopher

Institute of Archaeology, Tel-Aviv University
69978, Tel-Aviv (Israel)
agopher@post.tau.ac.il

Iván González

Universidad Complutense de Madrid
Ciudad Universitaria
28040 Madrid (Spain)

Marjorie E. Th. de Grooth

Aspelweg 49
53902 Bad Münstereifel (Germany)
grooth@t-online.de

Laurent Juhel

Inrap Bretagne,
UMR 6566-CReAAH (Université de Rennes I, Rennes II, Nantes, CNRS, MCC)
37 rue du Bignon, 35577 Cesson Sévigné (France)
Laurent.juhel@inrap.fr

Iosu Junguitu Íñiguez de Heredia

Dpto. de Geografía, Prehistoria y Arqueología de la Universidad del País Vasco
C/ Tomás y Valiente s./n., 01006, Vitoria-Gasteiz (Spain)
iosujun@gmail.com

David Larreina García

Institute of Archaeology UCL
31-34 Gordon Square, WC1H 0PY, London (United Kingdom)
dplauto@gmail.com

Roel C. G. M. Lauwerier

Cultural Heritage Agency
P.O. Box 1600
3800 BP Amersfoort (The Netherlands)
R.Lauwerier@cultureelerfgoed.nl

Ana Lázaro

Dep. Prehistoria y Arqueología
Universidad Autónoma de Madrid
Campus de Cantoblanco
28049 Madrid (Spain)

Hanna and Jacek Lech

The Institute of Archaeology and Ethnology Polish Academy of Sciences
Al. Solidarności 105
00-140 Warsaw (Poland)
lech@iaepan.edu.pl

Walter Leitner

Institut für Archäologien. Universität Innsbruck
Langer Weg 11
A-6020 Innsbruck (Austria)

Pedro José Lobo Urrutia

Servicio de Cartografía de la Universidad del País Vasco (SGIKER)
C./ Miguel de Unamuno s./n., 01006, Vitoria-Gasteiz (Spain)
lobourrutiap@gmail.com

José López Mazz

Dpto. de Arqueología, Facultad de Humanidades, Universidad de la República.
Magallanes 1577,
Montevideo 11200 (Uruguay)
lopezmazz@yahoo.com.ar

José A. Lozano

Universidad de Granada, Facultad de Ciencias
Instituto Andaluz de Ciencias de la Tierra (CSIC-UGR)
Avda. Fuente Nueva s/n
E- 18071 Granada (Spain)
jalozano@ugr.es

Ludomir R. Lozny

Department of Anthropology, Hunter College, CUNY (USA)
Ludomir.Lozny@hunter.cuny.edu

Jolanta Małecka-Kukawka

Nicolaus Copernicus University
Institute of Archaeology
Szosa Bydgoska 44-48
PL 87-100 Toruń (Poland)
jkukawka@uni.torun.pl

Laurence Manolakakis

CNRS, Protohistoire européenne-Arscan,
Maison de l'Archéologie et de l'Ethnologie,
21 allée de l'Université
92023 Nanterre cedex (France)
laurence.manolakakis@mae.cnrs.fr

Carmen Manzano

Dep. Prehistoria y Arqueología
Universidad Autónoma de Madrid
Campus de Cantoblanco
28049 Madrid (Spain)

Cyril Marcigny

Inrap Basse-Normandie,
UMR 6566-CReAAH (Université de Rennes I, Rennes II, Nantes, CNRS, MCC)
Le Chaos, 14400 Longues-sur-Mer (France)
cyril.marcigny@inrap.fr

Raúl Márquez

Arquex S.L.
Sector Embarcaciones, 10 Local 2
28760 Tres Cantos, Madrid (Spain)
http://www.arquex.es

Javier Martínez López

Departamento de Ciencias de la Tierra. Facultad de Ciencias.
Universidad de Cádiz.
Campus Rio San Pedro. 11510.
Puerto Real, Cádiz (Spain)
javier.martinez@uca.es

Henrique Matias

MA Student in Geoarchaeology
Faculdade de Ciências da Universidade de Lisboa (Portugal)
hamatias@gmail.com

Fabrizio Mazzarocchi

Studio Tecnico Laboratorio di Geofisica Applicata e Ambientale
Via P. Sarcoli 28
58024 Massa Marittima, Grosseto (Italy).
fabri.mazzarocchi@tiscali.it

Antonio Morgado

Universidad de Granada,
Facultad de Filosofía y Letras,
Dpto. Prehistoria y Arqueología
Campus universitario de Cartuja s/n
E- 18071 Granada (Spain)
morgado@ugr.es

Martin Oliva

Moravian Museum – Anthropos Institute
Zelny trh 6, 659 37, Brno (Czechoslovakia)
moliva@mzm.cz

Irene Ortiz

Dep. Prehistoria y Arqueología
Universidad Autónoma de Madrid
Campus de Cantoblanco
28049 Madrid (Spain)

Alfredo Pérez-González

Centro Nacional de Investigación sobre
Evolución Humana (CENIEH)
Paseo Sierra de Atapuerca s/n
09002 Burgos (Spain)

Teresa Pérez

Universidad Complutense de Madrid.
Ciudad Universitaria
28040 Madrid (Spain)

Gustavo Piñeiro

Dpto. Geología, Facultad de Ciencias,
Universidad de la República
Iguá 4225
Montevideo (Uruguay)
gpineiro@g.mail

José Polo

Arquex S.L.
Sector Embarcaciones, 10 Local 2
28760 Tres Cantos, Madrid (Spain)
http://www.arquex.es

Laurent Quesnel

CNRS- National Center of Scientific Research
UMR 6566 CReAAH, Bât. 24-25
263, avenue du Général Leclerc
35042 Rennes (France)

José Ramos Muñoz

Departamento de Historia, Geografía y Filosofía.
Universidad de Cádiz.
Avda. Gómez Ulla s.n.
11010. Cádiz (Spain)
jose.ramos@uca.es

Marta Roca
Dep. Prehistoria y Arqueología
Universidad Autónoma de Madrid
Campus de Cantoblanco
28049 Madrid (Spain)

Pedro Rodríguez
Arquex S.L.
Sector Embarcaciones, 10 Local 2
28760 Tres Cantos, Madrid (Spain)
http://www.arquex.es

Daniel Rubio
Dep. Prehistoria y Arqueología
Universidad Autónoma de Madrid
Campus de Cantoblanco
28049 Madrid (Spain)

Inmaculada Rus
Dep. Prehistoria y Arqueología
Universidad Autónoma de Madrid
Campus de Cantoblanco
28049 Madrid (Spain)

Alan Saville
Senior Curator, Earliest Prehistory
Department of Archaeology. National Museums Scotland
Chambers Street
Edinburgh EH1 1JF
Scotland (United Kingdom)
a.saville@nms.ac.uk

Muuk E. ter Schegget
Cultural Heritage Agency
P.O. Box 1600
3800 BP Amersfoort (The Netherlands)
M.ter.Schegget@cultureelerfgoed.nl

Ana Catarina Sousa
Grupo de estudos sobre as antigas sociedades camponesas.
Centro de Arqueologia da Universidade de Lisboa (UNIARQ)
Faculdade de Letras, Alameda da Universidade
P-1600-214 Lisboa (Portugal)
sousa@campus.ul.pt

Markus Staudt
Institut für Archäologien. Universität Innsbruck
Langer Weg 11
A-6020 Innsbruck (Austria)

Massimo Tarantini
Department of Archaeology and Arts History, Prehistory Section
University of Siena
Via Roma 56
53100 Siena (Italy)
tarantini@unisi.it

Andoni Tarriño Vinagre
Centro Nacional de Investigación sobre la Evolución Humana (CENIEH)
Paseo Sierra de Atapuerca s/n,
09002, Burgos (Spain)
Antonio.tarrinno@cenieh.es

Xavier Terradas
Consejo Superior de Investigaciones Científicas
Departamento de Arqueología y Antropología, Inst. «Milá y Fontanals»
c/ Egipcíaques, 15
08001 Barcelona (Spain)
terradas@imf.csic.es

Peter Topping
English Heritage
37 Tanner Row
York, YO1 6WP (United Kingdom)
Pete.Topping@english-heritage.org.uk

Gerhard Trnka
Institut für Ur- und Frühgeschichte
Der Universität Wien
Franz Klein-Gasse 1
A – 1190 Wien (Austria)
Gerhard.Trnka@univie.ac.at

Rodrigue Tsobgou Ahoupe
Inrap – National Institute of Preventive Archaeological Research
37 rue du Bignon
CS 67737
35577 Cesson-Sévigné (France)
tsar500@hotmail.com

Dagmara Werra
The Institute of Archaeology and Ethnology Polish Academy of Sciences
Al. Solidarności 105
00-140 Warsaw (Poland)
dagmarawerra@yahoo.com

Scientific Committee (UISPP Commission)

Françoise Bostyn

Jacek Lech

Anne Hauzeur

Alan Saville

Pierre Allard

Organizing Committee

Susana Consuegra (CCHS, CSIC)

Pedro Díaz-del-Río (CCHS, CSIC)

Xavier Terradas (IMF, CSIC)

Scientific Secretariat (CCHS, CSIC)

Cristina Casas

Enrique Capdevila

Marta Capote

Nuria Castañeda

Cristina Criado

Aurora Nieto

Acknowledgements

The 2nd International Conference of the UISPP Commission on Flint mining in Pre- and Protohistoric Times was funded through the following projects and institutions: Proyecto de Investigación Arqueológica en el yacimiento de Casa Montero (Madrid)" (Autopista Madrid Sur Concesionaria Española S.A., Consejería de Cultura y Deportes de la Comunidad de Madrid, Consejo Superior de Investigaciones Científicas - CSIC); Acción Complementaria HAR2009-06777-E7HIST del Ministerio de Ciencia e Innovación; Centro de Ciencias Humanas y Sociales (CSIC); Ayuda 2009 del Consejo Superior de Investigaciones Científicas para la organización de Congresos y Reuniones de carácter nacional e internacional.

The present edition has been entirely funded by *"Proyecto de Investigación Arqueológica en el yacimiento de Casa Montero (Madrid)"* (Autopista Madrid Sur Concesionaria Española S.A., Consejería de Cultura y Deportes de la Comunidad de Madrid, Consejo Superior de Investigaciones Científicas - CSIC).

The following persons have participated in the elaboration of these Proceedings:

Edition of English texts: Tatiana Andronova and Adrian Burton.

Adequacy of manuscripts to norms: Verónica Balsera (CCHS, CSIC) and Rosa Millán (CCHS, CSIC)

Design and text formating: Laura Paz (CCHS, CSIC)

Setting the context

A brief introduction to the Proceedings of The 2nd International Conference of the UISPP Commission on Flint mining in Pre- and Protohistoric Times

Participants of the Madrid conference at the CCHS conference hall.

The *Union International des Sciences Préhistoriques e Protohistoriques* Commission on Flint mining in Pre- and Protohistoric Times was established during the UISPP XV Congress, held in Lisbon (Portugal, September 2006). It was during the following year in Paris (September 2007) when the Commission held its first Conference. It appointed Dr. Françoise Bostyn and Dr. Jacek Lech as president and vice-president respectively, and agreed to maintain regular activities in-between quinquennial meetings, aiming at specific research topic of the group's general interest. These conferences would be held previous to the XVI UISPP Congress at Florianopolis (Brazil, August 2011), the first to be hosted in Spain (Madrid 2009), the second in Austria (Vienna 2010), closing the five year cycle in France (Paris 2012).

The triggering cause for setting the second conference in Madrid was the 2003 discovery of the Early Neolithic flint mine of Casa Montero, and the by then ongoing five year long research project on the site *(Casa Montero, Madrid. Neolithic production and circulation of flint in central Iberia)*. This project had been made possible by the Agreement signed on February 2007 by the Spanish National Research Council (CSIC), Autopista Madrid Sur Concesionaria Española S.A. and the regional government of the Comunidad de Madrid, and contemplated the organization of both national and international meetings in order to publicize the site on a scientific level. The Conference would be hosted at the newly created Centre for Human and Social Sciences of the Spanish National Research Council (CCHS, CSIC), while the 'Casa Montero Team' would act as both hosts and organizers.

The *2nd International Conference of the UISPP Commission on Flint mining in Pre- and Protohistoric Times* was held at the main meeting hall of the CCHS between the 14th and 17th of October 2009, under the title *Flint Mining and Quarrying Techniques in Pre- and Protohistoric Times.* Two specific topics where suggested, following the guidelines originally set by the Commission: mining strategies, and technical procedures and mining tools. The former would include topics such as the organization of mining exploitations, the spatial organization of mining episodes, or the time and timing of extractive activities and its intensity, among others. The latter was oriented towards studies on technical decisions involved in shaft/gallery mining, extraction methods, variability in mining tools, their use, efficiency and final discard patterns.

The call was reasonably satisfactory, both in terms of its distribution among specialists and of the scientific interest of the papers presented at the conference: 72 enrolled researchers, 65 authors from 14 different countries, 26 papers, 9 posters and 3 workshops. These figures reflect the

A moment of the workshop. Marta Capote explaining a set of hammerstones and other pounding tools recovered from the Casa Montero flint mine.

agility and effectiveness of an interconnected "social network" linking individuals interested in prehistoric mines and quarries throughout an important part of the world. This network is a result of both individual and collective efforts, such as the creation and maintenance of the UISPP Commission and the SAA Prehistoric Quarries and Early Mines Interest Group, or the infamous Flint Symposia. Networks are not easy to build, and their maintenance frequently demands time consuming individual commitments. Meetings such as the Madrid Conference should encourage members of this active prehistoric mining interest network to keep up their enthusiasm.

The Madrid sessions were officially opened with some welcoming words by the head of our host institution, the CCHS of the CSIC, Eduardo Manzano, the President of the UISPP Commission, Françoise Bostyn, and Pedro Díaz-del-Río on behalf of the Organizing Committee. This was followed by a tribute to the late P.J. (Sjeuf) Felder, who unfortunately passed away on May 15th 2009. Gillian Varndell and Jacek Lech glossed his activities, stressing the pioneering nature of his work and the importance of his intellectual legacy.

Conversations during the poster session at the CCHS.

The present volume would wish to be a modest contribution to the memory of both P.J. (Sjeuf) and Werner M. Felder, who also passed way on December 15th 2008.

The sessions were organized taking into account previously proposed topics, beginning with those papers relating to mining strategies and followed by those more focused on specific aspects of mining techniques and mining tools. Two workshops were organized during the second day of the meeting, both related to different aspects of the Casa Montero Research Project: mining tools and flint operative chains. The complete set of archaeological remains from Casa Montero was by then stored at the actual venue, all of which allowed the participants to have a firsthand look at a good selection of objects. Finally, members of the team presented the first online version of SILEX, the Casa Montero Spatial Data Infrastructure (SDI).

The poster session was held on the afternoon of Friday 16th, followed by the Plenary Meeting of the Commission. Along with several organizational aspects, Françoise Bostyn and Jacek Lech pondered the development of the meeting, opening a short debate on the need to adjust contributions to the chosen topics in future conferences, something that was not exactly the general trend of the Madrid Conference. Other members of the commission considered that these gatherings were the result of the existence of some shared interests in certain general topics (e.g. prehistoric flint mining), so that limiting them would be somehow counterproductive. Finally, the organizing committee suggested the possibility of publishing the conference proceedings in the British Archaeological Report international series, all of which was agreed on by the commission.

Saturday 17th was devoted to our field trip, touring the participants to some of the flint outcrops of the Madrid region, including the site of Casa Montero and some close-by XVIII-XIX century flint mines. We had the pleasure to accompany the geologist José Luis Pérez-Jiménez in his explanations on the origin and formation of the different flint outcrops in the area. Some great weather accompanied the enthusiasm of all the participants, who, in some cases, where able to expand their own lithological collections. Finally, the event was closed at a local restaurant, with a typical paella.

The present volume includes 27 of the 35 original papers presented at the Conference. Considering the variety of topics, the editors decided to organize them following a geographical criterion. Hopefully this order will give a certain sense of continuity to the frequently diverse issues at stake. All in all, the volume represents the tireless activity of a reasonably large group of researchers that consider the social and economic context of flint mining as a key source for understanding prehistoric and protohistoric societies. It will seem obvious that many papers do not deal directly with the two topics chosen for the conference, mining strategies and technical procedures and mining tools. Actually only 10 out of 27 fall clearly into the subject matter. Out of

Members of the UISPP commission sampling Casa Montero's flint for their lithoteque.

During the closing lunch, participants sat at different tables. This group obviously had a good time...

the remaining 13, 10 contain new evidence for prehistoric flint procurement, while the remaining 7 focus on specific geological, chronological or methodological issues related to prehistoric flint mining. These figures should stand out as some 'food for thought' for future conference calls. It is nevertheless the opinion of the editors of this volume that this variety can only add to our previous knowledge on the subject matter, since much of the advance in the most recent research has come from both new discoveries and renewed insights on previously known sites.

As leading members of the organizing committee and co-editors of this volume we would like to finish this introduction and short chronicle by sincerely thanking each and every one of the wholehearted, enthusiastic and unconditional members of the organizing team, and to all the participants who came to Madrid and made the 2nd International Conference possible, and a success for all.

¡Muchas gracias a todos!

Susana Consuegra
Pedro Díaz-del-Río

Residues at the Neolithic flint extraction site at Den of Boddam, Aberdeenshire, Scotland

Alan SAVILLE

Abstract

As at all extraction sites, the flint residues are the main evidence for the processing of the resource. This paper examines the residues recovered from excavations at the Den of Boddam Neolithic quarry site in north-east Scotland. Inferences are drawn about the homogeneity of assemblages from various contexts and their significance in terms of primary processing or more resolved knapping. The few implements recovered include enigmatic prismatic rods, two of which are published here for the first time.

Keywords

Flint. Quarry. Scotland. Neolithic. Rods. Levallois cores. Flakes.

1. Introduction

In north-east Scotland, close to the harbour town of Peterhead, is found the only major inland source of flint in northern Britain (Figure1). This source, known as the Buchan Ridge Gravel Formation, is the result of deposits formed millions of years ago in the Tertiary era, which have become buried inland as a result of subsequent processes of erosion and deposition (Bridgland 2000; Bridgland *et al.* 1997; Merritt *et al.* 2003). As it survives today at Den of Boddam, the variant of these deposits – referred to here as the Buchan Ridge Gravel (and abbreviated to BRG) – consists largely of cobbles and pebbles of flint and quartzite (and similar very robust rocks), interpreted as the remains of a fossil marine beach. Most other, less resilient, lithic components of the BRG at Den of Boddam have decomposed as a result of deep weathering (Hall 1986) and form the kaolinized matrix for the surviving flint and other clasts, giving what was in origin an open-framework deposit the character of a matrix-supported one. The flint-rich BRG at Den of Boddam is overlain by a metre or so of later glacial deposits, which also contain abundant flint.

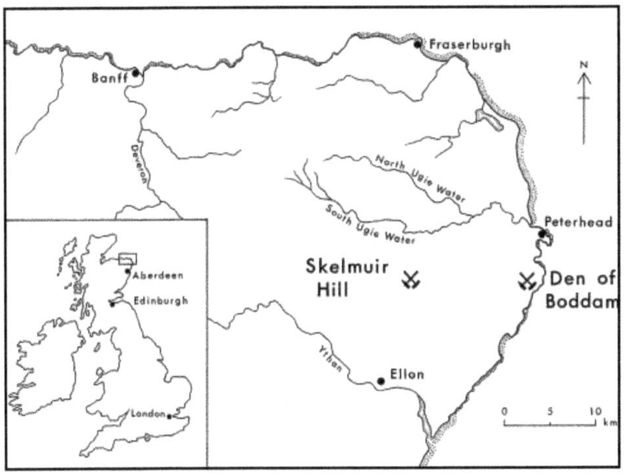

Figure 1. The flint-rich Buchan Ridge Gravel occurs in a small area south of Peterhead, in Aberdeenshire, north-east Scotland. The two known flint extraction sites shown, Den of Boddam and Skelmuir Hill (see Saville 1995), are approximately at the east and west edges respectively of the occurrence of in situ Buchan Ridge Gravel deposits.

Beyond the Den of Boddam the BRG survives as a very localized phenomenon within the region, forming the non-continuous capping to the higher parts of undulating low-relief land, extending west from the modern coast over an area about 13km (eight miles) across, although very flinty soils containing material derived from the Buchan Ridge Gravel Formation can be found spread over a wider area (Gemmell and Kesel 1979, figure 2; Kesel and Gemmell 1981, figure 2).

The existence of the flint-rich deposits at Den of Boddam was recognized by prehistoric people, probably following on from their observation of flint cobbles in the stream running through this location, which is a relict glacial meltwater channel (in which the stream was dammed in the

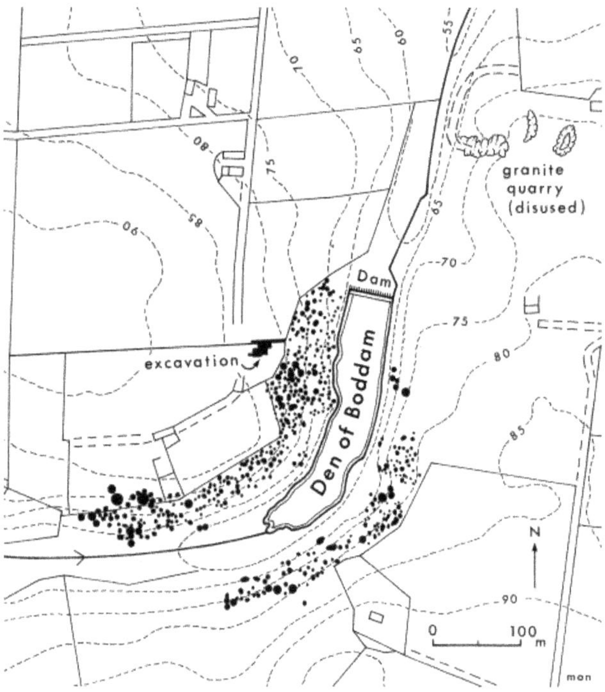

Figure 2. Den of Boddam location map. The black dots represent all the extraction pits visible on the surface, as plotted by surveyors of the Royal Commission on the Ancient and Historical Monuments of Scotland (RCAHMS 1994, 14-15). Cultivation of the fields to the west of the Den (including the excavation site), has removed surface indications there of the underlying extraction pits.

19th century to create a small reservoir). Prehistoric quarry pits are still visible on the surface at Den of Boddam as pronounced hollows on the steep unploughed slopes of the channel, making this a unique survival of a Neolithic industrial monument in Scotland (Figure 2).

2. Character of the resource and its exploitation

At Den of Boddam flint clasts sized 50mm or larger constitute about 35 per cent (by weight) of the total content of the BRG deposit down to the level penetrated by prehistoric extraction. These clasts are rounded pebbles and cobbles of grey flint, with chatter-marked external surfaces closely resembling the type of flint cobbles to be found on a marine beach. The mean maximum dimension of the larger cobbles is only 64mm and, although cobbles up to about 130mm are occasionally present, the relatively small average size of the clasts, together with the fact that the flint is rarely clear-structured and free of flaws, means that it is only suitable for the manufacture of small tools, not for anything larger such as axeheads. Nevertheless, the grey BRG flint was clearly preferred in prehistory to the similarly-sized brown/yellow-brown flint from the overlying glacial deposits, which was largely ignored by the flint knappers.

The extraction method employed involved the repeated excavation of roughly circular, cylindrical pits through the overlying topsoil and glacial gravel and down into the BRG. The deepest pits (amongst the very small sample of pits investigated thus far) reached just over 4m below the surface. The prehistoric exploitation of the BRG was thus a matter of simple extractive technology, but extraction which would have required careful management of spoil if extraction capacity was to be controlled and in any sense maximized. In some cases the fact that there are gaps between what appear to have been perfectly productive pits may indicate the position of former spoil heaps; in other cases the pits overlap or were contiguous.

The scale on which extraction took place at Den of Boddam was considerable. There is still the surface evidence for at least 458 pits (Figure 2). Following archaeological investigations (Bridgland and Saville 2000; Saville 1995), it is estimated that perhaps as many as 1000 pits were originally dug at this location over an area of almost 12 hectares (30 acres). Very approximate calculations have been made to estimate the yield from the BRG removed from the Den of Boddam quarry pits, with the conclusion that some 3000 or more cobbles sized 50mm or larger would have been obtained from each quarry pit.

Dating evidence for the quarrying activity at Den of Boddam is limited, both because of the absence of diagnostic lithic tool-types or other material culture, such as pottery, and because of the lack of organic materials for radiocarbon dating (the acidic deposits do not preserve items of bone, antler, or wood). Of the few well-stratified occurrences of charcoal,

Lab code and no.	Site coding and sample no.	Sample type	Context	$\delta^{13}C$ ‰	^{14}C years BP	Cal date BC OxCal v3.10 @ 95.4%
OxA-13102	DB'91/357	birch charcoal	'ditch' Area A west; Pit 3 infill (knapping debris)	-26.9	4372 ± 35	3090–2900
OxA-13103	DB'93/323	birch charcoal	Area 3, Pit 46, base of context 48 (knapping debris)	-25.1	4387 ± 34	3100–2900

Figure 3. Den of Boddam: radiocarbon dates.

two from quarry pit infills were sampled for radiocarbon dating and gave *termini ante quos* of c. 3000 cal BC for the quarrying activity in one part of the site (Figure 3).

The position of one of these samples (from Pit 46), at the base of an infill horizon of knapping debris, is probably at the surface at what was a relatively swift initial collapse and infill of the lower part of the pit, implying that the date for the digging of the extraction pit itself might not be very much earlier than the radiocarbon date.

Further information on various aspects of the Neolithic exploitation of the BRG at Den of Boddam can be found in previous publications (Saville 2005; 2006; 2008). This paper will focus on the flint residues which have been recovered during the excavation, looking at the residues themselves but also aspects of their analysis.

3. Residues

Although this was only a small-scale excavation project, well over one million pieces of lithic debris were removed from the site. Analysis has so far examined some 65,000 pieces to a lesser or greater degree, and the post-excavation study is still in progress. Knapped flint residues do of course exist everywhere at Den of Boddam – flint artefacts are densely strewn throughout this location in every context, including the topsoil – and this ubiquity and numerosity is itself a problem, given that constant movement of residual debris would likely be the norm on such sites. Nevertheless, these residues are the principal resource for examining what has gone on at Den of Boddam in prehistory and, whilst diagnostic pieces would be studied irrespective of their context, when it comes to the *chaîne opératoire* one is seeking more coherent, self-contained sub-assemblages from which meaningful conclusions can be drawn about production and organization.

In practice isolating sub-assemblages of lithic debris which are of high integrity is not straightforward. There are flint-rich stratified deposits within the backfilled pits, but their formation processes are often obscure. In the form in which they are found they are not the *in situ* debris from specific knapping events, and will have been through at least one redistributional transform since initial production, and in many cases probably through several transforms. Further

Type	BRG flint numbers	BRG flint weight in grams	Till flint numbers	Till flint weight in grams	Total numbers	Total weight in grams
Cores	235	35182	5	803	240	35985
Core fragments	80	9053	1	46	81	9099
Tested cobbles	132	34794	1	347	133	35141
Split cobbles	321	28036	3	404	324	28440
Chunks	217	8083	-	-	217	8083
Flakes, primary	1487	18490	7	126	1494	18616
Flakes, secondary	1551	13977	18	144	1569	14121
Flakes, tertiary	745	2779	1	1	746	2780
Piercers	3	32	-	-	3	32
Scrapers	-	-	-	-	-	-
Misc retouched pieces	24	1449	1	24	25	1473
Chips, spalls 20-4mm	-------	---------	------	-------	8897	4156
Unclassified burnt pieces	-------	---------	------	-------	87	745
Flint anvils	1	1023	-	-	1	1023
Flint hammerstones	-	-	-	-	-	-
Flint anvil/hammers	-	-	-	-	-	-
Non-flint anvils	-------	---------	------	-------	7	7263
Non-flint hammerstones	-------	---------	------	-------	4	1400
Non-flint anvil/hammers	-------	---------	------	-------	2	1728
Non-flint debitage	-------	---------	------	-------	38	2159
Totals	***4796***	***152898***	***37***	***1895***	***13868***	***172244***
raw material percentages	*99%*	*99%*	*1%*	*1%*		

Figure 4. Den of Boddam, Pit 46, Context 48: total assemblage.

complexity is added by the fact that not all such contexts were fully excavated, or were excavated at different stages by different excavation staff applying slight variations in recovery criteria.

Apart from within the pit fills, stratified flint-rich deposits were recovered from areas of buried soil protected from incorporation in subsequent ploughsoil by being covered by upcast from contemporary mining and further protected more recently by upcast from 19th-century ditch-digging. The buried soil might be expected to potentially retain the residues from *in situ* knapping activity, but in the areas excavated this did not appear straightforwardly to be the case. Rather, the buried soil residues showed some signs of having themselves been redeposited and mixed, and artificially concentrated by the heaping up of flint debris in places, rather than representing undisturbed knapping floors. Although the sheer numbers of artefacts prevented the general use of refitting ratios as an indicator of homogeneity or otherwise, it was clear from the obviously low presence of refitting halves of split cobbles in most excavated contexts, including the buried soil, that the assemblages recovered were disaggregated and/or incomplete.

However, one of the pit infill contexts (Pit 46, Context 48; Saville 2008, figure 6), which has an associated radiocarbon date (see above), comprised a substantial and virtually matrix-free deposit of knapping debris. This knapping debris had the potential to have been derived more or less directly from a single knapping area or event, and seemed appropriate for detailed study, especially as this context had been fully excavated. The typological categorization of this material is shown in Figure 4. All the flint, apart from the chips, is divided in raw material terms between flint from the BRG and that from the overlying till. In this assemblage the till flint represents only one per cent, and in all other assemblages so far examined from Den of Boddam the presence of till flint is never more than nine per cent by number or weight, showing that the exploitation of the superficial flint was casual and insignificant. The categories used in Figure 4 are 'broad-brush' and largely self-explanatory, but for clarity it can be explained that: primary flakes are those with a completely, or nearly completely, cortex covered dorsal surface; secondary flakes are those which retain some cortex; and tertiary flakes have no cortex. It should also be noted from Figure 4 that, in addition to the flint, there are anvils and hammerstones, usually in non-flint materials such as quartzite, and that these are a constant presence at Den of Boddam. Anvil flaking, in the sense of bipolar anvil core reduction, is not a major feature of the *chaîne opératoire* at Den of Boddam, where most of the flaking is freehand hard-hammer style, but anvils were invariably used for the initial splitting and testing of smaller flint cobbles.

For further characterization and comparative analysis of the struck flint residues from various excavated contexts at Den of Boddam it is convenient to group together those typological categories shown in Figure 4 into the major components as shown in Figure 5. This is even more 'broad brush' but does make it easier to demonstrate certain features of the Context 48 assemblage, such as the high proportion of chips (mean weight 0.5 grams), which permit the suggestion that these residues are close to what might be expected from a chipping floor. The retouched element of the assemblage is very low, as is typical of a quarry site where for the most part it is only the primary processing of the raw material which is taking place.

When comparisons are drawn with the other assemblages so far available from the analyses of the Den of Boddam residues, some obvious differences emerge in the relative presence of chips, flakes, and cores, although the low representation of retouched pieces stays the same (Figures 6 and 7).

Area 1, Context 20, is an area of buried soil adjacent to the excavated pits. The profile of its major categories would seem to approximate the most closely of all these assemblages to that of a chipping floor, with a very high component of chips, both by number and by weight, a low mass of tested and split cobbles, and a high mass of flakes. By comparison, and assuming the excavation recovery to have been equal and consistent for each of the assemblages, the profile of the pit infill contexts suggests they represent either a more primary phase of processing, or more human interference and/or natural intervention in the formation process. Context 48 in Pit 46 is probably one of the least contaminated in this respect, in that it was almost exclusively composed of knapping debris in a largely open framework condition, and could even represent knapping at the edge of that pit. The other pit contexts are of lesser integrity since the assemblages are samples, smaller in size, from partially excavated contexts and their profiles are accordingly more difficult to interpret. For example, there is no obvious explanation for the dominance of tested and split cobbles in the Pit 25, Context 12 assemblage.

If the sub-20mm fraction (i.e., the chips) is omitted, then the core-to-flake ratios are reasonably constant (Figure 8). Again taking the buried soil assemblage from Context 20 of Area 1 as the best proxy for that of a chipping floor, the other assemblages all diverge from this in having lower ratios (i.e., fewer flakes per core piece). This reinforces the notion that these assemblages are further removed from chipping-floor status.

Further confirmation of this comes from consideration of the purpose of the process which produces the residues. For the most part the flint residue is indicative of the primary processing stages of reduction, not of the intended end product of the knapping activity. Most instructive in the latter respect are the cores, amongst which Levallois ('tortoise') types are the most clear-cut as to the intended outcome. Frequently using half of a split cobble as the blank – cortex covering the dorsal surface, a single, flat, flake surface forming the ventral face – the knappers

Category	No.	%	Wt in g	%
Cores & core fragments	321	*2.4*	45084	*30.1*
Tested & split cobbles	457	*3.4*	63581	*42.4*
Flakes (>20mm)	3809	*28.2*	35517	*23.7*
Chips (<20mm)	8897	*65.8*	4156	*2.8*
Retouched	28	*0.2*	1505	*1.0*
Totals	***13512***		***149843***	

Figure 5. Den of Boddam, Pit 46, Context 48: assemblage major categories.

Category	Area 1 (20)	Pit 46 (48)	Pit 1 (1/2, 1/3, 2)	Pit 25 (12)	Pit 67 (39/68)	Pit 19 (sub-peat)
Cores & core frags	*1.3*	*2.4*	*2.3*	*3.8*	*4.8*	*2.7*
Tested & split cobbles	*2.1*	*3.4*	*6.4*	*19.8*	*15.8*	*15.5*
Flakes	*25.9*	*28.2*	*31.0*	*44.6*	*53.2*	*42.3*
Chips	*70.4*	*65.8*	*59.9*	*31.1*	*24.8*	*39.0*
Retouched	*0.2*	*0.2*	*0.4*	*0.6*	*1.5*	*0.6*
n=	**33551**	**13512**	**6806**	**4812**	**1407**	**1316**

Figure 6. Den of Boddam, percentage comparison of major categories in selected assemblages (by number).

Category	Area 1 (20)	Pit 46 (48)	Pit 1 (1/2, 1/3, 2)	Pit 25 (12)	Pit 67 (39/68)	Pit 19 (sub-peat)
Cores & core frags	*21.4*	*30.1*	*18.0*	*11.8*	*16.9*	*11.1*
Tested & split cobbles	*38.7*	*42.4*	*59.8*	*74.0*	*70.7*	*70.9*
Flakes	*32.2*	*23.7*	*19.6*	*12.3*	*10.0*	*16.3*
Chips	*7.0*	*2.8*	*1.0*	*0.4*	*0.4*	*0.6*
Retouched	*0.8*	*1.0*	*1.6*	*1.4*	*2.0*	*1.1*
Weight in kg	**153.7**	**149.8**	**123.9**	**202.9**	**66.5**	**42.2**

Figure 7. Den of Boddam, percentage comparison of major categories in selected assemblages (by weight).

Area 1 (20)	Pit 46 (48)	Pit 1 (1/2, 1/3, 2)	Pit 25 (12)	Pit 67 (39/68)	Pit 19 (sub-peat)
1:20	1:12	1:13	1:12	1:11	1:15

Figure 8. Core:flake ratios in selected assemblages (by number). Cores include core fragments; flakes include all types greater than 20mm in size.

Figure 9. Den of Boddam: 'Levallois' cores from Area 1, Context 20 (photo: A. Saville).

prepared the core by peripheral retouch from the bulbar face through the edge of the dorsal cortex, then used the retouched edge as the platform for removals across the bulbar surface. More than one such removal would result in subsequent flakes having multi-directional dorsal flake scars. Cores of this type are very numerous at Den of Boddam (Figure 9), but successful examples of resulting flakes are almost non-existent, for the obvious reason that they would have been taken off-site, either as blanks or after further modification. Examples of unsuccessful flakes from Levallois cores do occur, however, exhibiting the characteristic attributes of faceted platforms, multi-directional dorsal scars, and flat profiles (Figure 10). The assumption is made that the successful Levallois flakes are the blanks for manufacturing transverse (chisel and oblique) arrowheads (Saville 2005), even though no examples of arrowheads of any kind are present amongst the excavated finds at Den of Boddam.

Figure 10. Den of Boddam: 'Levallois' flakes from Area 1, Context 20 (photo: A. Saville).

Area 1 (20)	Pit 46 (48)	Pit 1 (1/2, 1/3, 2)	Pit 25 (12)	Pit 67 (39/68)	Pit 19 (sub-peat)
64.9	*31.1*	*48.1*	*24.2*	*29.5*	*37.0*
n=322	**235**	**104**	**99**	**44**	**27**

Figure 11. Den of Boddam: percentage comparison of 'Levallois' core types in selected assemblages (by number).

Area 1 (20)	Pit 46 (48)	Pit 1 (1/2, 1/3, 2)	Pit 25 (12)	Pit 67 (39/68)	Pit 19 (sub-peat)
32.7	*18.5*	*18.6*	*13.6*	*13.7*	*11.3*
n=4646	**1493**	**909**	**821**	**322**	**142**

Figure 12. Den of Boddam: percentage comparison of faceted platform flakes amongst bulbar flakes (secondary and tertiary flakes and BRG flint only) in selected assemblages (by number).

Category	Area 1 (20)	Pit 46 (48)	Pit 1 (1/2, 1/3, 2)	Pit 25 (12)	Pit 67 (39/68)	Pit 19 (sub-peat)
Piercers	5	3	2	-	-	1
Scrapers	-	-	-	2	2	1
Rods	-	-	-	1	-	-
Edge-trimmed flakes	1	-	-	-	-	-
Microliths	1	-	-	-	-	-
Miscellaneous retouched	66	25	29	28	19	3
Assemblage n=	***33551***	***13512***	***6806***	***4812***	***1407***	***1316***

Figure 13. Den of Boddam: numerical presence of retouched pieces in selected assemblages.

The relative occurrence of Levallois cores and faceted platform flakes within the various assemblages analysed so far is shown in Figures 11 and 12, from which it can again be interpreted that the buried soil assemblage from Area 1 has a higher proportion of the later, more resolved stages of prepared flaking than do the assemblages from the pit infills.

4. Retouched pieces

It has already been emphasized that, it being a quarry site, the residues recovered at Den of Boddam would be expected to be low on implements and other secondarily retouched pieces other than cores. Indeed, almost all the implements represented are ad hoc tool forms, such as mi-

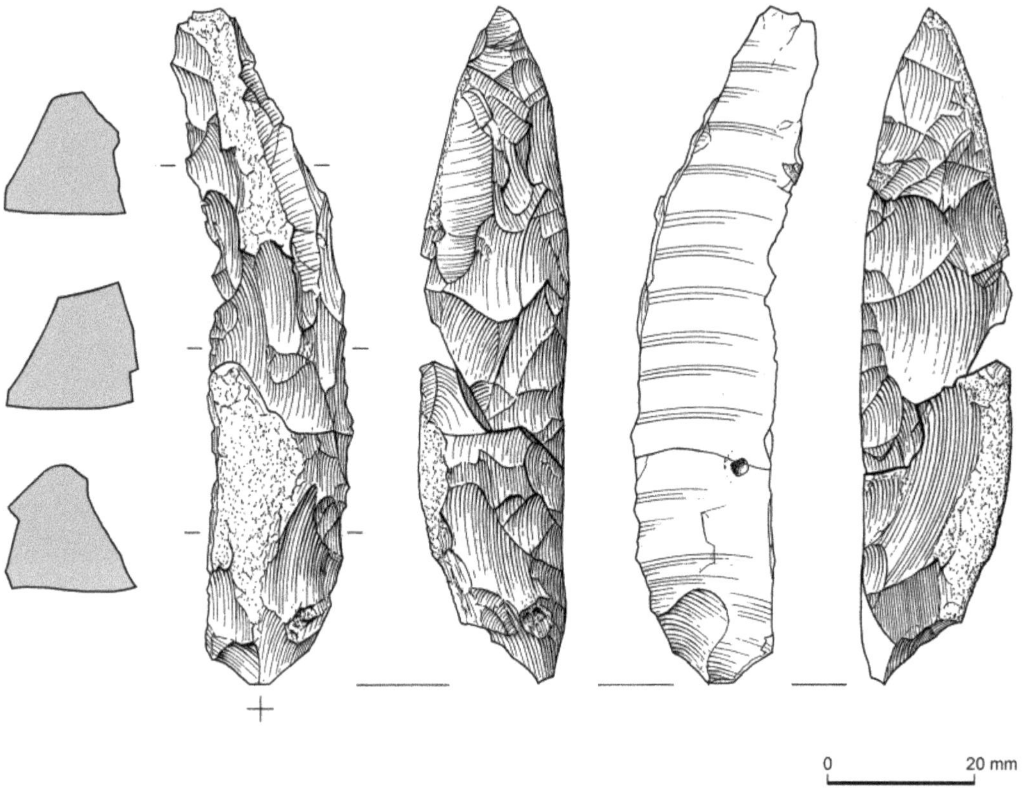

Figure 14. Den of Boddam: prismatic flint rod implement from Pit 25, Context 12 (drawn by Marion O'Neil).

nimally retouched scrapers and lightly modified, naturally pointed piercers.

Tabulation of the retouched pieces within the assemblages already encountered confirms that the residues are mainly of an industrial, primary processing nature (Figure 13), though the caveats must be added that only a very small part of the overall quarry site was excavated (Figure 2), and that the residues reflect what was left behind, not what may have been taken away. The majority of pieces with identified retouched zones are in the miscellaneous category and are not assignable to recognized implement types, either because they are fragmentary or because they are non-standard. The microlith is a residual item in a much more abraded and discoloured condition than any of the other struck flint, and indicates some Mesolithic activity at Den of Boddam well before the quarrying phase.

The most interesting implement type found is the rod, of which one example was refitted from two fragments found in the same location (Pit 25, Context 12; Figure 14). This implement is complete (91mm long, 20mm wide, 21mm thick; weight 43 grams) and is made on a bulbar flake. It has broken as a result of a blow at a point along one side. Another rod, slightly broken at the bulbar end of the flake blank, and again refitted from two pieces from the same context, was found amongst a deposit of knapping debris within the fill of Pit 3, about 15 metres to the north of Pit 25 (Figure 15). This example is smaller and lighter (80mm long, 18mm wide, 19mm thick; weight 28 grams), but shares the same prismatic form with steep bilateral retouch and trapezoidal to triangular cross-section. Neither of these rods has terminal retouch or any evidence for use abrasion at the terminal except that one (Figure 15) has heavy impact damage inversely at the distal end; the indications are that, if used at all, it has been along the steeply retouched lateral edges.

Within the UK there are precise parallels for the rod implement type at Grimes Graves flint mine site in Norfolk (Saville 1981, 10 figures, 76-85), although there rods seem to be associated with the Middle Bronze Age activity rather than with the late Neolithic mining (Saville 1981, 62-63). Nevertheless, there does appear to be an association of this tool type with flint source locations in southern England, including, for example, at the Easton Down flint-mine, Wiltshire (Saville 1981, 63; Stone 1935, 72). Despite the 600km or so separation between the Den of Boddam and the flint-mine zone of southern England, this little understood tool type provides another cultural connection between north and south Britain.

Acknowledgements

I am grateful to Marion O'Neil for the drawings in Figures 1-2 and 9-10 and to Craig Angus for their digitization.

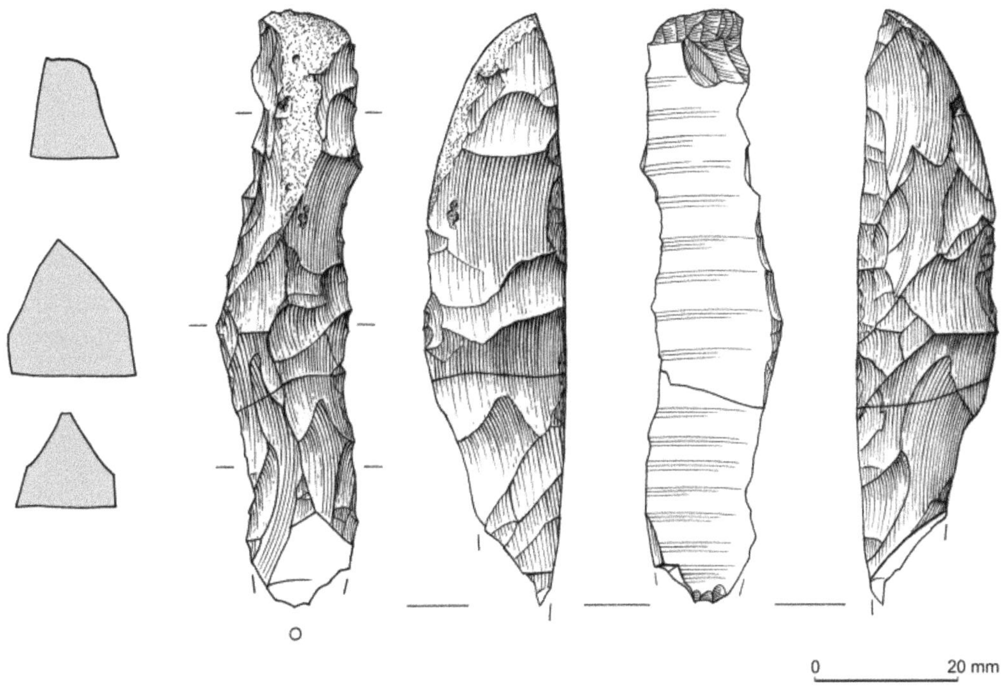

Figure 15. Den of Boddam: prismatic flint rod implement from Pit 3 (drawn by Marion O'Neil).

References

Bridgland, D. R. 2000. Discussion: the characteristics, variation and likely origin of the Buchan Ridge Gravel, in J. W. Merritt, E. R. Connell and D. R. Bridgland (eds.), *The Quaternary of the Banffshire Coast & Buchan: Field Guide*, 139-143. London, Quaternary Research Association.

Bridgland, D. R. and Saville, A. 2000. Den of Boddam, in J. W. Merritt, E. R. Connell and D. R. Bridgland (eds.), *The Quaternary of the Banffshire Coast & Buchan: Field Guide*, 102-115. London, Quaternary Research Association.

Bridgland, D. R., Saville, A. and Sinclair, J. M. 1997. New evidence for the origin of the Buchan Ridge Gravel, Aberdeenshire. *Scottish Journal of Geology* 33(1), 43-50.

Gemmell, A. M. D. and Kesel, R. H. 1979. Developments in the study of the Buchan flint deposits. In L. M. Thoms (ed.), *Early Man in the Scottish Landscape*, 66-77. Scottish Archaeological Forum 9. Edinburgh, Edinburgh University Press.

Hall, A. M. 1986. Deep weathering patterns in north-east Scotland and their geomorphological significance. *Zeitschrift für Geomorphologie* 30, 407-422.

Kesel, R. H., and Gemmell, A. M. D. 1981. The 'Pliocene' gravels of Buchan: a reappraisal. *Scottish Journal of Geology* 17(3), 185-203.

Merritt, J. W., Auton, C. A., Connell, E. R., Hall, A. M. and Peacock, J. D. 2003. *Cainozoic Geology and Landscape Evolution of North-East Scotland.* Edinburgh, British Geological Survey.

RCAHMS 1994. *Monuments on Record: Annual Review 1993–4.* Edinburgh, Royal Commission on the Ancient and Historical Monuments of Scotland.

Saville, A. 1981. *Grimes Graves, Norfolk, Excavations 1971–72: Volume II, The Flint Assemblage.* Department of the Environment Archaeological Reports No.11. London, Her Majesty's Stationery Office.

Saville, A. 1995. Prehistoric exploitation of flint from the Buchan Ridge Gravels, Grampian Region, north-east Scotland. *Archaeologia Polona* 33, 353-368.

Saville, A. 2005. Prehistoric quarrying of a secondary flint source: evidence from north-east Scotland, in P. Topping and M. Lynott (eds.), *The Cultural Landscape of Prehistoric Mines*, 1-13. Oxford, Oxbow Books.

Saville, A. 2006. Flint technology and production associated with extraction sites in north-east Scotland, in G. Körlin and G. Weisgerber (eds.), *Stone Age – Mining Age*, 449-454. Der Anschnitt 19. Bochum, Deutsches Bergbau-Museum.

Saville, A. 2008. Flint extraction and processing from secondary flint deposits in the north-east of Scotland in the Neolithic period, in P. Allard, F. Bostyn, F. Giligny and J. Lech (eds.), *Flint Mining in Prehistoric Europe: Interpre-*

ting the Archaeological Records, 1-10. Oxford, BAR Publishing. British Archaeological Reports International Series 1891.

Stone, J. F. S. 1935. Excavations at Easton Down, Win-terslow, 1933–34. *Wiltshire Archaeological Magazine* 47, 68-80.

The origins of flint extraction in Britain

David FIELD

Abstract

Excavations at well-known Neolithic flint mines in Britain have concentrated on shafts that represent a mature stage of site development. Using evidence from several mine sites in southern England and Grimes Graves in Norfolk, this paper discusses the limited evidence for the origins of extraction and the reasons for adopting certain locales for this activity over others that are equally well-placed.

Keywords

British flint mines. Grimes Graves. Swallets. Swallow holes. Ritual shafts. Mesolithic extraction.

1. Introduction

Natural flint deposits are extremely widespread in southern England. Flint is ubiquitous – it's everywhere. Not only on the parent chalk formations, for there are also very widespread secondary deposits that occur, not only in Tertiary strata, but also as Pleistocene river and beach accretions. Material can easily be obtained on the surface. In some places seams of flint outcrop just below the turf and all you need to do is strip off the grass. It can be obtained from seams in cliff faces situated along the coast and where rivers cut through the chalk formations, or from around the flanks of deep coombs that are engraved in the chalk. Where such seams outcrop nodules can be easily extracted with very little effort. Such locations may have been used in an ad-hoc way for obtaining material for millennia and many knapping sites, for example, around Stonehenge, where nodules fall from valley sides and collect under gravity in the bottom of coombes, or at Ringland in Norfolk, where the River Yare has cut through deposits of chalk leaving flint seams exposed, occur in these places.

Ever since our study of the English mines in the 1990s (Barber *et al.* 1999) we have posed the question as to why it should be necessary to mine for flint. Given the great extent of the flint seams why one location should be chosen over another is not immediately apparent. And why are mines hafts clustered together in compacted areas instead of being spread across the chalk deposit as historic mines are?

The sparse, but nevertheless important, evidence from Upper Palaeolithic sites such as Hengistbury Head on the south coast (Barton 1992), or much further inland at Fisherton, near Salisbury (e.g., Roe 1981), or the caves of Mendip in the west (e.g., Jacobi 2004), indicates a significant human presence in the region from very early in the post-glacial period. Flint artefacts discovered, certainly at the latter sites, which occupy a non-Cretaceous geology, imply considerable movement through the countryside, at least in so far as the raw material for these tools might only be obtained from the chalk deposits of central southern England. Implicit in this movement is knowledge among travellers of the position of different resources. Flint was being repeatedly obtained from somewhere on the chalk and carried reasonable distances from it.

2. Mesolithic extraction

Unlike elsewhere in Europe (e.g., Bando *et al.* 1992; Negrino *et al.* 2006), there is no direct evidence of Palaeolithic or Mesolithic extraction in Britain. The greater number of picks and axes dredged from along the River Thames appear to be of gravel flint obtained from secondary sources and this is also true of material from many of the excavated sites. However, a series of pits excavated in 1938 at Farnham, in Surrey, that had been dug into the gravel were associated with a large amount of Mesolithic flint work (Clark and Rankine 1939). They were considered by J G D Clark, the excavator, to be dwellings or crude huts, but no post holes were present and one of them cut through the gravel to the underlying clay. More recently Roger Ellaby (1987) has pointed out that the clay would have collected water and it is therefore unlikely that the pits were for living in. Instead he suggested that they were the result of quarrying for large gravel flint cobbles for the manufacture of flint tools. The large amount of knapped material on the site certainly originated somewhere and the suggestion appears to be quite reasonable. Thus simple pits of this nature might be expected right across the areas of secondary deposits.

Such simple pits on chalk take on a little more mechanical structure. While aware of the need to be cautious because of the way that simple pits were used at Grimes Graves and elsewhere, it may be that the pits at Martins Clump in Hampshire are a development of this simple process. Of four small pits exposed by an electricity cable trench (Ride 1998; Field 2006), one produced a ^{14}C date of 4230-4190cal +BC (Barber *et al.* 1999), not particularly early in Europe as a whole, but within Britain providing the earliest date for a flint mine. As a complex, the site differs from others in comprising a large number of relatively small pits and of all British sites is probably most deserving of further attention. Although there are relatively few ^{14}C dates available for British sites there is nevertheless an indication that the Sussex flint mines are some of the earliest Neolithic sites in the country. With dates often preceding 3750 cal BC, they appear to pre-date the apparent monument horizon when many long barrows and causewayed enclosures were constructed. Church Hill has a date of 4490-3810 cal BC; Cissbury 4040-3780; Harrow Hill 4240-3700; Long Down 4250-3600 and Blackpatch 4310-3530 cal BC (ibid). Most of these dates were on material taken from locations at the heart of complex; that is, from shafts that appear to be positioned at a mature stage of the sites development and therefore may not reflect the origins of the site.

3. Flint quality

Chance finds of ground axes in the areas around the Sussex flint mines concentrate on the Coastal Plain rather than around the mines themselves (Field 2008, 89-90). This is perhaps of no surprise, but assuming that the distribution marks out the places of favoured occupation, it demonstrates that the extraction sites themselves may have been on the border of land-units, rather than a central feature. Indeed, any such land-unit may have changed significantly with the rising sea-level which will have deprived communities of much valuable farm land and shifted activity onto the chalk. A plot of submerged contours off the Sussex coast by Imperial College, London (Gupta *et al.* 2004; Gupta *et al.* 2007) revealed a former landscape, no doubt densely occupied, of valleys, bluffs, lakes, marshy areas, whereby the chalk hills provide a substantial backdrop at the limit of the territory. Bounded by the Greater Rhine or Channel River in the south and the chalk hills in the north, all of which were highly visible across the coastal plain, materials could be obtained by middle distance journeys and extensive expeditions were not necessary. But it is evident that such journeys would have been for more than mere raw material - for flint in one form or another, at least sufficient for most ad-hoc purposes, was available right across these surfaces. So why bother to go to specific locations to extract it? It is curious that the mining doesn't appear to relate to the 'quality' of the flint for Sussex mine sites are not necessarily situated on what we might consider being the best seams. What is considered to be poorer quality flint was mined at both Blackpatch and Harrow Hill when compared to Cissbury and Church Hill (Barber *et al.* 1999). All of these sites are situated within no more than a few miles of each other; and while Cissbury and Church Hill exploited the 1st and 2nd nodular seams of the Peacehaven Beds of the Newhaven Member, Blackpatch and Harrow Hill utilized material from between the Rottingdean and Old Nore Marls of the Newhaven Member (Mortimore 1982, 1986a, 1986b). In both cases, better quality flint is available nearby. It would appear that the topographic location of the mines was of greater importance than the quality of the raw material. So why should one place be chosen in preference to another?

It must be admitted that our perception of 'quality' is a modern one largely derived from gunflint miner terminology at Brandon, Norfolk. There, gunflint miners employed by Canon Greenwell on the early excavations at Grimes Graves used their own descriptions and the terms stuck (Greenwell 1870). In the gunflint trade 'quality' related to how well material produced sparks when struck with a hammer, an essential element of gunflints (Skertchley 1879). Thus the British army determined that the seam of 'floorstone' sparked better than the others and only purchased gunflints from the black flint of that layer. With time the Brandon knappers simply painted the grey flint from other layers, black (Forrest 1983). We can only guess at the properties prized by Neolithic communities which may have been very different from our own. An example of this is the Pipestone, traditionally quarried in Minnesota, United States. George Catlin an early explorer recorded how people would travel huge distances to the quarry location to obtain the material. Investigating the site more recently, Scott and Thiessen (2005) recounted how the various purification rituals and other ceremonies involved were carried out before quarrying could take place. The importance in the stone was not that it was a particularly good material for making pipes, but in its reddish pink colour which is seen in a supernatural way as representing flesh and the origin of the human race.

In Sussex, lenses of flint act as weather retarding agents in the soft chalk and the resulting breaks of slope marking the location of these seams often stand out on hillslopes. From such places material is easily obtained, however, the summit of the hills in southern Britain, i.e., where the mines are located, is usually capped with a tenacious deposit of Clay-with-flints and the position flint seams totally obscured. Careful study of the earthworks at a number of these sites, however, suggests that the mining process moved uphill rather than down, spoil from one shaft being thrown downhill (its easier) to fill a shaft that had been excavated previously (Barber *et al.* 1999). The lower levels of extraction appear to correspond to the presence of the Clay-with-flints capping and it seems likely that the junction of Chalk and Clay-with-flints was therefore also marked at the surface in some way. The underlying chalk at this point, differentially weathered and dissolved, might even have undercut the deposits of tenacious clay, making flint seams at this level easy picking. Such naturally occurring exposures might be expected to attract activity from the earliest times and, protected from peri-glacial activity by the Clay-with-flints, the nodules would be free of frost-induced imperfections – a definite advantage. Like the pipestone visible seams in the strata might easily be perceived within a metaphysical perspective; as bones of a supernatural being, especially as the red colour of the clay might be held to be flesh coloured. Once marked out by human activity such sites might easily be identified and confined to myth, tradition and memory, maybe even taking on 'sacred' or 'special' connotations. Thus vast areas of quite appropriate and useable flint seams might be ignored, as people consistently and repeatedly visit the locales that presented themselves in this way.

4. Grimes Graves

Despite the presence of extensive deposits of flint across East Anglia, formal flint extraction concentrates in one place - Grimes Graves. This stands out among British flint mines as dating to rather later in the Neolithic. It was exploited during the latter part of the 3rd millennium and the first part of the 2nd, at a time when elsewhere Stonehenge was under construction, bluestones quarried and transported huge distances and, crucially, metal work introduced. Metal axes were now available and a desirable product. So why was flint in such great demand? Few flint axes were found at Grimes Graves and equally there is no indication that the contemporary flint axes within the region are of mined flint. Alan Saville (1981) suggested that cores may have been exported. It would seem so, but at a time when the nature of flint-use was changing and the old cutting functions superseded, why go to the extent of digging for material when for ad-hoc purposes it was widely available on the surface. There is certainly a social imperative.

But did extraction start earlier, in doing so kick-starting a series of traditions, myths, memories and ultimately supernatural beliefs that embedded the place as an important component of the social structure? Saville's analysis identified a small quantity of residual Mesolithic flintwork and discovery of a flaked axe with differential staining and patina on flake scars during English Heritage fieldwork indicates some depth of time to site use. Similary, a few sherds of Early Neolithic pottery were present from excavations (Longworth *et al.* 1988).

The valley at Grimes Graves is beset with peri-glacial features, relict meanders, and dolines. On the slopes on both sides of the valley are peri-glacial soil stripes - cracks in the chalk surface resulting from the annual freeze / thaw and subsequent drainage chaos following the last glacial. These have subsequently been filled with wind-blown cover-sand, a deposit that masks much of the surface area and the resulting alternating chalk and sand deposits support different vegetation and stand out dramatically on the surface.

During the 1930s, Leslie Armstrong excavated trenches along the edge of the dry valley and found it to be filled with nodules that had rolled from positions on the slopes. Glacially contorted layers of flints were found along the edge of the site very close to the surface and by following this seam into the chalk the pits would need to become progressively deeper. Gale Sieveking's excavations during the 1970s also demonstrated that smaller pits and quarries cut into the soliflucted material around the edge of the valley, while excavation of a cable trench by Frances Healy (1985) left no doubt that the extraction pits continued into the dry valley.

It seems likely that the earliest recognition of flint deposits here during prehistory may have been of the soliflucted material exposed on the valley floor and sides; it would have been a relatively easy matter to obtain this. However, without extensive excavation, mapping the extent to which this might have been done presents a problem. Nevertheless, recent surveys using Ground Penetrating Radar have proved extremely promising in demonstrating the degree to which material in the valley was utilised (Linford *et al.* 2009). The plot indicates widespread pitting of the valley floor and sides, but it may be that it was not only the soliflucted material that was utilised. The peri-glacial stripes stand out proudly, as if they have had the sand cleared out of them as an easy method of accessing the flint seam. This needs further investigation, but the initial indication is that these peri-glacial features were being used as a method of easily accessing the flint seam.

5. **Swallets**

It is evident that there was a fascination with natural geological and peri-glacial features in the landscape elsewhere as well. Caves and swallets, alternatively termed swallow holes or solution hollows, in the Mendip Hills often have deposits of Neolithic and Beaker material in impossibly difficult to access ledges, in one case over 10m down (Lewis 2005). Neolithic visitors will also have encountered the stalagmite encrusted skulls of humans and animals that died in these places during an earlier period and it is easy to imagine how such places take on an important ancestral significance. Barrow cemeteries, for example that at Tyning's Farm in the Mendip Hills, often accumulate around and among such swallet holes and the practice has also been noticed in Dorset (Woodward 2000, 125-6). In a landscape stripped of its modern vegetation and paraphernalia, such peri-glacial features might be of considerable importance, appearing as though they were constructed by past occupants or, alternatively, as entry points into a supernatural underworld.

This brings us back to the Mesolithic site at Farnham. On excavation, one of the pits tuned out to be a natural swallet hole, within which prehistoric deposits had accumulated. And this is not the only occasion on which the relationship has been noticed. A similar situation occurred at the Fir Tree Field shaft, in Dorset, where Mesolithic, Neolithic and Beaker material was encountered in the silting of a massive natural shaft (French *et al.* 2007, 280-305). Mesolithic debris comprising some 3000 pieces of struck flint was also encountered in the upper fills of a series of solution hollows on Portsdown, close to the junction of the chalk and a 30m raised beach that overlooks Portsmouth Harbour (Hughes and ApSimon 1977).

There are other unexplained features enshrined in the literature. At Eaton Heath in Norfolk, not far from Grimes Graves, a series of 17 shafts excavated by Geof Wainwright during the 1970s, some of which reached a depth of 8m, were found lined with clay and thought to have been deliberately constructed (Wainwright 1973). The subsoil comprised sands and gravels and was very unstable while the chalk bedrock lay at an unknown depth. Neolithic settlement material lay scattered across the surface and the filling of the shafts contained occasional Neolithic artefacts, implying that they were open whilst the settlement was active. It was clear that they were not flint mine shafts for they did not encounter the chalk and their purpose remained unexplained, although it was felt that, on the analogy of similar Iron Age shafts, a ritual function might be inferred. That might be so, but with hindsight, these might now be alternatively considered to be solution hollows and part of the increasing fascination with holes in the ground. It may be that tradition and the memory of these somehow provided a catalyst for early activities at Grimes Graves.

6.Conclusion

Natural fissures and other holes and exposures in the ground provide a means of obtaining knowledge of subterranean stratigraphy and during the post-glacial period much greater opportunity for engaging with soliflucted and ice formed landscapes will have been available. Some sites such as Cissbury, in Sussex, may have originated as an entirely practical response to weathered geology and only later after generations of extraction, stacked up traditions and mythology, they may have taken on a more symbolic role. In other cases, features such as swallets provided a rare opportunity of inspecting the interior of a hidden world, one that is likely to have been highly symbolic from the outset. Certainly a number of such sites were later encapsulated into ritual or ceremonial landscapes and thus it may be that ritual shafts and ritual in mines derive from a common origin. While swallets in Mendip with their associated Neolithic deposits will not have encountered flint, those in Dorset, Wiltshire, Sussex and elsewhere will have. The peri-glacial stripes recorded at Grimes Graves and present widely across the chalk will have served a similar function.

At face value the relationship between natural exposures and quarrying is an obvious one, but the importance of swallet holes is less transparent. Nevertheless, the relationship has been observed in the United States (LaPorta pers comm.). In certain places the chalk in southern England is riddled with such natural holes and it may be worth paying them far greater attention.

References

Bando, C., Dagnan-Ginter, A., Holen, S., Kozlowski, J., Montet-White, A., Pawlokowski, M., and Sobczyk, K. 1992. Wolowice, province of Krakow (fl int extraction and processing site) in *Archaeological Investigations of 1990*, 5-24. Krakow, Institute of Archaeology, University of Cracow.

Barber, M., Field, D. and Topping, P. 1999. The Neolithic fl int mines of England. Swindon, English Heritage/Royal Commission on the Historical Monuments of England.

Barton, R. N. E. 1992. *Hengistbury Head, Dorset. Volume 2 .The Late Palaeolithic and Early Mesolithic sites.* Oxford, Oxford Committee for Archaeology.

Clark, J. G. D. and Rankine, W. F. 1939. Excavations at Farnham, Surrey (1937-38). *Proceedings Prehistoric Society* 5, 61-118.

Ellaby, R. 1987. The Upper Palaeolithic and Mesolithic in Surrey, in J. Bird and D. Bird (eds.), *The archaeology of Surrey to 1540,* 53-70. Guildford, Surrey Archaeological Society.

Field, D. 2006. Procurement and extraction of flint in Wessex, in G. Korlin and G. Weisgerber (eds.), *Stone Age-Mining Age,* 77-82. Bochum, Deutsches-Bergbau Museum.

Field, D. 2008. *Use of land in central Southern England during the Neolithic and early Bronze Age.* Oxford, BAR. British Archaeological Reports British Series 458.

Forrest, A. J. 1983. *Masters of flint.* Suffolk, Lavenham Press.

French, C., Lewis, H., Allen, M. J., Green, M., Scaife, R. and Gardiner, J. 2007. *Prehistoric landscape development and human impact in the upper Allen valley, Cranborne Chase, Dorset.* Cambridge, McDonald.

Greenwell, W. 1870. On the opening of Grime's Graves in Norfolk. *Journal Ethnological Society London* New Series 2, 419-439.

Gupta, S., Collier, J., Palmer-Felgate, A., Dickinson, J., Bushe, K. and Humber, S. 2004. Submerged *Palaeo-Arun River: Reconstruction of Prehistoric Landscapes and Evaluation of Archaeological Resource Potential* Report for English Heritage. Imperial College London.

Gupta, S., Collier, J. S., Palmer-Felgate, A. and Potter, G. 2007. Catastrophic fl ooding origin of shelf valley systems in the English Channel. *Nature* 448 (19 July 2007), 342-346.

Healy, F. 1985. The hunting of the fl oorstone. In A. J. Scofield (ed.), *Interpreting artefact scatters: contributions to ploughzone archaeology,* 29-37. Oxford, Oxbow.

Hughes, M. F. and ApSimon, A. M. 1977. A Mesolithic fl int working site on the south coast motorway (M27) near Fort Wallington, Farnham, Hampshire 1972. *Proceedings Hampshire Field Club Archaeological Society* 34, 23-36.

Jacobi, R. M. 2004. The Late Upper Palaeolithic lithic collection from Gough's Cave, Cheddar, Somerset and human use of the cave. *Proceedings Prehistoric Society* 70, 1-92.

Lewis, J. 2005. *Monuments, ritual and regionality: the Neolithic of northern Somerset.* Oxford, BAR Publishing. British Archaeological Reports British Series 401.

Linford, N., Martin, L. and Holmes, J. 2009. *Grimes Graves, Norfolk: report of geophysical survey.* English Heritage Research Department Report Series 64. Portsmouth, English Heritage.

Longworth, I., Ellison, A. and Rigby, V. 1988. *Excavations at Grimes Graves, Norfolk 1972-1976.* Fascicule 2. The Neolithic, Bronce Age and Later Pottery. London, British Museum Publications.

Mortimore, R. N. 1982. The stratigraphy and sedimentation of the Turonian-Campanian in the Southern Province of England. *Zitteliana* 10, 27-41.

Mortimore, R. N. 1986a. Controls on Upper Cretaceous sedimentation in the South Downs, with particular reference to fl int distribution, in G. de G. Sieveking and M. B. Hart (eds.) *The scientifi c study of fl int and chert,* 21-42. Cambridge, Cambridge University Press.

Mortimore, R. N. 1986b. Stratigraphy of the Upper Cretaceous White Chalk of Sussex. *Proceedings Geological Association* 97, 97-139.

Negrino, F., Martini, S., Ottomano, C. and Del Lucchese, A. 2006. Palaeolithic evidence for quarrying activity at 'I Ciotti' (Mortola Superiore, Ventimiglia, Imperia, Italy), in G. Korlin and G. Weisgerber (eds.), *Stone Age-Mining Age,* 153-162. Bochum, Deutsches-Bergbau Museum.

Skertchley, S. B. J. 1879. On the manufacture of gunflints, the methods of excavating for fl int, the age of Palaeolithic man, and the connextion (sic) between Neolithic art and the gunfl int trade. Memoirs of the Geological Survey of England and Wales. London, HMSO.

Ride, D. 1998. Excavation of flint mines and a linear boundary at Martin's Clump. *Proceedings Hampshire Field Club and Archaeology Society* 53, 1-23.

Roe, D. 1981. *The Lower and Middle Palaeolithic periods in Britain.* London, Routledge & Kegan Paul.

Saville, A. 1981. *Grime's Graves. Norfolk. Excavations 1971-72: volume II the flint assemblage.* Department of the Environment Archaeological Report no 11. London, HMSO.

Scott, D. and Thiessen, T. D. 2005. Catlinite extraction at Pipestone National Monument, Minnesota: social and technological implications, in P. Topping and M. Lynott (eds.) *The cultural landscape of prehistoric mines,* 140-154. Oxford, Oxbow Books.

Wainwright, G. J. 1973. The excavation of prehistoric and Roman-British settlements at Eaton Heath, Norwich. *Archaeological Journal* 130, 1-43.

Woodward, A. 2000. *British Barrows.* Stroud, Tempus.

The evidence for the seasonal use of the English flint mines

Peter TOPPING

Abstract

This paper presents evidence which confirms that certain flint mines in England were used on a seasonal basis. The chronology of several flint mines demonstrates that they are some of the earliest – if not the earliest - sites in the cultural landscape and therefore must have been regarded as special places. Evidence further suggests that the miners travelled some distance to the mines following the appearance of spring migrants (mostly bird species) and the annual collection of antler picks. The mines were temporarily abandoned around the time of the autumn equinox each year when reduced daylight hours and poor weather conditions would have become problematic. During the winter 'downtime' the miners would have led, or participated in, world renewal ceremonies, distributed significant artefacts to underpin kinship links, build alliances and maintain social narratives and belief systems. Such evidence demonstrates that flint miners followed a seasonal cycle at some English flint mines.

Keywords

Seasonality. Seasonal indicators. Antler collection. Downtime. Ritualisation.

1. Introduction

One of the interesting issues concerning the Neolithic flint mines in England is the question of how intensively they were used. Were they 'stone age weapons factories', as recently characterised by the Sunday Times, with the implication that they mass-produced artefacts on a modern 'industrial' scale, or were they more akin to a 'cottage industry' and comparatively small scale operations? Circumstantial evidence does, however, indicate certain trends which can be used to sketch the operational life of at least some of the flint mines.

2. The chronology of the English flint mines

The chronological information currently available for the English flint mines suggests that mining began around 4000BC at Harrow Hill, Church Hill and Blackpatch in Sussex, which arguably makes them amongst the earliest sites constructed in the landscape and therefore must have been places of special cultural significance. At Harrow Hill, where there are nine dates (five from antler picks, the remainder unidentified charcoal), the chronology of mining activity may be as long as a millennium, although is more likely to be some 200-400 years. The 1994 survey of this site recorded 245 'certain or probable' mine shafts which represent an early phase of deep shaft mining followed by a phase of shallow quarrying or pit digging on the southern perimeter of the complex (RCHME 1994, 6 and 11). Considering this chronology, the exploitation of 245 mines over a period of 200-400 years suggests that in an average year as few as one or two mines may have been excavated - perhaps none in some years. The radiocarbon chronology for both Church Hill and Blackpatch is restricted to single dates, so the duration of mining at these sites is unclear.

At Grime's Graves, which is the most comprehensively dated site in England and currently undergoing a re-evaluation sponsored by English Heritage using Bayesian modelling (inf. F Healy), the traditional Late Neolithic chronology suggests a period of use of some 400-500 years which overlaps with the introduction of the first metalwork. Although we clearly must await the detailed results of the new English Heritage project for a more accurate picture, if this chronology is broadly retained and considered in combination with the earthwork and geophysical evidence which suggests there was originally between 500 to 1000+ shafts and pits – this again implies that only one or two mines may have been dug each year. Such a level of exploitation can hardly be termed 'industrial' in a modern westernised sense, and cannot indicate mass-production, although each mine would have clearly provided enough raw material to craft many tools. It is all a question of scale, we appear to be seeing the exploitation of small numbers of mines over a period of several centuries, which in a given year may have produced artefacts in their hundreds or even thousands, but not in the 'production line' quantities we might expect of a more modern, intensive industrialised process. Flint mining as a skilled, specialist activity does not appear to have mobilised large numbers of people in the same way as the construction of communal monuments did, such as the great henges at Marden or Durrington Walls. Partly this was because henge building was undertaken on a massive scale spatially, and could accommodate a high percentage of unskilled workers, whereas flint mining was by its very nature and restricted workspace an activity which only required a small but highly-skilled crew who knew and understood the tolerances of the geology, and how to dig safely underground. Arguably, the size of the mining workforce and the quantity of raw material and artefacts produced at Grime's Graves is reflected in the small percentage of mined flint found in the local assemblages in the surrounding Brecklands and adjacent fenland basin (inf. F Healy). Consequently, the radiocarbon chronologies, although imperfect, when combined with these other factors do strongly suggest that flint mining was undertaken by relatively small groups of miners who dug only a few mines each year.

3. Evidence for long-distance journeys, and trade or exchange

If the weight of chronological and logistical evidence points to small scale exploitation of the English flint mines, what lies behind this scenario? An important theme in ethnography is ritualised journeys. Australian ethnography records journeys to important quarries of distances between 500km [311 miles] (Taçon 2004, 34) to 1000km [621 miles] (Brumm 2004, 153), illustrating the deep cultural value embedded in long-distance treks for the collection of particular minerals or stones (Boivin 2004, 10). The lack of permanent settlement at any extraction site so far excavated in the UK suggests that the miners here may also have travelled some distance and only used temporary

Figure 1. An aerial view of the Late Neolithic flint mines at Grime's Graves, Norfolk. Each mine or pit is surrounded by the irregular, mounded waste dumps of chalk dug from the shafts and galleries. © Peter Topping

shelters on site. Recent studies of human skeletal material have provided evidence of a range of Neolithic journeys, a group at Monkton Up Wimborne travelled roughly 60km [37miles] from the Mendips (Green 2000, 79); the Boscombe Bowmen from Cumbria or southwest Wales journeyed a maximum of 386km [240 miles] to Wessex (Fitzpatrick 2004), whereas the Amesbury Archer had travelled some 1046km [650 miles] from the Alps to Wessex (Fitzpatrick 2003). At Grime's Graves there is evidence that the miners may have been coastal inhabitants (see below) – the nearest coastline would have been roughly 80km [50 miles] distant, although the miners may equally have lived further away.

Long journeys and/or extensive exchange networks are also demonstrated by the often widespread distribution of stone axes, especially the abundant Group VI Langdale axes which are found not only throughout the UK but some 100+ crossed the Irish Sea to Ireland (Cooney and Mandal 1998, 111-173; Cooney pers comm.). In contrast, Group IX axes from Northern Ireland travelled in the opposite direction eastwards over the Irish Sea and were distributed throughout the UK (cf. Clough and Cummins 1988). Such examples illustrate the value placed upon certain tools which had been crafted from recognisable or mythologised raw materials sourced from culturally important locations. This tradition followed the epitome of long-distance exchange networks whereby exotic jadeitite axes derived from high-altitude quarries in the Italian Alps were transported throughout continental Europe, the UK and Ireland (cf. Pétrequin *et al.* 2008).

Further evidence is provided by a Cornish greenstone axe (Group IIIa) was discovered in Greenwell's Pit 'in the first gallery, 4 feet from the entrance' at Grime's Graves, demonstrating another example of a long distance, north-easterly journey of roughly 540km [335 miles]. A second Cornish axe of picrite (Group IV) was found in the assemblage on Floor 15. The fact that the first axe was carefully positioned in the gallery as part of a culturally significant placed deposit deep in the mines demonstrates its cultural importance to the miners. The gallery also had the impact marks of a polished axe in several places on the chalk walls, suggesting the Cornish axe had been periodically used to symbolically excavate the gallery - despite the main extraction tool being the antler pick (Clarke 1915,

Figure 2. A view of one of the underground galleries in Greenwell's Pit at Grime's Graves, Norfolk. The low ceiling stands roughly 1m high, and the black flint which the miners sought can be seen on the right hand side of the gallery entrance and preserved below a buttress. © Peter Topping

17; Greenwell 1870; Norfolk County Records Office, #21198). During mining symbolism clearly played a role alongside the deposition of carefully placed assemblages as part of the practice of ritualised mining.

Bird remains are rarely discovered in flint mines, but again in Greenwell's Pit the skull of a phalarope (*Phalaropus sp.*), a shorebird, was placed between a pair of antler picks with the Cornish axe described above positioned at the base of the picks (cf. Norfolk County Records Office, #21198). Placing bird remains below ground may have referenced engendered sky and earth deities, creating a metaphor for fertility and renewal. The presence of a carved chalk phallus in Greenwell's Pit may support this suggestion (cf. Longworth *et al.* 1991, 94 and 147, Figure 75 [C299a]). Curiously, 'the skull of what appears to have been a mouse' was discovered with an antler pick in Gallery III of Pit 21 at Harrow Hill (Curwen and Curwen 1926, 115), hinting at other forms of animal symbolism - the fact that again, as with the phalarope, only the skull was discovered suggests both careful curation and deliberate use.

In relation to seasonality, the presence of a shorebird at Grime's Graves offers other interesting possibilities, suggesting that this species may have had a symbolic association with flint mining – at least for the miners at this site. Phalaropes are a migratory species which arrives on the eastern coast of England from early March, peaking in mid-April (data from the British Trust for Ornithology; http://blx1.bto.org/bt-dailyresults/results/cov5-05.html). Considering the difficulties of traversing slippery chalk surfaces and the short daylight hours in the preceding autumn and winter months, the suggestion that flint mines were only used seasonally during the warmer months makes operational sense. Consequently, the reverential treatment of the phalarope skull implies it had a special place in the iconography of the miners. Did the appearance of these spring migrants trigger the annual mining operations at Grime's Graves? The fact that the phalarope is exclusively a coastal species implies that the miners at Grime's Graves may also have been coastal inhabitants – or at least had access to the coast to enable them to encounter the migrating phalaropes. The nearest coastline to Grime's Graves during the Neolithic period was at least 80kms from the flint mines.

The recovery of another 'possible' bird bone from Shaft 27 at Cissbury (Pull Archive, Worthing Museum and Art Gallery) may reinforce this symbolic link between mining, the miners, and birds. The fact that mortuary deposits in long barrows have produced remains of swan, crane, heron and geese -all migratory species- emphasises the importance of avian seasonal indicators to Neolithic communities. This evidence echoes the use of medicine bundles by Native Americans such as the Hidatsa, who celebrated the Big Bird ceremony which occurred during the spring migration of large raptors and was a ceremony that also provided the supernatural endorsement for flint knapping (cf. Topping 2005, 85).

The importance of journeys, exploration and experiencing different landscapes was to impart knowledge and identify and collect necessary resources for the Neolithic lifeway, both secular and spiritual. Gould (1977, 164) has observed that amongst the Australian Aborigines 'quarries occur at or near sacred sites – that is totemic "dreaming" places. People who believe themselves to be descended patrilineally from the particular totemic being at one of the sites will make special trips to the quarry to secure the stone there. A man places a high value on stone from a site of his dreamtime totem. Stone like this is often transported over long distances (as much as 500km) and is given to distant kinsmen of the same totemic patrilineage ... Because of his patrilineal relationship to the site, a man sees the stone as part of his own being ...'. Trade or exchange to distant kin thus created tangible links between people and place (cf. Brumm 2004, 153).

The seasonal hypothesis may also be strengthened by the paucity of evidence for Neolithic buildings in general, and the fact that little settlement evidence has been recovered at the mines, all suggesting that mobility was a major feature of the Neolithic lifeway. This can be seen at the South Downs flint mines, which appear to have been primarily extraction and workshop sites which concentrated upon the production of rough-outs and preforms but where permanent settlement was absent (Gardiner 1990, 121). However, caution is still needed as the destructive nature of mining may have destroyed any evidence of settlement, or excavation bias may simply have missed it. Nevertheless, the elusive nature of Neolithic settlement evidence in general in the UK compared to continental Europe is notable, and must reflect the temporality of occupation.

It is not only 'domestic' evidence that might be expected at the mines, but functional, mining-related structures should also be anticipated. At Harrow Hill, excavations near the opening of Shaft 13 recorded sub-circular depressions, two of which were 'deep and quite steep-sided' (McNabb *et al.* 1996, 28-9). Their location could suggest foundations for a winch; part of an earlier building; or possibly some form of roof structure to prevent flooding in the mine as has been postulated at Krzemionki - which could enable year-round use (W. Borkowski and W. Migal pers comm.; Borkowski 1997, 46-7). Recent surveys of both Grime's Graves and Harrow Hill recorded several level platforms scattered amongst the shafts which may have been storage, processing or activity areas with associated structures (cf. Barber *et al.* 1999, 33 and 43).

At Blackpatch, Pull (1932, 52-4) did not discover any structural evidence amongst the mines, nor to the east of the complex; some ambiguous features did contain contemporary assemblages but their character remains unclear. Pull also systematically explored Church Hill '... over an area of nearly half a square mile in the vicinity of the shafts' (Pull 1935), but these excavations similarly proved inconclusive in terms of discovering structures. Consequently, the weight of evidence does suggest that mining was a seasonal activity, which explains why the archaeological record contains little more than ephemeral traces of structures and cultural debris typical of short-term occupation (see below). The lack of permanent settlement might also imply that the mines were 'special' places where settlements may have been restricted by cultural taboos as was recorded at the Pipestone (Catlinite) Quarries in Minnesota (cf. Topping 2005, 87).

The question of why certain extraction sites achieved a special status in the cultural landscape is of interest. Extraction sites were not used lightly. The dangers of mining – both physical and metaphysical - would have heightened the social value of mined materials, becoming as Bradley (2000, 85-90) has suggested 'pieces of places', embedded with a deeper symbolism than more easily obtained materials. This is graphically illustrated at the dangerously exposed axe quarries on the near vertical faces of Top Buttress on Pike O'Stickle at Langdale. It was the degree of difficulty invested in procuring recognisable or mythologised types of stone from culturally significant locations that visibly labelled the context of the raw material and thus created its value to society. Identifiable varieties of stone –such as the banded flint from Krzemionki in Poland or jadeitite from the Italian Alps– embedded biographies into an artefact, which then transmitted cultural narratives to the community and beyond. These cultural narratives could then reinforce kinships, clans and alliances, and maintain belief systems.

4. Downtime at the mines

There are a number of practical reasons which emphasise why flint mines may only have been exploited seasonally. The location of many sites on or near downland crests would have introduced both operational and access issues, suggesting that they could only have been safely used seasonally during the warmer months of the year when there was longer daylight hours. This situation would have been exacerbated at the high altitude axe quarries such as Great Langdale in the UK which exploited Seathwaite Tuff, or Monte Viso in the Italian Alps where Jadeitite was quarried, both of which would have been closed by snowfall each winter. In addition, during the autumn and winter months daylight becomes restricted to some seven hours by mid-December, and as there is little evidence for artificial light in any of the English flint mines, such conditions will have significantly affected underground mining. At this time of the year reflected light would not have penetrated so far down the shafts and into the galleries because of the lower declination of the sun which would have limited direct natural light to the upper parts of the vertical shafts and therefore reduced the amount of reflected light into the galleries.

As an added complication, this period of the year also sees the wettest climatic conditions, turning chalky surfaces slippery, and offering the potential of flooding in unprotected workings. At Krzemionki Opatowskie in Poland,

Flint mines [Sussex data from Gardiner 2001; Grime's Graves from Saville 1981 & G Varndell pers comm]	Flint axes (maximum number of all types, including miniatures)	Largest non-axe tool type	Total number of tool types in each site assemblage
Early Neolithic sites			
Blackpatch v	149	35 [cores]	27
Church Hill	64+	266 [flakes/blades]	18
Cissbury	1052 [+1 stone axe]	79 [cores]	33
Harrow Hill	152	7 [cores]	10
Late Neolithic sites			
Grime's Graves	374+	1015 [points]	15

Figure 3. Assemblage variation at the English Neolithic flint mines.

evidence of flooding was recorded in the galleries of mine #795 and in the 'Great Chambers' (Bąbel 2008, 106-107), which clearly occurred before the shafts were backfilled –thus providing evidence that some shafts had remained open and uncovered during the wetter parts of the year. The fact that these mines were flooded would, by force of circumstance, have prevented underground working and thus arguably enforced seasonal exploitation in at least some of these Polish mines. However, other shafts are claimed to have had 'Sheds ... built over the chamber shafts to protect the miners and the mine from rain and snow' (Bąbel 2008, 98), and that the deeper mines were used in the 'cold season', although the evidence for this claim is not presented.

At other European mines where there is evidence for adjacent 'miners' settlements, such as the Michelsberg enclosure at Petit-Spiennes (Collet *et al.* 2008, 47-48), there is clearly the potential for year-round working if seasonal climate-related difficulties could be overcome. At Casa Montero near Madrid, however, the excavation data suggested seasonal working by small teams, and that the mining could be characterised as 'the result of sequential short term activities ... over a relatively short period of time', perhaps 13 shafts per year excavated over a period of some 300 years of mining activity (Capote *et al.* 2008, 129-130). In continental Europe, therefore, the overall picture of the intensity of flint mining is clearly variable. To date, the only potential evidence in England for structural remains adjacent to a mine was the fragmentary structure discovered next to Shaft 13 at Harrow Hill (see above). The conclusion must be that if buildings or roofing existed at the English flint mines, then much of this evidence has subsequently been destroyed by mining when successive shafts have disturbed by earlier workings.

However, excavation data records other evidence which demonstrates that the English flint mines were not roofed, thus providing additional support for the seasonal hypothesis. At Grime's Graves, both the Pit 1 and Pit 2 shaft sequences record at least ten phases of backfilling with evidence of stabilised horizons and windblown silts, suggesting episodic filling –the windblown silts would not have accumulated with a roof in place. Such an interpretation is confirmed by the presence of bats in galleries 1 and 8 in Pit 1 (Clarke 1915, 55 and 58) and 1, 3, 5 and 6 in Pit 2 (Clarke 1915, 90), which demonstrates that the mines and certain galleries remained open and were quiet long enough to be colonised by these timid creatures during their roosting season from October to March. Consequently, the presence of bat roosts at Grime's Graves demonstrates that this was a period of the year when some mines remained open, were quiet and temporarily abandoned. Similar episodic backfilling sequences have been recorded elsewhere, for example at Shaft 27 and the 'Cave Pit' at Cissbury, but without the presence of bats which 'may simply reflect a different biomass, a range of habitat constraints or excavation bias' (Topping 2005, 72). However, importantly, windblown silts in these fills demonstrate again both a lack of roofing over the shafts, and the presence of stabilised horizons provides further evidence for episodic backfilling. Such evidence thus confirms periods of downtime, in this case at certain flint mines in Sussex.

5. Artefact assemblages

Other circumstantial evidence also indicates the seasonal use of flint mines. Studies in the USA and Australia have found that assemblage variation not only reflects activity and depositional processes 'but also the occupation span at a particular place' (Doelman 2008, 145-146). This hypothesis suggests that the level of tool diversity correlates with the length of site use, thus the greater the range of tool types in an assemblage the longer the 'duration of occupation'. The logical conclusion of this model implies that a limited tool assemblage at a given site is a reflection of group mobility, whereas high diversity represents more sedentary occupation with a greater range of subsistence tasks.

The application of this model to the English flint mines is interesting, despite acknowledging the potential effects of excavation bias and taphonomic processes. The data - as a sample of the total assemblage - still provides an indication of trends and tool type priorities at the various mines. The four Early Neolithic flint mines near Worthing in Sussex mainly focussed upon axe production (see Figure 3). Even at Church Hill where, unusually, the number of axes is outnumbered by flakes/blades (some of which may be debitage from axe making), axes here still formed the second largest artefact group. The greatest assemblage variation occurs at Cissbury, but this is also the largest mine complex of the four and may simply reflect a longer period of use. The overall composition of the Worthing assemblages demonstrates that the priority product was the axe, complemented by a secondary range of tools.

By the Late Neolithic, the focus at Grime's Graves had moved away from axe production to the crafting of points (non-arrowheads). However, significant numbers of axes, discoidal knives, scrapers, rods and flake tools were still produced, alongside minority tool types such as burins, knives and fabricators; dressed nodules may also have been shipped off-site. The reduction in the importance of axes at Grime's Graves is paralleled by the small numbers of off-site axes attributed to these mines by trace element analysis (Craddock *et al.* 1983). Clearly, axes had begun to loose some of their cultural importance by the later Neolithic period, and had been eclipsed by more mundane tool types. However, despite this re-focus, the overall number of tool types represented at Grime's Graves is still the second lowest of any flint mine at 15, and would fit the hypothetical model for the short term use of the site and the restricted repertoire of Grooved Ware users.

Further evidence for seasonal working is provided by the use of antler picks in the mines. Critically, antlers are shed between March to May, and the detailed study of antler picks from Grime's Graves has shown that over 80% were shed, less than 20% came from dead stags, thus proving that the majority were collected rather than being products of the hunt (Clutton-Brock 1984, 13 and 16). Deer shed their antlers in the same place each year (Clason 1981, 122) and this predictability may have assisted the seasonal collection of antlers. The fact that antlers are only carried by stags limits their availability, and their dense woodland habitats make them difficult –if not impossible– to herd. This prompted Clutton-Brock to state that the antlers from Grime's Graves '.. retain the lack of variability ... which reflects a population that was still untampered with by man' (Clutton-Brock 1984, 23).

An additional perspective is provided by deer behaviour. Population studies have shown that each individual red deer stag inhabits a territory of roughly 10 hectares of mature forest, and excavations have suggested that an average of 140+ antler picks were used in each mine at Grime's Graves. These statistics suggest that the process of collecting sufficient antlers would have involved the miners in foraging over a minimum area of roughly seven square kilometres or 1,400 hectares (allowing for two antlers per 10ha); a greater area would have been required if the numbers of shed antlers was erratic. Consequently, the logistics of procuring enough antlers for use in the mines may have been a considerable time commitment and another reason why mining was a seasonal activity to allow the miners time to collect enough antler picks before mining could begin.

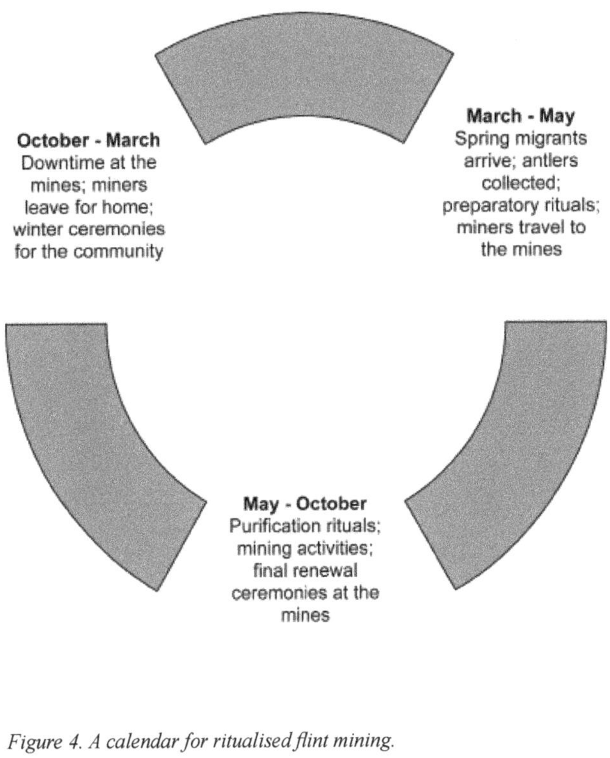

Figure 4. A calendar for ritualised flint mining.

6. Conclusions

In summary, the evidence for the seasonal use of the English flint mines includes the possibility that the appearance of spring migrants may have initiated mining activities, particularly colourful birds in their breeding plumage (cf. Colston and Burton 1988, 141, pl 22). This would have coincided with a general improvement in weather conditions, combined with increased daylight hours which would have created easier access to the mines: such safety considerations are survival-defining parameters at dangerous extraction sites.

The apparent lack of permanent settlement implies that only temporary encampments existed at the mines. The fact that some mines were clearly left open during the winter months which allowed bats to colonise the galleries provides further proof of periods of downtime, when the mines were temporarily abandoned during the season of shorter daylight hours and wetter climatic conditions. The radiocarbon chronology at Grime's Graves and elsewhere suggests that as few as one or two mines may have been open each year, which undermines any suggestion of mass production.

So, how does this all fit together? The various strands of evidence suggest the flint miners followed an annual round, much of which was influenced by seasonal indicators and climatic conditions loosely occurring around both the Spring and Autumn equinoxes, both of which were significant periods in Neolithic spiritual and ceremonial life. The clear evidence for downtime at the mines indicates that they were temporarily abandoned during the period when bats roosted between October to March. The appearance of colourful spring migrants such as the phalaropes between March to April arguably triggered the preparations of the miners; this also coincided with the shedding of red deer antlers during the period March to May when they could be collected for use in the mines. Consequently, following these events mining could then begin during April to May, and it would continue until the onset of significantly shorter daylight hours around the Autumn equinox in October. This would then bring mining to an end at the same time of the year as other annual ceremonies surrounding the harvest, world renewal and year-end festivities were being staged at a range of sites.

This conjunction of major events surrounding the autumn equinox would provide a social platform for the miners to return to their communities with their 'harvest' of flint. During this Autumn-Winter downtime, if ethnographic records are a reliable indicator, we should envisage renewal ceremonies taking place to ensure the future supply of flint. These ceremonies would create a social mechanism by which the miners, through the agency of gift giving or using the flint and their knapped tools as collateral, were able to play a central role in maintaining kinship links and alliance building, and reinforcing social narratives. In addition, during this period of ceremony and festivities, the miners may have introduced new initiates into the 'mysteries' of flint mining. Downtime was therefore as important as actual mining for both the miners and the social relations of their communities.

Acknowledgements

The author would like to thank Dave Field, Gill Varndell and Alan Saville for many fruitful discussions about flint mines, all of which have shaped the ideas presented here. Gill Varndell read and commented upon this paper and also helpfully provided information about the Grime's Graves assemblage in advance of the British Museum's own publication of Fascicule 6. To all go the author's grateful thanks. Any errors or misrepresentations are the sole responsibility of the author.

Bibliography

Bąbel, J. 2008. The Krzemionki flint mines latest underground research 2001-2004, in P. Allard, F. Bostyn, F. Giligny and J. Lech (eds.), *Flint Mining in Prehistoric Europe, Interpreting the archaeological records*, 97-110. Oxford, British Archaeological Reports International Series 1891.

Barber, M., Field, D. and Topping, P. 1999. *The Neolithic Flint Mines of England.* Swindon, Royal Commission on the Historical Monuments of England / English Heritage.

Boivin, N. 2004. From veneration to exploitation, human engagement with the mineral world, in N. Boivin and M. A. Owoc (eds.), *Soils, Stones and Symbols, Cultural Perceptions of the Mineral World*, 1-29. London, UCL Press.

Borkowski, W. 1997. Deposit Management System of the Krzemionki Exploitation Field, in R. Schild and Z. Sulgostowska (eds.), *Man and Flint, Proceedings of the VIIth International Flint Symposium*, 41-48. Warszawa, Institute of Archaeology and Ethnology Polish Academy of Sciences.

Bradley, R. 2000. *An Archaeology of Natural Places.* London, Routledge.

Brumm, A. 2004. An axe to grind, Symbolic considerations of stone axe use in Ancient Australia, in N. Boivin and M. A. Owoc (eds.), *Soils, Stones and Symbols, Cultural Perceptions of the Mineral World,* 143-163. London, UCL Press.

Capote, M., Castañeda, N., Consuegra, S., Criado, C. and Díaz-del-Río, P. 2008. Flint mining in early Neolithic Iberia, A preliminary report on 'Casa Montero' (Madrid, Spain), in P. Allard, F. Bostyn, F. Giligny and J. Lech (eds.) *Flint Mining in Prehistoric Europe, Interpreting the archaeological records*, 123-137. Oxford, British Archaeological Reports International Series 1891.

Clarke, W. G. (ed.) 1915. *Report on the Excavations at Grime's Graves, Weeting, Norfolk, March - May 1914.* London, Prehistoric Society of East Anglia.

Clason, A. T. 1981. The Flintminer as a Farmer, Hunter and Antler Collector. In *Third International Symposium on Flint, 24-27 Mei 1979, Maastricht*, 119-125. Nederlandse Geologische Vereniging.

Clough, T. H. McK. and Cummins, W. A. 1988 (eds.), *Stone Axe Studies*, vol 2, The Petrology of Prehistoric Stone Implements from the British Isles. London, Council for British Archaeology, Archaeological Research Report 67.

Clutton-Brock, J. 1984. *Excavations at Grimes Graves, Norfolk 1972-1976*, Fascicule 1, Neolithic Antler Picks from Grimes Graves, Norfolk and Durrington Walls, Wiltshire, A Biometrical Analysis. London, British Museum Press.

Collet, H., Hauzeur, A. and Lech, J. 2008. The Prehistoric Flint Mining Complex at Spiennes (Belgium) on the occasion of its Discovery 140 Years Ago, in P. Allard, F. Bostyn, F. Giligny and J. Lech (eds.), *Flint Mining in Prehistoric Europe, Interpreting the archaeological records*, 41-77. Oxford, British Archaeological Reports International Series 1891.

Colston, P. and Burton, P. 1988. *A field guide to the Waders of Britain and Europe with North Africa and the Middle East.* London, Hodder and Stoughton.

Cooney, G. and Mandal, S. 1998. *The Irish Stone Axe Project*, Monograph 1. Bray, Wordwell Ltd.

Craddock, P. T., Cowell, M. R., Leese, M. N., and Hughes, M. J. 1983. The trace element composition of polished flint axes as an indicator of source. *Archaeometry* 25, 135-163.

Curwen, E. and Curwen, E. C. 1926. Harrow Hill flint-mine excavation 1924-5. *Sussex Archaeological Collections* 67, 103-138.

Doelman, T. 2008. *Time to Quarry, The Archaeology of Stone Procurement in Northwestern New South Wales, Australia.* Oxford, British Archaeological Reports International Series 1801.

Fitzpatrick, A. 2003. The Amesbury Archer. *Current Archaeology* 184, 146-152.

Fitzpatrick, A. 2004. The Boscombe Bowmen, builders of Stonehenge? *Current Archaeology* 193, 10-14.

Gardiner, J. P. 1990. Flint procurement and Neolithic axe production on the South Downs, a re-assessment. *Oxford Journal of Archaeology* 9, 119-140.

Gardiner, J. P. 2001. Catalogue of Surviving Flintwork from the Worthing Group of Mines. In M. Russell (ed.), *Rough Quarries, Rocks and Hills, John Pull and the Neolithic Flint Mines of Sussex*, 202-223. Oxford, Oxbow Books.

Garrow, D., Beadsmoore, E. and Knight, M. 2005. Pit clusters and the temporality of occupation, an earlier Neolithic site at Kilverstone, Thetford, Norfolk. *Proceedings of the Prehistoric Society* 71, 139-157.

Gould, R. A. 1977. Ethno-archaeology; Or, Where do Models Come From? In R. V. S. Wright (ed.), *Stone Tools as Cultural Markers*, 162-177. Canberra, Australian Institute of Aboriginal Studies.

Green, M. 2000. *A Landscape Revealed.* Stroud, Tempus.

Greenwell, W. 1870. On the opening of Grime's Graves in Norfolk. *Journal of the Ethnological Society of London*, New Series 2, 419-439.

Longworth, I., Herne, A., Varndell, G. and Needham, S. 1991. *Excavations at Grimes Graves, Norfolk, 1972-1976*, Fascicule 3. London, British Museum Press.

McNabb, J., Felder, P. J., Kinnes, I. and Sieveking, G. 1996. An archive on recent excavations at Harrow Hill, Sussex. *Sussex Archaeological Collections* 134, 21-37.

Pétrequin, P., Sheridan, A., Cassen, S., Errera, M., Gauthier, M., Klassen, L., Le Maux, N. and Pailler, Y. 2008. Neolithic Alpine axeheads from the Continent to Great Britain, the Isle of Man and Ireland. *Analecta Praehistorica Leidensia* 40, 261-279.

Pitts, M. 1996. The stone axe in Britain. *Proceedings of the Prehistoric Society* 62, 311-372.

Pull, J. H. 1932. *The Flint Miners of Blackpatch.* London, Williams and Norgate.

Pull, J. H. 1935. The stone age villages of Downland, no 1, the discovery of Neolithic settlements in West Sussex. *Sussex County Magazine* 9 (7, July), 437-439.

R. C. H. M. E. (Royal Commission on the Historical Monuments of England). 1994. *Harrow Hill, West Sussex [TQ 01 SE 23 and 17].* Swindon, National Monuments Record. Unpublished earthwork report.

Saville, A. 1981. *Grimes Graves, Norfolk, Excavations 1971–72*, vol 2, The Flint Assemblage. Department of the Environment Archaeological Report 11. London, HMSO.

Taçon, P. S. C. 2004. Ochre, Clay, Stone and Art, The Symbolic Importance of Minerals as Life-Force among Aboriginal Peoples of Northern and Central Australia, in N. Boivin and M. A. Owoc (eds.), *Soils, Stones and Symbols, Cultural Perceptions of the Mineral World*, 31-42. London, UCL Press.

Topping, P. 2005. Shaft 27 Revisited, an Ethnography of Neolithic Flint Extraction, in P. Topping and M. Lynott (eds.), *The Cultural Landscape of Prehistoric Mines*, 63-93. Oxford, Oxbow Books.

Territories and lithic resources in the Paris basin during the Middle Neolithic (4200-3600 BC)

Laurence MANOLAKAKIS and François GILIGNY

Abstract

Neolithic tool-makers of the Paris basic had access to a great diversity of lithic material resources, among them—secondary or tertiary flint, sandstone and orthoquartzite. Tertiary flint and secondary flint were most commonly used for domestic production. These materials can be found mainly in the Mesozoic Senonian limestone (Coniacian and Campanian), and in the Tertiary Saint Ouen (Bartonian) silicified limestone. Other materials are exogenous, such as Jurassic flint originating from the flint mines of Caen-Falaise plain of Normandy (the fringes of the Armorican Massif), used for the production of axes.

The presence of axes manufactured from non-local igneous and metamorphic rocks suggests that some axes were brought to this region from the Armorican massif and the Alps. Others may have come from Vosges, the Central Massif or the Ardennes. This evidence consequently points to the Seine valley's role as an important geographical centre of both distribution and exchange of Armorican and Alpine axes. Evidence points to a sharp difference in use of lithic resources between the Chasséen septentrional and Michelsberg.

Keywords

Flint-mine. Paris basin. Middle Neolithic. Procurement system. Chasséen, Michelsberg.

1. Introduction

Neolithic tool-makers of the Paris basic had access to a great diversity of lithic material resources, among them—secondary or tertiary flint, sandstone and orthoquartzite. Tertiary flint and secondary flint can be found mainly in the Mesozoic Senonian limestone of Secondary age (Coniacian and Campanian), and in the Tertiary Saint Ouen (Bartonian) silicified limestone. Other materials are exogenous, such as Jurassic (bathonian) flint originating in Normandy or materials coming from metamorphic rocks of the Alps (jadeite, eclogite), the Armorican Massif (dolerite) and less frequently of the Vosges (pelite-quartz), the Massif Central (both magmatic and volcanic) and the Ardennes (amphibolites), all of which were used for the production of axes.

The presence of axes manufactured from non-local igneous and metamorphic rocks suggests that some axes were brought to this region from the Armorican massif and the Alps. Others may have come from Vosges, the Central Massif or the Ardennes. This evidence consequently points to the Seine valley's role as an important geographical centre of both distribution and exchange of Armorican and Alpine axes.

During the second stage of the Middle Neolithic (4200-3600 BC), two different cultures were present in the Paris basin on either side of the Seine-Oise valleys: Chasséen séptentrional on the western side and Michelsberg on the eastern side. Both cultures are characterized by the same types of sites: enclosures, open settlements, mines and funerary sites.

In the west, in the Chasséen séptentrional, influences from the Chasséen méridional and the Burgundy Chasséen (Chassey) are evident in the pottery, but the lithic industry is clearly based in the local tradition, inherited from the Cerny. In the east, the Paris basin Michelsberg is similar to the Michelsberg of the Rhine both in pottery and lithic industry, with a few differences in the relative frequency of blade and flake debitage.

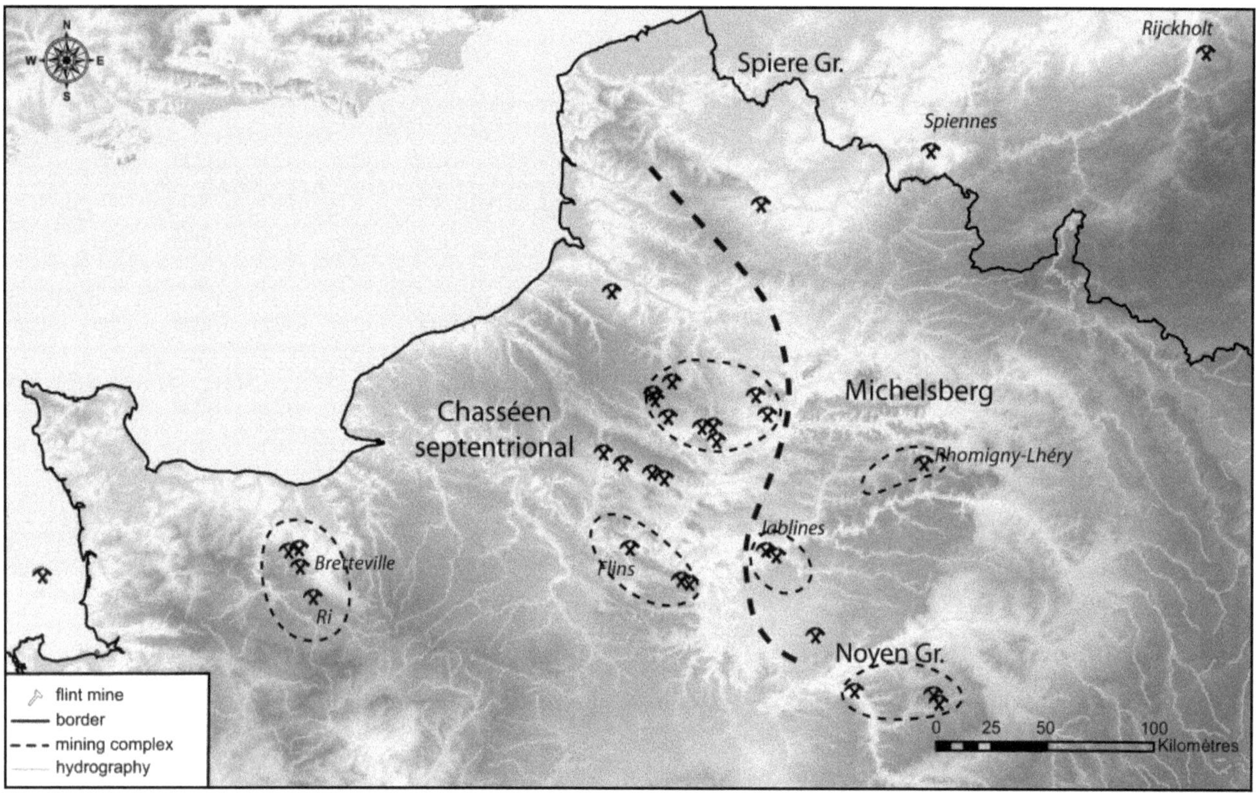

Figure 1. Flint mine distribution and cultural borders during the Middle Neolithic (4200-3600 BC).

2. Flint mines

The distribution of flint mines in the Paris basin is not uniform, whether or not they date to the middle Neolithic (Figure 1). Certain zones, characterized by a concentration of extraction sites, debitage and polishing workshops, correspond to mining complexes. From east to west, the main zones are:

- the Caen-Argentan plain, situated along the Armorican massif.

- the Seine valley, located downstream from Paris.

- the Picardy plateaus, the north and west of the Oise valley.

- the Marne valley.

- the Romigny-Lhéry and Ourcq plateaus.

- the Pays d'Othe between the Seine et the Yonne.

This distribution affected procurement networks, since not all settlements were located near mines or mining complexes. However, flint was accessible almost everywhere in diverse forms and could be obtained without digging a mine shaft, either through collection from colluvial or alluvial deposits, cleaning of natural sections on slopes or through digging shallow pits. In most cases these means of procurement remain undetected, in the absence of archaeological features.

There are many factors complicating the task of dating the start of mining. One of the most important ones is the long duration of their exploitation and the large standard deviations of radiocarbon dates. The earliest date, from Bretteville-le-Rabet (Ly-3680 : 5560+190 BP), covers the period 4826-3975 at 1 sigma and may correspond to the VSG or Cerny as well as to the Chasséen. The sites, for which the largest numbers of dates are available, show clustering by phases, the earliest of which lie between 4200 and 3800 BC (Jablines and Ri). Flint mines appear to be more frequent during this Middle Neolithic period (Figure 2).

The manufacturing activities at the mines were essentially linked to the production of axes. Those were made at the mines and finished near settlements and in sandstone polishers workshops located along rivers. In the Chasséen, the mines seemed to be separate from settlements, whereas in the Michelsberg, enclosures were located in the near proximity of the extraction sites (Spiennes, Jablines), and, as in the case of Spiennes, long blades were produced directly at the mine.

The axes (and Michelsberg blades) produced at the flint mines represent a specialized production and circulate inside their own culture. However, given that identical Bartonian and Senonian flints can be found in both culture areas, it is difficult to identify circulation of axes made of these materials between Chasséen and Michelsberg. This difficulty is clearly evident for the mines like Jablines, located on the border between the two cultures. In the Chasséen area, the products circulated between mining complexes,

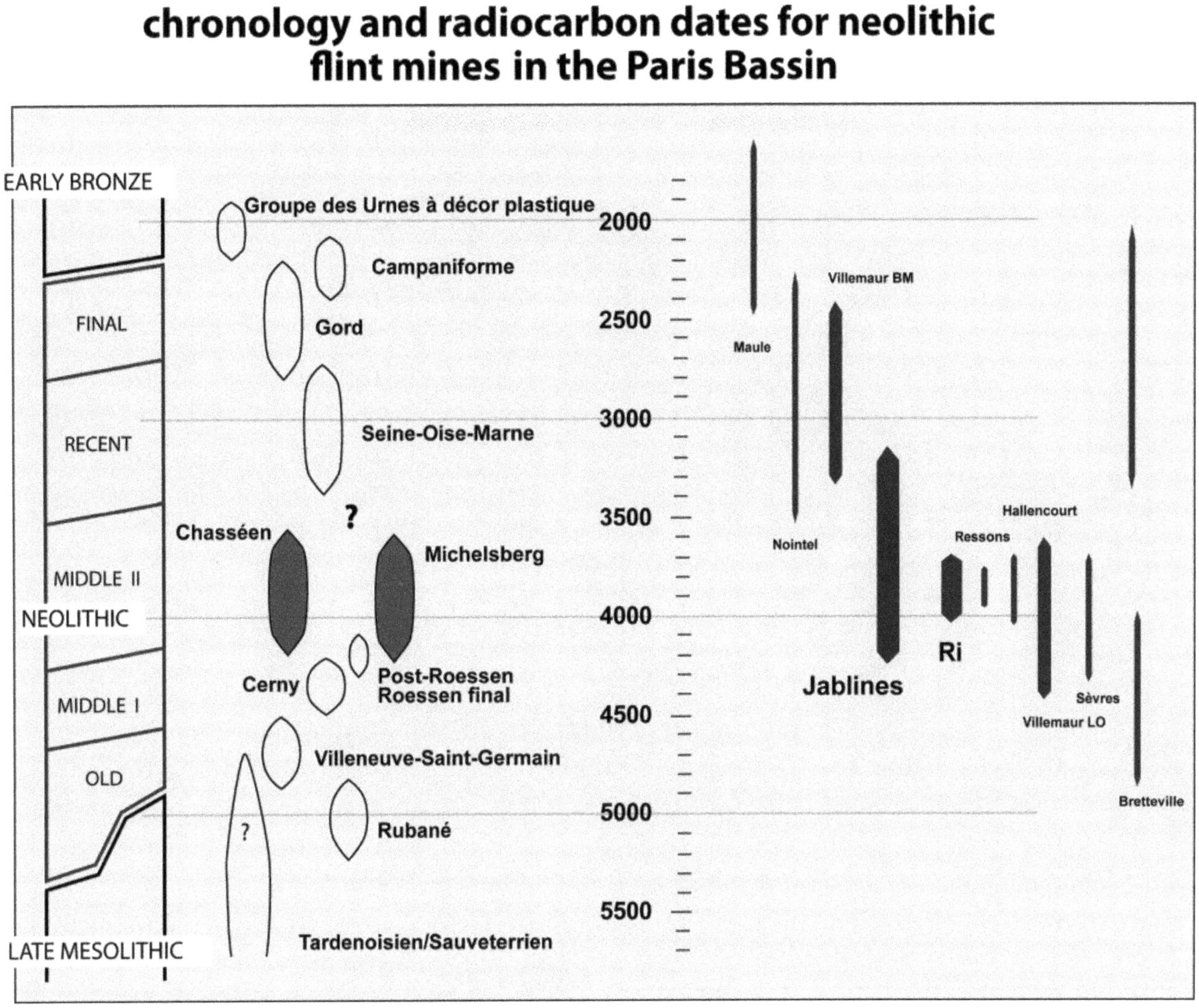

Figure 2. Chronology and radiocarbon dates for Neolithic flint mines in the Paris basin.

illustrated by the example of Bathonian and Bartonian flint axes, which moved in both directions between the Caen plain and the Seine valley. A small number of imported axes were even present in the flint axe production zone (Giligny *et al.* 2009).

Site rescarchers have identified several procurement distances, ranging from local scale up to distant products. We have distinguished three categories of procurement distance available simultaneously for domestic and specialized axe production: local (0-10km), regional (10-70km), extra-regional (>70km). The products made of long distance materials included in the extra-regional category are sometimes called 'exotic'. Here, we do not make such a distinction and consider even alpine rocks as extra-regional.

3. Settlement and enclosure location

Almost all the Chasséen settlements were located on flint sources, even when raw material was of poor quality (Figure 3). This pattern suggests that the presence of flint resources was an important factor in the choice of settlement location. Louviers is a flood-plain site next to a channel. The secondary flint was extracted from the side of the Eure valley and debitage waste discarded on the surface of the former channels around which the Chasséen settlement developed (Giligny *et al.* 2005). At Jonquières, the raw material came from the vicinity (2 to 4km) (Blanchet et Petit, 1972). At Chartres, the raw material used at the site was of poor quality and came from the area located at 1km distance (Garmond in press).

The frequency of axes in extra-regional materials is higher at Louviers (36%) than at the other Chasséen sites, such as Boury-en-Vexin and Jonquières, where it reaches only 9 to 10%. The site's geographical location, adjacent to a major communication axis, the Seine valley, is the probable explanation.

In the Michelsberg, although debitage is mostly of flakes, genuine domestic production of two kinds of blades existed: one, in soft direct percussion, employed Senonian and

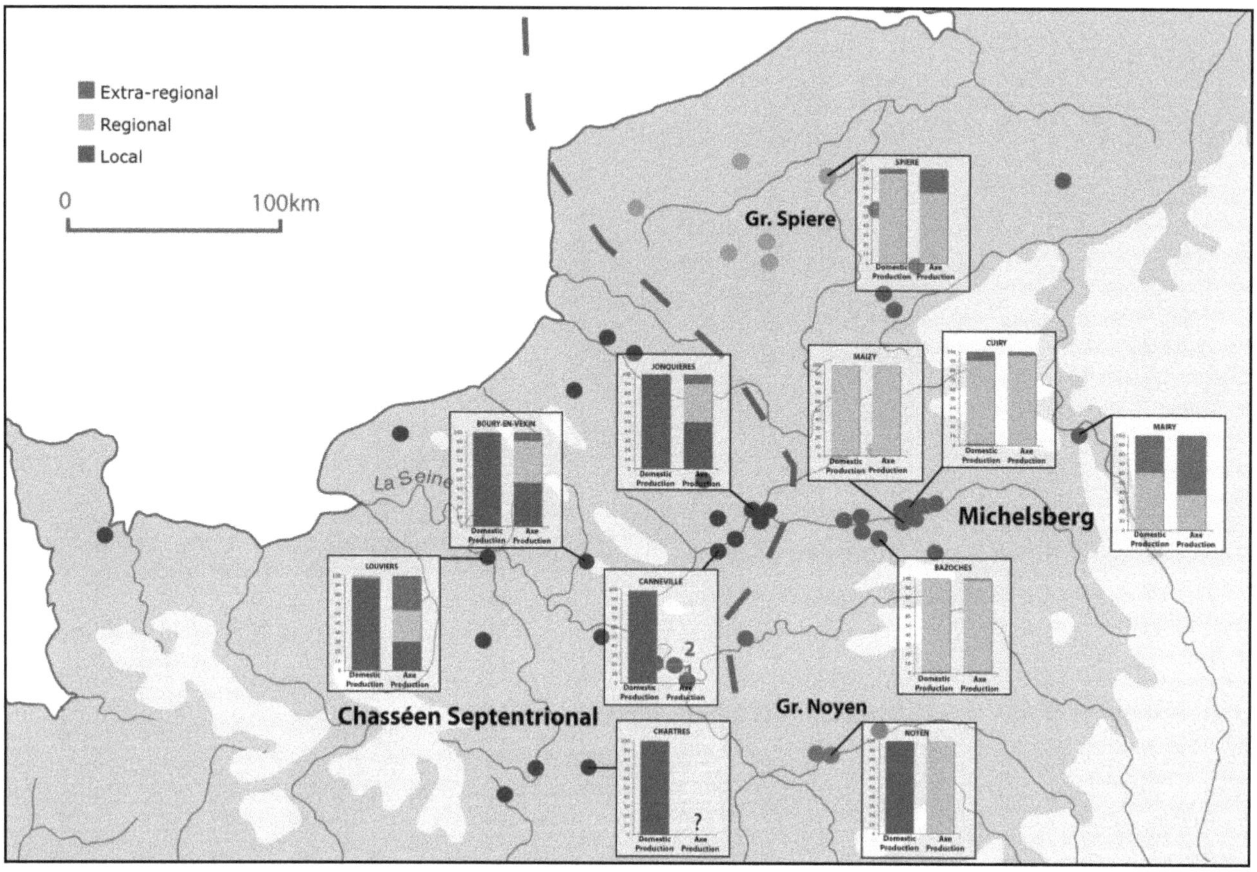

Figure 3. Domestic production versus axe production in Michelsberg and Chasséen settlements and enclosures.

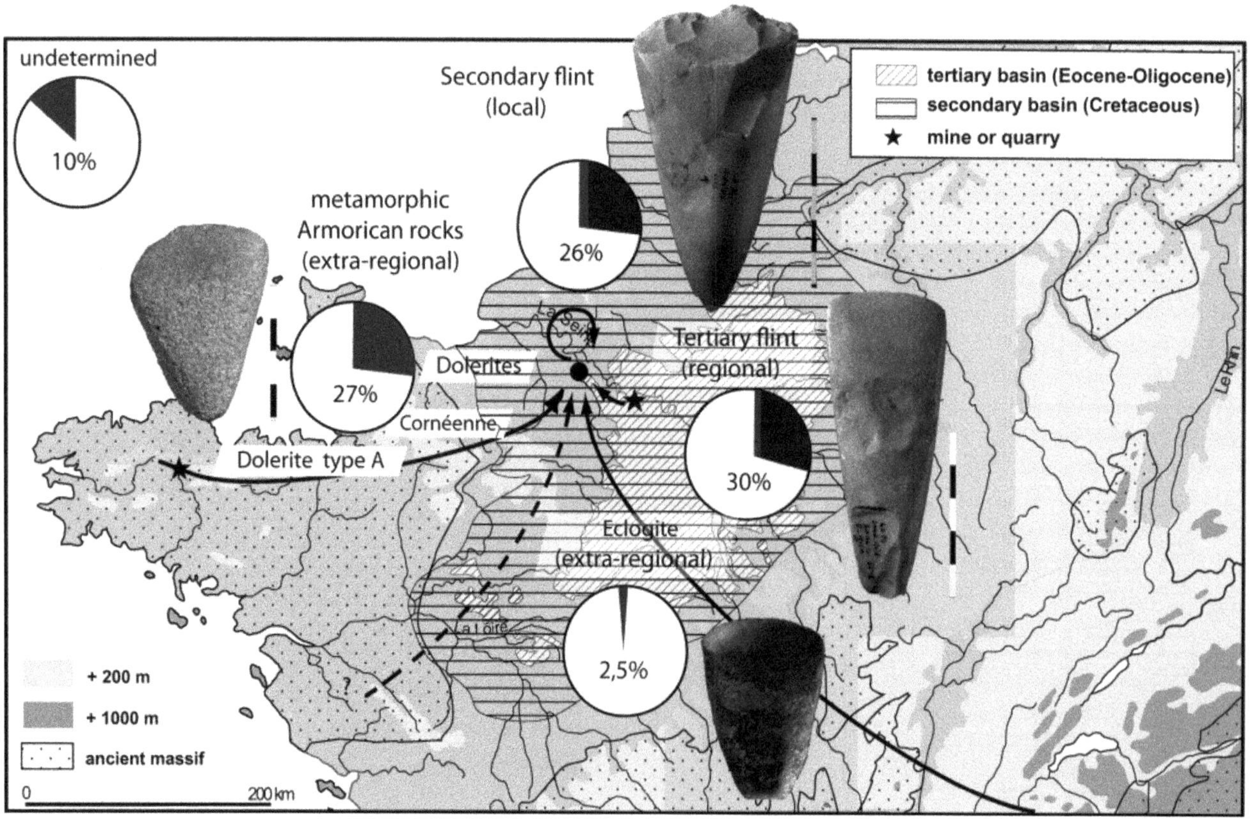

Figure 4. Raw material procurement for axes at Louviers (Chasséen, 3900 BC).

Bartonian flint, and the other, in indirect percussion, was of long blades, employing Bartonian flint from Romigny, Campanian flint from Spiennes, or Maastrichtian flint from Rijckholt (De Grooth 1991). The long blades seem to have been produced at the flint mines. In the settlements located in areas without any flint sources such as those in the Moselle valley, distant materials could be easily obtained (Figure 3) from Champagne, Belgium or the Netherlands. In the Michelsberg, flint axes were made of the same materials as those used in the domestic production. Metamorphic and igneous axes are quite rare or absent. The exchange networks do not seem to extend beyond the culture area.

4. Axe procurement system on Chasséen sites

In the Chasséen séptentrional, domestic production and common tools were made of strictly local flint (Figure 4). They are made on flakes, sometimes on irregular blades, or long flakes, unlike the Chasséen méridional in southern France. Axe production follows a different pattern : exogenous materials are more numerous than local ones. For example at Louviers, only 1/3 of axes are in local flint, 1/3 are in exogenous flint and 1/3 in exogenous rocks. These metamorphic and igneous rocks come from the Amorican Massif or the Alps.

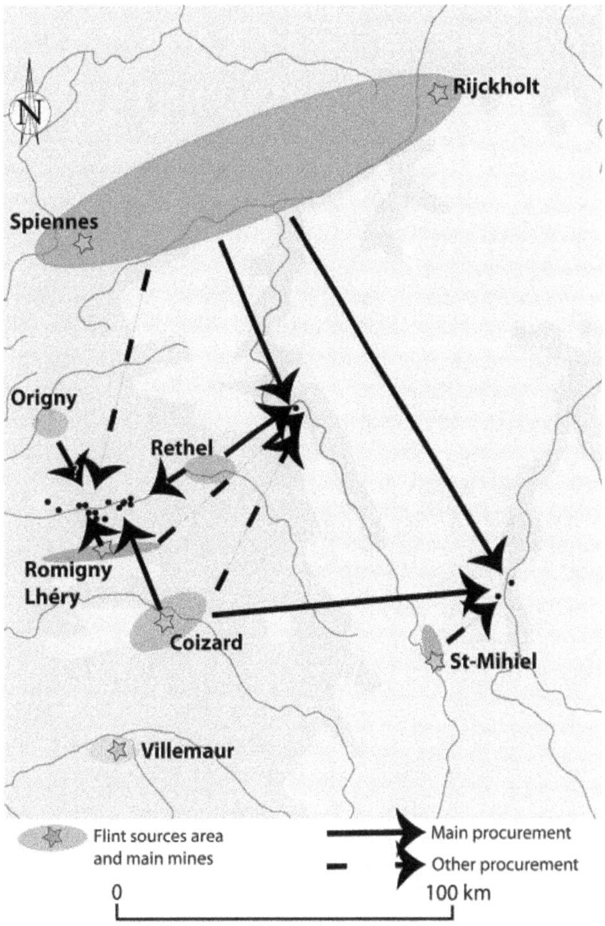

Figure 5. Main flint procurement for Michelsberg settlements and enclosures.

5. Flint procurement system on Michelsberg sites

In the Michelsberg area, some enclosures like Spiennes were located on flint deposits and used the Campanian flints from the mining complex. But settlements were not always located on flint sources. The procurement patterns demonstrate a preference for good quality materials (Figure 5). In the Aisne valley, mainly Bartonian flint from the Romigny and Lhéry zone (at 20-40km distance) was procured and Senonian flint, probably from the Petit Morin valley or the Aisne-Oise confluence, was used as a complementary resource. The extra-regional flints from the Mons basin appear in very small quantities relative to other raw materials, as do the regional flints of medium quality (Turonian from Rethel). The local flints (Thanetian) of small size and medium quality were not used.

On sites located in zones where raw materials were scarce, such as Mairy, flint procurement was mainly regional (60%) or extra-regional (40%), particularly in the production of the axes. In the Moselle valley, procurement was entirely extra-regional, almost equally divided between Paris basin flints and flints from the Belgian-Dutch zone.

The Chasséen séptentrional and Michelsberg are contemporaneous cultures having distinct patterns of flint procurement and production. In the Michelsberg, raw material procurement was characterized by the search for good quality flint for domestic production, whereas in the Chasséen, the producers employed local flint material, even that of poor quality. This probably implies varying difficulties of access to flint and/or products from the mines: the communities in the Chasséen area did not have dependable access to good quality flint from the mines and thus needed the guarantee of a local source. *A contrario*, Michelsberg communities had a reliable access to flint and/or products from the mines, regardless of their distance.

References

Augereau, A. and Hamard, D. 1991. Les industries lithiques du Néolithique moyen II des vallées de la Petite-Seine, de l'Aisne, de l'Oise. In A. Beeching *et al.* (Dir.), *L'identité du Chasséen*, Actes du colloque international de Nemours 1989. Mémoires du musée de préhistoire d'Ile-de-France, 4. 235-250.

Bostyn, F., Lanchon, Y. (dir.). 1992. *Jablines, Le Haut Château (Seine-et-Marne): une minière de silex au Néolithique.* Paris, Maison des Sciences de l'Homme, Documents d'Archéologie Française 35.

Collet, H., Hauzeur, A. and Lech, J. 2008. The prehistoric flint mining complex at Spiennes (Belgium) on the occasion of its dicovery 140 years ago, in P. Allard, F. Bostyn, F. Giligny and J. Lech (dir.), *Flint Mining in Prehistoric Europe. Interpreting the archaeological records,* 41-78. Oxford, Hadrian Books. British Archaeological Reports International Series 1891.

de Grooth M. 1991. Socio-economic aspects of neolithic flint mining: a preliminary study. *Helinium* XXXI (2), 153-189.

Desloges J. 1999. Une mine de silex au Néolithique. L'exemple de Bretteville-le-Rabet, in G. San Juan and J. Maneuvrier (dir.), *L'exploitation ancienne des roches dans le calvados : Histoire et archéologie*, 53-77. Caen, Couleurs Calvados, Service départemental d'Archéologie du Calvados & Société Historique de Lisieux.

Garmond, N. i. p. Typologie et technologie du débitage dans le Chasséen Septentrional : l'exemple du Locus 1 du « Parc d'Archevilliers » à Chartres (Eure-et-Loir), Actes du 28e Colloque Interrégional sur le Néolithique, Le Havre, 9-10 nov. 2007, Presses Universitaire de Rennes.

Giligny, F., Augereau, A., Billiou, D., Binet, C., Bocherens, H., Delattre, N., Guillon, M., Lebret, P., Limondin, N., Léon, G., Monchablon, C., Morzadec, H., Perrault, C., Philibert, S., Reckinger, F., Sidera, I., Théron, V. *et coll.*, 2005. *Louviers "La Villette". Un site Néolithique moyen en zone humide.* Rennes, Document Archéologique de l'Ouest.

Giligny, F., Bostyn, F., Couderc, J., Le Maux, N., Lethrosne, H., Lo Carmine, A. and Riquier, C. 2009. Axe Production and Exchange in the Seine Valley (France), *Internet Archaelogy,* Issue 26, September 2009

http://intarch.ac.uk/journal/issue26/giligny_toc.html

A New Approach for analysing Mining Production and Management combining Geomorphological, Geological and Physical approaches: the case of Ri/Fresne Neolithic flint mine, France

Rodrigue TSOBGOU AHOUPE, Cyril MARCIGNY, With collaborations of Valérie DELOZE, Emmanuel GHESQUIÈRE, François CHARRAUD, Erik GALLOUIN and Laurent QUESNEL

Abstract

In this paper we present an analysis of the organization and strategies of extraction of materials through the application of geo-studies. In Europe, the scholars of Polish flint mines pioneered the research on the methods used by prehistoric miners and the nature of their products. At Ri, we have looked at the relationship among the stratigraphy, the fabric of flint, and its physical qualities. We correlate the type of product, the nature of raw material, rock properties, limestone layer characteristics and evolution of mining strategies. This approach has turned out to be very productive and resulted in a number of important findings about the situation and nature of archaeological structures of extraction and axes production. We compare these data with the results obtained at other European Neolithic mines and suggest a number of new interpretations of mining economies and European production systems.

Keywords

Flint. Geomorphology. Petrography. Micropalaeontology. Physical properties. Mining strategies. Mining production. Neolithic. Normandy.

1. Introduction

The analyses of rock mining and quarrying have been an important part of many research projects in European prehistory. These studies have considered an array of raw materials extracted, the best examples being the materials exploited during the Neolithic such as the metadolerite of Plussulien (Le Roux 2002) and jadeites and eclogites of the Italian Alps (Pétrequin *et al.* 2003).

Flint mine locations during the Neolithic in Europe were conditioned by the nature of sedimentary rocks from the Mesozoic and Cainozoic. To compare the results obtained at Ri (Orne, France) with those of others flint mines of Europe, we have chosen a number of well-studied and known Europeans mines. Spiennes mining complex in Belgium (Hubert 1978, 1988; Gosselin 1986), Kleinkems near Efringen-Kirchen in Germany (Lais 1948; Diethlem 1997), Grime's Grave, Windover, and Harrow Hill in England (Greenwell 1870; Mercer 1976; Curwen 1928; Sheperd 1980a, 1980b; Sieveking 1980) are great examples of Neolithic flint mining economies.

In France, an important example of flint mining is Jablines 'Le Haut Château' in Seine-et-Marne (Bostyn and Lanchon 1992). In Normandy, the research on flint mining was initiated by the discovery of the Neolithic mine of Bretteville-le-Rabet (Desloges 1986, 1999), dating to the beginning of the middle Neolithic (Cerny Culture), and later enriched with the finings from other sites such as Pontigny,

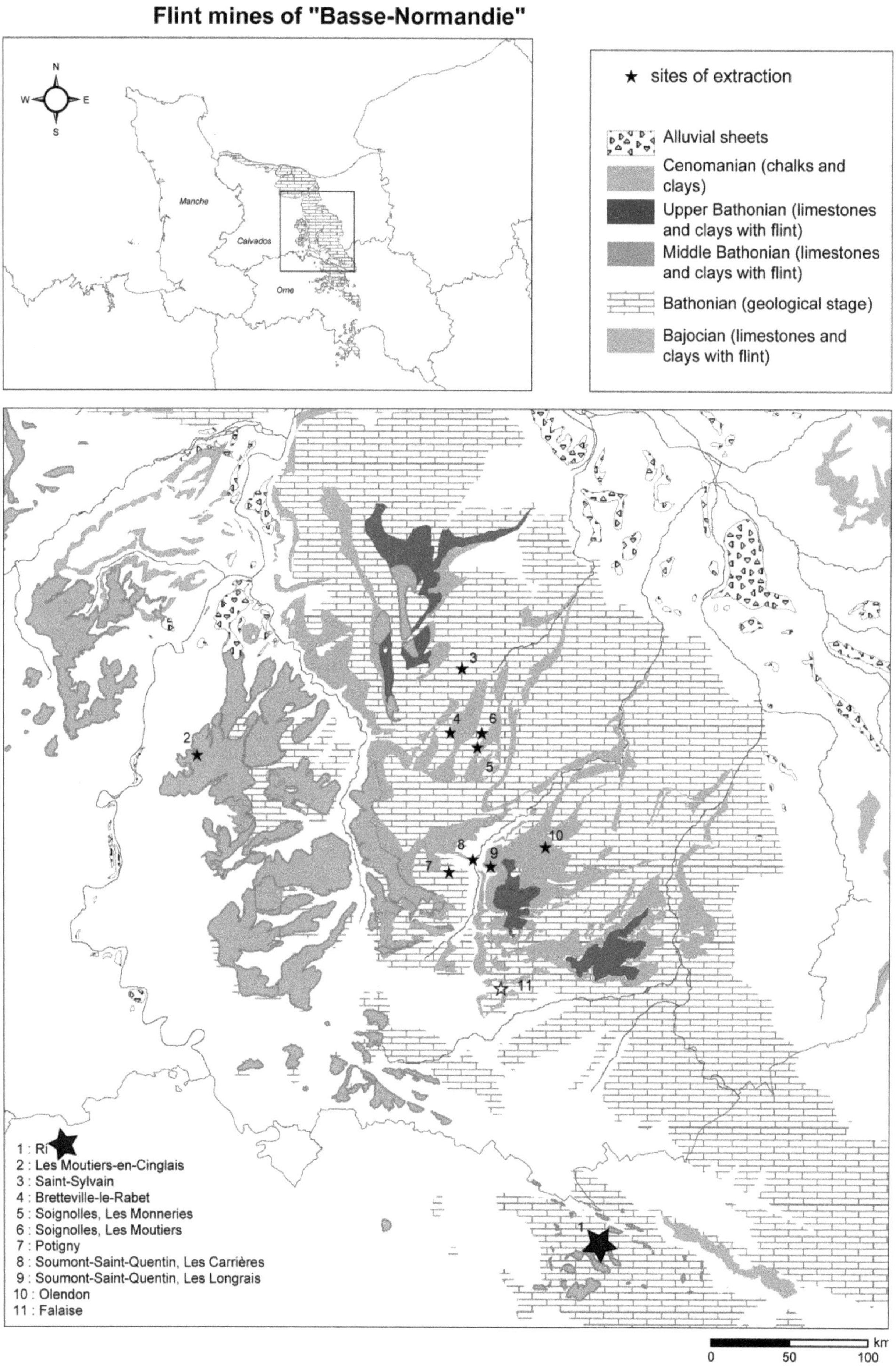

Figure 1. Geological map of 'Basse-Normandie' and location of regional Neolithic flint mines (data from the MCC, modified).

Soignolles, and Soumont-Saint-Quentin (Calvados), and the mine of Ri (Orne) (Figure 1).

Flint mines were bored in the substratum of clays and limestones. Flint mined from clays was important in earlier Neolithic industries in Normandy and Brittany, but the exploitation of limestone resources characterized the start of the middle Neolithic. Found in the course of the construction of the A88 highway in France, the Neolithic flint mine of Ri is dated at 4000-3500 cal BC (Middle to Recent Neolithic: Chasséen Culture).

The mine is located at an altitude of 200 to 250m NGF ('Nivellement Général de la France'), on a slope oriented south-westward, at the border between the Armorican Massif and the Parisian Basin. The Jurassic marine transgressions on this part of the continental basement were possible because of the lower altitudes of the area (Kuntz *et al.* 1989). The mine is then, located on a Jurassic limestone formation (Limestone Formation of Caen), and dated to the lower and middle Bathonian (Gigot *et al.* 1999). The mine's surface area is estimated approximately at 30ha, of which only 2ha have been excavated. It is marked by a series of holes, pits and galleries, at different depths, depending on the nature of strata and their distribution. Many knapping accumulations were discovered all over the surface of the mine, and the extraction structures contain a large quantity of deer antlers.

The mining activities and production play a key role in the Neolithic settlement implantation and socio-economic evolution. The type of production, organization of the mine, procurement of digging tools, and exchange of production output actively shaped habitats, group organization and social relations. The research projects on prehistoric mining have deployed numerous approaches. Geo-studies, combining geological, geomorphological and geochemical studies, and examination of technologies (digging and debitage techniques) constitute the principle scientific approaches used within this line of investigation. But the scholars of prehistoric mining rarely draw on physics or applied mechanics data. In this work, we have attempted to enhance the understanding of mining conditions and products by combining geo-studies and physics analyses of flint and technologies of its extraction.

2. Geomorphologic and Stratigraphic contexts of Ri exploitation

The limestone or clay formations containing flints present different outcropping characters. In the south of Ri, the Houay's valley is filled with fine siltoargillaceous and siltoarenaceous alluvial deposits from the Holocene. The color of these sediments is grey to brown. They are sometimes hydromorphic at high depths and contain thin geological layers rich in organic material or peaty levels. Silty colluviums from the Holocene are located strictly in small dales, near the Houay's Valley. These colluviums are formed by the destruction of the covering loessic sediments and are melted to gravels from the weathering of autochthonous rocks. The loessic covering (Weschelian loess) is decarbonated and 2 to 4-meters-thick on plates, to 5 meters-thick on the steep valley sides.

From the bottom to the top, the 15 geological profiles or drillings (Figure 2) are present at Ri:

- The bottom of the stratigraphy is characterized by the presence of a pinkish fresh sandy limestone (L5). The thickness of this layer is unknown, but it is closest to the surface south of the mine, at 1.2-1.6m of depth. In the north, it did not appear at the depth greater than 2 meters. The limestone of Caen is dated from lower to middle Bathonian (Gigot *et al.* 1999). This limestone is a pelmicrite to biopelsparite (Figure 3b- S15, C3-C4) and fined.

- A stratified and hardened limestone facies of the Limestone of Caen appearing in the south of the mine (L6). Its hardness is the result of silica transfer from the upper sandy loam layer L3. This layer disappears in the north because of the increase of weathering. A carbonated layer (L7) is present at D8 and intercalated in L3 sandy loam. It seems to be the equivalent of the upper L6 level.

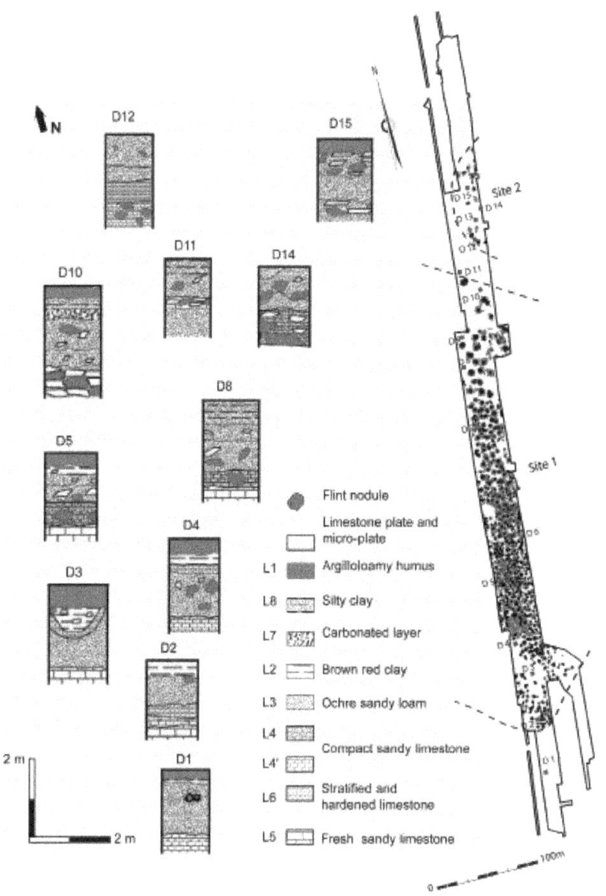

Figure 2. Stratigraphic columns of Ri flint mine and the sketch map of mines structures.

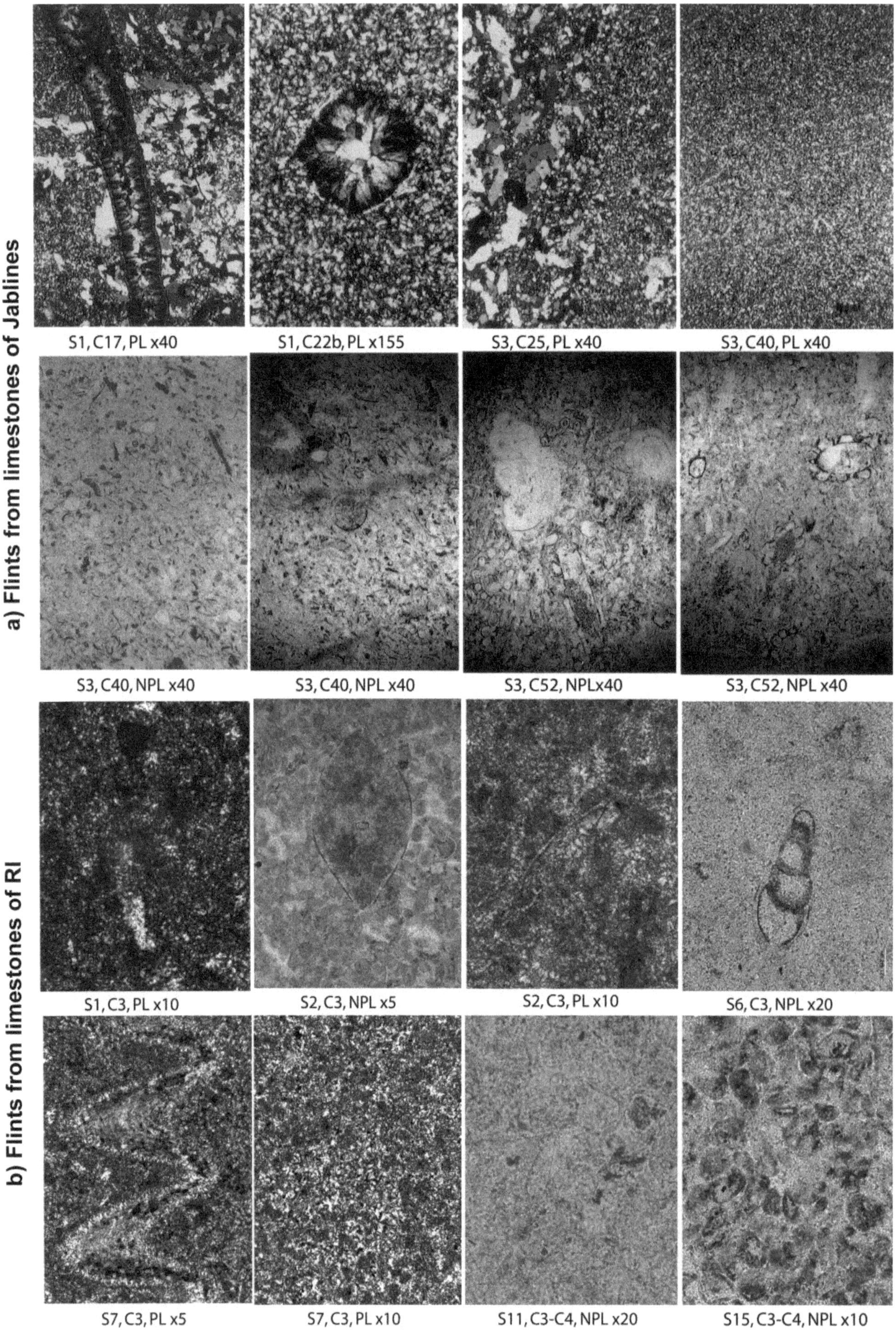

Figure 3. Flints microstructures from Ri and Jablines Neolithic flint mines in France (data from Tsobgou et al. in review; Bostyn and Lanchon 1992).

- Two layers of the four facies of the Limestone of Caen (L4 and L4'). These compact sandy limestone layers are less exploited than the first one L3. The Layers L4 and L4' are less weathered than L3, and contains less flint nodules in the south-west of the mine (D1 to D4), and voluminous spherical blocks in the north-east (D8 and D14). L4 is affected by frost and the plates of limestones are larger than those observed in the ochre sandy loam L3.

- An ochre sandy loam (L3). It results from weathering of limestone of Caen (L5). Layer L3 is underlying of colluviums layer L8 on the major part of geological profiles. Ochre sandy loam depth varies from 1.5-2m and it contains microplates of limestones and flint nodules. Its thickness varies from 0.4m at D1, 0.8m at D6, 0.9m at D14 to 1.5m at D3. The top of this level is poor in silicified limestone plates, but rich in small blocks of 4-5cm of thickness. Transfers of iron oxide and hydroxide figures are frequently observed. They are mainly coloured in red ochre (hematite, goethite and limonite).

- The silty clay layer (L8) mostly in the northern and north-eastern sides of the site. It is constitute by the melting of parts of the weichselian loess (alluvial formations of the area) and the brown-red clay layer (L2). This argilloloamy amalgamation is the consequence of the solifluction phenomena.

- A brown-red clay (L2) at the bottom of L1. It derives from the weathering of limestones layers L5, L4, and L3. Holes filling material is observed in this stratum, particularly at D1 and D3 drillings.

- An argilloloamy humus layer (L1) at the summit of the stratigraphy. The thickness of this stratum is 0.16 to 0.3m. It contains numerous fragments of limestone. It loamy fraction comes from the covering loess from the Holocene.

The stratigraphic profiles demonstrate the existence of series of anticlines and synclines oriented NE-SW. The deposition of calcareous sediments during the middle Jurassic is represented by a series of limestone strata, constituting the Limestone of Caen geological complex. The intensity of weathering of this layers, and silica transfer and accumulation varies along the mine surface. At Ri, layers L3 and L4 are those which are exploited by Neolithic miners (Figure 2).

3. Flint microstructures determination and classification

In this paper we draw on Dunham's (1962) classification of limestones to characterize the microstructures of flints. Along the mine, the mineralogy is constant. The rock is colored with hematite (flocculated mass) and goethite (pearls). Quartz is the most common mineral found in phenocrystal (megaquartz) and microcrystal (microquartz and fibrous chalcedony type calcedonite) form. The xenomorph quartz is mainly present in the matrix (0.5 to 2µm in size). Megaquartz is rare, but can reach 20 to 50µm in size. We have noted some detritic quartz with grain size of 100µm. At the boundary of megaquartz, calcedonite growths are observed. Muscovite is present in microflakes less than 5µm, but some crystals can reach 210µm.

The fossils and micro-remains of organisms which are contained in flint are pellets, foraminifera, spongiae, ooliths and brachiopods. The observation of *Rynchonella* (Figure 3 b- S7, C3) and bryozoans *Fenestra* (Figure 3 b- S2, C3), uniseriate foraminifera (Figure 3 b- S6, C3) confirms the age of the original limestone (Limestone of Caen). The proofs of the midoacen limestone character of the Limestone of Caen are given by the presence of oogonia of *Characea* (fresh water algae) (Figure 3 b- S11, C3-C4).

Flints from the fifteen drillings made at Ri (from layers L3 and L4) are classified in three macroscopic categories noted C1, C2, and C3 (Figure 4). These categories are distinguished by the nature and number of cortex and core of flint nodules.

- The nodules with a single cortex and two cores are classified as C1. We have noted two cores because the contact zone between the siliceous core (inner core) and the cortex (sandstone-like to chalky) is sandstone-like to siliceous (outer core). Flint blocks contained in this category are highly frost-fractured and contain numerous geodes of prismatic microquartz. D1, D10, D11, D13, D14, and D15 geological profiles contain flints of this category. Their structure is scattered, from mudstone, wackstone, to packstone (Figure 3 b-S1, C3; S11, C3-C4).

- In the category C2, the flints have two cortex and two cores. The outer cortex is sandstone-like to chalky, and the inner one is gritty. The outer core is sandstone-like, and the inner core is siliceous and more homogeneous. Only two drillings compose this category (D6 and D12). They are located at the transition areas between the flints of C1 and C3 categories. Even if they are homogeneous structurally, they are wackstone to packstone because of the size and junctions of minerals (Figure 3 b- S6, C3).

- The nodules contained in the group C3 are observed in drillings D3, D4, D5, D7, and D8. As samples of C2, those of C3 have two cortex and two cores, but their inner cortex is gritty to chalky, and thinner. The flints of C3 category possess an outer core thinner than the one of C1 flints. The flints of category C3 have essentially a mudstone structure (Figure 3 b-S7,C3).

4. Physical properties determination and interpretation

To determine the effect of rock properties on production features we have applied fracture mechanics and technology of brittle solids to prehistoric materials. These properties are elastic and resistance qualities. The methods and parameters used are described by Tsobgou (2009) and Tsobgou and Dabard (accepted). The structural and stratigraphic descriptions of flints and layers of Ri mine are completed with series of physical properties determination (Hardness H_v, Toughness K_{IF}, Young's modulus E, Shear modulus μ, Poisson's ratio v, Bulk modulus K, Density ρ, Edge toughness M_I, Fracture energy G_{IF}). We have used a microindenter for resistance and ultrasound echography for elastic properties. The results are reported in Figure 4. We have used Principal component Analysis (PCA) and Upward Hierarchical Classification (UHC) to discriminate categories of flint depending on their properties and structural features.

Two zones of concentration of flint samples are noted by PCA and UHC classifications, at probability $p \leq 0.05$. These areas of concentration are valuable only at 82.78% (sum of F_1: 34.5%, F_2: 30.15% and F_3: 18.13% factorial axes). On the PCA diagram (Figures 5-1 and 2) the distinction between the two areas is more detectable than on UHC profile (Figure 6). Samples from drillings D1 and D6 are plotted far from the others. We should note that flint samples from drillings D1 to D6 have shear moduli (μ) higher than bulk moduli (K). They are more resistant to shear deformations (Gercek 2007). They should then be more difficult to knapp or cut using bifacial debitage technique, with a tangential force applied.

The higher values of their toughness (K_{IF}) also indicate a big resistance to fracture propagation in the flint. The UHC diagram shows well that flints from these drillings aggregate at distances ≥ 0.02. Sample D5 is an exception in this area. The constancy of the isotropy coefficient (v) is due to it's closely link with the structure of flints (Figure 5). This explains well the distribution observed for flint samples from drills on Figure 5.

The area delimited by these drillings (south-eastern end of the mine zone excavated), is the zone where flint layers L3-L4 and less deep. From D7 to D14 rocks are less resistant to shear. But D7 to D10 constitute the transition between very resistant flint samples and mid-resistant ones (D11 to D15). Even if D7 and D8 flints have a $\mu < K$, they are very resistant to fracture and need a lot of energy for knapping (G_{IF}). So, flints from drillings D2 to D10 are better for axes used for farming or wood cutting. The mean value of the ratio G_{IF}/M_I of flints of these drillings is 1.7-1.8. This ratio has no physical significance but represents well the production of the mine of Ri.

The northern end flints are less resistant than those of the central zone of exploitation. The number of mining structures (holes, pits, or galleries) is abundant in the zone delimited by drillings D3 to D10. The mudstones microstructures (fine) are examined for their best properties.

When comparing flint axes from Ri with some in others important materials present during the middle Neolithic in France (metadolerite of Plussulien in Brittany and Alps metamorphic rocks), we note that Ri flint axes mostly likely to be contemporaneous with the eclogite and jadeite axes from

Drillings	Samples	Samples category	Textures	ρ (g.cm^{-3})	E (GPa)	υ	μ (GPa)	K (GPa)	H_v (GPa)	K_{IF} (MPa.m$^{-1/2}$)	M_I (Pa)	GIF (J.m^{-2})	Layer
D1	Cortex	C1	Mudstone	2.5	54	0.1	25	22	6.2	1.4	81.7	36.3	3-4
D2	Core	C3	Mudstone	2.6	74	0.1	37.5	31	7	3.8	62.6	195	3-4
D3	Core	C3	Mudstone	2.6	91	0.1	43	34.5	7.3	2.9	47	94.4	3
D4	Core	C3	Mudstone	2.6	91	0.1	43	34.5	7.3	2.9	47	92.4	3
D5	Core	C3	Mudstone	2.6	85	0.2	35	50	6.4	3.9	36.3	179	3-4
D6	Core	C2	Mudstone	2.6	78.8	0.1	37	30	8.3	3.5	92.1	155.5	3
D7	Core	C3	Mudstone	2.6	83.5	0.2	35	43.5	7.4	3.2	58	122.6	3
D8	Core	C3	Mudstone	2.6	83.5	0.2	35	43.5	7.4	3.2	58	122.6	3
D11	Core	C1	Mudstone-Wackstone	2.5	79	0.2	34	41	7	2.4	55	73	3-4
D12	Core	C2	Wackstone-Packstone-	2.6	80	0.1	35	37	8.2	1.7	86	36	
D13	Core	C1	Wackstone-Packstone	2.5	75	0.2	32	40	8.3	2.6	101.6	90	3-4
D14	Core	C1	Wackstone-Packstone	2.6	88	0.2	38	49	8.2	2.5	71	71	3-4

Figure 4. Synthesis of physical properties of flints from geological drillings of Ri mine.

the Alps. Bradt *et al.* (1973) have calculated the toughness of those minerals or rocks (KIc ≈ 3.5-4 MPa.m$^{-1/2}$). Even if the method used to measure this property is different (fracturing by notches) from the one used in this work (indentation), toughness results could be compared. According to Tsobgou (2007, 2009) the toughness of metadolerite of Plussulien is KIc ≈ 2.25 MPa.m$^{-1/2}$. Ri flint axes are then, more resistant than metadolerite ones, and closer to jadeites and eclogites from Alps. Thus, they could be valuable for cultural groups of the 'Chasséen' in Normandy or neighbouring zones of the Armorican Massif.

How then do the observations made at Ri compare with the findings from other important flint mines of Europe? At Jablines for example, three levels with flint are observed, at different depths: 1.5-1.6m, 3.5-4m, and 5.5m. The more exploited layers are powdery (layers 39 and 41) at 3.5-4m of depth. Jablines mine is located on marly calcareous limestones of Saint-Ouen. Limestones of Saint-Ouen are a biomicrite formed by a succession of layers either calcareous or marly (soft and crumbly), and compact layers of indurated lithographic to sublithographic limestones (Bostyn and Lanchon 1992). The best microstructure is a mudstone (isogranular to cryptocrystalline) (Figure 3a). Wackstone and packstone are also observed (Figure 3a- S1; S3, C40; S3, C52), on flints of middle and upper limestones levels. The original limestones nature at Ri (Limestone of Caen, layer L5) and Jablines (limestone of Saint-Ouen, layers 39 and 41) could explain differences observed in microstructures.

No microstructural study of flint was performed at Bretteville-le Rabet (Normandy), and this absence of data is a significant handicap for the comparative analysis of flint mines. Some similar characteristics of the two mines suggest that qualities of Bretteville-le Rabet and Ri flint must be comparable. Desloges (1986) has noted that the depth of limestones with the most exploited flint at Bretteville-le-Rabet is 2.5m. Nevertheless, there is a first discontinuous level 1.5m deep. The depth difference between best flint layers exploited at Ri is 2-3m with Jablines, and 0.5-1m with Bretteville. We will see below that this difference is not a hazard.

As compared to the French flint mining sites, the sites of Spiennes in Belgium displays a more complex exploitation system. Extraction structures at Petit-Spiennes are not as deep as those at Camp-Cayaux. This difference of limestones with exploited flint layers is conditioned by the geomorphology of the terrains. But, superficial holes or pits (3-4m depth) are recorded at Camp-Cayaux and Petit-Spiennes (Verheyleweghen 1953; Lefort 1954; Hubert 1976).

At the mine of Krzemionki in central Poland, exploited between 3500-2800 cal BC (Babel *et al.* 2005), limestones layers are oolithic, coral and politic from the upper Oxfordian. Stratigraphic profiles show a synclinal form oriented NW-SE. Limestones layers with flint are located at 8 and 8.5m depth in deep zones. Structures of extraction vary depending on the depth of the limestone with flint layers. Shallow cavities are noted at 2m depth, niches at 4-5m, and chamber-pillar and chamber mines at 8-9m. This classification is correlated to limestone strata with flint. There, the difference between the best flint layers of Ri and Krzemionki is of 6 to 7m.

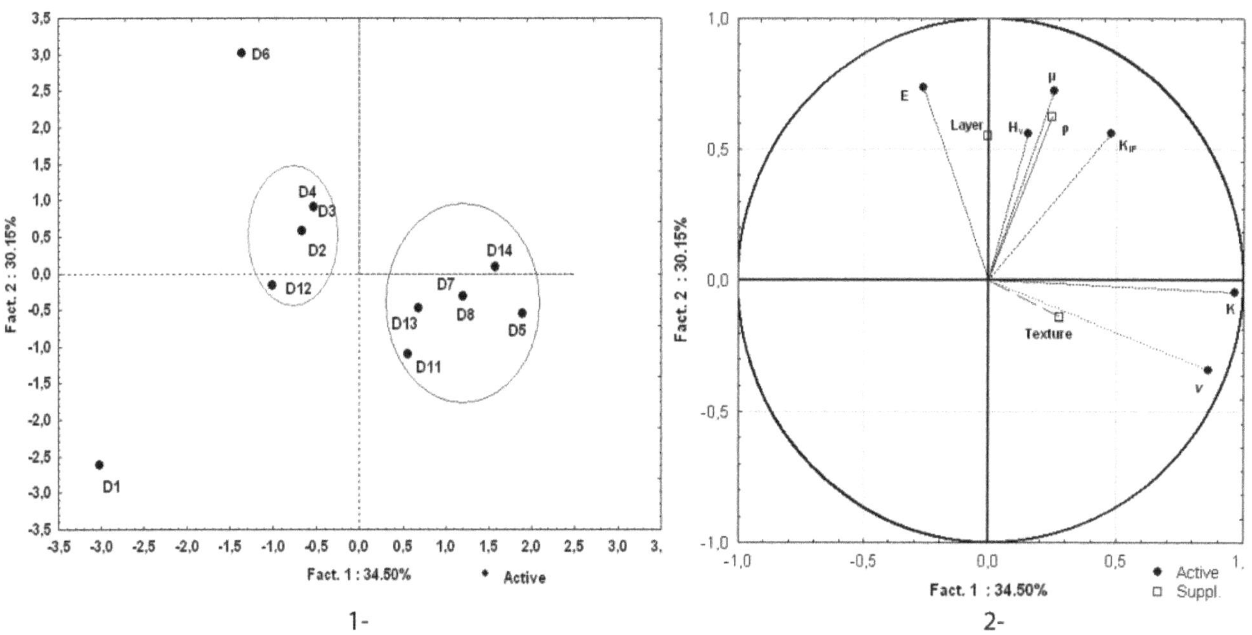

Figure 5. Principal Component Analysis of drillings from Ri flint mine F1/F2 1: Individuals projection (observations) 2: Properties projection (variables).

5. Evaluation of mining, digging and axes production techniques

To compare mining techniques, we have decided to simplify the classification of bone tools observed in mines. At Ri, the most observed deer tool is a pick. Its predominance has to do with the uncemented nature of the deepest layers of flint. Picks constitute 45 % of deer tools extracted at Ri (192 deer antlers tools, Figure 7).

At Jablines, a great number of deer antlers tools (quarantine) were also recorded. Bostyn and Lanchon (1992) have noted the high proportion of type c2 (picks) and type h tools (multifunctional tools). The multifunctional tools combine crowbar, wedge, chisel, and hammer functions. This multifunctional type h tools are found there in greater numbers than picks.

The extraction tool sets at Bretteville-le-Rabet and Spiennes mines are mostly composed of picks (Hubert 1988). Picks double the total number of crowbars at Bretteville-le-Rabet. Unlike at Jablines and Ri, a great number of picks were made of flint. Desloges (1986) has recorded four flint picks for every deer antler one (200 picks in flint and 50 in deer antler). At Jablines, not a single flint pick was found in the mining structures. The more restricted use of flint picks at Jablines and Ri suggests that the exploited sediment layers are not as hard as that at Bretteville-le-Rabet. However, this analysis should take into account the presence and availability of deer antlers.

Is there a relationship between the abundance of flint picks and the economization of antlers? If so, is it in turn affected by the quality of limestones microstructure? Is the rarefaction of antlers consequently linked to the deer population size at a moment? In flint mines where the quality of rock is less homogeneous, we expect to find an increase in the number of picks. The expected depth of pits required to access to the best flint layers and mined surfaces could also be criteria for the choice of mining tools (Figure 8).

At Spiennes, the exploitation is marked by series of holes, pits and galleries, depending on the depth of layers of flint of the best and most homogeneous quality. Some pits reach 7-8m in depth at Petit-Spiennes and 15-16m at Camps-à-Cayaux (Hubert 1988; Collet 2002; Collet and Wim 2002). As at Bretteville-le-Rabet, flint picks are abundant, which maybe explained by the fact that in order to reach the best

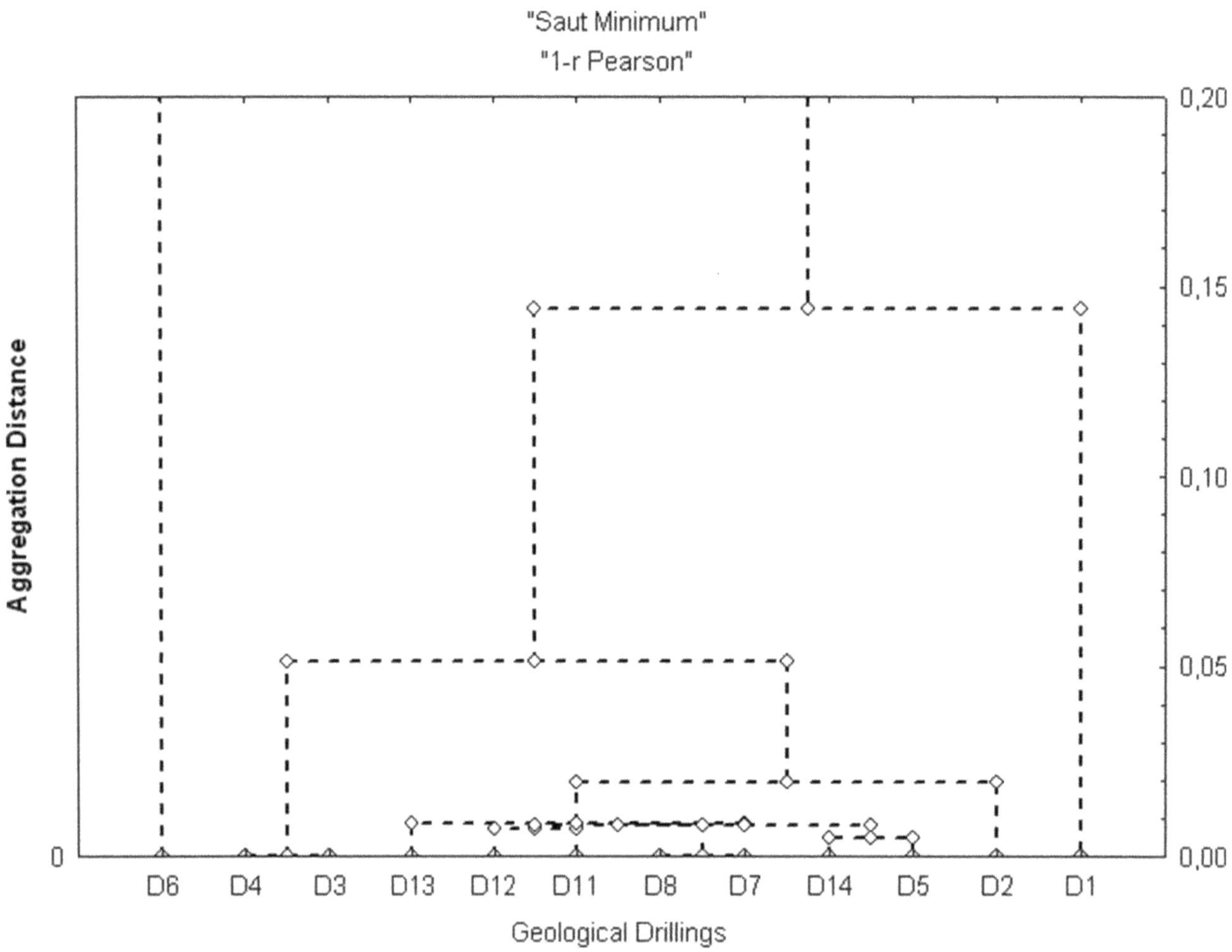

Figure 6. Dendrogram of Ri flint mine geological drillings based on Upward Hierarchical Classification (UHC).

flint layers numerous strata had to be drilled. Depth difference between Spiennes and Bretteville-le-Rabet could also be due to the period of exploitation of the mine. Does the 1.5-2.5m depth of pits and the hardness of the limestone at Bretteville-le-Rabet explain the great quantity of flint picks? We suggest that the miners' economization of deer antlers (due to limited access to biological resources at Bretteville-le-Rabet) may provide the best explanation of the choice of mining tools.

In Poland's Krzemionki mines the majority of mining tools are made of stone, antlers and wood (Babel 2008). Stone and flint picks, as well as roe deer and deer antlers chisels were deployed for wall extraction. Some bonfires traces (10-15cm layers of coal) were recorded at Krzemionki Opatowskie, evidencing rock hewing by fire. The use of this technique and the abundance of antler chisels and flint picks probably had to do with the hardness of the limestone layers. Flint picks were made of poor quality flint. The selection of flint for the production of mining picks or axes seems to reproduce the same pattern as in Ri, where some atypical roughouts are noted only in mine areas with poor flint quality.

What factors explain the absence of flint picks at Jablines and their abundance at Bretteville-le-Rabet? At Jablines they include the high depth of best flint layers (layer C40), the possible abundance of deer in the region, and the miners' desire to economize flint extracted in the mine, and their adaptation to geological conditions (hence the choice of multifunctional tools). However, the example of Krzemionski shows that mining tools choice is a compromise between their availability and strata characteristics (Borkowski, 1990). The analysis of deer antlers in European mines shows predominance of deer deadwood. At Ri 78% of deer antlers are deadwoods. The figure 7 shows some specimen of these tools. The percentage of deadwoods was evaluated at 81.3% at Grimes Gaves, and 87.3% at Durrington Walls mines in England (Clutton-Brock 1984). Deer antlers at Jablines (Bostyn and Lanchon 1992) and Bretteville-le-Rabet (Desloges 1986) are represented mainly by deadwoods. In Neolithic habitats of Jura, Billamboz (1977) notes that 70 to 80% of antlers are deadwoods (probably linked to mining activities). The choice of deadwoods is justified by their resistance and elasticity, and the facility of procurement (not requiring deer capture).

The constant mineralogy and the predominance of mudstones structures in flints of Ri explain the concentration of excavation structures. The ovoid form of flint nodules of Ri limit the direct removal of flakes. Gallouin has noted that the nodules are broken into two parts using the technique of pier percussion (Figure 9). The two reduced cores are then used as nuclei. The reduced cores are also extracted directly from layer L3. They are produced by the effect of gel. On unspherical blocks, thick flakes are obtained by direct percussion with heavy stone hard-hammer (mainly flint, Figure 10). Some hard-hammers in sandstone were also found. The use of hard-hammers is justified by the volume of nodules, their toughness and hardness, and the thickness of their cortex.

In contrast to the phase core reduction, the phase of roughout obtainment shows the use of small flint hard-hammers, recovered from the substratum of the mine. But, in both phases, the chalky character of the cortex of flint hammers facilitates the removal of thin flakes. The big size of hammers used in core reduction allows the production flakes that are thick but not too thick, so as to preserve sufficient volume for the second phase of knapping. The maximization of the length of roughouts using this method makes possible the production of big axes (Figure 10). The chalk hammers were used in the mine for the last stage of roughout manufacturing, allowing the removal of small and controlled size flakes by soft-hammer percussion.

6. Mining activities and cultural behaviours interpretation

The presence of good quality flint at low depths explains the absence of very deep structures at Ri and Bretteville-le-Rabet. The use of flint picks at Bretteville-le-Rabet may be due to the superior quality and abundance of flint. But, as we have already noted, the choice of tools may also have to do with the chronology of the mine's exploitation. Being an older mine and belonging to the Culture of Cerny, Bretteville-le-Rabet was initially exploited, when in Normandy deer antler was saved for fulfilling other productive needs. During the second phase of the middle Neolithic (Culture of Chasséen), the intensification of mine exploitation and growing demand for flint may have forced the miners to use others natural resources such as deer antlers.

The use of deer and roe deer antlers in mining was also influenced by seasonal factors. In other words, there must have been a close connection between the periods of antler gathering from January to March (Billanboz 1977) or February to April (Clason 1979) and the periods of smoother climactic conditions necessary for drilling activities (autumn and summer). At Ri the loamy character of flint layers and the argillaceous facies of upward terrains were disadvantageous for mining in rainy and cold weather of winter and autumn. It was also the case of Jablines mine, hypothesized by Bostyn and Lanchon (1992), where the mining took place during dry seasons.

The status of flint axes from different European mines depended on their structural and physical qualities, size, colour, and accessibility. Over time their position in cultures changed, depending on these principal criteria. The resistance of flint axes produced at Ri made them ideally suitable for farming using and wood cutting, and also may have given them prestige in Brittany or other neighbouring areas. The importance of Normandy's flints in the industries of Brittany since the earlier Neolithic (RRBP : Rubané Récent du Bassin Parisien, and VSG : Villeneuve-Saint-

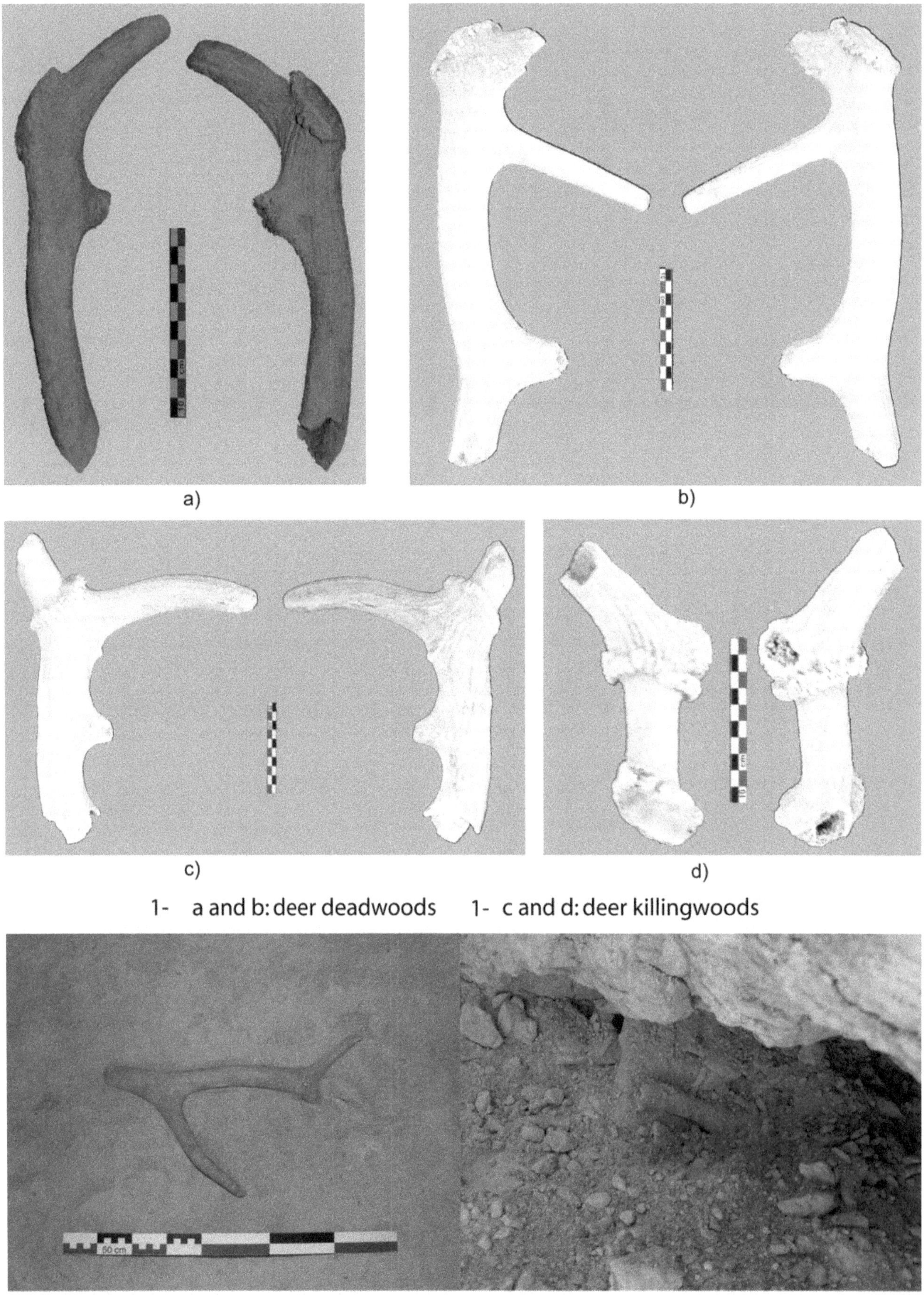

Figure 7. 1- Deer antlers from Ri showing deadwoods and killingwoods 2- Examples of picks (photographies, E.Gallouin).

Germain) supports this hypothesis (Cassen *et al.* 1998; Blanchet 2006; Marchand *et al.* 2006; Tsobgou 2009).

If the flints from Normandy in earlier Neolithic occupations came principally from clay formations (at the top of limestones formations), they were only used for blade production. The example of Soumont-Saint-Quentin in Calvados shows that clay extraction mines contain mainly blade and bladelet cores (Ghesquière *et al.* 2008). The use of flints from limestones of Ri is then not excluded during the early Neolithic and more probably during the middle Neolithic, when flints from limestones layers were exploited. The measures obtained for physical properties, discussed above, are key for elaborating the theory of Neolithic mining activities and production evolution in Normandy.

The access to clay layers is easier than to limestone layers. This characteristic of this form of exploitation makes it comparable to the exploitation of erratic block (from rivers) at the beginning of the middle Neolithic in France: metadolerite of Plussulien in Brittany (Le Roux and Giot 1965; Delibrias and Le Roux 1975), fibrolite in Auvergne (Surmely *et al.* 2001; Goër de Herve de *et al.* 2002). This also gives us some clues about differentiation of activities and the time spent by miners at the extraction sites. Few numbers of common lithic products of the Neolithic (blades, knifes) have been recorded at Ri and Jablines. At Bretteville-le-Rabet, none of these tools was discovered. What do this finding then tell us about the nature of relations between flint mining zones and neighbouring habitats?

At Bretteville-le-Rabbet, miners may have returned home for the night. The exploitation of flint of Krzemionki was performed by specialists, notably during the phase of chamber-pillar and chamber construction. Miners of Ri could be considered specialists since the start of exploitation. However, the model of exploitation at Ri is distinct from that of Krzemionki, where during the first stages; low deep pits were available to all the persons needing flint (Babel 2008). The difference between the strictly vocational specialization of Ri miners (since the beginning of excavation) and the gradually transforming status of miners at Jablines and Krzemionki mines had to do with the duration of their exploitation. At Jablines and Krzemionki mines the development of underground techniques of flint extraction eventually resulted in monopolization of economic production and exchange by the miners' societies or cultural groups.

Axes roughouts are the predominant kind of product found at Ri, which means that the smoothing and polishing was performed in habitats or production settlements. As for the axes from Krzemionki flint, there is abundant evidence of similar production management, particularly in river basin of Kamienna (Babel 2008). In Normandy the data on flint dispersion in habitats or ritual contexts is lacking, inhibiting our understanding of production economies and exchange patterns. One exceptional example of flint axe fragment was noted in the middle Neolithic (Cerny, 4521-4259 cal BC) occupation of Mondeville in Calvados (Chancerel *et al.* 2006, 77), made of flint from Bretteville-le-Rabet mine. The second example is an accidental heated flint discovered in the lithic industry of a flint carrier at Argentan (Ghesquière and Marcigny 2004, 47). In sum, this data is insufficient for evaluating the economic potential of flints of Ri.

Mines	Dates cal. B.C.	Depths m	Layers	Layers nature	Flint structures	Mining tools
Ri	4000-3500	1.5-2	C3-C4	Sandy loam	+Mudstone -Wackstone	Deer picks
Bretteville-le-Rabet	4800-3975	1.5-2.5	C1-C2	Indurated sandy loam?	-Mudstone? +Wackstone?	Flint picks
Jablines	4300-2700	3.5-4	C40	Powder limestone	+Mudstone -Wackstone	Deer picks
Spiennes	4400-2450	15-16	?	Indurated sandy limestone?	?	Flint picks
Krzemionki	3500-2900	8-8.5	?	Cemented sandy limestone	Banded flint (mudstone?)	Flint picks Roe deer and deer chisels

Figure 8. Synthesis of exploitation of flint layers in some Neolithic mines in Europe.

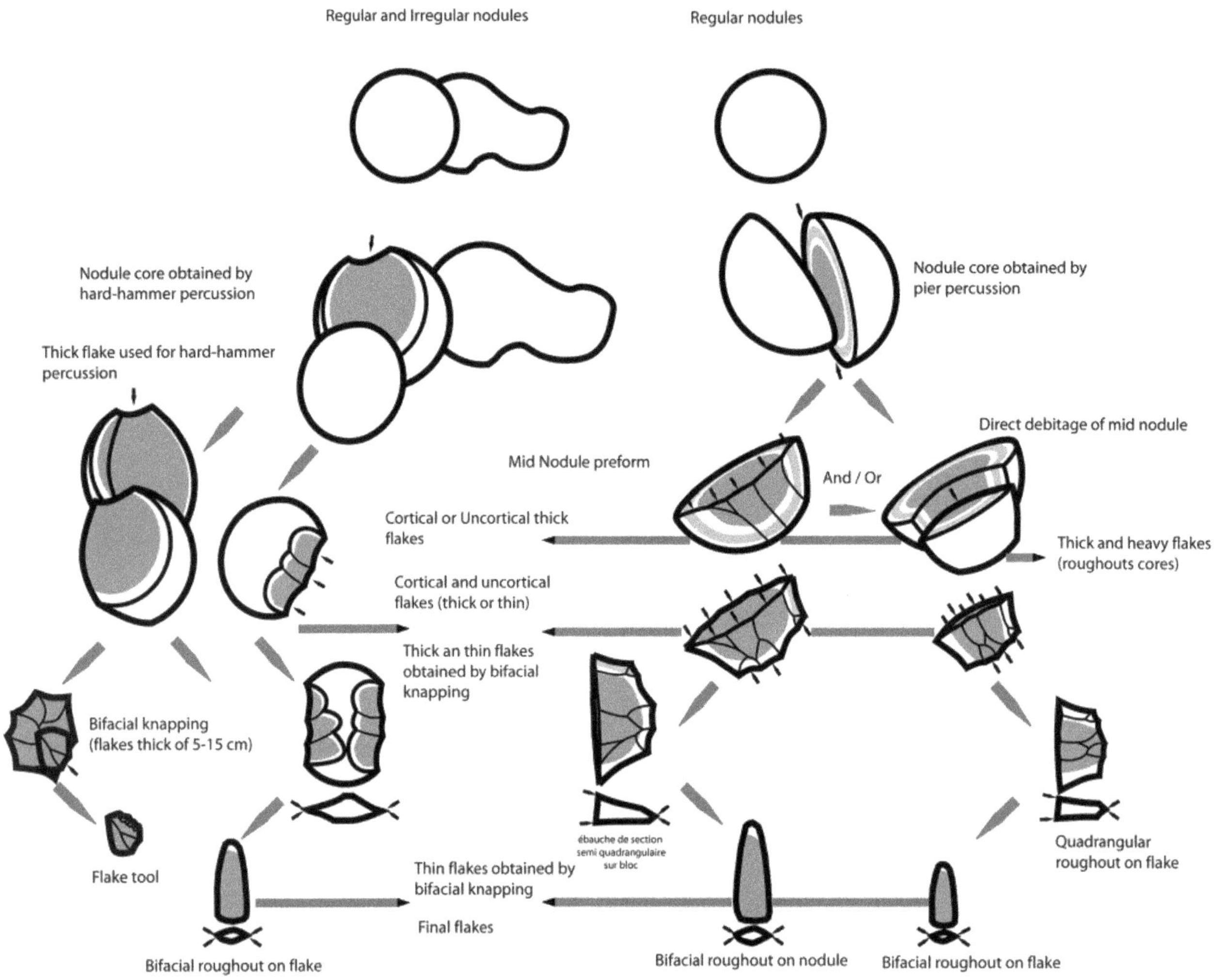

Figure 9. Knapping protocol for the production of flint axes in the mine of Ri, France (Draw, L. Juhel).

The flint axes produced at Krzemionki in Poland are considered common tools for cutting down trees and clearing land during the existence of the Funnel Beaker Culture (3900-2900 cal BC), and which finally had religious uses during the Globular Amphorae Culture (2900-2500 cal BC) (Babel 1986, 27; Babel *et al.* 2005). Giving the current lack of data, these uses cannot be compared to the uses of Normandy's flint. As the main material employed for axe production, the first products used in nearby areas had most of the time the function of common tools. The second phase of exploitation, which represents the maximum period of exploitation and distribution, these goods became items of value and prestige. Their growing prestige was not only due to the increase in demand, but also to the high value ascribed to them by populations located far away from production centres. The ritual character of a product is then, a consequence of its difficulty of acquisition (cost, difficulty of producing), and its beauty (banded character of flints of Krzemionki, colour of jadeites and eclogites of Alps, etc.). If during the first phases of production intensification, the value assigned to new products increased, later the same intensification caused the status of these products to diminish. Products, which are manufactured in mass and at low cost, become common tools because their acquisition is easy and democratic. This phenomenon is well illustrated by the case of jadeites axes of Alps. While nothing can be said about the ritual properties of flint axes from Ri, they were still prestigious for farmers and woodcutters capable of purchasing them. The resistance of the material to damage during activities, i.e., longer lifetime, is an essential quality for the attribution of prestige value. The status of prestige or the importance accorded to goods is then, not only attached to symbolic (ritual) or aesthetic characteristics. The quality required for effective utilization is also important.

7. Conclusion

The application of multidisciplinary studies to flint and substratum layers of the middle Neolithic mine of Ri demonstrates the miners' nuanced knowledge of their geological environment. The selection or concentration of mining structures is closely related to the nature of the sediment and its richness in good flint nodules. Considering their abilities to analyze physical properties of rocks allowing the quantification and qualification of the density and resistance to fracture of flint axes, and their choice of microstructures, the mine workers can be justifiably considered specialists for their superior knowledge and good management of the geological potential of their territory. The methods and technique of debitage employed during axes manufacture were essential for minimizing the loss of big masses of raw materials and maximizing the size and quantity of goods. The quality of limestone flint products at Ri was conditioned by two major factors—their structural and physical qualities and the technical skills of a flint-knapper. Thus the choice of this territory for the implantation of a flint mine during the second phase of the middle Neolithic contributed to the valorisation of its flint products.

Digging techniques vary throughout the Neolithic, and therefore mines established and exploited during this period are considerably different. The deep of geological layers, the thickness, density and hardness of layers containing good flint qualities for axes production, are the principal parameters controlling the adaptation modalities applied by Neolithic miners. At Ri the use of deer antlers tools, mainly picks, is the consequence of the powdery character of the sandy loam stratum which contains flint, and its low depth position. But, the great number of that special tool supposes the possibility of access to it. The acquisition of mining tools during the period of antler shedding by animals shows that the livestock was significant or the organization of acquisition was well-developed and efficient. The organization of activities of miners and farmers' societies was then closely linked with axes production.

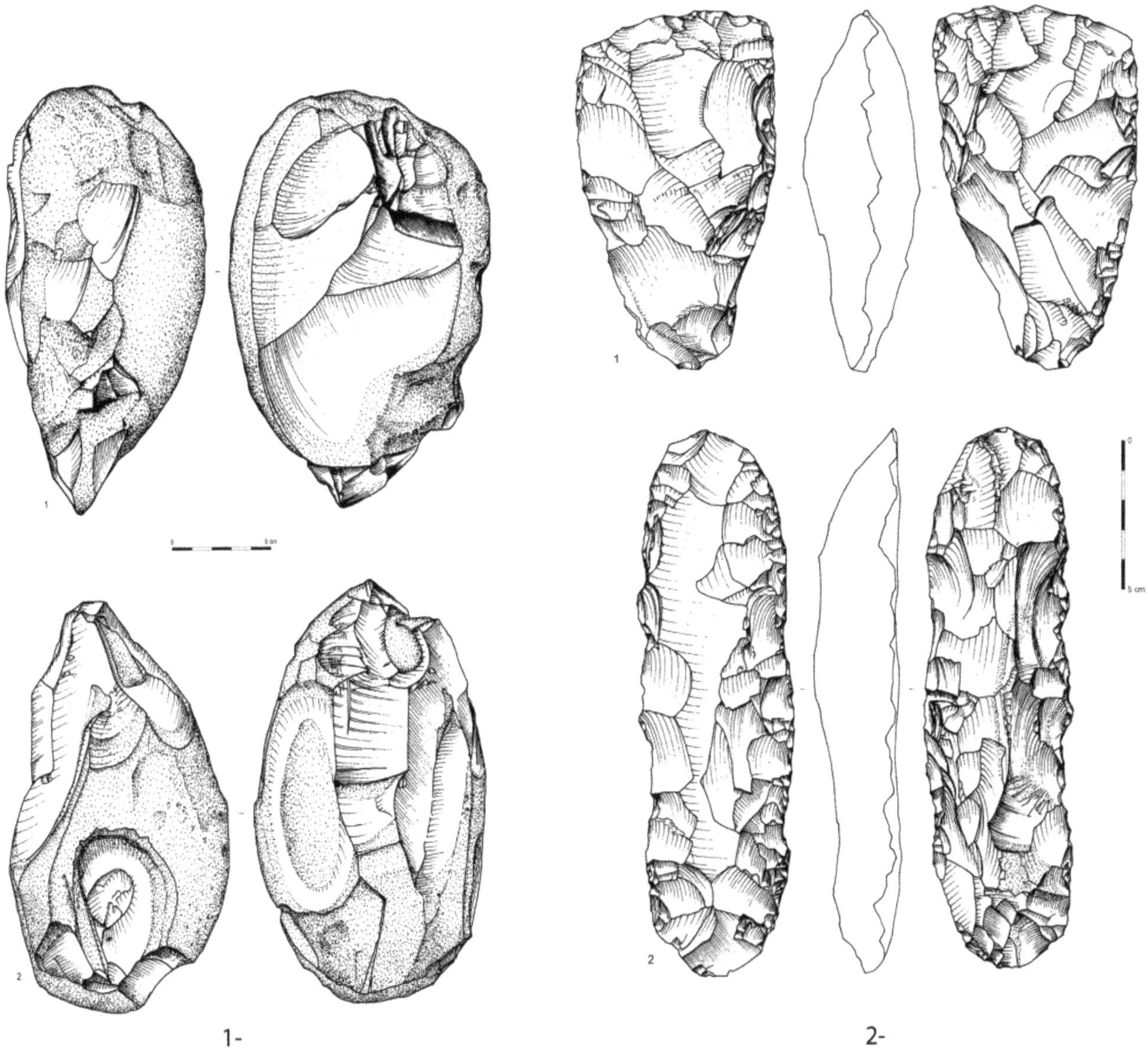

Figure 10. 1- Flint hard-hammers 2- Flint roughouts from the mine of Ri (Draw, L Juhel). +: high presence, -: low presence.

The analysis of differences in mining toolsets deployed at various Neolithic flint mines in France and others European sites demonstrates that the geological nature of their substratum and the long-term intensification of economic exchanges played a critical role.

References

Babel, J. 1986. The Problems of the Investigations of the flint mine at Krzemionki near Ostrowiec Świętokrzyski, Kielce and Tarnobrzeg Voivodeships. In K. T. Biro (ed.), *International Conferrence on Prehistoric Flint Mining and Lithic raw Material Identification in the Carpathian Basin, Budapest – Sümeg*, 27. Budapest, KMI Rota.

Babel, J. 2008. Krzemionki Opatowskie. The earliest beginnings of modern mining. In *The mine – history and monument of technology.* World Mining Congress, Krakòw 2008, 85-107. www.krzemionki.pl/files/Krzemionki_Opatowskie.

Babel, J., Braziewicz, J., Jaskóła, Kretschmer, W., Pajek, M., Semaniak, J., Scharf, A. and Uhl, T. 2005. The radiocarbon dating of the Neolithic flint mines at Krzemionki in central Poland. *Nuclear Instruments and Methods in Physics Research* 240, 539-543.

Billamboz, A. 1977. L'industrie en bois de cerf en Franche-Comté au néolithique et au début de l'âge du Bronze. *Gallia Préhistoire* XX(1), 31-34.

Blanchet, S. 2006. Un habitat du Néolithique ancien en Bretagne orientale – Pluvignon à Betton (Ille-et-Vilaine), in G. Marchand, C. Manen, and A. Tresset (eds.), *La néolithisation de la façade atlantique de l'Europe – interactions culturelles, transferts techniques, implications des milieux naturels.* Rapport terminal 2003-2005, 309-315. Rennes, CNRS.

Borkowski, W. 1990. Diverseness of the flint mines at Krzemionki Opatowskie, Poland; preliminary results of microgeomorphological analyses. *Cahiers du Quaternaire* 17, 225-231.

Bostyn, F. and Lanchon, Y. 1992. *Jablines, Le Haut Château (Seine-et-Marne): Une Minière de silex au Néolithique.* Paris, Maison des Sciences de l'Homme. Documents d'Archéologie Française, 35.

Bradt, R. C., Newnham, R. E. and Biggers J. V. 1973. The Toughness of Jade. *American Mineralogist* 58, 727-732.

Cassen, S., Audren, C., Hinguant, S., Lannuzel, G. and Marchand, G. 1998. L'habitat Villeneuve-Saint-Germain du Haut-Mée (Saint-Etienne-en-Coglès, Ille-et-Vilaine). *Bulletin de la Société Préhistorique Française* 95 (1), 41-75.

Chancerel, A., Marcigny, C. and Ghesquière, E. 2006. *Le plateau de Mondeville (Calvados) du Néolithique à l'âge du Bronze.* Paris, Maison des Sciences de l'Homme. Documents d'archéologie Française, 99.

Clason, T. 1979. The flintminers as a farmer, hunter and Antler collector. Troisième Symposium International du Silex. *Stragia*, 6 (1981), 119-125.

Clutton-Brock, J. 1984. *Excavations at Grimes Graves, Norfolk 1972-1976, Fascicule 1, Neolithic Antler Picks from Grimes Graves, Norfolk and Durrington Walls, Wiltshire, A Biometrical Analysis.* London, British Museum Press.

Collet, H. 2002. Un centre minier néolithique inscrit au patrimoine mondial en belgique. *L'Archéologue, Archéologie nouvelle* 58, 41-46.

Collet, H. and Wim, V. N. 2002. Stratigraphie et faune d'un puits d'extraction de silex néolithique à Petit-Spiennes. *Anthropologica et Praehistorica* 113, 73-104.

Curwen, E. C. 1928. The antiquities of Windeover Hill. *Sussex Archaeological Collections* 69, 93-97.

Delibrias, G. and Le Roux, C. T. 1975. Un exemple d'application des datations radiocarbone à l'interprétation d'une stratigraphie complexe : la fouille des ateliers de Plussulien (Côtes-du-Nord). *Bulletin de la Société Préhistorique Française* 72, 78-81.

Desloges, J. 1986. Fouilles de mines à silex sur le site néolithique de Bretteville-le-Rabet (Calvados). *Revue Archéologique de l'Ouest*, Supplément 1, 73-101.

Desloges, J. 1999. Une mine à silex au Néolithique: L'exemple de Bretteville-le-Rabet, in G. San Juan and J. Maneuvrier (dir.), *L'exploitation ancienne de roches dans le Calvados, Histoire et Archéologie, Couleurs Calvados*, Molay-Littry, p. 53-78.

Diethelm, I. 1997. Neolithic flint mining in Three-country Corner (Basel region), Kleinkems (Germany) and Löwenburg (Switzerland), in R. Schild and Z. Sulgostowska (eds.), *Man and Flint: Proccedings of the VIIth International Flint Symposium*, 63-64.Warsaw, Institute of Archaeology and Ethnology at the Polish Academy of Sciences.

Dunham, R. J. 1962. *Classification of carbonate rocks according to depositional texture. In W. E. Ham (ed.), Classification of carbonate rocks. American Association of Petroleum Geologists Memories* 1, 108-121.

Gercek, H. 2007. Poisson's ratio values for rocks. *International Journal of Rock Mechanics and Mining Sciences* 44, 1-13.

E. Ghesquière, S. Clément-Sauleau and C. Marcigny. 2004. Carrières de calcaire du Néolithique moyen II à Argentan (Orne). *Internéo* 5, 45-62

Ghesquière, E., Marcigny, C., Desloges, J. and Charraud, F. 2008. La production de lames en silex Bathonien dans la Plaine de Caen: redécouverte de la minière des Longrais à Soumont-saint-Quentin (Calvados). *Internéo* 7, 103-119.

Gigot, P., Dupret, L. and Le Gall, J. 1999. *Carte géologique de la France (1/50 000), feuille de Falaise (176).* Orléans, Editions BRGM.

Goër de Herve de, A., Servelle, C. and Surmely, F. 2002. Les haches polies du site de Chastel-sur-Murat (Commune de Chastel-sur-Murat, Cantal, France). *Comptes Rendus Paleovol* 1, 123-128.

Gosselin, F. 1986. Un site préhistorique d'exploitation du silex à Spiennes (Ht) au lieu-dit Petit-Spiennes. *Vie archéologique* 6 (2), 34-159.

Greenwell, W. 1870. On the opening of Grime's Grave in Norfolk. *Journal of Ethnological Society of London* 2, 419-439.

Hubert, F. 1976. Puits de mines à la tranchée du chemin de fer à Spiennes. *Archaelogia Belgia* 186, Conspectus MCMLXXV, 16-20. Brussel.

Hubert, F. 1978. Une minière néolithique à camp-cayaux de Spiennes. *Archaeoklogia Belgica* 210, 5-44.

Hubert, F. 1988. L'exploitation du silex à Spiennes. *Archaeologicum Belgi Speculum* 15, 63.

Kuntz, G., Menillet, F., Le Gall, J., Rioult, M. Callier , L. Pellerin, J., De La Querrière, P., Vautrelle, P. and Verron, G. 1989. *Carte géologique de la France(1/50 000), feuille d'Argentan (212).* Orléans, Editions BRGM.

Lais, R. 1948. *Die Höhle an der Kachelfluch bei Kleinskems im badischen Oberland: Eine jaspisgrube und Grabstätte der jüngeren Steinzeit.* Mit Beiträgen von R. Bay und H. G. Stehlin. Freiburg, Urban-Verlag.

Lefort, M. 1954. *Les cahiers de Spiennes 3* (fascicule polycopié).

Le Roux, C. T. 2002. Plussulien et la diffusion des haches polies armoricaines. In J. Guilaine (dir.), *Matériaux, productions, circulations du Néolithique à l'Age du Bronze, Séminaire du Collège de France*, 101-112. Paris, Editions Errance.

Le Roux, C. T. and Giot, P.-R. 1965. Etude pétrographique des haches polies de Bretagne. VI, découverte des ateliers de la dolérite type A. *Bulletin de la Société Préhistorique Française* 62, 143-160.

Marchand, G., Pailler, Y. and Tournay, G. 2006. Carrément à l'Ouest ! Indices du Villeneuve-saint-Germain au centre de la Bretagne (Le Dillien à Cléguerec et Bellevue à Neuillac, Morbihan). *Bulletin de la Société Préhistorique Française* 103 (2), 519-533.

Mercer, J. R. 1976. Gime's Graves Norfolk. An interim statement on conclusions drawn from the total excavations of a flint mine shaft and a substantial surface area in 1971-1972, in C. Burgess and R. Miket (dir.), *Settlement and economy in the 3rd and 2nd millennia BC*, 101-111. Oxford, BAR Publishing. British Archaeological Reports 33.

Pétrequin, P., Errera, M., Cassen, S. and Croutsch, C. (2003). De la pétrographie aux approches sociales – la circulation des grandes hachjes alpines en Europe occidentale pendant le Néolithique. In *Les matières premières lithiques en Préhistoire, Table ronde internationale d'Aurillac, du 20 au 22 juin 2002.* Préhistoire du Sud-ouest, supplément 5, 253-274.

Sheperd, R. 1980a. Harrow Hill bie Findon, West Sussex. *5000 Jahre Feuersteinbergbau, die Suche nach dem Stahl der Steinzeit* 517-521. Bochum. Deutschen Bergbau- Museum Bochum 22.

Sheperd R. 1980b. Windover Hill bie Eastbourn, East Sussex. *5000 Jahre Feuersteinbergbau, die Suche nach dem Stahl der Steinzeit* 525-527. Bochum. Deutschen Bergbau- Museum Bochum 22.

Sieveking, G. de G. 1980. Weeting village «Grime's Grave», Norfolk. *5000 Jahre Feuersteinbergbau, die Suche nach dem Stahl der Steinzeit* 528-540. Bochum. Deutschen Bergbau- Museum Bochum 22.

Surmely, F., Goêr de Herve de, A., Errrera, M., D'Amico, C. Santallier, D., Forestier, F.-H. and Rialland, Y. 2001. Circulation des polies en Auvergne au néolithique. *Bulletin de la Société Préhistorique Française* 98 (4), 675-691.

Tsobgou, A. R. 2007. *Matières et Techniques de la Préhistoire récente du Massif armoricain, pétrographie-géochimie-mécanique-technologies.* Unpublished PhD thesis, University of Rennes.

Tsobgou, A. R. 2009. Mapping Mesolithic and Neolithic cultures behaviours and interactions with nature and pro-perties of rocks in Western France. *Journal of Archaeolo-gical Science* 36 (7), 1615-1625.

Tsobgou, A. R. and Dabard, M.-P. 2010. Petrographical, Structural and mechanical Analysis of Armorican Phtanites: a key raw Material for Mesolithic in Western France. *Geoarchaeology* 25 (3), 327-351.

Tsobgou, A. R., Marcigny, C., Deloze, V., Juhel, L., Charraud, F. and Ghesquière, E. i.p. Mining production and management by geomorphological, geological and physical approaches at Ri 'le Fresne' Neolithic flint mine, France. *Journal of World Prehistory.*

Verheyleweghen, J. 1953. Découverte sur le territoire d'une phase d'occupation des hommes de Michelsberg, antérieure à celle du plateau du « Camps-Cayaux » (compte-rendu des fouilles entreprises en 1953). *Bulletin de la Société royale belge d'Anthropologie et de Préhistoire* 64, 141-162.

The flint mine of Ri «Le Fresne»

Cyril MARCIGNY, Emmanuel GHESQUIÈRE, David GIAZZON, Rodrigue TSOBGOU AHOUPE, François CHARRAUD, Laurent JUHEL and Sébastien GIAZZON

Abstract

The flint mine of Ri is located at +200 to +205m NGF (*Nivellement Général de la France*) on a south-eastward slope, in the north of the Houay's Valley, whose eponym river is a tributary of the Orne. The raw material mined at this site has the shape of spherical mid-grain-sized pebbles, measuring 20cm in diameter and 50cm in length. They outcrop in a powdery limestone layer, 1.2m in depth at the southern end of the excavated area to 4.5m at the northern end. From shallow to deep areas, the infrastructure of the mine varies from incurved holes to 30 square meters pits radiating a network of small galleries and extensions. While some of these galleries remained unfilled, most pits were refilled immediately after the extraction of flint. Their walls contain a great number of digging tool traces. Twenty eight radiocarbon dates have been obtained on deer antlers recovered from the entire surface of the excavation, situating the mine in the second half of the fourth millennium BC.

Keywords

France. Lower Normandy. Neolithic middle II. Flint mine. Flint axes.

1. Introduction

The Neolithic mine of Ri is one of a series of flint mines built in the Plain of Caen, a Jurassic limestone plate. This group of mines is oriented north-southward, along a thirty kilometer long line (Figure 1). Although the archaeological field survey estimated the surface of the mine to be 30 hectares, only two hectares have been excavated. The exact area of excavation measures 22,300 square meters, extending 730m in length and 40 to 50m in width (Figure 2).

During 6 months, from June to November 2007, the excavation made by Inrap has allowed the discovery of five thousand and fifty structures of flint extraction, along the area of extension of the future A88 highway transept (Caen-Sées), cutting a part of the mining complex. Five thousand and ten of these structures have been excavated using a backhoe. Forty were excavated by technicians. A team of 10 persons collected a great quantity of data during the fieldwork at Ri.

Assuming that the nature and the depth of the geological layers containing exploited flint determined different types of extraction structures at Ri, one of the foci of the project has been to improve our understanding of their morphology and distribution, correlating the layers to specific microstructures and qualities of the flint (see Tsobgou *et al.* in this volume). We have recorded many geological profiles—15 geological drillings, measuring 3m by 2m, and one 300 meters long stratigraphic transept (Figure 3).

In the course of this study we hope to enrich our knowledge of the economy of siliceous raw materials and their exploitation in European Prehistory, principally in mining activities, while drawing on the existent scholarship on this topic initiated at the end of the 19th century. In France, and principally in Normandy, the excavation of the Neolithic mine of Bretteville-le-Rabet marked the start of this research line (Desloges 1986, 1990).

2. The Substratum Characteristics

The flint extracted from de mine of Ri comes from the Limestone of Caen formation (J3ca), a fine bioclastic limestone located at a depth 10 to 12m. This limestone has white to beige color and is porous and poor in fossils. The age of the limestone formation has been assigned to the Lower and Middle Bathonian (Gigot, *et al.* 1999), but a more recent micro fauna examination has showed the presence of

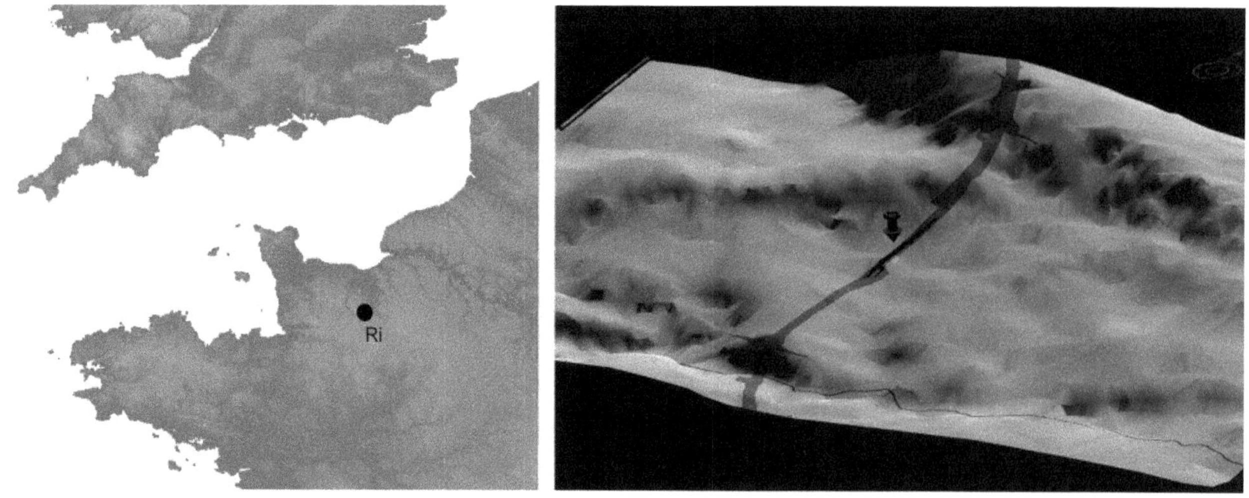

Figure 1. Location map of the road and the area of excavation on a 3D projection of the topography (DAO, L. Vipard, Inrap).

Rynchonella which extends its deposition to the upper Jurassic (Oxfordian) (see Tsobgou *et al.* in this volume).

The deposition of carbonaceous sediments occurred during many phases, followed by weathering process. In addition to weathered sediments of the limestones layers, we have also observed brownish-red clay and a sandy loam layer. These weathered layers are covered by silty Holocene colluviums and loess, sometimes mixed with fragments of altered limestones. The thickness of this cover increases from 0.5m in the southern to 3m in the northern parts of the excavation.

The flint extracted in this mine has the shape spherical nodules. It is associated with both a sandy loam layer, called 'chaussin' by current miners, and a mixed colluvium in the southern part of the mine. The rarefaction of flint in the surface layers of the northern section of the mine is not only due to a northward thrust, but also to the presence of good quality flint in deeper layers.

In the central area of the mine, where mining structures are more concentrated, flint has a mudstone texture, whether the structure of the limestone is packstone to wackestone. The external areas of the mine are poor in structures of exploitation because there the flint has the texture similar to that of the enclosing bedrock. We have identified three categories of flint nodules at Ri (see Tsobgou *et al.* in this volume). The nodules of category 3 present in the axial area of the excavation possess a double cortex (gritty to chalky) and a double core. Two sub-categories 3a and 3b of these nodules have a less weathered inner core, and are the most exploited. The mechanical properties of flint extracted in the fifty drillings are related to the substratum sedimentological characteristics and their topographic position. The quality of flint for knapping and use decreases from surface to deeper layers (see Tsobgou *et al.* in this volume). Although the best quality flint of the mine of Ri was suitable for the manufacture of tough and resistant axes, the size and the shape of blocks restricted their production in large volumes.

3. Mining Structures

Pits or shafts

The 550 excavated extraction pits are concentrated in two areas (Figure 3)—the widest area located in the southern part of the excavation and the narrow area of the northern section. Drawing on Bostyn and Lanchon's terminology (1992), we have identified five types of extraction structures (Figure 4):

Open holes

The first type of structures observed is located in the south end of the mining zone, at an altitude of 197m (NGF). These open holes vary in size and are mainly circular, with diameters ranging from 1.3 to 1.7m. Some of them are oblong or double spherical ('8' shaped), reaching 5 meters in length and 2 to 3 meters in width. These open exploitation structures cover an extended surface, forming a network of holes interconnected by thin trenches. The mean depth of open holes is 0.6m, but some reach 1m. The top of the layer which contains flint is at the depth of 0.4m from the surface. The walls of the holes are generally vertical, but sometimes they are bell-shaped.

Holes with chambers

The second category of structures is located at an altitude of 198m (NGF). This category is composed by circular holes, with diameter of 2 to 2.8m. The exploitation of these holes is extended in one or many directions by 1.2 meter-deep subsurface cavities. In this type of hole a miner could work on his knees. The mean height of the chambers is

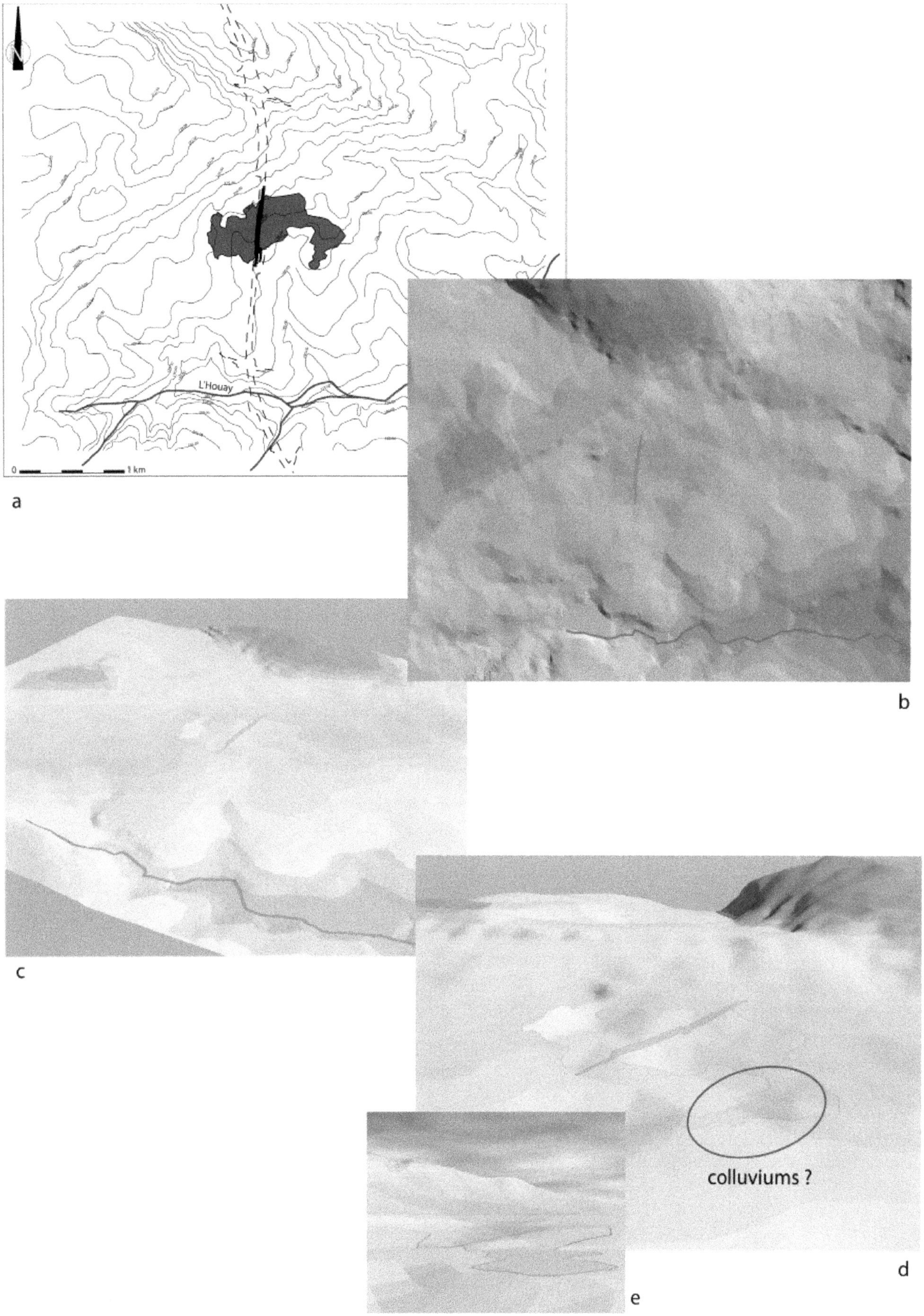

Figure 2. Estimated surface of the mine by land prospection (a and b). The 3D model of the artefacts position shows an area which could be disturbed by colluviums (d and e) (DAO, L. Vipard, Inrap).

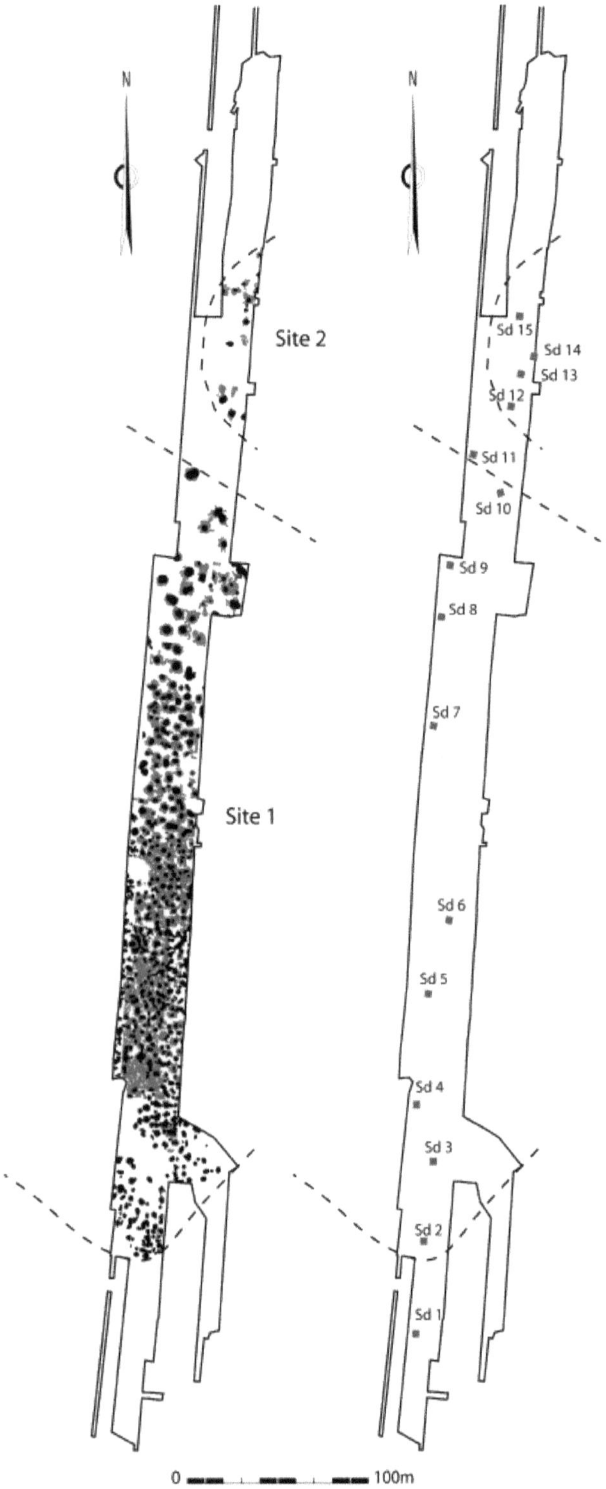

Figure 3. General scheme of the mine showing the two concentrations and the distribution of stratigraphic drillings (DAO, J.M. Palluau, Inrap).

0.8m, with the diameter of vertical access measures less than 0.4m. The layer with flint containing these holes with chambers is located at 0.8m from the surface of the mine.

Pits at medium depth

The third category corresponds to pits of medium depth, located at a 200 meters altitude (NGF). The pits of this category are generally situated at a 1.3 to 1.7m depth and are bell-shaped. As in the holes with chambers, the access is vertical, but here, its mean diameter is 2m. The exploited surface is 5m in diameter and 0.5m high (corresponding to the thickness of the layer with flint). At the bottom the pits contain some flint nodules, suggesting that the extraction took place primarily there, and their height simply facilitated miners' activities. Some pits of this type are characterized by a trench. These pits with trenches form a network of open galleries. This configuration of the structures of extraction permitted the optimization of production (Figure 5). This type of pits is dominant in the areas accounting for 90 percent of the exploited bedrock flint.

Deep pits

Deep pits constitute the fourth category of structures. While some of them are observed at a 200m and 201m NGF, others are situated in the north (200m from the first ones) at a 204m NGF. Their distribution between two different areas marks one of their distinctions. In the north area, the pits are scattered and the flint nodules are very gelifracted. The form of pits is circular to sub-circular, and their diameter and depth vary from 2m to m and 1.7m to 2m respectively. The thickness of the exploited layer of flint ranges from 0.5m to 0.6m. The access of pits is conical and serves as a central space connecting numerous galleries.

Very deep pits

The last category of pits is observed at considerable depths. These sub-circular structures, located at 201m to 203m (NGF), have a mean diameter of 5m. The mean depth of these pits is 2.3m, but some reach 4m (example of St. 643). The 0.6 meters- thick exploited layer is situated at the depth of 1.6m. The access of the pit is conical because of the instability of the substratum due to weathering. The exploitation of the layer occurred in different chambers connected to large central spaces. Sometimes, two layers of flint have been exploited, and the depth of pits then reaches 4m. To reach this second layer of flint, an access of 2 to 3m in diameter was made following the same axis as the first one.

The network of structures

The main parts of different structures are connected in a subterranean network. They still, however, have the appearance of separate structures. The design of bell-shaped pits connected to chambers or galleries apparently functioned to maximize the yield of exploitation and allowed the workers to meet during the extraction process. A 3D representation of the network shows a structure with a flat bottom, which corresponds to the exploited flint layer, and many arches working as pillars for the roof support. The management of the mining spoil sometimes consisted in stacking up small limestone plates. This layout has the profile of a dry fortification which may be used to prop up the roof, but most importantly to reduce the quantity of spoil that has to be removed from the chambers or galleries to the surface of the mine. The pits are systematically filled after their exploitation as represented by the cone of dust visible on the sections of structures. This choice of backfilling the pits seems to be related to the objective of minimizing the amount of dust removed from chambers and galleries.

Small sized pebbles of flint were left on the bottom floor of these structures, probably discarded after the initial selection of bigger size or better quality nodules. Some of these nodules had first been tested before their abandonment.

The debitage accumulations

The debitage accumulations are abundant surrounding the mouth of pits at Ri. Although they are present in different forms, some rare and small accumulations are found in the loamy upper horizons, situated at 200m to 202m NGF. To understand the conditions of their conservation we selected an area of 1200 square meters for a detailed excavation. The report of this area shows the presence of large heaps and empty areas. These empty areas may be the prints of some accumulations after their disturbance.

One of the thoroughly excavated debitage heaps contained a superposition of layers of flint flakes. Unlike the debitage accumulations present at the pits' bottom and in the spoil heaps, the disposition of accumulations at the top is controlled by the topography of previous heaps (Figure 6). The quality or the type of flakes found here differs from the quality of flakes present in the pit's bottom accumulations.

4. The Mining Tools

The ways in which pits were dug are identified through the types of tools left by miners in the mine, but also through the traces documented on the walls of chambers, galleries and pits. The most frequent traces found on the walls are thin grooves, 1 to 2cm wide and less than 40cm long. Their section is sub-circular and their extreme end circular, sometimes dissymmetric. The traces are generally grouped and their features are indicative of the deployment of a pointed tool.

In addition to giving us clues about the model of excavation, these traces allow us to reconstruct or imagine the miners' gestures. At the end of galleries, numerous traces cover the walls of excavations, showing the technique deployed for digging mining pits. The miners used here an angled pick, as evidenced by the short traces around flint pebble prints, and their movements or strokes were downward. Some differences however appear in digging techniques, notably in the pit St. 457, where the gestures alternate first from right to left, and towards the end from left to right (probably traces of a right-handed miner) (Figure 7).The mining tool here could have been a deer antler,

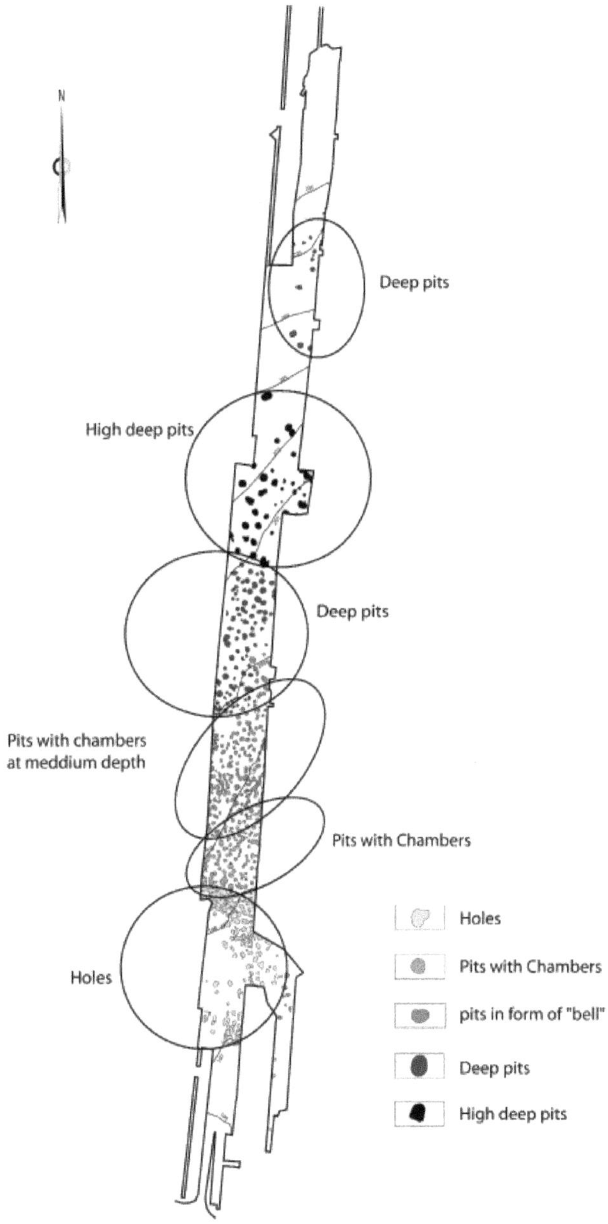

Figure 4. Types of pits or shafts with topographic curves in m; DAO, J.M. Palluau, Inrap).

as confirmed by our experiments and the great quantity of this type of tool in the pits.

Other traces are observed on some walls and flint pebbles in very deep pits. These traces are thinner and shorter (fine grooves) than those described above. They are mostly observed on hard surfaces (flint cortex or hard limestone) that are more resistant for a digging using antler tools. Our experiments have showed that the tool was probably a flint pick. However, the use of this tool is restricted to the bottom of very deep pits and of secondary importance to the mining activity, which is also the case for the use of cortical pieces of flint. At the bottom of very deep pits, two concentrations of charcoal evidence the use of torches for the lighting. However, it is still difficult to confirm this hypothesis or correlate it with the traces of torch flames.

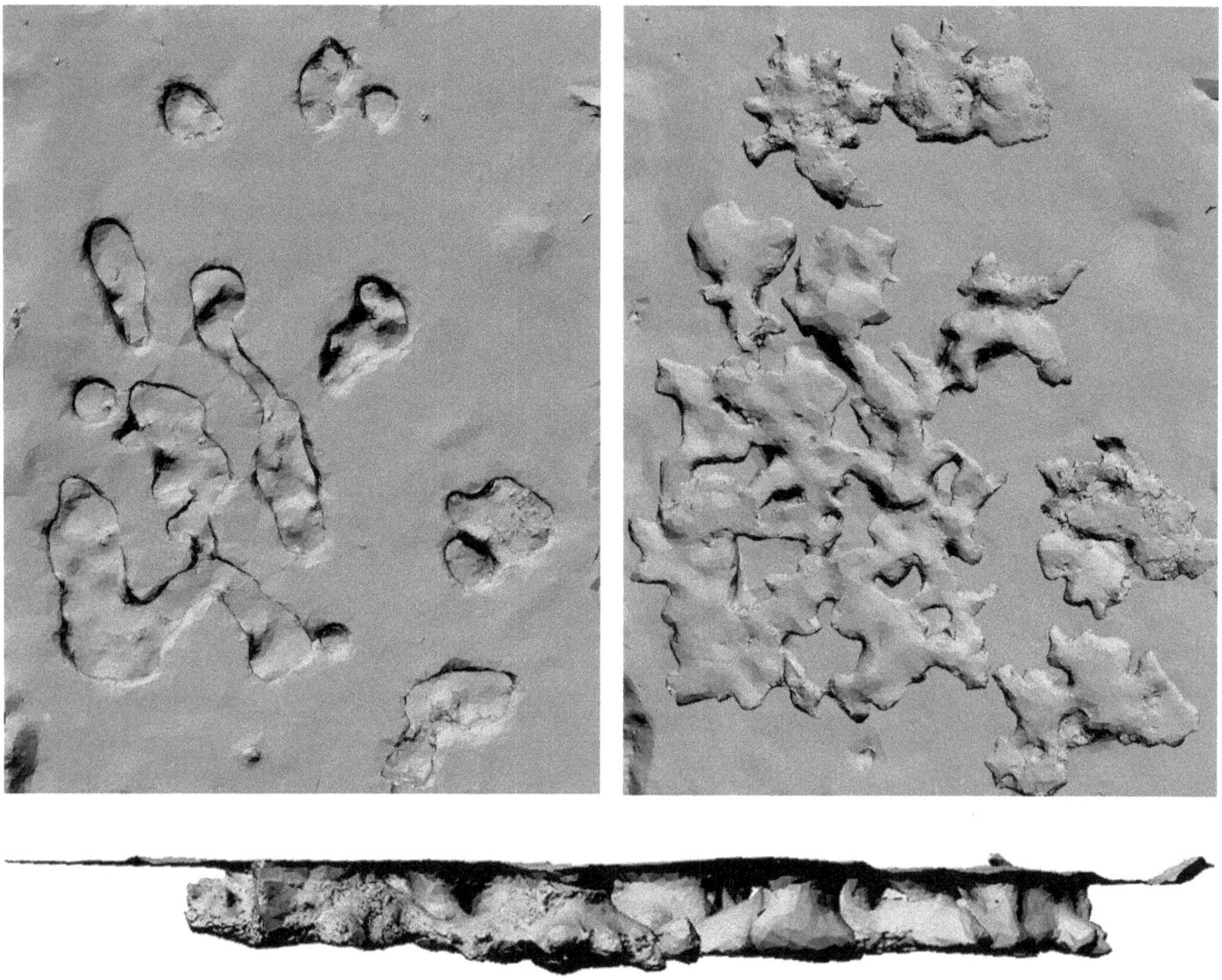

Figure 5. 3D scan of an area of the mine. At the left: view of the opening of pits. At the right, a view of the bottom of structures showing the density of the exploitation surface. Down: side view of the area with pits of 1.6 m depth in the limestone (ATM3D profile).

The hypothesis of the possible use of wooden mining tools is not excluded either. The presence of elongated grooves (more than 50cm length) on the walls of galleries, principally on roofs, may attest to the use of stakes as digging tools. These traces are straighter and wider than the traces left by antler tools. The plaster cast of these traces and the restitution of positive forms using a polyurethane rubber were successful just for two pieces from the structure St. 361. The analysis of the positive shows that the traces are of branches or small tree trunks ranging from 1.3 to 0.5m in length. The extremities of these wood pieces are broken. Whether they were used as stakes or were parts of a ladder still remains an open question.

The main digging tool: red deer antlers

In the course of the excavation of Ri, the team recovered 427 fragmented red deer antlers, which represents the largest quantity of deer antlers discovered in the flint mining complexes of the north-western Europe to date.

The antlers were well preserved as a consequence of their abandonment in depths of 1.5m to 2.5m. Deer tools have not been recovered in all mining structures: deer antler remains are found in only one third of the pits (with a mean of 10 tools per pit). The analysis of the preservation of these tools and their layout at the bottom of the pits suggests that some tools were lost while others (2 or 3 antler pieces) were probably purposefully discarded.

Twenty-eight percent of the recovered deer antlers have observable traces of cutting using the fire brake technique, currently recorded only at the mining sites of the Durrington Walls and Grimes Graves (Clutton-brock, 1984) and absent in the others contemporaneous sites in France. The typology of tools (Figure 8) gives little new information about the mining activities. Their shapes seem to be strictly related to the structure and characteristics of the substratum layers. Hence, the preferential use of short or very short tools (picks, tips) and the occasional presence of pickaxes and crowbars.

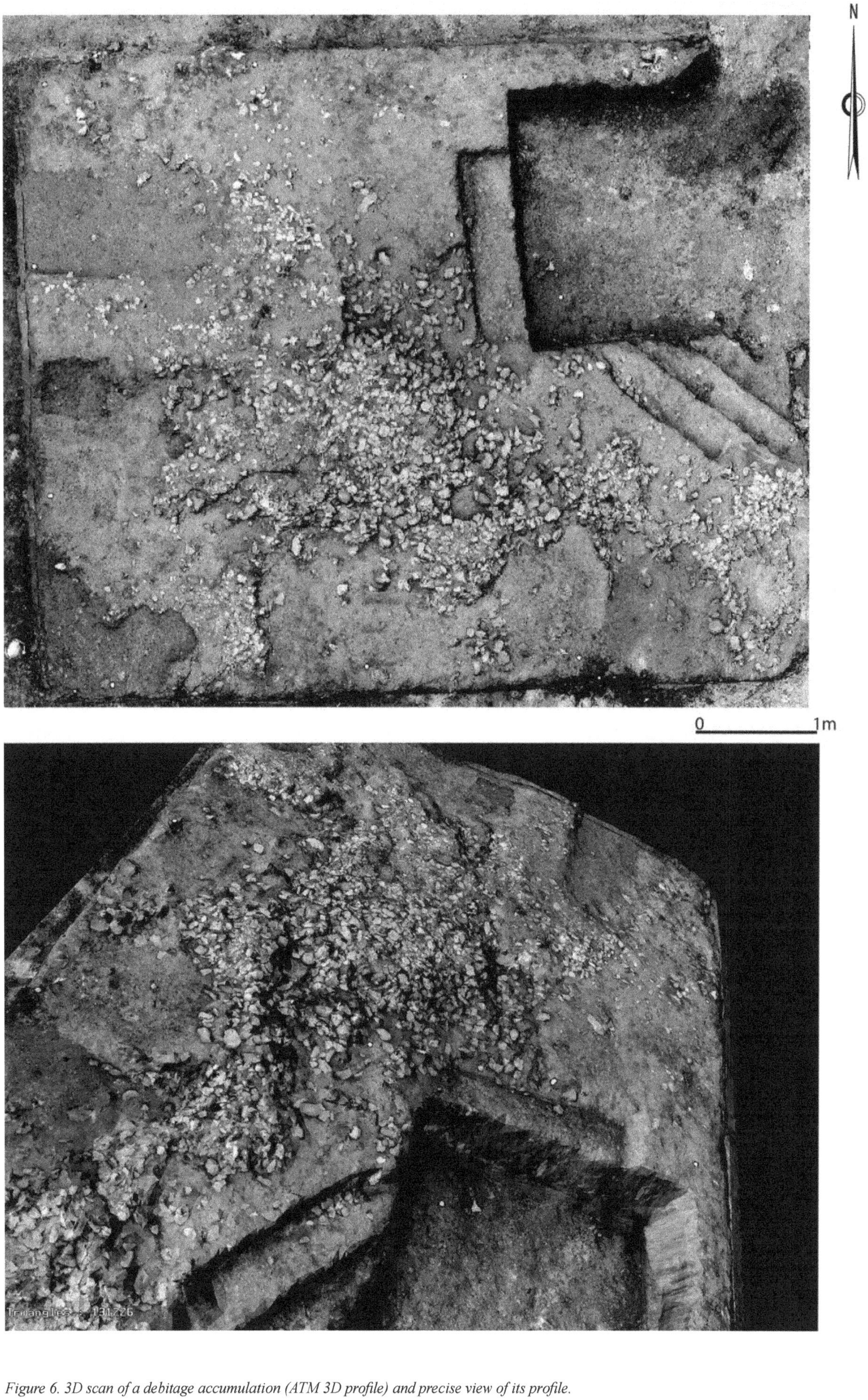

Figure 6. 3D scan of a debitage accumulation (ATM 3D profile) and precise view of its profile.

The use of red deer antlers in the mining activities at Ri constitutes the site's principal similarity with other mining sites. The percentage of deadwood versus killingwood antlers, the ratio between right and left antlers, the weakness of tools and the technique of wood cutting are also statistically similar to those of known flint mining sites. The observable differences among some characteristics such as the shape and the size of tools are closely related to the specificities of each flint mine and bear no cultural significance. The most substantial difference between the tools found at the mining site of Ri and those recorded at other mines in France is the aforementioned use of fire cutting technique, pointing to the possible connection to the English examples.

5. The Chronology of Exploitation

The chronology of the exploitation of the mine has been one of the main questions of this project. Since the start of the excavation at Ri, we have been studying the elements characterizing the phases of exploitation of the mine—its initial construction, evolution and abandonment. Although only a small part of the estimated mining complex has been excavated, the recovery of an important number of faunal remains allowed us to address the question of the mine's chronology.

The nature of the recovered artifacts and their stratigraphic significance complicates the task of establishing dates. Some artifacts, such as the ceramic fragments found in one of the pits at Ri, cannot be used for diagnosing the start or the end of the pit's exploitation. In most cases, ceramic fragments are present in trapped soils during the packing down of sediments in partially filled structures or in filling sediments of totally overflowed pits.

For radiocarbon dates, we have chosen to use only deer antler remains deposited at the bottom of pits. This choice is meaningful because the type of material directly bears on the interpretation of the link between the 'radiocarbon event' and the 'human activity event' (i.e. anteler decay and antler gathering). Although killingwood is the best type of antler, we have chosen deadwood samples because of their potential to provide us with the data which would allow a more nuanced spatial examination of the evolution of mining activities.

A total of 28 radiocarbon dates were obtained on deadwoods recovered from 7% of pits containing antlers. At Site 1, 27 date measurements were performed on this material (Figure 9, 1 date by AMS: Lyon-4370-OxA), allowing the dating of the entire surface of the excavation, while on Site 2 only one date was obtained, due to lack of good material suitable for analysis.

The dates obtained at Ri reflect the use of two types of mining structures. The date obtained on site 2 is 500 years (in BP) younger than those of the site 1 (Figure 9). The old part of the mine, on the surface of excavation, appears to have been exploited between 4100 and 3700 cal BC. The exploitation lasted three centuries, excluding the dispersion due to the dates between 5255 and 4985 BP. Although this figure sounds impressive, the number of structures of extraction (more than 5000 pits) could explain the longer time of the mine exploitation (considering that the surface of the excavation is representative of the total surface of the mining complex). This means that 20 to 30 pits were opened every year (2.5 in the area of excavation). The figure falls close to those obtained in others mines such as Jablines (Bostyn and Lanchon 1992). The site 2 is dated to the second half of the 4th millennium—the date obtained from the sample of a deer antler recovered from pit St. 1094. This result has yet to be correlated with the rest of the mine's surface.

6. The Production

The analysis of the debitage material from Ri and of the correlation between the limits of technical possibilities (in the production of flint axes) and the results obtained have led us to conclude that the technical system was not specialized. The examination of the technical protocol of axe production shows that such activity does not require specialization (Figure 10 and 11). The extraction and manufacture of flint axes are then characterized by a low level of skills, which each member of the community could obtain and develop. Sometimes the miners had to adjust to environmental changes, but such adaptations were most certainly spontaneous or transmitted to the rest of the group by the more qualified workers. However, these qualifications did not immediately give these individuals the status of specialists. As Augereau and Pelegrin (2005), we argue that the knowledge of mining techniques and practices was transmitted through copying.

Calculation of the mean surface of pits allows us to estimate the quantity of flint extracted in each structure of the mine. Based on this estimation, we have concluded that 200kg of flint can be dug out from small holes, 500kg to 1 ton from pits of medium depth. This quantity could reach 3.5 tons in the deepest pits. The experiments realized to estimate the ratio of flint axe pieces to the weight of material mined at Ri have showed that a total of 15 axes could be produced using 194kg of flint raw material. Taking into account the total number of pits in the mine, it was possible to manufacture 15 flint axes from the material extracted from each small hole, and more than 250 pieces in the case of deeper pits. The time needed to obtain a roughout was experimentally estimated at 30 minutes to one hour, but it is still difficult to estimate the time necessary to transform it into a finished product, especially since there are many techniques of time optimization and improvement of productivity such as the pressure during polishing operations.

The number of flint axes produced at Ri, estimated through our calculations or predictions, cannot be treated as an accurate figure representing the actual volume of production in the mine. There is no guarantee that the all flint extracted from the mine was intended for flint axe production. The percentages of the area of the mine that underwent exploitation vary from 40 % in the south end to 60 % in the north, and to more than 90 % in the central part of the mine. Therefore, our estimations are based on the mean of 70 % of the mine and given that 2010 square meters surface was excavated, resulting in approximately 2000 tons of flint mined and 1.5 million of flint axes produced.

The question of the circulation of products manufactured at Ri cannot be currently addressed. It is bound to be an exciting and ambitious but also very difficult to realize project, considering the difficulties of flint facies identification of all the substrata (limestones with flint) in the Plain of Caen such as those of Bretteville-le-Rabet (Desloges 1986; 1999). Nevertheless, we know that the polished axes in Bathonian flints from the Plain of Caen have been found across Basse-Normandie, in small quantities in Haute-Normandie and Ile de France, and have not been traced in Brittany. On the other hand, in Basse-Normandie, Bathonian flint axes are less frequent than the axes made of tougher raw materials such as dolerite.

A large number of flint workshops were located at a two kilometer distance of the mine, and their number became scarce at a distance greater than five kilometers. This fact suggests that different phases of flint axes production (shaping, polishing) could have been completed by different communities.

7. Conclusion

The flint mine of Ri was discovered many years ago by ground survey. Although its total surface was initially estimated at 30 hectares and then adjusted to 25ha (based on taphonomic analysis of remains and dispersion by slope colluviums calculation), the construction of the A88 highway in this area of Normandy enabled us to excavate 2ha of the mine, oriented north-southward in width. In the sampling area, which represents 10 % of the estimated total surface, more than 550 pits have been discovered. Most of these pits have been excavated mechanically or manually. The mining structures are pits, characterized by an open and somewhat deep circular access area branching out into chambers or galleries in the flint layers exploited. These pits are classified according to their morphology and depth into in 4 sub-types –holes, bell-shaped pits or pits with chambers, mid-depth pits, and very deep pits.

Twenty-eight radiocarbon dates have allowed us to identify two phases of exploitation of the mine. The first phase of exploitation is estimated at 4100 and 3700 cal BC, corresponding to the second phase of the middle Neolithic, and the second begins one and a half millennium later, as observed at Jablines, after the abandonment.

Different experimental approaches applied during this study have allowed us to improve our understanding of the techniques and methods employed by the miner of Ri.

References

Augereau, A. 2005. *L'industrie du silex du Ve au IVe millénaire dans le sud-est du Bassin Parisien. Rubané, Villeneuve-Saint-Germain, Cerny et groupe de Noyen.* Paris, Maison des Sciences de l'Homme. Documents d'Archéologie Française, 97.

Bostyn, F. and Lanchon, Y. 1992. *Jablines, Le Haut Château (Seine-et-Marne): Une Minière de silex au Néolithique.* Paris, Maison des Sciences de l'Homme. Documents d'Archéologie Française, 35.

Clutton-Brock, J. 1984. Neolithic antler picks from Grimes Graves, Norfolk, and Durrington Walls, Wiltshire : a biometrical analysis. *Excavations at Grimes Graves Norfolk 1972-1976, Fascicule 1, Dorchester* 1984, 55.

Desloges, J. 1986. Fouilles de mines à silex sur le site néolithique de Bretteville-le-Rabet (Calvados). *Revue Archéologique de l'Ouest,* 1er suppl. Actes du Xème colloque sur le Néolithique 1983,73-101. Caen, Pôle éditorial de l'Ouest.

Desloges, J. 1990. L'extraction minière du silex au Néolithique et l'exemple de Bretteville-le-Rabet (Calvados) EHESS. Polycopié.

Desloges, J. 1999. Une mine de silex au Néolithique, l'exemple de Bretteville-le-Rabet. In *L'exploitation ancienne des roches dans le Calvados, Histoire et Archéologie.* 52-77. Caen, Conseil Général du Calvados.

Gigot, P., Dupret, L. and Le Gall, J. 1999. Carte géologique de la France 1/50 000, feuille Falaise (176), notice explicative. Editions BRGM, 154.

New ^{14}C dates from the Neolithic flint mines at Rijckholt-St. Geertruid, the Netherlands

Marjorie E. Th. de GROOTH, Roel C. G. M. LAUWERIER and Muuk E. ter SCHEGGET

Abstract

During the 20th century, various studies were conducted at the Neolithic flint mines at Rijckholt-St. Geertruid in the southern Netherlands. To give us more idea of how long the mines were in operation, a series of new AMS ^{14}C dates have been performed. The results show that underground mines were already operational at the beginning of the Michelsberg culture (*c.* 5320 ± 40 BP), and were still in use at the time of the Stein group, at the end of the fourth millennium (4470 ± 35 BP).

One of the most intriguing finds – the human skull 'Rijckholt 1' – was also dated. Surprisingly, this find turned out to be not Neolithic but of recent date. This skull, an isolated find from one of the mine galleries, was originally interpreted as a ritual deposit. This paper attempts to place the recent date of the skull in the history of the mine, from its Neolithic origins to the present day.

Keywords

Netherlands. Rijckholt. Neolithic. Flint mine. ^{14}C date. Rijckholt 1 skull.

1. Introduction

The 'Prehistoric Flint Mines Working Group' of the Dutch Geological Society's Limburg branch excavated the Neolithic flint mines at Rijckholt-St. Geertruid (Figure 1) from 1964 to 1972 (Felder *et al.* 1998; Rademakers 1998a, 1998b) (we use the official spelling 'Rijckholt' in this paper; the traditional spelling – 'Ryckholt' – may be encountered elsewhere). A horizontal exploratory tunnel was dug into the limestone in the slope to the east of the *Groot Atelier*, at the level where Van Giffen had found the first galleries in 1925 (Van Giffen 1926, 107 note 1). This passage eventually formed an underground link between the *Groot Atelier* and the shafts that Waterbolk encountered on the plateau some 150m to the east in 1964 (Waterbolk 1994). In this underground excavation, on either side of the tunnel the prehistoric galleries were examined over a width of 10 meters. A total of 75 shafts and 1526 square meters of galleries were examined, the entire area measuring 2436 square metres (Figure 2). Underground mining is thought to extend over *c.* 8 hectares, comprising an estimated total of 2000 mines. The flint extracted came mainly from layer 10 of the Lanaye limestone (Gulpen Formation, Upper Cretaceous).

The Working Group commissioned five ^{14}C datings on the small amount of organic material found below ground (Felder *et al.* 1998). Four of the samples were charcoal, and the fifth dating was carried out on a fragment of red deer antler. As part of a study of the social and economic context of flint mining in Rijckholt facilitated by the Bonnefantenmuseum in Maastricht, in 2004 eight further ^{14}C datings were performed on samples with a negligible (or low) intrinsic age, known find circumstances and the greatest possible spread over the area exploited in prehistory (Figure 3). The spread was nevertheless fairly limited: two samples came from the Working Group's excavation, three from Waterbolk's research on the plateau, two from the western slope of the *Schone Grub*, a dry valley bisecting the mining complex; and a final sample excavated by Van Giffen in all probability also originates from the plateau (Figure 1,2). A full report on this research will be published shortly (De Grooth *et al.* i. p.).

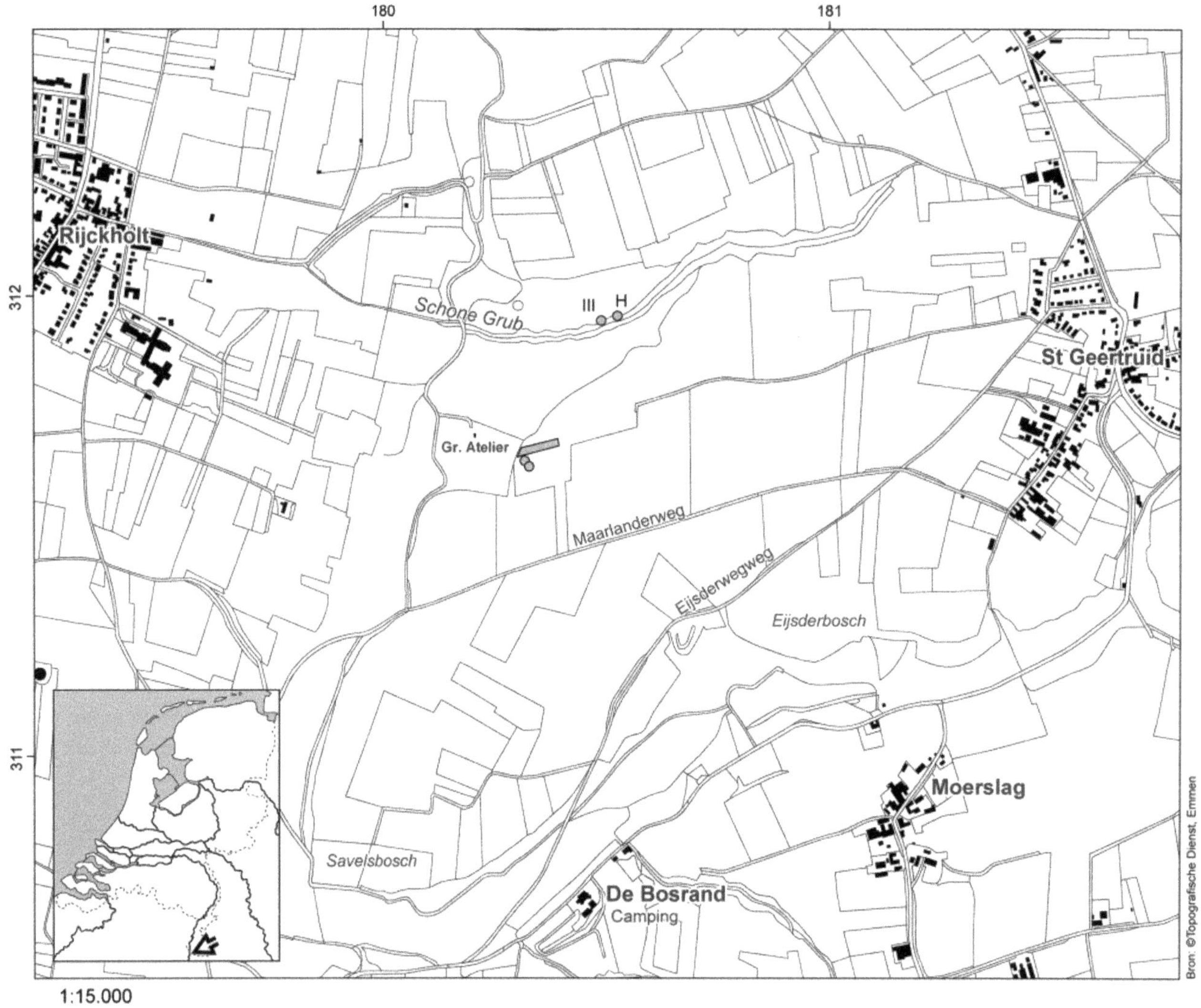

Figure 1. Map showing the location of the Rijckholt-St. Geertruid mining area and of the ^{14}C samples from Van Giffen's, Hamal Nandrin's and Waterbolk's excavations.

The samples were taken at the archaeozoological laboratory of Groningen Institute of Archaeology (GIA) in April 2004, and ^{14}C AMS dated by the University of Groningen's Center for Isotope Research (CIO). For the purposes of this paper, the results were calibrated using the CALIB 5.0 calibration programme (Stuiver and Reimer 1993) and the INTCAL04 ^{14}C calibration curve (Reimer *et al.* 2004). Six of the dates conformed to expectations; these are discussed in the next section. The analysis of the sample from the Rijckholt 1 skull (GrA-26905) produced a completely unexpected date of 210 ± 30 BP. After calibration, in calendar years, this gives a 2 sigma date of AD 1645–1684, 1734–1806 or 1929–1951. Since the skull is one of the most intriguing finds from the Working Group's excavations (Rademakers 1972, 1998b; Van Vark 1971), this result will be discussed in more detail in the second part of this paper.

2. Discussion of the new data

The new dates from mine 48 (Working Group excavation; GrA-26907) and from the plateau (fill of Waterbolk's shafts 7 and 10; GrA-26913, GrA-26903) give roughly the same age as the antler fragment from mine 67 (beneath the floor of the *Groot Atelier*), dated previously to 5065 ± 45 BP (Gr-N9058). In calendar years, they may represent a period of some 200 to 250 years, between *c.* 3970 and *c.* 3675 or *c.* 3720 cal BC. If we also consider the charcoal dates from the mines, taking into account the old wood effect, this part of the mine area might have been in production for at least 300 years. This coincides with Lüning's (1967) phases III and IV of the Michelsberg culture (Lanting and Van der Plicht 1999/2000). The two dates of Van Giffen's finds provide interesting new information on the duration of production. The sample from location III offers the earliest date so far for underground mining in Rijckholt, at 5320 ± 40 BP (GrA-26909), or 4316–4042 cal BC

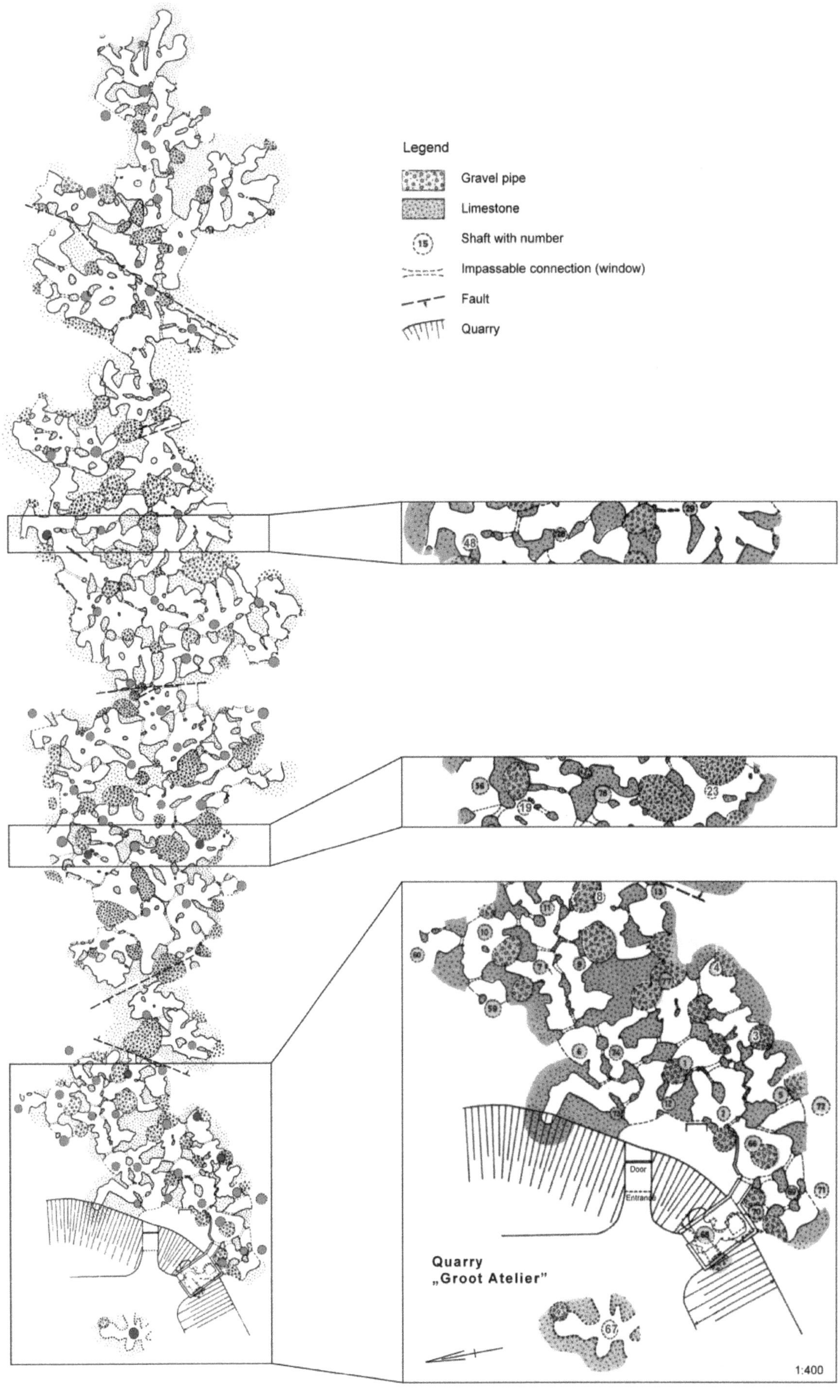

Figure 2. The locations of ^{14}C samples from the Working Group's excavation (map based on Felder and Bosch in Rademakers 1998).

(2 sigma): contemporaneous with the earliest phases of the Michelsberg culture, in other words (cf. Lanting and Van der Plicht 1999/2000). The pick is considerably younger, with a date of 4470 ± 35 BP (GrA-26904), or between 3340 and 3025 cal BC. This mining tool, therefore, is contemporaneous with the only sherd excavated at Rijckholt, a base fragment that can be assigned to the Stein group (Van Giffen 1925, Pl. 4; VII, 35). Finally, the 'Hamal Nandrin skull', with its date of 3840 ± 35 BP (GrA-26908), or between 2459 and 2201 cal BC, points to the presence of humans in the vicinity of Rijckholt at the time of the Bell Beaker Culture.

3. Evaluation of the duration of exploitation

A validation of these data on the duration of exploitation may obtained through a survey of the occurrence of mined Rijckholt flint in dated settlement and burial sites, as well as by comparing Rijckholt with similar flint mines in the region. Of the latter, Spiennes (Hainault, Belgium) is of major importance, because to date it is impossible to distinguish the flints mined at both sites by macroscopic means.

a. Comparison between Rijckholt and Spiennes

According to a series of 15 radiocarbon dates of samples with a negligible intrinsic age, deep shaft mining at Spiennes may have started somewhere before 4200 cal BC and continued to at least 2700 cal BC (Collet *et al.* 2006; Collet *et al.* 2008a; Collet *et al.* 2008b). The new Rijckholt dates lay well within this range. Given the important amounts of Michelsberg pottery found at the site, the main period of exploitation is thought to have occurred during the Michelsberg period, even though the majority of ^{14}C dates are coeval with the younger Seine-Oise-Marne and Stein group period.

b. Evidence from non-mining sites

In an alternative approach, the duration of exploitation of the mines may be estimated through a survey of mined flints in dated settlement and burial sites. In this, we are confronted with the afore-mentioned impossibility to distinguish between Rijckholt and Spiennes flint. In a previous study (De Grooth 1991), as an interim solution, only sites in regions for which Rijckholt would have been the closest and most conveniently situated source area were considered, thus excluding the well-dated and well-stratified sites in the Dutch coastal and western river areas. In the present survey, these sites will be included, as their assemblages provide interesting information on changing patterns in the use of mined flint on a more general level. Areas where Spiennes would have been the most probable source are not considered. Vermeersch and Burnez-Lanotte (1998) and Vanmontfort *et al.* (2008) provide a survey of the situation in Belgium. The distribution in Northern-France is still little-known (Collet *et al.* 2008a).

An analysis of the debitage from flaking floors at Rijckholt and Spiennes indicates that the main output consisted of semi finished or polished flint axes with an oval cross section, chisels, robust blades and large flakes. Production waste associated with the youngest mine shafts at Spiennes, however, shows that there axes were the only artefacts manufactured at that time (Collet *et al.* 2008a).

Apart from the axes and chisels, pointed retouched blades (*Spitzklingen*) and wide end-scrapers on flakes seem to be the most characteristic tool types. At present, size is the only suitable characteristic to distinguish between artefacts made from flint extracted from a primary geological context at the mining sites and those made from secondary flints, i.e., material collected from residual loams or river gravels. Following Wansleeben and Verhart (1990), 'mined' blades and blade tools should have a minimal width of 25mm, and a minimal length of 80mm. Flakes should be larger than 50mm.

Such artefacts are known from a large part of Western and Central Europe (Figure 4). Most of the well-dated sites are located in the southern and western parts of the Netherlands, the Rhineland and Westphalia, i.e., at a distance of up to 200km from the mines. Mined flint of the Rijckholt type also is encountered farther afield, e.g., in Lower Saxony and Hesse (*c.* 300km from Rijckholt) and Baden-Württemberg, 500–550km from Rijckholt.

Sites dated before c. 4000 cal BC

For the earliest period, i.e., before *c.* 4000 cal BC, evidence for the use of mined Rijckholt flint is rare, but not altogether absent. From the Rhineland, some (fragments of) polished axes are known from settlements of the Grossgartacher, Rössener and Bischheim Cultures (Dohrn-Ihmig 1983; Fiedler 1979; Gehlen *et al.* 2009; Schwellnus and Arora 1983). Further to the north, a number of macrolithic blades and axe fragments are reported for the Bischheim phase at Hüde I (Kr. Diepholz). However, because of the complicated post-depositional situation at this site, most researchers assign these finds on typomorphological grounds to a subsequent Michelsberg phase (Stapel 1991). At the other extreme of the distribution area, finds with a similar age are published from the Schwieberdingen settlement at Sindelfingen-Hinterweil (Kieselbach 2000). More evidence is known from the early stages of the Michelsberg Culture (phases MK I and II according to Lüning 1967) and the Middle Swifterbant Culture (*c.* 4200–3800 cal BC, Raemaekers 1999). Mined Rijckholt flint was surprisingly rare at both Maastricht-Watermolen Vogelzang (Brounen 1995) and Heerlen-Schelsberg (Schreurs and Brounen 1998), sites situated in the immediate vicinity of the mines. In the Rhineland, sites such as Koslar 10, Inden 9 and Bochum-Altenbochum, however, provided important amounts of macrolithic pointed blades and (fragments of) polished axes (Höhn 1997a, 1997b; Willms 1982). In the Dutch Coastal area, several macrolithic blade tools from Rijckholt or Spiennes flint have been published from

Lab nr.	Type and location of sample; comments CIO november 2004	BP	Cal BC 1 sigma	Cal BC 2 sigma	Probability 2 sigma
New dates					
GrA-26909	bone indet, large mammal, Van Giffen location III (GIA Rijckholt 1923/VIII 22)	5320 ± 40	4233 - 4222 4210 - 4154 4133 - 4060	4316 - 4298 4262 - 4042	0.021 0.979
GrA-26907	red deer antler, Working Group mine 48 (pdb 3313A-11796)	5100 ± 40	3962 - 3933 3875 - 3806	3972 - 3797	1.000
GrA-26913	tibia red deer, Waterbolk shaft 7 (GIA Rijckholt 1964-7, nr. 3); Soxhlet extraction for contaminated samples	5090 ± 40	3958 - 3928 3877 - 3804	3968 - 3794	1.000
GrA-26903	new-born calf, Waterbolk shaft10 (GIA Rijckholt 1964-10)	5060 ± 40	3943 - 3893 3883 - 3853 3847 - 3832 3825 - 3799	3962 - 3766 3721 - 3720	0.999 0.001
GrA-26904	pick, red deer antler, Van Giffen plateau (GIA Rijckholt 1925/VII 29; Clason Rijc-71)	4470 ± 35	3330 - 3215 3182 - 3157 3125 - 3090 3042 - 3039	3340 - 3202 3199 - 3079 3070 - 3025	0.526 0.358 0.116
GrA-26908	human skull 'Hamal- Nandrin', (BFM 2969A); Soxhlet extraction	3840 ± 35	2398 - 2383 2346 - 2271 2259 - 2206	2459 - 2416 2411 - 2201	0.109 0.891
GrA-26911	tibia pig, Waterbolk shaft 2 (GIA Rijckholt 1964-2); Soxhlet extraction insufficient	345 ± 30	1487 - 1524 1558 - 1631	1466 - 1637	1.000
GrA-26905	human skull 'Rijckholt 1', Werkgroep mijn 8 (pdb 3313A-1); 'peculiar as regards the result'	210 ± 30	1651 - 1677 1765 - 1772 1776 - 1800 1940 - 1951	1645 - 1684 1734 - 1806 1929 - 1951	0.321 0.527 0.152
Existing dates					
GrN-4544	charcoal gallery between shafts 3 en 4	5070 ± 70	3951 - 3893 3882 - 3799	3972 - 3748 3745 - 3712	0.949 0.051
GrN-5549	charcoal bottom shaft 23	5000 ± 40	3905 - 3880 3800 - 3709	3943 - 3854 3847 - 3830 3825 - 3695 3676 - 3675	0.301 0.022 0.675 0.002
GrN-5962	charcoal bottom shaft 19	5090 ± 40	3958 - 3928 3877 - 3804	3968 - 3794	1.000
GrN-9058	red deer antler, shaft 67	5065 ± 45	3944 - 3904 3898 - 3896 3880 - 3800	3966 - 3763 3723 - 3715	0.991 0.009
GrN-9085	charcoal 67	5080 ± 45	3953 - 3912 3878 - 3803	3971 - 3775	1.000

Figure 3. Overview of new and existing ^{14}C dates for the Rijckholt mines.

Brandwijk-Het Kerkhof, from a layer dated between 4220 and 3940 cal BC (Van Gijn 2008), whilst they are absent in the Middle Swifterbant layers at the Hazendonk (Raemaekers 1999) and at Hoornaar-Lage Giessen (Van Hoof 2008). Once more, some sites in Baden-Württemberg may be mentioned (Kieselbach 2000): Aichbühl (Aichbühl Culture, starting at 4260 BC) and Eberdingen-Hochdorf (Older Schussenried with imported MK II pottery).

Sites dated from c. 4000 – c. 3500 cal BC

The ^{14}C dates clustering around 3970–3700 cal BC indicate that an important part of the mining activities took place during the first half of the fourth millennium, i.e., contemporaneous with the younger part of the Michelsberg Culture (Lüning's MK III-V) and the Dutch Hazendonk Group (originally thought to be a local Michelsberg derivative, but nowadays seen as a Swifterbant development, cf. Raemaekers 1999). At many sites the majority of tools were made of imported Rijckholt (and/or Spiennes) flint, although locally available flint was worked as well. This holds true for sites such as Maastricht-Klinkers (Schreurs 1992), and for Hazendonk sites in the Dutch Eastern river area, e.g., Linden-Kraaienberg (Louwe Kooijmans and Verhart 1990), Wijchen-Het Vormer (Louwe Kooijmans 1980), Beers-Gassel (Verhart and Louwe Kooijmans 1989) and Grave-Pater Berthierstraat (Verhart 1989). In the Dutch Coastal area, mined flint from either Rijckholt or Spiennes was of minor importance. At Schipluiden only five artefacts were characterised as such, i.e., less than 0.1% of the assemblage (Van Gijn *et al.* 2005/2006). At Wateringen 4 and Hazendonk (layer 3) mined flint was only present as reworked axe fragments (Raemaekers 1999), and it was absent among the 5000 artefacts studied at Ypenburg 4 (Houkes 2008). Given the origin of the other lithics recovered at these sites, their inhabitants are thought to have had contacts with both Hainault and North-western France and with the Middle Meuse valley and the Ardennes (Louwe Kooijmans 2005/2006).

Only very few sites dated to this period are known in the Rhineland, notably a cache with 13 in part conjoinable blades from the MK IV or V settlement at Garzweiler (Arora *et al.* 1988). More evidence is available for Westphalia, from sites such as Osterwick, Nottuln, Coesfeld-Harle, and Soest (Knoche 2008; Willms 1982), where macrolithic blade tools and (reworked) axes occur in considerable numbers. Moreover, Wallbrecht (2000) mentions some 30 Michelsberg sites with mined Rijckholt flint in the region between Göttingen and Hannover, east of the Weser. In Baden-Württemberg imported blade tools have been published from a number of sites (Behrends 1991; Hoffstadt and Maier 1999; Kieselbach 2000). Some, such as Bruchsal, Ilsfeld-Ebene (Kr. Heilbronn) and Heilbronn-Klingenberg belong to the Michelsberg Culture. Eberdingen-Hochdorf (Kr. Ludwigsburg), Allershausen-Hartöchsle (Kr. Biberach) and Leonberg-Hofingen (Kr. Böblingen) are from a Schussenried context, with dendrochronological dates of 3955 –3871 BC (Billamboz 1998). Wangen-Hinterhorn (Kr. Konstanz) belongs to the Pfyn Culture (dated between 3870 and 3799 BC, Billamboz 1998), and Hornstaadt-Hörnle1 and Bodman-Weiler represent the Hornstaadt Group (from 3913 to after 3890 BC, Billamboz 1998).

Sites dated after c. 3500 cal BC

Sites with artefacts made from mined Rijckholt and/or Spiennes flint from the second half of the fourth and the first part of the third millennium exist as well. There are only a few them, and their distribution is annoyingly lacunal. This, however, is in accordance with the settlement pattern prevailing in this period, as depicted by Van Gijn and Bakker (2005) for the Netherlands, Richter (1997) for the Rhineland, and Raetzel-Fabian (2000) for Westphalia and Hesse.

Both macrolithic blade tools and axes are known from sites belonging to the Stein Group or the Vlaardingen Culture, such as Maastricht-Hoogenweerdt (Brounen *et al.* 1990), Geistingen-Huizerhof (Heymans and Vermeersch 1983; Vanmontfort *et al.* 2008) or Echt-Koningsbosch 27 (Van Haaren and Modderman 1973) in the Meuse valley, and Beuningen-Ewijkse Veld (Jansen 1989) or Nijmegen-Ressen (Verhart and Van den Broeke 2002) in the Eastern river area. The wetland settlements of the Vlaardingen Culture in the western part of the Netherlands, however, yielded almost exclusively fragments of damaged and reworked axes, e.g., the Vlaardingen occupation at the Hazendonk (Raemaekers 1999), Hekelingen III (Van Gijn 1990), Ypenburg-Gavi (Bulten i.p.; Houkes and Dorenbos 2004), Leidschendam (Hamburg 2005; Van Gijn 1990), and Hellevoetsluis-Ossenhoek (Goossen 2009). Only at the last-mentioned site several macrolithic blade tools were recovered as well.

In Germany, relevant finds occur mainly in Westphalia, Hesse and the southern part of Lower Saxony, regions where the Wartberg Culture was present between *c.* 35/3400 and 28/2700 cal BC (Knoche 2008; Raetzel-Fabian 2000, 2002). In this period, no axes made of Rijckholt flint are known, but macrolithic blades and blade tools occur repeatedly in both collective graves and in settlements (Knoche 2008; Raetzel-Fabian 2000; Schwellnus 1979).

4. Further consideration of the 'Rijckholt 1 skull'

In the following section we will consider how we should place the date of the skull in the history of the mine, from its Neolithic origins to the present day.

From the outset, it did not occur to those involved to dispute the Neolithic date because the entire context – which was regarded as sealed – also dated to this period. Since only a skull was found, without the rest of the skeleton or any accompanying finds, the excavators originally interpreted it as a ritual deposit (Rademakers 1998b, 237). Its interment in an operational underground flint mine suggested a link between the death and the mineworkers. It appeared to be one of them.

The first conclusion concerning the age on physical anthropological grounds, which put the skull in the Neolithic, was drawn a week after the discovery. During a visit to Rijckholt, Professor De Wilde, who was involved in the research in his capacity as professor of anatomy at the Anatomy and Embryology laboratory of the University of Groningen, determined that the skull came from a male aged between 16 and 30. He regarded the skull as highly valuable, since it displayed several features that were primitive even for the Neolithic (Felder and Rademakers 1971, 52). He also observed a number of severe cranial traumas in the form of a deep round hole in the frontal bone, a shattered nasal bone and a long horizontal suture running from the frontal bone, over the sphenoid bone, to the centre of the occipital bone (Felder and Rademakers 1971), and in addition remarked that, given the advanced regrowth, a lengthy healing process had occurred, and the man had survived his injuries. Dr Van Vark, assistant and subsequently successor of De Wilde,

Figure 4. General distribution of artefacts made of mined Rijckholt flint.

reports (1971) most of these observations and conclusions, which are largely confirmed by the recent analysis of the skull (Ter Schegget i.p.). The skull consists only of a *calvarium* with an upper jaw (*maxilla*); the lower jaw (*mandibula*) is missing. The most distinctive and therefore most valuable features for determining sex on the basis of the skull are the *glabella*, the relief of the nuchal plane and the zygomatic process of the *maxilla*. These features all score +1, so the skull is male. The mastoid process – only the left side of which has been preserved – is relatively small on this skull, scoring 0 (indifferent). Most of the less sex-specific characteristics also score +1, with only the *tubera* scoring -1 (slightly female). The skull is therefore without a doubt male. Since only three dental elements remained preserved, and age determination on the basis of fusion of the cranial sutures is highly imprecise, the age at death can only be roughly estimated to lay between 35 and 55.

The very recent date of the skull immediately prompted debate. The CIO remarked that it was a 'curious result', and suggested the following explanation (letter by Van der Plicht; CIO/557-2004-H). 'Sample 3313A1 was subjected to the standard pre-treatment (collagen extraction), rather than the extended version (like sample 2969A, which was treated with resin), which did not appear necessary. Curiously, the stable isotope numbers in this sample, which we also measure in the ^{14}C process, deviate markedly from the values normally found in bone collagen. This applies to ^{13}C, and most especially to ^{15}N. This suggests a reservoir effect (e.g., due to a diet consisting mainly of fish). However, this would make the bone appear older, so the skull would in fact be of recent origin. We have no explanation for this at the moment. I would not expect a repeat of the dating to solve the mystery'. Given these comments, it would not seem reasonable to doubt the validity of the ^{14}C date. The ^{13}C and ^{15}N values might have slightly influenced the interpretation of the ^{14}C age, but this does not detract from the fact that the skull has been dated to the modern period and not, as expected, to the Neolithic. Clearly, the skull's find spot cannot be regarded as a closed Neolithic context.

To obtain a clearer picture, we reconsidered the circumstances of the find. In retrospect, the report of the finding of the skull issued by those involved in the excavation in December 1965 and published in 1972 gives us the following picture of the circumstances (Felder *et al.* 1965; Felder *et al.* 1998, 54-56; Rademakers 1998b). On 5 March 1965, after a large flint nodule was removed, gallery C became visible, cross-cutting the exploratory passage almost at a right angle. The part to the left of the exploratory passage was entirely filled with limestone grit, blocks of limestone, blocks and splinters of flint, flakes and picks. The section to the right was 60% filled, and the roof had caved in here to some extent .When the gallery was emptied and investigated some difficulty arose at the right end, which came to a dead end at a gravel pipe filled with loose sand and gravel. After some of the fill in gallery C had been removed, the fill of the gravel pipe began to show signs of imminent collapse. It was therefore decided that a protective wall of limestone blocks should be built at the point where gallery C met the gravel pipe.

Owing to circumstances, the work did not continue until eight months later, starting with the reinforcement of the exploratory passage. Part of the side wall and floor had to be dug out to allow an iron prop to be positioned. The wall built in March and a small part of the gravel pipe also had to be removed. Once a small part of the wall had been removed, the floor that was revealed was dug out. This weakened the remainder of the wall, pushing out loose limestone and erosional clay from the edge of the gravel pipe. At that moment, the skull became visible at the edge of the gravel pipe, just below the floor of gallery C (Figure 5). Several 30-40cm chunks of limestone had already fallen on the skull, and there was a danger that it would be crushed. Therefore all excavators were called to come and help remove the skull. Two people supported the large block of limestone, while the third released the skull by scratching out loose material from beneath it. Once the skull had been safely recovered, the blocks were released, and rolled into the gallery together with more blocks and part of the fill of the gravel pipe. According to the excavators the skull was moved no more than a few centimetres by the blocks in the wall. It lay with its right side against a mix of limestone, gravel and erosional clay, at the junction between gallery C and the gravel pipe, just beneath the floor of the gallery. The find spot was approximately eight meters below the current surface, and some six meters from the associated shaft (no. 8).

So the question remains: how did this recent skull end up at this spot? One particularly important factor would appear to be the fact that precise observation of the circumstances was hampered because the skull had to be removed under threat of imminent collapse. The above description suggests that the skull was in fact discovered at precisely the moment when it was dislodged by the pressure of the gravel pipe fill as it collapsed. A factor that might be even more important is the eight-month delay between the time when the researchers first reached the find spot and the resumption of work, leading to the discovery of the skull.

Its recent date clearly makes the hypothesis of a Neolithic ritual deposition untenable. Two possibilities remain: either it somehow fell into the mine in the modern period, or it

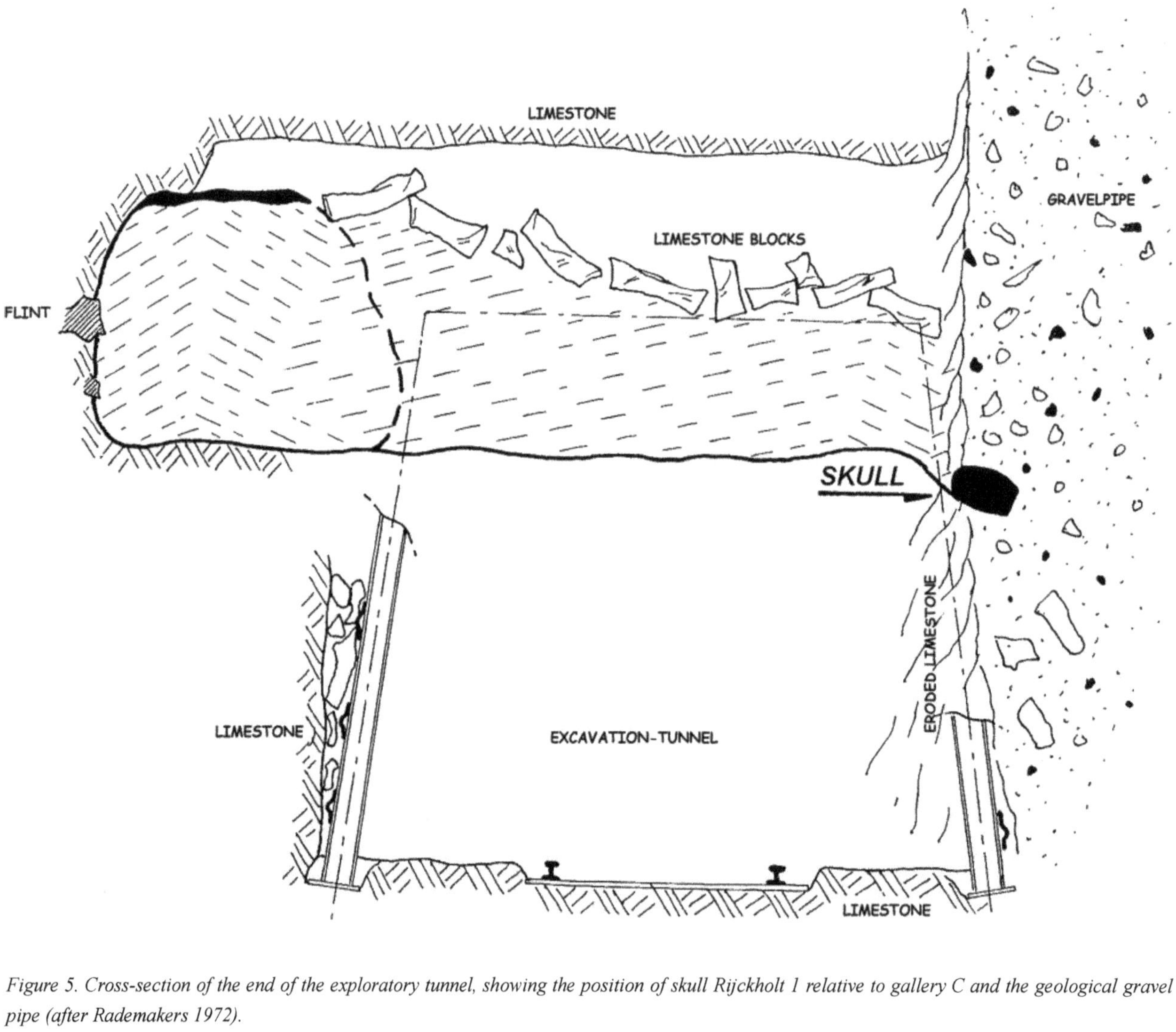

Figure 5. Cross-section of the end of the exploratory tunnel, showing the position of skull Rijckholt 1 relative to gallery C and the geological gravel pipe (after Rademakers 1972).

was placed in the mine at the time of the excavations. The skull cannot have found its way there via the mine shaft, because both the shaft and the gallery had already been back-filled with mining waste in the Neolithic. Neither can it have travelled down the solution pipe, as these karstic features were filled with gravel, sand and loess during the Quaternary, long before the mines were dug (Felder and Felder 1998). Once the recent date of the Rijckholt 1 skull was announced, it was suggested that the vertical feature in reality was not a geological solution pipe, but a shaft dug in the modern period for extracting marl, or a trial shaft to see if the flint was suitable for the flintlock industry – big business in the 18th and 19th centuries, in particular. The skull would not then be from 'gallery C' but from the fill of this shaft. There is, to date, no evidence to support this hypothesis. An investigation of the surface of the shaft field, however, might settle the question of the true nature of this vertical feature.

The final suggestion is that the skull must have been placed there in the context of the excavation. Given their reputation, we think it extremely unlikely that the Working Group was involved in such an activity, e.g., in an attempt to enhance the importance of their investigations. The deposition could have been performed by an intruding prankster, possibly between 5 March 1965, when the excavators closed gallery C and temporarily stopped work, and 5 November 1965, when work resumed and the skull was found. Despite the makeshift fence, several unauthorised visitors gained access to the excavation. There is however no evidence that the block wall was tampered with in this period, and the skull was not revealed until part of the limestone floor had been dug out. In short, the situation was so unstable that inexpert manipulation could have caused great danger. At any rate, the skull is no longer relevant to our knowledge either of the Neolithic or of the history of flint mining.

5. Conclusion

The majority of the Rijckholt ^{14}C dates suggests that the most important part of the exploitation took place during the first half of the fourth millennium. Its products are, indeed, encountered at many contemporary settlements. However, artefacts made of mined Rijckholt flint have also been found at older and younger sites. These data tended to be regarded with a certain amount of criticism, because no corresponding dates from the mines themselves existed. Therefore, many researchers thought them to reflect palimpsest situations or scavenging activities. Given the new ^{14}C dates, we think the evidence from settlements and burial sites presented in this study may be regarded as support for the idea that mining activities at Rijckholt lasted the same time as those at its counterpart at Spiennes. Of course the earliest new ^{14}C date does not represent the beginning of underground mining in Rijckholt, but it is consistent with the hypothesis that the underground workings began in the slopes of the *Schone Grub*. The other new dates do not provide unequivocal evidence to support the common hypothesis (Waterbolk 1994; Felder *et al.* 1998) that the small, shallow mines in the west are more recent than the large deep mines higher up the slope and on the plateau. Nor can they be used to dismiss this hypothesis, however. Finally, the direct dating of a mining tool can be seen as concrete evidence that underground mining was still going on at the time of the Stein group.

The Rijckholt 1 skull, which was recovered from the mine, and was originally interpreted as a ritual burial, was found to be recent. The past interpretation of the morphology of the skull, suggesting a Neolithic date, appears to have been the result of a foregone conclusion. The original assumption that this was a find from a closed Neolithic context is belied by the recent facts. How the modern skull ended up in this Neolithic context can no longer be established. It was probably placed in the mine in 1965, in the eight months between the point at which the excavators reached the find spot, and the resumption of work. It is therefore no longer relevant to our knowledge either of the Neolithic or of the history of flint mining.

Acknowledgements

We would like to thank the following individuals and institutions for their contributions to this paper: Béatric de Fraiture, Gemma Janssen, Fun Horbach (Provincial Repository for Archaeological Finds, Maastricht); the Bonnefantenmuseum (Maastricht); Thom Jacobs, Robert Kosters, Jan Lanting and Wietske Prummel (GIA Groningen); Hans van der Plicht (CIO Groningen); Jos Deeben (RCE Amersfoort); Marjolein Haars (BCL-Archaeological Support); and Leendert Louwe Kooijmans (Leiden University) for his comments on an earlier draft.

Special thanks go to Werner Felder, Sjeuf Felder, Fun Horbach, Jan Nilessen and Jo Willems of the Prehistoric Flint Mines Working Group for their frank and constructive discussion of the mysterious dating of the skull. Werner and Sjeuf Felder are unfortunately no longer with us. We hope they would have acquiesced in the deliberations presented in this paper.

References

Arora, S. K., Franzen, D. and Franzen, J. 1988. Ein Klingendepot des Jungneolithikums von Garzweiler (Gem. Jüchen, Kr. Neuss). *Archäologie im Rheinland 1987*, 27-28. Köln, Rheinland-Verlag.

Behrends, R. -H. 1991. *Erdwerke der Jungsteinzeit in Bruchsal. Neue Forschungen 1983-1991.* Stuttgart. Archäologische Informationen aus Baden-Württemberg 22. Landesdenkmalamt Baden-Württemberg.

Billamboz, A. 1998. Die jungneolithischen Dendrodaten der Pfahlbausiedlungen Südwestdeutschlands als Zeitrahmen für die Einfl üsse der Michelsberger Kultur in ihrem südlichen Randgebiet, in J. Biel, H. Schlichtherle, M. Strobel and A. Zeeb (eds.), *Die Michelsberger Kultur und ihre Randgebiete - Probleme der Entstehung, Chronologie und des Siedlungswesens*, 159-168. Stuttgart. Materialhefte zur Archäologie in Baden-Württemberg 43. Landesdenkmalamt Baden-Württemberg.

Brounen, F. T. S. 1995. Watermolen: Vogelzang. In T. A. S. M. Panhuysen (ed.) Archeologische Kroniek van Maastricht 1994. *Publications de la Société Historique et Archéologique dans le Limbourg* 131, 218-223.

Brounen, F. T. S., Knippels, B. and Orbons, J. 1990. Neolithische afvalkuilen en sporen van bijlproductie bij Hoogenweerth. In H. Stoepker (ed.), Archeologische Kroniek van Limburg over 1988 en 1989. *Publications de la Société Historique et Archéologique dans le Limbourg* 126, 212-214.

Bulten, E. i. p. Vlaardingen in de Hofstad. *Westerheem.*

Collet, H., Hubert, F., Robert, C. and Joris, J. P. 2006. The fl int mines of Petit-Spiennes (province of Hainaut, Belgium): an update, in G. Körlin and G. Weisgerber (eds.), *Stone Age – Mining Age*, 67-73. Bochum. Veröffentlichungen aus dem Deutschen Bergbau-Museum Bochum Nr. 148. Der Anschnitt Beiheft 19.

Collet, H., Hauzeur, A. and Lech, J. 2008. The prehistoric fl int mining complex at Spiennes (Belgium) on the occasion of its discovery 140 years ago, in P. Allard, F. Bostyn, F. Giligny and J. Lech (eds.), *Flint Mining in Prehistoric Europe. Interpreting the archaeological records*, 41-78. Oxford, BAR Publishing. British Archaeological Reports International Series 1891.

Collet, H., Jadin, I. and Woodbury, M. 2008. Apport à la chronologie radiocarbone des minières néolithiques. *Bulletin d'Information de la La Société de Recherche préhistorique en Hainaut 2008*, 2-4.

De Grooth M. E. Th. 1991. Socio-economic aspects of Neolithic fl int mining: a preliminary study. *Helinium* 31, 153-190.

De Grooth, M. E. Th., Lauwerier, R. C. G. M. and Ter Schegget, M. E. i p. Nieuwe ^{14}C dateringen van de vuursteenmijnen te Rijckholt-St. Geertruid, in J. Deeben and J. W. De Kort (eds.), *Het archeologisch onderzoek van de omgeving van het prehistorische vuursteenmijnveld te Rijckholt-St. Geertruid: de resultaten van 2008 en 2009.* Amersfoort. Rapportage Archeologische Monumentenzorg, 2011. Rijskdienst voor het Cultureel Erfgoed.

Dohrn-Ihmig, M. 1983. *Neolithische Siedlungen der Rössener Kultur in der Niederrheinischen Bucht.* Materialien zur Allgemeinen und Vergleichenden Archäologie 21. München, Beck.

Felder, P. J. and Felder, W. M. 1998. De geologie van de omgeving van het prehistorisch mijnveld te Ryckholt-St. Geertruid, in P. C. M. Rademakers (ed.), *De prehistorische vuursteenmijnen van Ryckholt-St. Geertruid*, 111-136. Beek, Nederlandse Geologische Vereniging, Afd. Limburg.

Felder, P. J. and Rademakers, P. C. M. 1971. 5 jaar opgraving van préhistorische vuursteenmijnen te Ryckholt-St. Geertruid. *Grondboor en Hamer* 3, 38-57.

Felder, P. J., Rademakers, P. C. M. and De Grooth, M. E. Th. (eds.) 1998. *Excavations of prehistoric flint mines at Rijckholt-St. Geertruid (Limburg, the Netherlands) by the 'Prehistoric Flint Mines Working Group' of the Dutch Geological Society, Limburg section*, Bonn, Habelt. Archäologische Berichte 12.

Felder, W. M., Felder, P. J., Rademakers, P. C. M., Kraayenhagen, F. C., Hensgens G. and Meessen, J. 1965. Rapport van de schedelvondst op 5 november 1965. In P. C. M. Rademakers (ed.), *De schedel Ryckholt I, Eindverslag van het onderzoek der neolithische vuursteenindustrie te Ryckholt – St. Geertruid, 1964-1972,* 11-21, bijlage 1. Werkgroep voor het onderzoek van prehistorische vuursteenmijnbouw, Nederlandse Geologische Vereniging, Afd. Limburg, Serie 1, Rapport 1.

Fiedler, L. 1979. Formen und Techniken neolithischer Steingeräte aus dem Rheinland., *Rheinische Ausgrabungen* 19, 53-190.

Gehlen, B., Langenbrink, B and Gaffrey, J. 2009. Die Gesteinsinventare der Rössener Siedlungen Aldenhoven 1 und Inden 1, in A. Zimmermann (Hrsg.), *Studien zum Alt- und Mittelneolithikum im Rheinischen Braunkohlerevier,* 287-374. Rahden/Westf., Marie Leidorf. Beiträge zur neolithischen Besiedlung der Aldenhovener Platte VII. Kölner Studien zur prähistorischen Archäologie 1.

Goossens, T. A. (ed.) 2009. *Opgraving Hellevoetsluis-Ossenhoek (Voorne-Putten). Een nederzetting van de Vlaardingen-groep op een kwelderrug in de gemeente Hellevoetsluis.* Leiden, Archol rapport 87.

Hamburg, T. 2005. *Neolithische bewoningsresten te Leidschendam. Begeleiding/IVO/DO Leidschendam-Voorburg – Prinsenhof.* Leiden, Archol Rapport 59.

Heymans, H. P. and Vermeersch, P. M. 1983. Siedlungsspuren aus Mittel- und Spätneolithikum, Bronzezeit und Eisenzeit in Geistingen (Provinz Limburg). *Archaeologia Belgica* 255, 15-64.

Höhn, B. 1997a. Das Steinmaterial der Michelsberg Siedlung Koslar 10, Gem. Jülich, Kr. Düren. In J. Lüning (Hrsg.), *Studien zur neolithischen Besiedlung der Alden-*

hovener Platte und ihrer Umgebung, 399-472. Köln, Habelt. Rheinische Ausgrabungen 43.

Höhn, B. 1997b. Das Michelsberger Erdwerk Inden 9, Gem. Jülich, Kr. Düren. In J. Lüning (Hrsg.), *Studien zur neolithischen Besiedlung der Aldenhovener Platte und ihrer Umgebung*, 473-598 Köln, Habelt. Rheinische Ausgrabungen 43.

Hoffstadt, J. and Maier, U. 1999. Handelsbeziehungen während des Jungneolithikums im westlichen Bodenseeraum am Beispiel der Fundplätze Mooshof und Hornstaad Hörnle I A. *Archäologisches Korrespondenzblatt* 29, 21-34.

Houkes, R. A. 2008. Vuursteen, in H. Koot, L. Bruning and R. A. Houkes (ed.), *Ypenburg-locatie 4. Een nederzetting met grafveld uit het Midden-Neolithicum in het West-Nederlandse kustgebied*, 213-247. Leiden, Hazenberg Archeologie.

Houkes, R. and Dorenbos, O. 2004. *Aanvullend Archeologisch Onderzoek op het Gavi-kavel, Ypenburg deelplan 25, Nootdorp. Rijswijk*, Gemeente Rijswijk.

Jansen, A.J. 1989. Een neolithische nederzetting in het Ewijkse Veld. *Westerheem* 38, 133-145.

Kieselbach, P. 2000. *Metamorphose des Steins – vom Rohmaterial zum Kulturgut. Versorgingsaspekte und technische Prozesse der Silexverarbeitung von jungneolithischen Silexinventaren aus Südwestdeutschland.* Unpublished PhD thesis, Dissertation Tübingen.

Knoche, B. 2008. *Die Erdwerke von Soest (Kr. Soest) und Nottuln-Uphoven (Kr. Coesfeld). Studien zum Jungneolithikum in Westfalen.* Rahden (Westf.), Marie Leidorf. Münstersche Beiträge zur Ur- und Frühgeschichtlichen Archäologie 3.

Lanting, J. N. and J. van der Plicht 1999/2000. De ^{14}C-chronologie van de Nederlandse pre- en protohistorie, III: Neolithicum. *Palaeohistoria* 41/42, 1-110.

Louwe Kooijmans, L. P. 1980. De midden-neolithische vondstgroep van Het Vormer bij Wijchen en het cultuurpatroon rond de zuidelijke Noordzee, ca. 3000 v. Chr. *Oudheidkundige Mededelingen uit het Rijksmuseum van Oudheden te Leiden* 61, 113-208.

Louwe Kooijmans, L. P. 2005/2006. Schipluiden: a synthetic view, in L. P. Louwe Kooijmans and P. F. B. Jongste (eds.), *Schipluiden. A Neolithic settlement on the Dutch North Sea Coast c. 3500 cal. BC*, 485-517. Leiden, Faculty of Archaelogy Leiden University. Analecta Praehistorica Leidensia 37/38.

Louwe Kooijmans, L. P. and Verhart, L. B. M. 1990. Een middenneolithisch nederzettingsterrein en een kuil van de Stein-groep op de voormalige Kraaienberg bij Linden, gem. Beers (N.-Br.). *Oudheidkundige Mededelingen uit het Rijksmuseum van Oudheden te Leiden* 70, 49-108.

Lüning, J. 1967. Die Michelsberger Kultur, ihre Funde in zeitlicher und räumlicher Gliederung. *Berichte der Römisch-Germanischen Kommission* 48, 1-350.

Rademakers, P. C. M. (ed.) 1972. *De schedel Ryckholt I, Eindverslag van het onderzoek der neolithische vuursteenindustrie te Ryckholt – St. Geertruid, 1964-1972.* Werkgroep voor het onderzoek van prehistorische vuursteenmijnbouw. Beek, Nederlandse Geologische Vereniging, afd. Limburg. Serie 1, Rapport 1.

Rademakers, P. C. M. (ed.) 1998a. *De prehistorische vuursteenmijnen van Ryckholt-St. Geertruid.* Beek, Nederlandse Geologische Vereniging, afd. Limburg. Serie 1, Rapport 1.

Rademakers, P. C. M., 1998b. De schedels van Ryckholt. In P. C. M. Rademakers (ed.) *De prehistorische vuursteenmijnen van Ryckholt-St. Geertruid,* 231-243. Beek, Nederlandse Geologische Vereniging, afd. Limburg. Serie 1, Rapport 1.

Raemaekers, D. C. M., 1999. *The Articulation of a 'New Neolithic'. The meaning of the Swifterbant culture for the process of neolithization in the western part of the North European Plain.* Leiden, Faculty of Archaelogy Leiden University. Archaeological Studies Leiden University 3.

Raetzel-Fabian, D., 2000. Calden. Erdwerk und Bestattungsplätze des Jungneolithikums. Architektur – Ritual – Chronologie. Bonn. Universitätsforschungen zur Prähistorischen Archäologie 70.

Raetzel-Fabian, D. 2002. Absolute Chronology and Cultural Development of the Neolithic Wartberg Culture in Germany (*www. jungsteinSite.de*).

Reimer, P. J., Baillie, M.G. L., Bard, E., Bayliss, A., Beck, J. W., Bertrand, C. J. H., Blackwell, P.G., Buck Caitlin, E., Burr, G. S., Cutler, K. B., Damon, P.E., Edwards, R. L., Fairbanks, R.G., Friedrich, M., Guilderson, T. P., Hogg, A. G., Hughen, K. A., Kromer, B., McCormac, G., Manning, S., Ramsey, C. B., Reimer, R. W., Remmele, S., Southon, J. R., Stuiver, M., Talamo, S., Taylor, F. W., van der Plicht, J. And Weyhenmeyer, C. E. 2004. IntCal04. Terrestial radiocarbon age calibration, 26–0 ka BP. *Radiocarbon* 46, 1029-1058.

Richter, J. 1997: *Neolithikum.* Köln, Rheinland-Verlag. Geschichtlicher Atlas der Rheinlande Beiheft II/2.1 - II/2.2.

Schreurs, J. 1992. The Michelsberg site Maastricht-Klinkers: a functional interpretation. *Analecta Praehistorica Leidensia* 25, 129-173.

Schreurs, J. and Brounen, F. 1998. Resten van een Michelsberg aardwerk op de Schelsberg te Heerlen. Een voorlopig bericht. *Archeologie in Limburg* 76, 21-32.

Schwellnus 1979. *Wartberg Gruppe und hessische Megalithik: ein Beitrag zum späten Neolithikum des Hessischen Berglandes.* Wiesbaden, Landesamt für Denkmalpflege Hessen.

Schwellnus, W. and S.K. Arora. 1983. *Wichtige jungsteinzeitliche Neufunde im rheinischen Braunkohlengebiet, Ausgrabungen im Rheinland 1981/1982,* 85-88.

Stapel, B. 1991. *Die geschlagenen Steingeräte der Siedlung Hüde I am Dümmer.* Hildesheim, Lax Veröffentlichungen der urgeschichtlichen Sammlungen des Landesmuseums zu Hannover 38.

Stuiver, M. and Reimer, P. J. 1993. Extended ^{14}C database and revised CALIB calibration program. *Radiocarbon* 35, 215-230.

Ter Schegget, M. E. i.p. De Rijckholt1-schedel fysisch antropologisch bekeken, in J. Deeben and J. W. de Kort (eds.), *Het archeologisch onderzoek van de omgeving van het prehistorische vuursteenmijnveld te Rijckholt-St. Geertruid: de resultaten van 2008 en 2009.* Amersfoort. Rapportage Archeologische Monumentenzorg. Rijskdienst voor het Cultureel Erfgoed.

Van Giffen, A. E. 1926. De ligging en aard van de overblijfselen der vóórhistorische vuursteenindustrie bij Ryckholt in Limburg. *Verslagen der Geologische Sectie van het Geologisch-Mijnbouwkundig Genootschap voor Nederland en Koloniën. Derde deeel,* 101-108.

Van Gijn, A. L. 1990. *The wear and tear of flint. Principles of functional analysis applied to Dutch Neolithic assemblages.* Leiden, Faculty of Archaeology. Analecta Praehistoria Leidensia 22. Universiteit Leiden.

Van Gijn, A. L. 2008. Exotic flint and the negotiation of a new identity in the 'margins' of the agricultural world: the case of The Rhine-Meuse delta. *Analecta Praehistorica Leidensia* 40, 193-202.

Van Gijn, A. L. and Bakker, J. A. 2005. Megalith builders and sturgeon fishers. Middle Neolithic B: Funnel Beaker culture and Vlaardingen group, in L. P. Louwe Kooijmans, P. W. van den Broeke, H. Fokkens and A. L. van Gijn (eds.), *The Prehistory of the Netherlands,* 281-307. Amsterdam, Amsterdam University Press.

Van Gijn, A. L., Van Betuw, V., Verbaas, A. and Wentink, K. 2005/2006. Flint: procurement and use, in L. P. Louwe Kooijmans and P. F. B. Jongste (eds.), *Schipluiden. A Neolithic settlement on the Dutch North Sea Coast c. 3500 cal. BC,* 129-166. Leiden, Faculty of Archaeology Leiden University. Analecta Praehistorica Leidensia 37/38. Van Haaren, H. M. E. and Modderman, P. J. R. 1973: Ein mittelneolithischer Fundort unter Koningsbosch, Prov. Limburg. *Analecta Praehistorica Leidensia* 6, Leiden, 1-49.

Van Hoof, L. G. L. 2008. *Hoornaar-Lage Giessen. Neolithische bewoning op een donk in de gemeente Giessenlanden.* Leiden, Archol rapport 96.

Vanmontfort, B., Collet, H. and Crombé, Ph. 2008. Les industries lithiques taillées des IVe et IIIe millénaires dans les bassins de l'Escaut et de la Meuse (Belgique), in M.-H. Dias-Meirinho, V. Léa, K. Gernigon, P. Fouéré, F. Briois and M. Bailly (eds.), *Les industries lithiques taillées des IVe et IIIe millénaires en Europe occidentale,* 11-40. Oxford, BAR Publishing. British Archaeological Reports International Series 1884.

Van Vark, G. N. 1971. Voorlopig onderzoek van de schedel, gevonden bij Ryckholt, Zuid Limburg, op 5 november 1965. In P. C. M. Rademakers (ed.), *De schedel Ryckholt I, Eindverslag van het onderzoek der neolithische vuursteenindustrie te Ryckholt – St. Geertruid, 1964-1972,* bijlage 1-2. Werkgroep voor het onderzoek van prehistorische vuursteenmijnbouw, Beek, Nederlandse Geologische Vereniging, afd. Limburg. Serie 1, Rapport 1.

Verhart L. B. M. 1989. Nederzettingssporen uit het Midden-Neolithicum langs de Pater Berthierstraat te Grave. *Westerheem* 38, 190-197.

Verhart, L. B. M. and Louwe Kooijmans, L. P. 1989. Een midden-neolithische nederzetting bij Gassel, gemeente Beers (N.Br.). *Oudheidkundige Mededelingen uit het Rijksmuseum van Oudheden te Leiden* 69, 75-117.

Verhart, L. B. M. and Van den Broeke, P. W. 2002. Vuursteen. In P. van den Broeke (ed.), *Van Mesolithicum tot Romeinse Tijd in Nijmegen-Ressen. Archeologisch onderzoek in een persleidingtracée door het monument Zuiderveld-West,* 25-27. Nijmegen, gemeente Nijmegen Bureau Archeologie Archeologische Berichten Nijmegen, rapport 2.

Vermeersch, P. M. and Burnez-Lanotte, L. 1998. La culture de Michelsberg en Belgique: état de la question, in J. Biel, H. Schlichtherle, M. Strobel and A. Zeeb (eds.), *Die Michelsberger Kultur und ihre Randgebiete – Probleme der Entstehung, Chronologie und des Siedlungswesens,* 47-54. Stuttgart, Landesdenkmalamt Baden-Württemberg. Materialhefte zur Archäologie in Baden-Württemberg 43.

Wallbrecht, A. 2000. *Die Höhensiedlung der Michelsberger Kultur auf dem Salzberg bei Höckelheim, Landkreis Northeim und der westeuropäische Flint östlich der Weser.* Oldenburg, Isensee. Veröffentlichungen der urgeschichtlichen Sammlungen des Landesmuseums zu Hannover 48.

Wansleeben, M. and Verhart, L. B. M. 1990. Meuse Valley project, the Transition from the Mesolithic to the Neolithic in the Dutch Meuse Valley, in P. M. Vermeersch and P.

van Peer (eds.), *Contributions to the Mesolithic in Europe,* 389-402. Leuven. Leuven University Press.

Waterbolk, H. T. 1994. Opgravingen in het vuursteenmijnbouwgebied van Rijckholt–St.-Geertruid, Zuid-Limburg. *Archeologie in Limburg* 61, 33-52.

Willms, C. 1982. Zwei Fundplätze der Michelsberger Kultur aus dem westlichen Münsterland, gleichzeitig ein Beitrag zum neolithischen Silexhandel in Mitteleuropa. Hildesheim, Lax. Münstersche Beiträge zur Ur- und Frühgeschichte 12.

Traces of Earliest Prehistoric Flint Mining Activity in high alpine region of Western Austria

Walter LEITNER, Thomas BACHNETZER and Markus STAUDT

Abstract
The article treats the earliest flint mining traces in the inner alpine region. This work is part of a special international research project of the University of Innsbruck called HIMAT (History of Mining Activities in the Tyrol and Adjacent Areas). In the course of investigation we were able to detect a concentration of primary layers of radiolarite in the Gemstel Valley in the county of Vorarlberg (Western Austria). An archaeological survey enabled us to deduce that this site was an open air mining zone where people exploited this raw material as early as the Stone Age. Since the quality of the local radiolarite variety is very good, this valley, situated at an altitude of 1600m above sea level, may have been a small supply centre in the west of Austria at that time and one of the oldest flint mining sites in the inner Alps.

Keywords
Radiolarite. Alps. Open cast mining. Mining tools.

1. Introduction

Flint is one of the most important raw materials used by the Stone Age tool-makers and has many well-known exploitation zones scattered all over Europe. The alpine arc, however, constitutes a large gap. The discovered exploitation zones are found outside the Alps and date mainly to the Neolithic period.

We know that hunters and gatherers already densely populated the inner alpine region during the postglacial period and used higher alpine regions as hunting areas at regular intervals. Very little, however, is known about the origin of the flint material they used. Was its source regional or local or were its users 'dependent on imports'? When did the systematic exploitation of flint begin? Are there any indications of the early exchange and circulation of this raw material?

As a part of the international research project mentioned above, numerous archaeological surveys and geophysical surveys have been conducted in the region of the northern Limestone Alps of western Austria, where the most significant flint formations of this area are located. Several years of investigation permitted us to conclude that alpine flint layers are abundant in this zone but were not always exploited due to the material's poor quality. The major challenge of this study consisted in locating real traces of mining activity in open terrain.

2. The site

The most obvious findings in this context became evident in the Gemstel Valley, a small side valley of the *Kleinwalsertal* in the *Allgäuer* Alps, in the country of *Vorarlberg*, Austria (Figure 1). In this region several mesolithic campsites of hunters and gatherers are known (Leitner 2003, 2005, 2009; Gulisano 1994). Nearly all of the stone tools used by the early inhabitants of these sites were manufactured from red and green radiolarite—the raw material of high quality that can be found in the immediate vicinity in secondary gravel deposits of the main stream of the valley. In other words, stone material was transported in a natural way from the mountains down to the valley floor. So, in their search for primary deposits the prehistoric hunters needed only to walk upstream and find a section of the creek where radiolarite was no longer visible. This precise spot was the location of flint layers situated on both flanks of the

Figure 1. Map of Western Austria with the investigation area in the Kleinwalsertal (white square).

valley. And indeed it is not difficult to notice long heaps of weathered radiolarite descending from steep and rocky slopes. This area constitutes the centre of the radiolarite deposits in the valley.

The focus of the study was a woodland hilltop called *Bärenkopf*, rising over the left side of the valley and not at all surprisingly known among the locals as 'Am Feuerstein' (Flintstone-Hill). On this slope, layers of radiolarite reach the surface in several spots.

At an altitude of 1550m a woodless area of about 25x40m is situated. Here flint outcrops were clearly visible. Transversely superimposing and having an average width of eight meters their emergence in the foreground of a steep slope is truly striking. The appearance of the adjacent plains makes the surface of these protrusions stand out even more and highlight their distinctiveness. Many small terraced spots and shallow cavities structure the zone and do not appear to be natural. After the top soil of the radiolarite formation was excavated, it showed evidence of intentional extractions (Figure 2, Figure 5). This finding was the first suggestion of the possible prehistoric exploitation.

In 2005 the researchers of the Institute of Archaeology of Innsbruck University, Austria, began the systematic excavation of the site. By 2009, the team uncovered seven sections, each containing thousands of scattered radiolarite fragments.

The principle colours of the material are red and green, in some cases laced with fine black bands. On the opposed side of the valley dark grey and black varieties are also common. Recent knapping experiments have demonstrated that the green variety is the best. Geological and petrographical analyses confirm this particular quality.

Geologist A. Binsteiner (2008) identified five types of radiolarite:

Type I- 'Walser Jasper'

Red and green radiolarite. The amount of radiolarian inclusions exceeds 40%. The width of the radiolarian skeletons is discernable measuring 100μ in diameter. Red and light red quartz veins are included; under magnification they illustrate structures similar to jasper or carnelian. The common term 'Walser Jaspis' (Walser Jasper) originates from this variegated appearance.

Type II – Banded radiolarite

Mostly red or green radiolarite, containing parallel, light and dark quartz veins, effecting foliation of the rock.

Type III – Green radiolarite

Radiolarite characterized by alternating bluish-green, greenish-blue, turquoise and moss colour.

High density of radiolarian skeletons with an amount over 70%, partly they are undirected. Single specimens are dissolved and only their indistinct outlines are preserved. Very homogenous and compact, high quality material with excellent knapping features.

Type IV – Red radiolarite

Red – brown radiolarite, often containing clefts and rock veins, sometimes filled

with quartz or calcite. It has a characteristic red – green, grained or speckled appearance.

Type V – Black radiolarite

Black – grey radiolarite. Heavily cleft. Rock veins filled with quartz or calcite.

Often black radiolarian skeletons, dense packed and showing a width up to 100μ.

Macroscopically a light foliation is visible. The quality in terms of knapping features is rather low.

Sectors five, two and one were most optimally suited for our study. These three sectors were initially treated separately, to be later linked together into a 21m long section.

3. Artefacts

Among the lithic artefacts found on the site we have identified two main categories—flakes and cores, constituting 95% of the pieces. The flakes are small in size and are rarely or only partially retouched at the edges. In most cases cores show irregular and arbitrary negative marks and are not worked systematically. These characteristics of the appearance of flakes and debris evidence the repeated testing of the quality of flint material. Since the site is situated on a steep slope and the ascent to the hilltop must have been difficult and exhausting, we presume that people came here solely to exploit the raw material and not to produce specific artefacts or settle down. Typical stone implements found here are most likely to be lost pieces. Therefore, we did not expect to find either camps or lodging structures.

All things considered, the tools recovered here do not facilitate the task of dating the site to a specific culture period, which partially explains why we have geared our inquiry towards questions about mining practices in the Stone Age.

4. Mining tools and mining technique

No conclusions can be made about mining activity in general or about the specific techniques or quarrying methods before the recovery of mining implements. The steepness of the terrain made their discovery very difficult and the first significant breakthrough happened only in 2007, when a spherical stone was found in sector five, at a depth of 1.10m. This nodular object was very different from the sharp-edged flint debris, because it had the traces of intentional shaping on its surface (Figure 3.1). It is made up of green radiolarite, has the size of a small fist and weighs 650 grams. This objects falls into a general category of simple pounders, but considering the context of the find, it can be interpreted as a mining tool, more specifically—a stone hammer. It could have been simply hand-held by a miner or hafted into a wooden stick.

The location of the find is particularly important, since it corresponds to the most horizontal area of this steep terrain, and the only place where the mining tools could have been laid down without the risk of rolling down the slope. Apparently people came upon an ideal layer of radiolarite knocking out a small cavity in the bedrock at this spot.

We found two other nodular stone objects having an apparent functional connection with the first mining tool located close to where the hammer stone was recovered. One is composed of sandstone, the other of limestone (Figure 3.3-4). Both have flattened and smoothed areas and appear to be grinding stones, employed for grinding or sharpening bones or antler artefacts.

In 2009, in the course of straightening and cleaning the profile in sector five, the second hammer-stone appeared. The tool is about 11cm long, weighs 868 grams, and has an oval shape (Figure 3.2). In comparison with the hammer found in 2007 its surface is more roughly worked, but traces of impact marks are clearly visible and the material is the same.

The above-mentioned artefacts (grinding- and hammerstones) make up a mining tool set which was intentionally deposited at an ideal spot in the miners' workspace.

Figure 2. Outcrops of radiolarite layers on the Bärenkopf-hill in the valley of Gemstel (Vorarlberg, Western Austria). 1600 m above sea level.

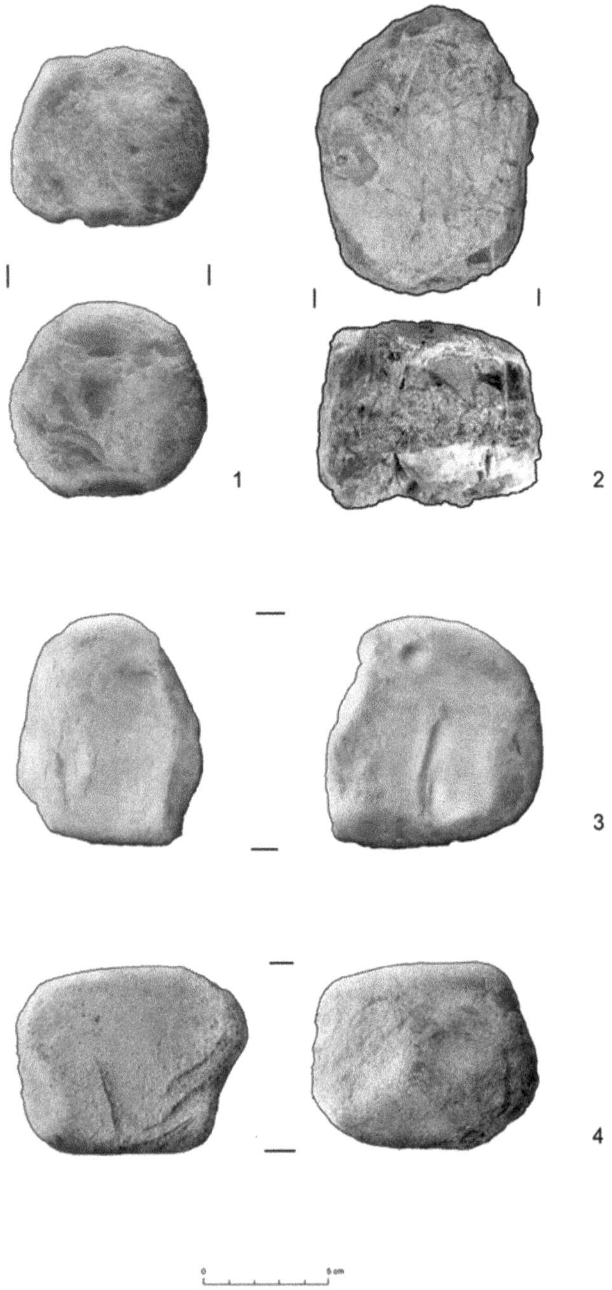

Figure 3. Types of mining tools from sector five. Among them two hammer-stones (1-2) and two grinding stones (3-4).

Why then was this mining equipment, used during the prehistoric exploitation, left behind? Usually the tools are stored when a worker has an intention of coming back. It could also be the case that the mine was abandoned abruptly.

The recovered tools have provided us with some clues about the techniques of radiolarite extraction. According to experiments, this material can be mined best by continuously hammering down on the surface of the layers, which loosens the structure of the rock so that it could be easily removed. The chunks obtained in this manner are reduced to smaller pieces, thus producing a great number of flakes and angular debris. This way the miners tested the material on the spot in order to finally carry away only the best nodules. However, we do not exclude the possibility that other quarrying methods were also practiced, such as levering the rock with antlers. This alternative, however, cannot be supported with relevant archaeological evidence and, additionally, the space between the radiolarite layers is undoubtedly too narrow to fit in an antler pick. Neither have we encountered the traces of the use of wooden wedges for cracking flint layers or firesetting often used for facilitating the task of extraction.

5. Stratigraphy

Slopy terrain made it difficult to accurately describe the stratigraphic conditions. To do so, we dug a 11m long test trench in the flattest area. The profile in sector five can be interpreted as follows (Figure 4).

In the test trench several different layers could be observed. Beneath a 5-20cm thick humus-layer (layer 1) three different layers can be observed along the 11m stretch. Upslope to the north a yellow-brown, slight reddish, dense loamy layer containing numerous flakes, debris and retouched artifacts became visible (layer 2). This material is likely to have come from the stockpile left over from the Stone Age mining activities located further up. We cannot say with certainty if this layer was relocated due to human activities or is the result of a land slide. To the south layer 2 ends downhill after 4 meters. Under layer 2 a 30cm thick layer of similar contents but lower density can be found (layer 3). Pieces of radiolarian rock tend to fall out of the profile of this layer almost by themselves. In the southern area this layer meets the humus. This is a stockpile from a small quarry two to three meters above. Most of the retouched artifacts and cores are in this layer. Numerous preparation flakes, cores and many retouched artifacts, tested for their quality, suggest that the Stone Age miners seem to have processed the quarried radiolarian lumps on the spot. The top of this layer forms an almost 5m long level, that is likely to be the original Stone Age working place surface. The mining waste fills a knocked out cavity in the radiolarian bedrock, constituting the best proof of Stone Age mining activities so far, and demonstrating that there must have been two periods of mining, since the gap was filled during later quarrying. Lower, below the hollow, a 4th layer consisting of consolidated loam and of apparently natural origin rests directly on the bedrock.

Except for the artificially worked mining areas that can be easily differentiated due to their fresh appearance the radiolarian rock is weathered.

In summer of 2009, we also found a thin charcoal layer between layer 2 and 3, along the extension of the west-profile. Radiocarbon dates date the site to the Late Neolithic (VERA-5109: 3895 +/- 40 BP, 2480-2270 cal BC, 91.7% probability; VERA-5110: 3755 +/- 35 BP, 2290-2110 cal

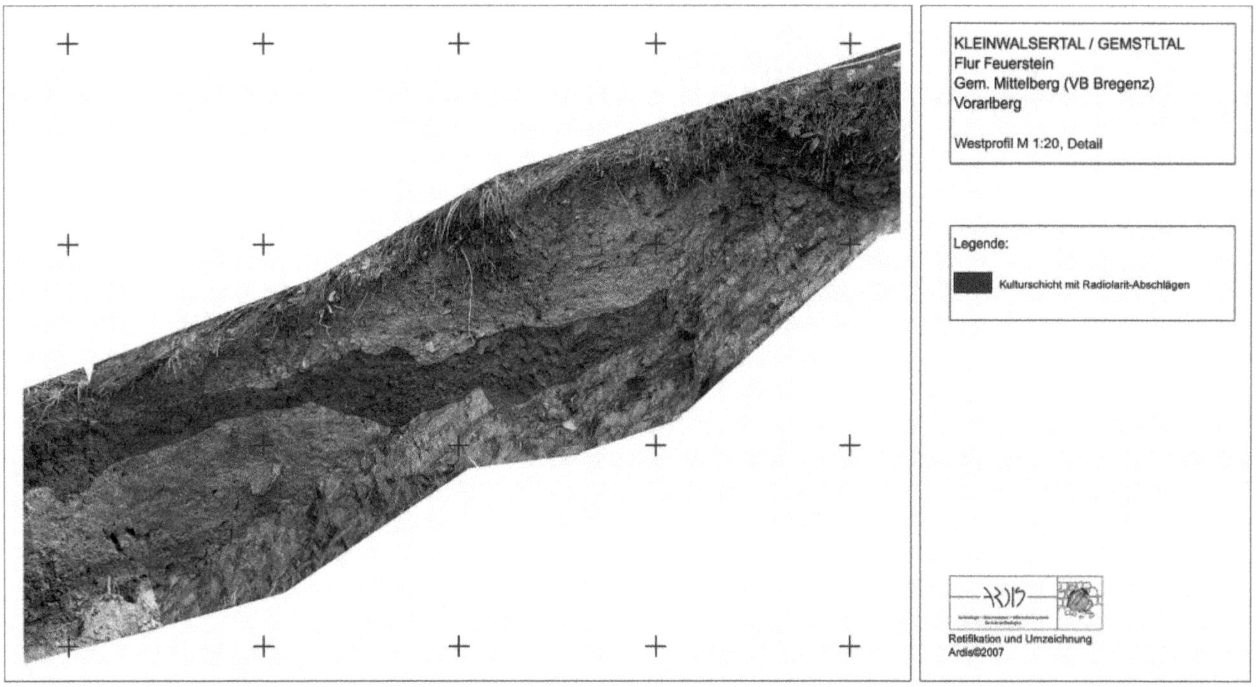

Figure 4. Stratigraphic picture in the exploitation area of sector five.

BC, 76.3% probability), which means that the subjacent layers of mining waste date back to at least the Neolithic or are even older. Unfortunately the artifacts offer no definite clue that would allow a more exact dating of the site. The origin of the charcoal is still unknown. It may well derive from a fireplace. Further excavations will hopefully give us additional answers.

5. Chronology

We have both direct and indirect evidence concerning the age determination of this open cast mining site.

First of all, it should be noted that all archaeological sites excavated in the valley to date are situated in the relative vicinity of the radiolarite exploitation area and all date to the Mesolithic (as determined by several types of geometric microliths and relative radiocarbon dates) (Leitner 2003). Thus we can assume that that people took the raw material not only from the river gravels but also possibly from primary deposits.

Both the mining tools found here and modified flake containing characteristic marks of prehistoric flint knapping technique provide solid evidence about the existence of a prehistoric mining activity in this valley.

Only one radiocarbon date is available so far. The AMS-measuring of scant traces of charcoal from the profile in sector five resulted in a calibrated age between 2500 and 2100 BC corresponding to the late Neolithic age. This does not automatically mean that mining work only took place during that period. As we have already mentioned, the inventory of the main findings all point us to the Mesolithic age. We do, however, hope to find addition datable samples.

Finally, it is worth mentioning that the historical tradition of flint mining does not exist anywhere in the valley.

6. Conclusion

When taking into account all the factors, it can be argued that the *Bärenkopf*-hill must have been an open air radiolarite exploitation area during the Stone Age. We do not know its actual dimensions, since only a small section of the prehistoric mining area has been excavated (Figure 5). A three dimensional model of the hilltop, developed from circa 16,000 measuring points showing the very uneven and bumpy morphology resulting from mining operations, helped us to get an idea of the width of the expansion of the district and estimate the area at approximately 400 square meters. Since recoverable flint material can also be found on the opposed valley flank, the area may have been much bigger.

Several surveys demonstrated that, in terms of quality and quantity, no comparable radiolarite deposits can be found in the surroundings of the *Kleinwalsertal*. This is perhaps the reason why the small valley of *Gemstel* can be considered as a regional supply centre in Mesolithic and Neolithic times. The evidence pointing to the distribution of stone tools made of comparable radiolarite has been found in the immediate vicinity of the exploiting

Figure 5. In sector five dimensions of exploitation can be seen clearly. The deepest knocked out point in the cavity is filled up with water, coming down continuously from the slope after a rainy period. The immediate vicinity of the cavity corresponds to the deposition place of the mining tools.

area, the alpine Rhine valley, and the south-east coast region of the Lake Constance corresponding to a radius of about 50km.

The question about the possible reason for the mine's abandonment is still waiting to be addressed. There may have been several contributing factors. It is possible that the top radiolarite layers had been exhausted, or a bigger flint mining site had been found along the Danube, or the demand for flint had fallen when the extraction of copper ore had intensified.

The Bärenkopf-hill is the second discovery of a flint mining site in Austria, following the 1924 discovery of the Neolithic mine in *Mauer*, near Vienna (unlike the *Gemstel*-valley site, it is located below the surface and consists of a few pits and galleries) (Ruttkay 1970). Situated at an altitude of 1600m above the sea level, it is the highest and one of the oldest open air exploitation zones in the inner Alps.

In the course of other investigations in the northern Limestone Alps in 2009 more traces of flint mining activity were found in the *Allgäuer*-Alps and in the *Rofan*-mountains in the Tyrol. Excavations on these sites are under way and promise new revealing results (Bachnetzer *et al.* 2009; Kompatscher 2005).

References

Bachnetzer, T., Leitner, W., Staudt, M. 2009. Radiolarit, Hornstein und Bergkristall. Steinzeitliche Bodenschätze aus den Tiroler Alpen, in K. Oeggl and M. Prast (eds.) *Geschichte des Bergbaus in Tirol und seinen angrenzenden Gebieten.* Innsbruck University Press, Conference Series 261-267.

Binsteiner, A. 2008. Steinzeitlicher Bergbau auf Radiolarit im Kleinwalsertal/Vorarlberg (Österreich). Rohstoff und Prospektion. *Archäologisches Korrespondenzblatt* 38(2), 185-190.

Gulisano, G. 1994.Neue mittelsteinzeitliche Fundplätze im oberen Illertal und im Kleinwalsertal. *Archäologische Informationen* 17(1), 79-84.

Kompatscher, K., Kompatscher, N. 2005. Steinzeitliche Feuersteingewinnung. Prähistorische Nutzung der Radiolarit- und Hornsteinvorkommen des Rofangebirges. *Der Schlern* 79(2), 24-35.

Leitner, W. 2003. Der Felsüberhang auf der Schneiderkürenalpe. Ein Jäger- und Hirtenlager der Vorzeit. Die ältesten menschlichen Spuren im Kleinwalsertal. *Bergschau* 1(2003).

Leitner, W. 2005. Eine mittelsteinzeitliche Fundstelle bei Riezlern im Kleinwalsertal (Vorbericht). *Jahrbuch des Vorarlberger Landesmuseumsvereins* 148, 15-20.

Leitner, W. 2008a. Steinzeitlicher Silexabbau im Kleinwalsertal. *Archäologie in Deutschland 4*, 28-29.

Leitner, W. 2008b. Steinzeitlicher Bergbau auf Radiolarit im Kleinwalsertal/Vorarlberg (Österreich). Archäologische Ausgrabungen. *Archäologisches Korrespondenzblatt* 38(2), 175-183.

Leitner, W. 2009. Scant structural evidences of Mesolithic sites in high alpine regions. In F. Cavulli (ed.), *Proceedings XV UISPP World Conference, Lisbon 2006. Vol. 32*, 5-11. Oxford, BAR Publishing. British Archaeological Re-ports International Series 2009.

Ruttkay, E. 1970. Das jungsteinzeitliche Hornsteinbergwerk mit Bestattung von der Antonshöhe bei Mauer (Wien). *Mitteilungen der Anthropologischen Gesellschaft in Wien* 100, 70-83.

Chert Mining in the Krumlov Forest (Southern Moravia)

Martin OLIVA

Abstract

Since 1994, the Anthropos Institute of the Moravian Museum in Brno has been surveying the prehistoric mining of Jurassic chert in the region of the Krumlovský les, in southern Moravia. The oldest traces of extraction date to the early and Late Mesolithic. During the LBK, a large manufacturing center was located directly in the region of the source, near Nové Bránice. The Early Eneolithic (late Lengyel) extraction system can be characterized by deep and narrow vertical mining pits, while the Late Eneolithic extractions are parallel horizontal terraces on the slope. Mining pits of the Late Únětice or Early Věteřov EBA cultures predominate in most areas while chipped Urnfield Culture industries made of extremely poor quality chert have been recovered only on mining heaps, platforms and under the edges of "seat-like" boulders. Mining activity was renewed in Hallstatt times. The disproportions between the large quantities of chert mined but scarcely used, suggests that the reasons behind the long term mining activity at Krumlov Forest were other than the procurement of raw materials for tool manufacture, probably relating to the spiritual significance of the place.

Keywords

Krumlovský les. Chert extraction. Mesolithic. Late Lengyel. Bell beakers. Early Bronze Age. Hallstatt Age. Sacred landscape.

Since the early 1990s, the Anthropos Institute of the Moravian Museum in Brno has been surveying the prehistoric mining of Jurassic chert in the region of *Krumlovský les* (Krumlov Forest) in southern Moravia. This research follows the recent identification of this area as one of the largest mining areas in prehistoric Europe, due to both the extent of its distribution, and its excellent state of preservation (Figure 1). Much of the acquired data have become the subject of further research initiatives (for instance Oliva 2003, 2004, 2005, 2006, Oliva *et al.* 1999), and only some of the basic findings will be mentioned here.

The actual source of Jurassic chert at the mines are Miocene (*Ottnang – Eggenburg*) sands and weathered Quaternary loams. Further sources include blocks of various chert breccias, primarily located in mining fields II and III, which are frequently brightly colored and highly suitable for flaking. Even though some of these sources reached large proportions (chert variety KL I up to 1/2m, KL II up to 20cm, weathered blocks of breccias up to 3m), their complicated genesis and frequent re-deposition resulted in numerous fissures within the chert mass. This is the reason why the Krumlov Forest chert was never chosen for the manufacture of prestigious tools, as was done with siliceous rocks in other mining areas.

With the exception of prestigious tools, the Krumlov Forest chert was nevertheless used for a wide range of implements, and with the exception of the two most advanced Palaeolithic cultures (Gravettian and Magdalenian), remained the dominant choice of stone in southern Moravia (Oliva *et al.* 1999; Oliva 2008, 2010). Overlooking for the moment, the presence of Krumlov Forest chert in the Lower and Middle Palaeolithic assemblages (often only tentatively dated), the use of this source climaxed in the early phase of the Upper Palaeolithic, in terms of both quantity and quality. However, a similar peak is also observable during the Mesolithic. Chert from Krumlovský les represents the main raw material of all Mesolithic chipped industries throughout South Moravia, and predominates in all collections within the range of 60km away from the outcrops (Smolín, Přibice, Břeclav – Pohansko). The site of Dolní Věstonice - Písky lies under the Pavlov Hills and has provided a large quantity of debitage of the Jurassic hornstone and a limited quantity of tools. It is also dominant in the Mikulčice assemblage recovered some 80km away, and at several sites in Lower Austria (Horn – Galgenberg, Kamegg, less frequent in Wien - Bisamberg). A considerably high occurrence is also detected in Bratislava (110km), and a sporadical representation in Sereď in the Lower Váh Basin (130km). The most distant occurrence

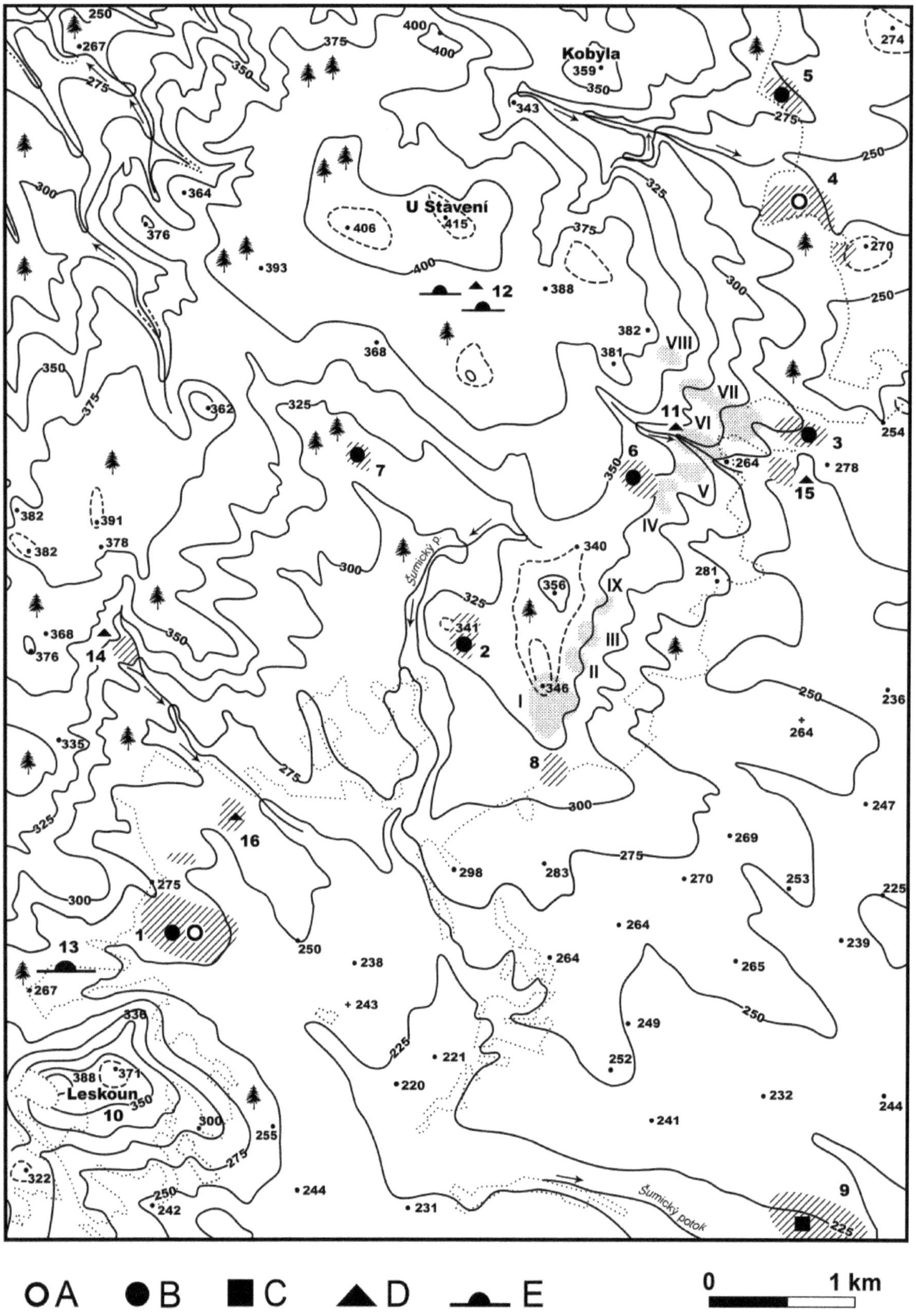

Figure 1. Prehistoric settlements and extraction areas (I-IX) in the Krumlov forest. A LBK, B Lengyel culture, C Early Bronze Age, D Hallstatt. 1 Vedrovice (settlement and burial site LNK, 2 rondels MMK), 2 Moravský Krumlov – Vysoká hora, 3 Jezeřany-Maršovice – Na Kocourkách, 4 Nové Bránice – V Končinách, 5 Nové Bránice B, 6 Moravský Krumlov – traces of a MMK settlement over area V, 7 Moravský Krumlov – Dlouhá louka, 8 Moravský Krumlov – polycultural settlement near area I, 9 Kubšice - Nad Lukama, 10 Olbramovice –Leskoun hillfort, 11 Moravský Krumlov – Horákov settlement in mining area VI, 12 Hallstatt mounds near Stavení (not visible today), 13 Urnfield mound below the Leskoun hillfort. The dotted line indicates the forest edge.

(200km) could be registered in Hořín near Mělník at the confluence of the Vltava and Elbe Rivers in Bohemia (1 core and 2 flakes). This raw material also regularly appears in South Bohemia (150-175km), both in the Upper Palaeolithic (5 sites) and in the Mesolithic (9 sites, Vencl *et al.* 2006). A regular occurrence could be recently detected in Early Mesolithic industries of Glatz (e.g., Lawica 8: 15 pcs), where the high-quality erratic flint is available to utilize. Regarding the low quality of the Krumlovský les chert, the frequent occurrence mentioned above can indicate that the distribution of this raw material resulted from the emission of a production centre.

Jurassic chert, frequently with remains of black cortex, predominates in three main Mesolithic settlements, out of which two (Přibice and Smolín) lie in the immediate vicinity of Krumlovský les. The mass distribution of the above-mentioned sort of chert continues in the Linear Pottery Culture (LBK) (Mateiciucová 2008), when a large manufacturing centre was located directly in the region of the source, near Nové Bránice. An even wider distribution, although indicated by a much smaller quantity of artefacts, occurred during the Stroked Pottery Culture (StK) (Čižmář and Oliva 2001). During the Moravian Painted Ware Culture (belonging to the Lengyel Culture, LgK) the geographic distribution of artefacts made using Krumlov Forest Chert was limited to the area between the region of Znojmo (for example Těšetice - Kyjovice) and the northern boundary of the Brno region. A significant change occurred in the late phase of this culture, when the hitherto dominant use of Krumlov Forest chert type KL II, began to be replaced by KL I type (which is coarser and permitted the manufacture of larger blanks), and which thereafter remained the dominant source. With the exception of the actual mining areas, most of the flaked industry was distributed to hillforts on the Jihlava (Kramolín) (Oliva 1990; Oliva *et al.* 1999, 279) and Oslava Rivers, that is, within a distance of about 30km to the northwest. The identified inventory comprised a large portion of cores and cortical flakes, indicating these elevated settlements were supplied with unprepared raw material. Analysis of the flaked industry found in 9 pits attributed to the terminal LgK from Jezeřany-Maršovice (directly within the vicinity of KL region) however, indicate the export of manufactured blades away from the settlement area (Přichystal and Svoboda 1997; Oliva 2001). Within the region of Brno, Krumlov Forest chert was both dominant (Brno - Obřany, Brno - Kníničky), and submissive to local Jurassic chert of the Stránská Skála type (Brno - Líšeň, Popelákova street). Further to the north it was completely overtaken by quality Jurassic flints of Polish origin (Mokrá) or erratic flint (Opatovice u Vyškova, Drnovice – within the Jordanów Culture assemblage).

A similar situation is seen during the Funnel Beaker Culture, when the Brno region was supplied primarily by the source found at Stránská Skála (Svoboda and Šmíd 1994). However, directly at the Stránská Skála source, more than 7 thousand flaked stone tools were found in a large clay pit, along with blades manufactured of various sources, Krumlov Forest chert among them. Its southern Moravian provenience is, however, uncertain, since similar cherts are also available beneath the nearby Hády hill. Krumlov Forest chert was not commonly used during the Early and Late Eneolithic. Its use has been documented during the Bell Beaker Culture and Corded Ware Culture, and constitutes 14% of Moravian assemblages (Přichystal 1999, 223). The use of the Krumlov Forest chert again gradually increases, in spite of the availability of higher-quality flints from the north, and reaches dominance in the Moravian Proto-Únětice Culture at 37% (Kopacz and Šebela 1998). It is noteworthy that it was only employed for common (non-prestigious) artefacts.

In the Early Bronze Age, Krumlov Forest Chert type KL I and to a certain degree also the brightly coloured breccias, were present only at one site in significant quantities, approximately 5km SE from the southern mining fields. In an area of 700 x 250m, east of Kubšice, various concentrations of flaked stone tools have been identified, in addition to a unique core reduction technique whereby the cores were worked into 'disks', or gradually thinning 'tablets' (Oliva *et al.* 1999, 288-298; 2003). In a similar position in the terrain, but 3km further east, near a Věteřov Group settlement, a ditched enclosure and a 57m long timber house were identified (Stuchlík and Stuchlíková 1999). Pit number 3, situated directly in the centre of the circular ditch, yielded a depot of 7 blades with sickle sheen. Although now broken, the largest of these blades probably represents the highest-quality item ever made from Krumlov Forest breccias (Oliva *et al.* 1999, Figure 43: 1).

The large quantity of Krumlov Forest chert tools at settlement sites near Kubšice sharply contrast with their scarcity at other early Bronze Age sites (Oliva 2003; 2010). Small amounts of chert artefacts were found in Únětice Culture pits on the hillfort Cezavy u Blučiny. Excavations, conducted between 1983-1998 by M. Salaš, yielded 34 chipped stone artefacts with an overall weight of about 0.2kg. Unlike the stone tools found at the source in the Krumlov Forest, many of the tools at the hillfort were retouched. Among these was a distinctive 86mm long knife made from Krumlov Forest chert, with a sharply retouched edge. It was found alongside the cut and burnt bones of 11 children and 1 adult in pit number 5 (Salaš 1990a). The slightly serrated edge extends along the entire length of the knife, and shows signs of polish.

Within the region of Brno, at Brno-Černá Pole, approximately 40 chert artefacts are known from settlements of the Únětice Culture and the Middle Danube Tumulus Culture. However, the direct import of chert from the source region near the Krumlov Forest cannot be confirmed. This material was also absent within the early Únětice Culture toolkit at Brno–Slatina, acquired by M. Salaš (1987, 59, 61; Oliva 2003). Of vital significance in understanding the distribution of Krumlov Forest chert is a series of 7 sickle blades with serrated edges, from a late Únětice Culture pit

No.	Labor/no.	^{14}C-Age BP	cal. BC 68%	cal. BP 68%	trench	depth	probe origin
Mezolitická těžba - Mesolithic mining							
KL 24	GrA-34410	9410+-50	8690+-60	10640+-60	I-12-1	200 cm	mouth of shaft
KL 30	GrA-38110	6775+-40	5680+-30	7630+-30	II-19-1	120-130 cm	shaft
KL 33	OxA-18595	6612+-32	5560+-40	7510+-40	I-13-1	300 cm	shaft
Neolithic mining (LBK)							
KL 35	GrA-45664	6270+-40 BP	5260+-40	7210+-40	I-13-2	300-320 cm	shaft
Pozdně lengyelská těžba - Late Lengyel mining							
KL 5	GrN-27500	5490+-60	4350+-70	6330+-70	VI-9-1	200 cm	shaft 4, above LGK bowl
KL 6	GrA-22839	5380+-50	4210+-90	6160+-90	VI-9-1	600 cm	shaft 4, human bone
KL 25	GrA-34265	5380+-40	4230+-80	6180+-80	VI-9-1	120-130 cm	mouth of shaft 4
KL 16	GrA-30350	5325+-40	4160+-70	6110+-70	VI-9-2	180 cm	mining terrace
KL 13	GrA-28033	5395+-40	4260+-60	6210+-60	VI-9-1	170-200 cm	mouth of shaft 10
KL 19	GrA-30369	5630+-40	4450+-50	6400+-50	VI-9-2	350 cm	shaft 17
KL 27	GrA-38082	5920+-35	4790+-50	6740+-50	I-11-2	280-310 cm	shaft
Final Eneolithic mining (Bell Beakers?)							
KL 2	GrN-27497	4100+-50	2700+-120	4650+-120	VI-1-3	170-180 cm	mining terrace
KL 3	GrN-27498	4070+-120	2650+-170	4600+-170	VI-1-3	190-210 cm	mining terrace
KL 4	GrN-27499	3870+-50	2350+-80	4300+-80	VI-1-3	270-300 cm	mining terrace
MK 01	VERA-3038	3920+-35	2410+-60	4360+-60	VI-1-3	270-300 cm	mining terrace
KL 10	GrN-28874	3900+-40	2390+-60	4340+-60	VI-9-1	750 cm	shaft 1, bone
KL 22	GrA-34364	4360+-35	2980+-50	4930+-50	I-10-1	130 cm	shaft
KL 23	GrA-34266	3875+-35	2370+-70	4320+-70	I-10-1	250 cm	shaft
KL 12	GrA-27034	4060+-40	2620+-90	4570+-90	II-10-2	280 cm	shaft
KL 34	GrA-45662	4065+-35	2620+-90	4570+-90	I-13-1	270 cm	shaft
Early Bronze Age mining							
KL 7	GrA-23556	3630+-50	2020+-80	3970+-80	I-1-1	70-90 cm	pit
KL 8	GrA-23559	3490+-50	1820+-60	3770+-60	III-1-1	130-135 cm	shaft, sherds of Early Bronze Age
KL 1	GrA-22835	3340+-45	1620+-60	3570+-60	IV-2-1	60-70 cm	heap
KL 15	GrA-29162	3510+-50	1840+-70	3790+-70	V-5-1	140 cm	shaft
KL 18	GrA-30352	3310+-40	1590+-50	3540+-50	VII-4-1	180-190 cm	shaft
KL 21	GrA-34262	3490+-35	1820+-50	3770+-50	II-10-3	125 cm	shaft margin
KL 28	GrA-38107	3640+-30	2020+-50	3970+-50	I-13-1	80 cm	workshop
KL 31	GrA-38112	3365+-30	1670+-40	3620+-40	II-20-1	100 cm	mouth of shaft
KL 32	GrA-38113	3255+-30	1540+-50	3490+-50	II-22-1	100 cm	mouth of shaft
Urnfield period							
KL 14	GrA-28034	3170+-35	1450+-30	3400+-30	V-2-2	120 cm	shaft, sherds of Unrnfield culture
KL 11	GrN-28875	2840+-50	1020+-70	2970+-70	II-9-1	90 cm	shaft, sherds of Unrnfield culture
KL 26	GrA-38081	2755+-25	890+-40	2840+-40	I-11-2	105-133 cm	pit in shaft refill
KL 29	GrA-38108	2760+-30	900+-40	2850+-40	II-2-2	120 cm	pit in shaft refill
Hallstatt Age mining							
KL 17	GrA-30351	5210+-40	4020+-40	5970+-40	VI-9-1	300-400 cm	shaft 5
Late Hallstatt Age pithouse							
KL 20	GrA-30370	2445+-35	580+-130	2530+-130	VI-6-1	130 cm	hearth
Late Hallstatt/Early LaTene							
KL 9	GrN-28873	2465+-55	600+-120	2550+-120	II-11-1	90 cm	shaft, second. filling

Figure 2. Krumlovský les mining area - radiometric dates.

at the elevated settlement Zelená Hora near Radslavice u Vyškova (Staňa 1986; Kopacz and Šebela 2006, 153-155). Krumlov Forest chert, nor any other Moravian chert, was employed in the manufacture of flat retouched daggers (dated mainly to the Proto-Únětice Culture), which were instead made from northern flint (Šebela 1998).

The distribution of Krumlov Forest chert during the Urnfield Culture was insignificant, even when considering a wider area. Nevertheless, it was the only type of chert used for flaked stone tool manufacture, within the minimal stone tool assemblage at the hillfort Cezavy u Blučiny. The stone tool assemblage may have been poor due to the unique ritual character of the site. Common settlement features were absent, replaced instead with shallow pits containing unusual fill in the form of burnt grain, ceramics and stone (Salaš 1989, 122). Two such pits (48 and 50) yielded 3 flaked stone tools. Therefore the majority of material was found in surface concentrations 1, 2 and 4, alongside Velatice-type ceramics, arranged animal and human remains, as well as coarse stone (Salaš 1990b, concentration 2 is visible in his figures 3-4). Not surprisingly, cores were not identified. Instead larger quantities of blades were present at the site. From the overall count of 31 artefacts, 17 were retouched or showed significant sings of use (Oliva 2000).

Judging from the significantly varying distribution and use of Krumlov Forest chert in the Holocene, its mining climax can be anticipated in the Mesolithic and Neolithic. The oldest documented mining falls indeed within the Mesolithic. The date GrA-34410: 9410±50 BP comes from a small fireplace with red-fired sand in the narrowed mouth of shaft I-12 with even bottom, so that with regard to mining it undoubtedly represents a *terminus cum quem* or *ante quem*. Radiometric dates from the complicated system of shafts I-13-1 (KL 33 and 35, see Figure 2) prove that mining at this place survived from the Late Mesolithic up to Early Neolithic (LBK). The latter date is namely contemporary to the middle phase of the LBK cemetery at Vedrovice, which is situated directly beneath the Krumlovský les area (Podborský *et al.* 2002; Pettitt and Hedges 2008, 127). In the pits I-12 and I-13 mining was carried out using the method of horizontal undercuttings into solidified granodioritic detritus abundant in cherts (Figure 3). The loose fill of the inclined up to almost horizontal undercuttings splits clearly from the intact seam. The height and length of the undercuttings go over 1m, at some places a subterranean interconnection of the shafts can be detected. The chipped stone industry is very indistinct, relatively small-shaped for local conditions, with irregular as well as parallel cores. In its size it differs from the inventories of all hitherto investigated shafts.

We have no evidence of mining in the Middle and Late Eneolithic. The Early Eneolithic (in our terms, i.e., late LgK) can be characterized by mining from rather narrow and very deep pits (up to 8 meters) recessed vertically into the slope. One of the pits (VI-9-1, no. 4) contained a Late Lengyel pedestal bowl surrounded by charcoals (GrN-27500: 5490±60 BP) at a depth of 2m. Skeletal remains of two females were found at a depth of 6 and 7m (Figure 4). The latter skeleton was deposited along with a newborn child. The baby was aligned so that the head was placed on the woman's chest, and the baby's hip bone lay within her pelvis. Radiocarbon samples taken from the skeletal remains yielded a date of 5380±50 years BP. According to DNA analysis (O. Šerý in Oliva, 2010), both of the females are closely related, and none of them is mother of the discovered child.

Figure 3. Mesolithic to LBK shaft I-13-1.

Another type of mining was identified in the other two trenches found on the same slope. Trench VI-8-1, which lies in the western part of the area VI, contained oblique shafts undercut to a depth of 5m below the steep slope. Trench VI-9-2 yielded the same mode of extraction, overlain by a 3m thick heap of Hallstatt debris. In the latter case, Miocene sand had been undermined to access the granodiorite bedrock. The lithic industry found in these oblique shafts is more numerous than the lithics found in the vertical shafts. It consists primarily of cortical flakes and unsuccessfully shaped cubic pre-cores, with very limited amounts of true blades. These areas were clearly used also as primary workshops, where the raw material was tested for workability, and roughly prepared as part of the extraction process.

The upper section of the incline encompasses a subtle drop in the terrain, which may indicate the remains of a mined wall. A trench sunk in the year 2000 confirmed that this

Figure 4. The female skeleton near the bottom of the shaft 4, trench VI-9-1 (Late LgK).

section was mined on two terraces (VI-1-3). The upper one yielded a 3.5m high extraction face covered with 1m of debris from the upper part of the slope. Chert had been extracted from quaternary sediments overlaying tertiary sands. At a depth of 2.7 – 3.4m on the upper terrace, a cavity had been opened and filled with accumulations of flaked stone industry alongside a significant quantity of charcoal (mainly *Quercus sp.*). Charcoal and indistinct ceramic sherds were also identified at a depth of depth of 200cm in the same profile. The base of the parallel lower terrace was found 2.9m beneath today's surface, and had been covered with Miocene sand, shoveled from the upper terrace. Dates from charcoals found at various depths within the fill of mined shaft VI-1-3 suggest the mines were used before the Early Bronze Age (Figure 2). Calibrated values of these samples place the mining into the late Eneolithic, probably into the Bell Beaker Culture. Studies at other sites have identified a decline in blade technologies and the introduction of the flake industries (Lech 1982-83, 53; Kopacz 2001; Kopacz and Šebela 1998). This radical change in the lithic technology can also be observed here. However, it is interesting that the beaker cultures made very few interventions into the earth. The up to now small amount of chipped stone industry documented (mainly of the graves), is not only of poor quality, but also did not require any mining.

In the Early Bronze age, improvements in the flaking technique resulted in many discoid, irregular and sometimes very flat cores (with both flake and blade scars). Mining pits of this period predominate in all areas except area VI. In trench VIII-1-2, which lies in northernmost area VIII, chert pebbles and artefacts were included in granodiorite detritus with sand up to 3m thick and lying atop bedrock of solid granodiorite. The extraction of area IV falls within the time span of the Late Únětice or Early Věteřov Cultures. Other areas within the southern mining fields yielded Únětice Culture sherds, both inside mining pits (I-1-1, III-1-1), but also near accumulated waste flakes found lodged underneath large flat boulders (III-3-1).

Únětice ceramics are always accompanied by a concentration of technologically distinctive flaked industry corresponding to the toolkit found at the Únětice settlement near Kubšice (Figure 5). Similar industry predominates in all excavated mining pits in areas I, II, III and IX in the southern group of mining fields, and in areas IV, V, VII and IX in the northern group of mining fields. The walls of Early Bronze Age mining shafts in the southern mining fields were sunk vertically down, sometimes with slight undercutting (pit III-1-1). However even at a depth of 7.2m of the so far largest mining shaft II-9-1, still did not yield a sterile base. At a depth of 2m the diameter of this funnel-shaped shaft was 3 meters.

Pit I-1-1 belonging to the southern group of mining fields was 270cm deep. Small flakes decrease near the base, and cores with only a few removed flakes increase, probably representing raw material tests or misused pre-cores. The periphery of this shallow pit encompasses an almost continuous 'pavement' of small chips several cm thick. At a

Figure 5. Kubšice, chipped industry of Early Bronze Age.

depth of 1.5m and a distance of 4m from the margin of the depression, we still had not reached a sterile substrate. Over 150 thousand items were acquired from 9.5 cubic meters of sediment with an overall weight of more than 1200kg. Of these, 7340 are cores. Floatation and careful sifting of sediments would surely increase the count of minute waste flakes. Merely by examining (without sifting) half a cubic meter of earth in pit II-2-1, more than 745 cores, and 14 thousand flakes with an overall weight of 137kg were acquired! Test trenches and surface sampling suggest the southern mining fields contain several tens of such mining pits with analogous quantities of flaked industry. Disproportions between the quantity of mined and exported chert are thus far more distinct during the Bronze Age than during the Eneolithic.

The southern group of mining fields however, contained distinctive assemblages, both in terms of technology (splinter technique) and sources used (extremely poor quality chert). Based on one accumulation, dated by graphited ware sherds (pit II-5-1), this industry dates to the Urnfield Culture. Similar industries have not been found in mining pits, but instead on heaps/ piles, platforms and under the edges of 'seat-like' boulders.

In the Late Bronze Age (Urnfield cultures), fragments of vessels, ash and burnt bones (including a bronze needle fragment) with many large stones were sunk into Early Bronze Age shaft II-9. This probably constitutes a secondary translation of cremated burials.

Excavations in 2002-2005 indicated that mining activity was renewed in Hallstatt times. In the eastern part of mining field VI (9-1) and underneath a large translocated boulder, 6 narrow shafts were found leading up to a depth of 8m. Most of them did not reach chert-bearing seam at all. After a several-years-long research we could detect a certain system in their arrangement. The oldest shafts from the turn between Neolithic and Eneolithic are disturbed by narrower shafts from the Early Iron Age, among which one can distinguish two generations: the younger ones penetrate the whole spoil heap (which becomes thicker down the slope), and they are usually overlaid only with strongly humous delluvial deposits containing many Late Hallstatt Period sherds. The bottom part of the spoil heap is dated to the Late LgK, the upper part comes from Hallstatt Period extraction that proceeded basically the slope up (so that new shafts hadn't to penetrate the spoil heaps from previous shafts). A few of these late shafts (5, 13 and 15) yielded isolated sherds of the Horákov Culture. Chipped industry occurs here only in certain deposits, but in a very large amount, which testifies to some one-shot chipping actions separated by longer time spans (in backfills of the Early Bronze Age extraction pits, chipped industry is always scattered more regularly). The brown to black-brown humous loams above the trenches VI-1 and mainly VI-9 contained sherds of the Late Hallstatt Period graphite pottery. The densest cluster was found inside the superimposed layers above the shaft 7 in the trench VI-9-1. The analysis by A. Majer showed that these delluvial deposits are considerably enriched with phosphorus (probably from dung), which is, on the contrary, absent at the hilltop above. After the same analysis, a higher content of phosphorus is typical of sediments from small Late Hallstatt Period hillforts at Krumlovský les, which are probably associated with sheep breeding. In the mining area VI, the humous layer with ceramic sherds was undoubtedly washed down from the hilltop onto the slope. Settlement activity (above all textile production) at the top of the promontory is documented by the discovery of a typical pit dwelling above the west part of the mining area (point VI-6-1). The dwelling sized 5 x 4m with a step-like entrance from the hillside contained Late Hallstatt Period graphite and non graphite pottery, in single cases with La Tène elements.

The described disproportions between the large quantity of chert mined but scarcely used, suggests that the reasons for mining in the Krumlov Forest were other than the procurement of materials to manufacture tools or weapons. Today it is evident that the mining of quality silicites at other mines also did not arise out of everyday needs, but from the desire to manufacture prestigious, and hence sought after, tools. Krumlov Forest chert however, concerns a low quality raw material source, which *cannot* be used to make complex tools. No Eneolithic axes or Early Bronze Age daggers (see for instance Šebela 1998) were made from it.

Organizers and participants in the mining of Krumlov Forest chert focused on the act of mining itself, without considering any further use of the product. Hence the various large accumulations of flakes and cores all over the mines. The social and symbolic aspects which motivated and rationalized the actual mining, could have fluctuated over time. However these repeated cooperative acts involved the expenditure of human energy, and undoubtedly contributed to the maintenance of traditional values and political stability within these prehistoric societies. It is not coincidental that the upsurge of Krumlov Forest chert mining occurs during the Early Bronze Age, when there was a significant amount of population growth and structured permanent settlements.

Very strange and significant is the method of mining in the Hallstatt Period. Here, the founding of shafts in dug-over zones need not to be a mistake but rather an intent – the uselessness of such a mining place would be namely revealed by the first or second trench already! Digging at the place of older activities below the megalith, and deposition of industry at the bottom of the shafts may have been caused by a common factor – both of these activities may have targeted the subterranean world of the ancestors. The shafts probably mediated the contact to this world, and the depositions then should have to endow, honour or conciliate the ancestors.

Returning the stone back to the earth was on spiritual level a logical continuance of mining, and had to not weaken the ancestral power of the place and to ensure its continuance

(Cooney 2005, 25). Search for these contacts occurred in a period when the population from surrounding agricultural regions took shelter in wooded uplands and founded small hillforts there (in Krumlovský les so far two of them are documented), attended to sheep husbandry, and conducted home textile production. The curious 'mining' in the late Hallstatt Period may have been a reaction to an external threat represented by the expansion of nomadic Vekerzug Culture (Golec 2003, 104 et seq.).

Therefore we can imagine the tradition of these mass actions was carried on by the spiritual significance of this 'extraction' area. Taking into account the incoherent course of the mining and, certainly, the scattered settlement of the area as well, there must have existed uninterrupted awareness of the unique character of these places which, from time immemorial, have yielded gifts of the earth. In acquiring them, contact was also established with the ancestral underground world.

Certainly, it was the interest of the *élite* that all free time – which in archaic societies, lacking home media, had to be mostly spent in public – was spent in the spirit of certain ideology. Collective participation in such actions helped the *élite* in showing that they can control a large number of persons; the participants, on the other hand, enjoyed a deepening of the spiritual experience and perhaps even resulted in collective ecstasy (cf. the hallucinations occurring during religious pilgrimages, or the 'ghost dance' of prairie Indians in the 19th century, etc.). It seems that the cycles of communal work played a significant role in the maintenance of *status quo* in archaic societies. The significance of these cooperative acts was in releasing energy and tension in times of increased social stress. Such activities are not therefore limited only to ancient kingdoms, and to higher levels of social organization which often already have effective mechanisms for enforcing the work. Instead they must have arisen out of voluntary participation, motivated by local social pressure to conform (such as the pressures on traditional country folk to participate in local religious ceremonies). The attraction of these acts surely involved the basic human need to congregate, to compete, to spread information, to exchange certain tools, to meet life partners, to break with routine and to partake in the collective experience of euphoria (irrespective of whether of a spiritual or earthly nature) and so on. Simultaneously, the surrounding countryside became symbolically domesticated. The periodic repetition of such meetings undoubtedly

Figure 6. Group of boulders near the mining area III.

contributed to the cyclical repetition of actions, reminiscent of 'archetypes' upon which many 'oral' societies are founded (compare various contributions in *Archeologické rozhledy* XLIX, 1997, number 2). Much of this may also apply to circular rondels and henges. Their circular shape somehow symbolised the cyclic character of Time.

The presence of rock outcrops, unusual in this part of South Moravian landscape, is an important factor aiding the mythicised memory of the ancestors' country (Field 1997, 68). In the Krumlovský les, such rocky formations appear in all three mining districts from the Bronze Age while they are missing from the older mining fields. In field I it is the 'U Kroužku' outcrop with elevation point 346; in its immediate environs some sherds of Early Bronze Age pottery were found. An obvious group of granodiorite boulders are found approximately 80m NW to the connected fields II and III; it is likely that one of them originally stood erect. It is oblong with an oval profile and it lies on the forest ground in front of a group of rocks with flat upper surfaces (Figure 6: at right). The upper part of this megalith bears two head-like shapes, undoubtedly artificially made. Several kilograms of shattered chert, as if hit against the boulders, were found at its foot. Due to their size, the fragments must have come from large compact blocks which very seldom occur in excavation pits.

In the course of time, these myth-making signs of natural origin were added with human artefacts such as spoil heaps, pits and concentrations of chipped industry. All such shapes thus became similar components of ritual landscape as were elsewhere megaliths, barrows and hillforts (Cooney 1998, 114). From the Late Eneolithic on, the value of Krumlovský les landscape already did not consist in its practical usability but in its past. This could be revived in various time periods, and used for manifold social and ritual purposes. It was just the situation at Krumlovský les that showed most convincingly that in some cases the term '*extraction of lithic raw materials*', within which the prehistoric flint mining usually falls, is completely erroneous.

References

Cooney, G. 1998. Breaking stones, making places: The social landscape of axe production sites, in A. Gibson and D. Simpson (eds.), *Prehistoric Ritual and Religion*, 108-118. Stroud, Sutton publ.

Cooney, G. 2005. Stereo Porphyry: Quarrying and Deposition on Lambay Island, Ireland, in P. Topping and M. Lynott (eds.), *The Cultural Landscape of Prehistoric Mines*, 14-29. Oxford, Oxbow Books.

Čižmář, Z. and Oliva, M. 2001. K ekonomii surovin štípané industrie lidu s vypíchanou keramikou na Moravě. In: *Pravěk NŘ, Supplementum 8, Otázky neolitu a eneolitu našich zemí*, 97-130. Brno, ÚAPP.

Gibson, A. and Simpson, D. (eds.). 1998. *Prehistoric Ritual and Religion*. Stroud, Sutton publ.

Golec, M. 2003. *Těšetice-Kyjovice 6/ Horákovská kultura v těšetickém mikroregionu*. Brno, Masarykova univerzita.

Field, D. 1997. The Worthing flint mine complex, in R. Schild, Z. Sulgostowska (eds.), *Man and Flint. Proceedings of the VII International Flint Symposium*, 65-69. Warszawa, Institute of Archaeology and Ethnology, Polish Academy of Sciences.

Kopacz, J. 2001. *Początki epoki brązu w strefie karpackiej w świetle materiałów kamiennych*. Kraków, PAN.

Kopacz, J. and Šebela, L. 1998. Chipped Stone Materials of the Moravian Proto-Únětice Culture. *Przeglad archeologiczny* 46, 37-57.

Kopacz, J. and Šebela, L. 2006. *Kultura unieticka i grupa wieterzowska na Morawach na podstawie materialów krzemieniarskich*. Kraków, Polska Akad. umięt.

Lech, J. 1982-83. Flint Work of the Early Farmers. Production Trends in Central European Chipping Industries from 4500-1200 BC. An outline. *Acta Archaeologica Carpathica*. 22, 5-63.

Mateiciucová, I. 2008. Talking Stones: The Chipped Stone Industry in Lower Austria and Moravia and the Beginnings of the Neolithic in Central Europe (LBK), 5700-4900 BC. *DABP* 4. Brno, Masarykova univerzita.

Oliva, M. 1990. Štípaná industrie kultury s moravskou malovanou keramikou v jihozápadní části Moravy. *Acta Musei Moraviae - Časopis Moravského muzea sc.soc.* 75, 17-37.

Oliva, M. 2000. Exploatační oblast rohovce v Krumlovském lese v době popelnicových polí a význam pozdních štípaných industrií. Pravěk NŘ 10, 335-364.

Oliva, M. 2001. Sídliště lidu s moravskou malovanou keramikou v okolí Krumlovského lesa a jejich štípané industrie. In Konf. o otázkách neolitu a eneolitu našich zemí, Mostkovice 1999. Pravěk Suppl. 8, 197-231.

Oliva, M. 2003. O nezanedbatelnosti neočekávatelného: štípané industrie starší doby bronzové na Moravě. *Archeologické rozhledy* 55, 10-46.

Oliva, M. 2004. Flint mining, Rondels, Hillforts... Symbolic works or too much free time? *Archeologické rozhledy* 56, 499-531.

Oliva, M. 2005. Výzkum pravěké těžby rohovce v Krumlovském lese. *Acta Musei Moraviae - Časopis Moravského muzea, sci. soc.* 90, 161-183.

Oliva, M. 2006. The Krumlovský les (Southern Moravia) exploitation area: mining techniques, chronology, symbolic meaning. *Der Anschnitt, Beiheft* 19, 163-172.

Oliva, M. 2008. Paleolitické osídlení litické exploatační oblasti Krumlovský les. *Acta Musei Moraviae - Časopis Moravského muzea*, sci. soc. 93, 3-38.

Oliva, M. 2010. Pravěké hornictví v Krumlovském lese – Prehistoric mining in the "Krumlovský les". *Anthropos Studies Vol.* 32 (N.S. 24). Brno, Moravian Museum.

Oliva, M., Neruda, P. and Přichystal, A. 1999. Paradoxy těžby a distribuce rohovce z Krumlovského lesa. *Památky archeologické* 90, 229-318.

Pettitt, P. and Hedges, R. 2008. The Age of the Vedrovice Cemetery: the AMS Radiocarbon dating Programme. *Anthropologie (Brno)* XLVI/2-3, 125-134.

Podborský, V. 2002. *Dvě pohřebiště neolitického lidu ve Vedrovicích.* Brno, FF MU.

Přichystal, A. and Svoboda, J. 1997. Výroba štípané industrie na sídlišti kultury s moravskou malovanou keramikou v Jezeřanech - Maršovicích. *Přehled výzkumů* 1993-1994, 15-25.

Salaš, M. 1987. Záchranný archeologický výzkum na sídlišti z doby bronzové v Brně - Slatině. *Acta Musei moraviae, sci. soc.* 72, 53-73.

Salaš, M. 1989. Der gegenwärtige Forschungsstand der Untersuchungen auf der jungbronzezeitlichen Höhenfundstelle Cezavy bei Blučina. In: Studia nad grodami epoki brazu i wczesnej epoki zelaza w Europie srodkowej, 113-130. PAN, Wroclaw etc.

Salaš, M. 1990a. Únětická sídlištní jáma s lidskými kosterními pozůstatky na Cezavách u Blučiny. *Památky archeologické* 81, 275-307.

Salaš, M. 1990b. To the Problem of Human Skeletal Remains from the Late Bronze Age in Cézavy near Blučina. *Anthropologie* 28, 221-229.

Schild, R. and Sulgostowska, Z. (eds.) 1997. Man and Flint. *Proceedings of the VII International Flint Symposium.* Warszawa-Ostrowiec Swietokrzyski Sept. 1995. Warszawa.

Staňa, Č. 1986. Výšinné únětické sídliště na Zelené Hoře u Vyškova na Moravě. *Archeologické rozhledy* 38, 46-61.

Stuchlík, S. and Stuchlíková, J. 1999. Die Erforschung des Věteřover Rondells in Šumice, in J. Bátora and J. Peška (eds.), *Aktualle Probleme der Erforschung der Frühbronzezeit in Böhmen und Mähren und in der Slowakei,* 169-182. Nitra.

Svoboda. J. and Šmíd, M. 1994. Dílenský objekt kultury nálevkovitých pohárů na Stránské skále. *Pravěk NŘ* 4, 79-125.

Šebela, L. 1998. Spätäneolithische und altbronzezeitliche Silexdolche in Mähren. *SASTUMA* 6/7, 1997/98, 199-226.

Šerý, O. 2010: DNA analýza tří kosterních pozůstatků z mladolengyelské těžební šachty v Krumlovském lese. In: Oliva. *Pravěké hornictví v Krumlovském lese – Prehistoric mining in the "Krumlovský les"*. 419-422. Brno, Moravian Museum.

Topping, P. and Lynott, M. (eds.) 2005. *The Cultural Landscape of Prehistoric Mines.* Oxford, Oxbow Books.

Vencl, S. (ed.). 2006. *Nejstarší osídlení jižních Čech. Paleolit a mesolit.* Praha, ARÚ Praha.

Extraction methods in the Bronze Age at the Wierzbica 'Zele' flint mine site (Central Poland): a model

Hanna & Jacek LECH, Kamil ADAMCZAK and Dagmara WERRA

Abstract
The flint mine at Wierzbica 'Zele' is one of several mines from the central cluster of chocolate flint exploitation points. Chocolate flint was one of the most common siliceous rock materials found in central Europe during the Stone Age. However, the 'Zele' mine operated in the Bronze Age and its largest dated shafts were exploited in the Late Bronze Age. At 'Zele' methods of flint extraction and production in the Early Bronze Age and in the Late Bronze Age differed significantly.

Keywords
Flint mining. Wierzbica 'Zele'. Chocolate flint. Flint work. Bronze Age. Mierzanowice culture. Urnfield complex. Archaeology of Poland. Backed blades of 'Zele' type.

The flint mine at Wierzbica 'Zele' was discovered some years before the Second World War by Stefan Krukowski (1890-1982), one of the most outstanding figures in the history of Stone Age flint mining studies in Europe. The site was not excavated for many years and little was known about it (Schild 1980) until the first systematic surface collection in the autumn of 1979 and further rescue excavations, which allowed the scholars to produce the first comprehensive description of the site (Młynarczyk 1983; Lech 1984).

1. Location of site

The prehistoric flint mine at Wierzbica 'Zele' is located in the north-east fringe of the Świętokrzyskie (Holy Cross) Mountains (51°14'47" N; 21°3'10" E). In the natural stratification of the site there is a layer of Pleistocene sands mixed with boulder clay under the topsoil. Underneath, there is a layer of clay containing flints of different shapes and sizes and, still deeper, a weathered Late Oxfordian limestone, also with flints (Figure 1). In some places, small flattened flint nodules and tabular flint plates occur just below the surface, but the large nodules of flint are found at deeper and the deepest exploitation levels (Figure 2).

2. Excavations and the extent of the mining field

The objective of the research program was to date mining activity in various parts of the site, characterize the mine field and final products of flint working, and to study the differences between exploitation units, mining techniques and prehistoric organization of labor. In the 1980s, T. Herbich (1993) tried, for the first time, to apply the resistivity method to measure the extent of pre- and protohistoric mining fields.

Rescue excavations began in 1980 and lasted until 1988, resulting in 2180 square meters of excavated area. They determined that the mining field within the site had the shape of an elongated ellipse, with a longer diameter (SE to NW) of about 270m and a shorter diameter (SW to NE) of about 50-60m. The 'Zele' mining field covered an area of about 1.4 - 1.8ha, which constitutes about 15 per cent of the site excavated to date (Lech 1984, 188-191; 1997, Figure 3).

3. Raw material

The raw material exploited here was black-brown Jurassic flint, a variant of 'chocolate colored flint' (Schild 1976, 147-150; 1987, 137-139 and 148). It has been given the

Fig. 1. Geological stratification of the limestone quarry near the "Zele" mining field: topsoil, layer of Pleistocene sands mixed with boulder clay, layer of carstic clay, weathered Late Oxfordian limestone, solid Late Oxfordian limestone.

general name of 'Zele' type chocolate flint, whose weak gloss and the appearance, at times, of sharp borders between the black and brown colors in the silica distinguishes it from other types of chocolate flint. The material appears in various forms: small, flat nodules; large plate shaped and bulbous nodules; very large bulbous nodules, reaching 100cm in diameter. The latter occur very rarely, because their deposits are situated deeper and were not exploited.

4. Extraction of flint

Excavations located 81 shafts and exploitation pits. Shafts were dug through Quaternary formations to the level of eluvial clays, penetrating heavily karstified thin plate limestone (Figure 1 and 2). The flint extracted from limestone and limestone weathering products was of various shapes and sizes. Exploitation units were either open shafts, with or without side workings, or large shallow pits. All shafts were wide, open-mouthed and sunk into the limestone (Figure 3 and 4).

The mining field was densely covered with exploitation units. Some of the shafts had a communication platform and their respective depths ranged from about 3m (shafts 6 and 17) to about 7m (shaft 19).

In addition to the shafts, there were also simpler exploitation units – exploitation pits. Their depth was about 1.5m below the level of the ploughsoil. The pits were usually large, with a diameter of several meters. None of these features have been dated. The type of exploitation unit in the 'Zele' mine depended on the geological conditions of the site, while their size and depth were determined by varying demand for flint.

After the shafts had been exploited, they were filled with material extracted when digging neighboring extraction units (Młynarczyk 1983, 104; Lech 1995, 468).

5. A few words about mining tools

The tools used to dig the shafts were mostly of organic material, such as wood, which has not survived, and red deer antler (*Cervus elaphus*). Antlers were used as hammers both in flint working and for breaking up the limestone rock. Examination has revealed numerous traces of use on the antlers but few traces of their having been worked. Glacial boulders with traces of use were frequently found as well. They ranged in size from small ones to specimens weighing several kilograms (Lech 1995, 475).

Fig. 2. Wierzbica 'Zele'. Mining field. Geological stratification, in cross-section. Measuring rods: 20 cm in the centre and 1 m from the right. Showing different levels of the 'Zele' flint: a - layer of thin tabular flint fragments; b – level of flint nodules of medium size; c – level of large flint nodules.

6. Dating and culture history

At the outset of the excavation, the researchers were expecting to find features that could be associated with activities of prehistoric communities ranging from the Final Paleolithic to the Early Bronze Age (Schild 1976, 171; 1980; Lech 1984, 196). ^{14}C analyses were obtained for two opposite parts of the mining field. All dates connect the examined shafts to the Bronze Age in the Vistula River catchment area (Figure 5). However, neither the samples from shafts in the central part of the site nor from the shallow pits have been dated yet. Therefore, one cannot exclude the possibility of earlier mining activity in the area.

An analysis of the morphology of flint artifacts confirms the Bronze Age dating of most of the features in the 'Zele' mining field. Unfortunately, no pottery has been recovered so far.

Exploitation of the mine in the Early Bronze Age was confirmed by the three oldest dates obtained from a small and shallow shaft, no 17, situated in the south-east part of the mining field (Figure 3). Three radiocarbon dates and small finds from the filling of the shaft connect it to the Mierzanowice culture. Flint from 'Zele' was found in small amounts in a cemetery belonging to community of this culture in Szarbia, Kazimierza Wielka district, about 110km SW from the mine.

Fig. 3. Wierzbica 'Zele'. Shaft 17. Cross-sections: west face. Measuring rod 3 m.

In the culture history of the Vistula catchment area, the Mierzanowice culture is followed by the Trzciniec culture, with which shaft no 20, lying in the north-west part of the mine, can be associated.

The largest dated shafts (19 and 28) were exploited in the Late Bronze Age by communities of the Lusatian culture, at the end of the 2nd and beginning of the 1st millennium (Figure 4). They can be associated with the early and middle phase of this culture. Small amounts of 'Zele' flint were discovered in material from excavations of a cemetery and settlement of the Lusatian culture community in Maciejowice, Siedlce district, about 60km to the north-east of the mine, on the other side of the Vistula River (Dąbrowski and Mogielnicka 1993, 87-89 and 97). The Lusation culture belongs to the Urnfield complex, widespread in Europe. Flint mining at 'Zele' continued at the beginning of the Iron Age.

7. Landscape of the mining field

In the Bronze Age the mining field was covered by mixed forest. Analysis of 3711 fragments of charcoal, from 224 samples obtained from shafts, showed that the dominant tree species were pine (*Pinus sp.*) and oak (*Quercus sp.*). The forest also included hazel (*Corylus avellana*), ash (*Fraxinus Excelsior*), hornbeam (*Carpinus betulus*), and

Fig. 4. Wierzbica 'Zele'. Shaft 19. Cross-sections: south face. Measuring rod 3 m.

birch (*Betulus*). The hazel and birch could have been the result of forest succession in places where trees had been cut down to allow the excavation of exploitation units.

The mining field was marked by a large number of shafts filled in to various degrees, accompanied by heaps of extraction waste with numerous pieces of plate limestone, and flint knapping workshops with thousands of waste pieces. We can also suppose there were camps, but their remains, apart from charcoal and single household flint tools, were not found. They occurred on the surface and were destroyed in medieval and modern times.

When flint mining activity ceased, it left behind a characteristic landscape made up of depressions, where the shafts had been, and heaps of waste. Unfortunately, this landscape did not survive to modern times, primarily because the forest, which had covered the area known locally as a "sacred place", had long ago given way to ploughland.

8. Two main directions of flint work: a model

In the Early Bronze Age, flint-knappers favored small, flat shaped nodules, which were used to produce axe and adze blades. In the Late Bronze Age, they were more interested in large bulbous nodules used to produce large, massive blades and flake blanks.

At the time when flint was being exploited by communities of the Mierzanowice culture, in the Early Bronze Age, small flat-shaped nodules and fragments of flint plates were the most popular. Such material was extracted from the shallower shafts, such as no17. Shafts of similar size, dated to the same period and providing similar raw material, are known from nearby chocolate flint mines Polany Kolonie II and Polany II (Schild 1976, 153-158; 1987; Chmielewska 1988; Lech and Leligdowicz 2003). The 'Zele' finds dateable to this period include numerous early and advanced roughouts for bifacial axe blades and headadzes produced from this type of raw material. Apart from bifacial headaxes and headadzes, sickles were produced using the same technique. Flake blanks were also obtained but whether they were produced at this time is still an open question.

Production was concentrated on bifacial axe and adze blades of various types (Figure 6). In the course of surface collections and excavations we recovered a rich array of early bifacial axe/adze roughouts and advanced bifacial axe/adze roughouts (preforms) with different profiles and in various stages of working. Preparation included the longer edges of the specimen and both sur-

Shaft number	Depth of sample	Laboratory number of sample	Conventional dating bp	Dating in calibrated years BC – level of probability 68%
Dating of north-west part of the mining field				
19	400-410 cm	OxA-5101	2780±45 bp	984-873 cal BC
20	340 cm	BM-2383	3150±80 bp	1509-1328 cal BC
Dating of south-east part of the mining field				
17	100-104 cm	GrN-11 852	3680±70 bp	2188-1980 cal BC
17	215 cm	GrN-11 853	3570±90 bp	2071-1799 cal BC
17	300-310 cm	GrN-11 854	3670±60 bp	2162-1978 cal BC
18	82-95 cm	GrN-11 856	2670±60 bp	894-810 cal BC
28	330-350 cm	BM-2385	2750±70 bp	989-847 cal BC
28	330-350 cm	BM-2385A	2780±80 bp	1054-863 cal BC
28	430-450 cm	BM-2386	2890±110 bp	1254-955 cal BC
28	430-450 cm	BM-2386A	2800±100 bp	1134-876 cal BC

Figure 5. Flint mine Wierzbica 'Zele'. Radiocarbon dating (charcoal samples only).

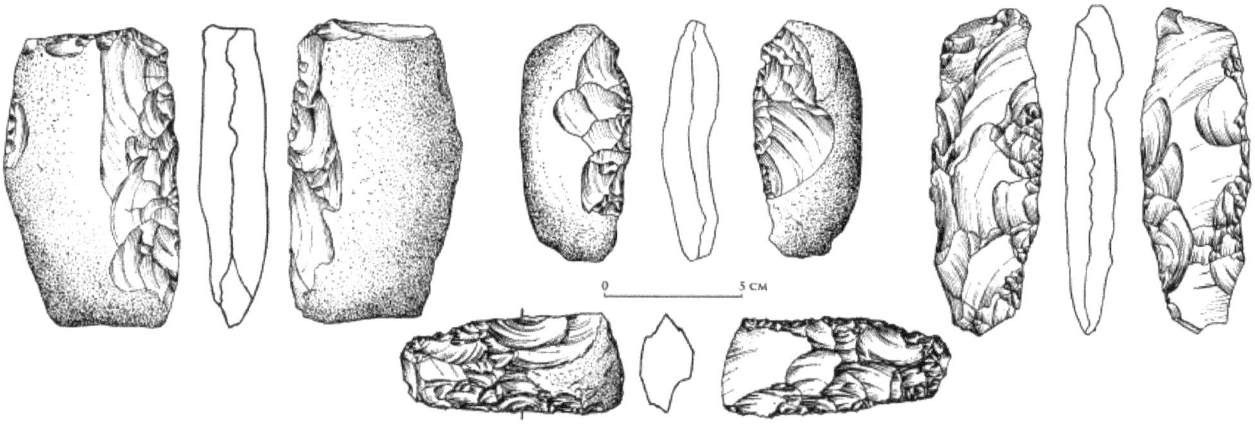

Figure 6. Wierzbica 'Zele'. Early Bronze Age. Bifacial headaxe roughouts from small flat flint nodules.

faces. The back and cutting edge were formed by removing small flakes (Lech 1983, 64-67; Młynarczyk 1983, 107-111).

The 'Zele' finds, which can be dated to the Late Bronze Age and communities of the Lusatian culture, include characteristic backed blades of 'Zele' type, large cores for massive flakes and blades, as well as flake and blade blanks. The flint industry of the Lusatian communities had the capacity of producing large knives which had long and sharp cutting edges and a blunt backs (Figure 7). They were produced from large, massive blade, blade/flake and flake roughouts obtained from irregularly shaped cores. These, in turn, were obtained from large bulbous nodules, sometimes broken up, of a type which probably did not interest knappers of the Mierzanowice culture. The blades, if they had a naturally blunt back, preferably with cortex, were used as knives without further working. In other cases, the back was shaped by a series of several careful retouchings. The massive backed blades (knives), obtained in this way, and occurring in several variants, are a characteristic product of the Lusatian industry from 'Zele' (Lech 1984, 195).

Comparing the exploitation of the 'Zele' mine in two different periods of the Bronze Age, it is possible to draw some conclusions as to the type of raw material which interested the flint-knappers, the extraction methods (types of exploitation units) which they used and the specific features of the flint industry. These conclusions have served as the basis of a model which needs to be verified through further studies.

The Early Bronze Age:

a) interest primarily in small flat flint nodules and fragments of tabular flint plates;

b) extraction of flint mainly from small shallow shafts and large surface pits;

c) production, primarily, of bifacial headaxes.

The Late Bronze Age:

a) interest mainly in large flint nodules and fragments of large tabular flint plates;

b) extraction of flint primarily from large and deep shafts;

c) preparation of large cores; production mostly of knives from large blade and flake blanks, including backed blades of the 'Zele' type.

This model points to the probable age of many hitherto undated shafts and large surface pits. It will be verified by planned radiocarbon determinations, still possible in the case of some features, and by further studies of flint material. We also need to determine when production of headaxes from flint came to an end, not only in the case of the 'Zele' mining field, but also in the Vistula catchment area and other regions of Europe.

9. Final remarks

The depth of shafts exploited by communities of the Lusatian culture demanded a considerable outlay of labor, often greater than in the case of shafts dated to the Early Bronze Age. Groups of several flint knappers/miners must have probably labored for many days. Procuring the flint must have required considerable knowledge of practical geology and mining skills. The size of the shafts indicates that in Central Poland in this time flint continued to be an important raw material.

Mining in the Late Bronze Age was not a casual and straightforward occupation. If we compare the exploitation units dated to the Early Bronze Age with the Lusatian shafts from 'Zele', it will be found that Lusatian mining was not a haphazard activity but possessed traditions, rules, and regulations (Lech 1997, 96-97; Högberg 2009, 19 and 272). The specialization of labor which occurred here,

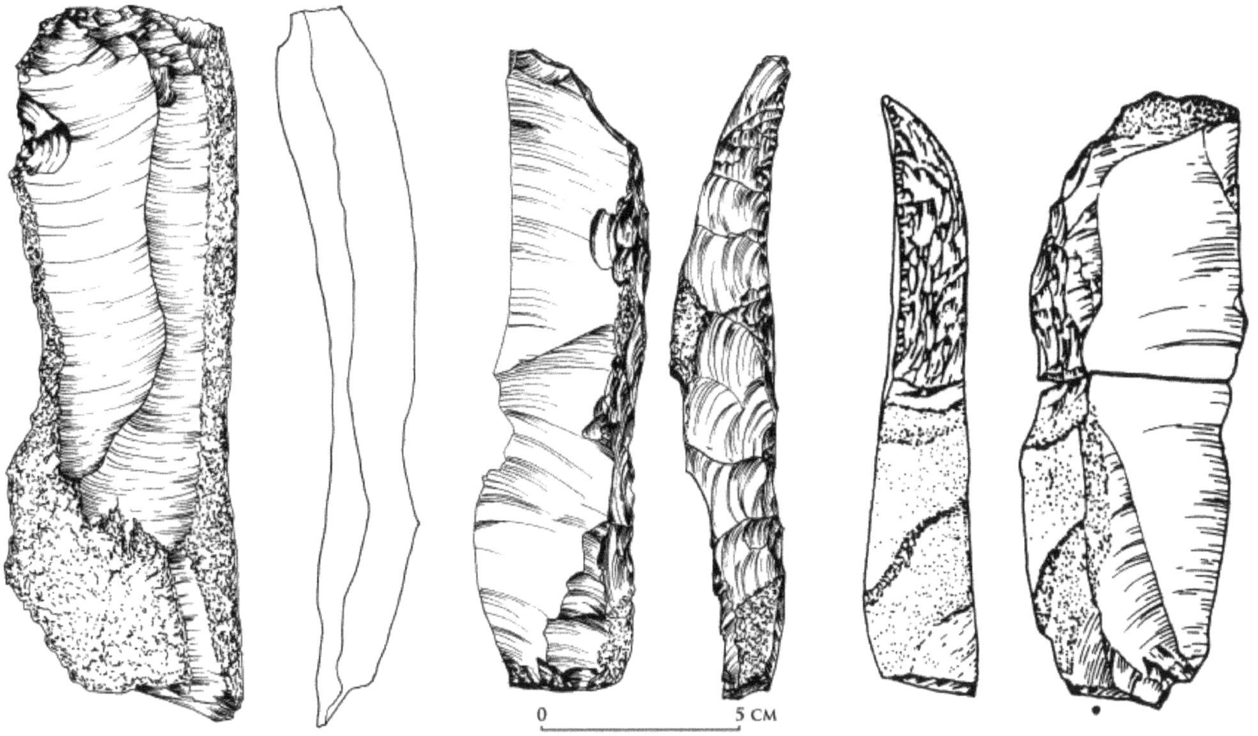

Figure 7. Wierzbica 'Zele'. Late Bronze Age. A blade blank and backed blades of 'Zele' type.

resulted from the needs of communities, their knowledge and skills, as well as from the conditions created by the natural environment in which people functioned (Högberg 2009, 218).

The archaeological material excavated at the 'Zele' mine field is now being studied. The results obtained so far do not confirm the view that flint was extracted, with intervals, from the Final Paleolithic to the Hallstatt D period. Archaeological finds and radiocarbon determinations do, however, definitely confirm the existence of flint mining in the Bronze Age and at the beginning of the Iron Age. What we now see is the result of a diversified model of exploitation of the flint deposits from the 'Zele' mining field.

References

Chmielewska, M. 1988. The Early Bronze Age flint mine at site II, Polany, Radom district. *Przegląd Archeologiczny* 35, 139-181.

Dąbrowski, J. and Mogielnicka-Urban, M. 1993. The radiocarbon dating of the three objects from the settlement of Lusatian culture at Maciejowice, Siedlce voivodeship. *Przegląd Archeologiczny* 41, 87-99.

Herbich, T. 1993. The variations of shaft fills as the basis of the estimation of flint mine extent: a Wierzbica case study. *Archaeologia Polona* 31, 71-82.

Högberg, A. 2009. *Lithics in the Scandinavian Late Bronze Age. Sociotechnical change and persistence.* Oxford, BAR Publishing. British Archaeological Reports International Series 1932.

Lech, H. and J. 1984. The prehistoric flint mine at Wierzbica 'Zele': a case study from Poland. *World Archaeology* 16(2), 186-203.

Lech, H. and J. 1995. PL 3 Wierzbica 'Zele', Radom province. *Archaeologia Polona* 33, 465-480.

Lech, H. and J. 1997. Flint mining among Bronze Age communities: a case study from central Poland, in R. Schild and Z. Sulgostowska (eds.), *Man and Flint*, 91-98. Warszawa, Institute of Archaeology and Ethnology Polish Academy of Sciences.

Lech, J. 1983. Flint mining among the early farming communities of central Europe. Part II – the basis of research into flint workshops. *Przegląd Archeologiczny* 30, 47-80.

Lech, J. and Leligdowicz, A. 2003. Studien zum mitteleuropäischen Feuersteinbergbau in der Bronzezeit, in T. Stöllner, G. Körlin, G. Steffens and J. Cierny (eds.), *Man and Mining – Mensch und Bergbau. Studies in honour of Gerd Weisgerber on occasion of his 65th birthday*, 285-300. Bochum, Deutsches Bergbau-Museum. 'Der Anschnitt' Beiheft 16.

Młynarczyk, H. 1983. Wstępne wyniki badań kopalni krzemienia czekoladowego Wierzbica 'Zele', woj. Radom, w latach 1979-1981. *Sprawozdania Archeologiczne* 35(1984), 87-115.

Schild, R. 1976. Flint mining and trade in Polish prehistory as seen from the perspective of the chocolate flint of central Poland. A second approach. *Acta Archaeologica Carpathica* 16, 147-177.

Schild, R. 1980. PL 3 Wierzbica, Fundplatz Zele, Wojew. Radom, in G. Weisgerber, R. Slotta and J. Weiner (eds.), *5000 Jahre Feuersteinbergbau* 581. Bochum, Deutschen Bergbau-Museum.

Schild, R. 1987. The exploitation of chocolate flint in central Poland, in G. de G. Sieveking and M.H. Newcomer (eds.), *The human uses of flint and chert,* 137-149. Cambridge, Cambridge University Press.

Danubian organization of flint mining in the southern part of the Polish Jura: a study from Sąspów near Cracow

Jacek LECH

Abstract

The flint mine of Sąspów is situated in the southern part of the Polish Jura. The mine is one of the better known sites of prehistoric flint mining in Poland. Jurassic-Cracow flint was exploited here from karstic clays in open, wide-mouthed shafts by Danubian communities of the Linear Pottery culture (LBK) and the Lengyel-Polgár cultural complex. The depth of shafts ranged from 2 to 5.2m. Flint working in the mining field centred on the preparation of pre-cores and early core forms, as well as on production of blade blanks. Different types of workshops were distinguished. The Sąspów mine lay outside the region of permanent settlement of Danubian communities and never formed part of the immediate zone of economic activity of the LBK or Lengyel-Polgár settlements. Miners organized expeditions for exploitation of flint, which took from at least several days to several weeks depending on the number of participants and demand for flint.

Keywords

Neolithic. Neolithic in Poland. Flint mining. Sąspów flint mine. Flint supply. Working of flint. Jurassic-Cracow flint. Danubian culture. Linear Pottery culture (LBK). Lengyel-Polgár cultural complex.

The first farmers came to the area of present day Poland from the south, through the Moravian Gate and the Carpathian and Sudeten mountain passes. In 1925 Gordon Childe coined the term Danubian Ia to denote these people, since they represented the first Neolithic culture to arise in the basin of the middle Danube (Childe 1925, 171-176; 1929, 36-47). Today, we use the term Linearbandkeramik (LBK) or Linear Pottery culture (Whittle 1996, 3; Milisauskas 2002, 162-165; Burnez-Lanotte ed. 2003). In the following period of the Neolithic, cultural influences and, to a certain degree, migration of new communities from the Danube and Tisa basins resulted in the rise of the Lengyel-Polgár cultural complex in the basins of the Vistula and Oder rivers. Both the first Danubians and the later Danubian communities used Jurassic-Cracow flint, exploited in the southern part of the Polish Jura, near Cracow. The most closely studied mining field of the Danubians can be found in the village of Sąspów (Lech 1981a, 2008).

The initial results of the Sąspów studies were published with material from the Second International Symposium on Flint (Lech 1975) and in the well-known Bochum book (Lech 1980). For the Madrid conference an up-to-date summary of what is known about the mining field has been prepared.

1. The Sąspów flint mine: geography, geology and archaeological investigations

The mine in Sąspów lies about 450m above sea level, in the southern part of the Polish Jura, approximately 25km north-west of Cracow, in the highest part of the region known as the Ojców Jura (Figure 1 in Lech 1981a). This area is composed of Upper Jurassic, Oxfordian limestone and their weathering products, covered by loess. The Ojców Jura is a waterless loess-covered plateau, cut across by deep valleys. The mine is situated in an area of palaeogenic planation surface. Under the layer of loess lies Tertiary red karstic clay, in some parts rich in nodules of Jurassic-Cracow flint. The mining field at Sąspów is situated on the hill-side at the beginning of the Sąspów valley. In May of 1960, W. Chmielewski (1929-2004) discovered here a prehistoric workshop site which turned out to be a Neolithic flint mine (site 1), the next flint mine of Danubian communities to be discovered after Vienna-Mauer (Ruttkay 1970). Site 1 at Sąspów was the second Neolithic flint mine excavated in Poland after the Second World War; the first was Krzemionki Opatowskie (Lech 1981a, 2008, 281-283; Bąbel 2008, 97).

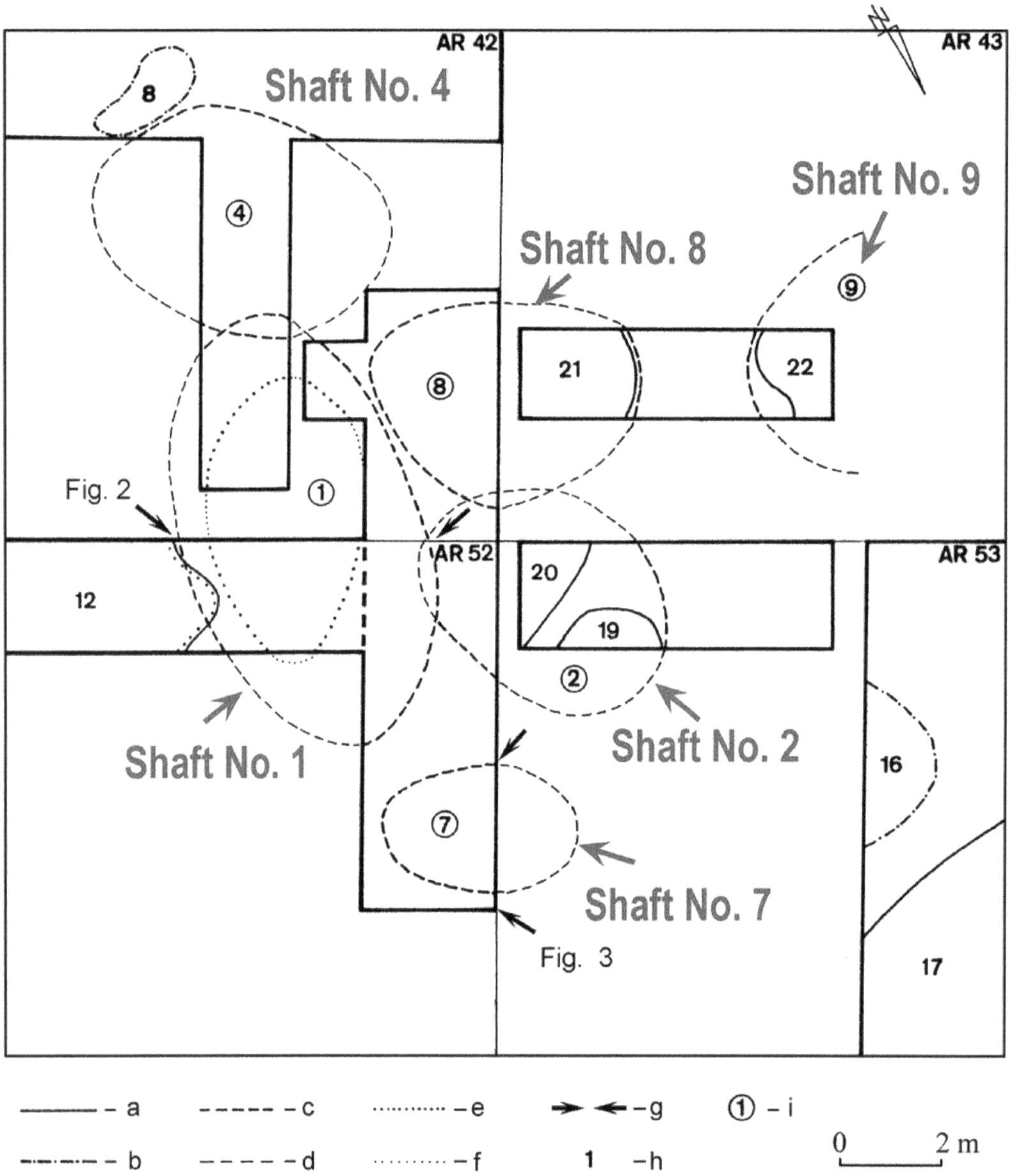

Figure 1. Sąspów, Cracow dist. Flint mine. Part of cuttings from 1970-71. Distribution of various features. a - borders of flint floors (krzemienisko); b – borders of other concentrations of flints; c – outlines of shafts; d – reconstructed outlines of shafts; e – outlines of waste heaps; f – reconstructed outlines of waste heaps; g – location of cross-sections from figures 2 and 3; h - numbers of flint floors and other concentrations of flints; i – numbers of shafts.

Rescue excavations at the complex of Sąspów sites were carried out in 1960, 1962, 1970-1973, 1994, and 1996. The mines were investigated together with nearby sites where flint was worked out in the open, as well as two sites located on the slope of the Sąspów Valley, near the headwaters from which springs the Sąspów Stream: the rock shelter 'Beside the mine' and the cave 'Below the Church'. Flint was brought there from the mining field (site 1). The cave also contained the remains of camps used by the miners, with traces of fires and numerous pottery sherds. In connection with the Sąspów Project, an archaeological survey of approximately 190 square kilometers was carried out in the Prądnik and Rudawa Rivers' catchment areas, in the years 1976-80 and 2005 (Lech 2001, with earlier literature in Polish).

In addition to excavation reports, other publications relating to the Sąspów mine have appeared: several short summaries outlining the state of knowledge at the time (see Lech 2001), a detailed study of two mine workshops excavated at the mining field site in 1960 (Dzieduszycka-Machnikowa and Lech 1976), and a monograph on flint mining in relation to settlement of Danubian communities in the Cracow Upland, according to the state of knowledge towards the end of the 1970s (Lech 1981b). At the Institute of Archaeology and Ethnology of the Polish Academy of Sciences, a monograph of the mine at Sąspów is being prepared and should be ready a few years from now. A large collection of flint material from excavations of the site is kept at the Archaeological Section in Igołomia, a part of the Cracow Branch of the Institute.

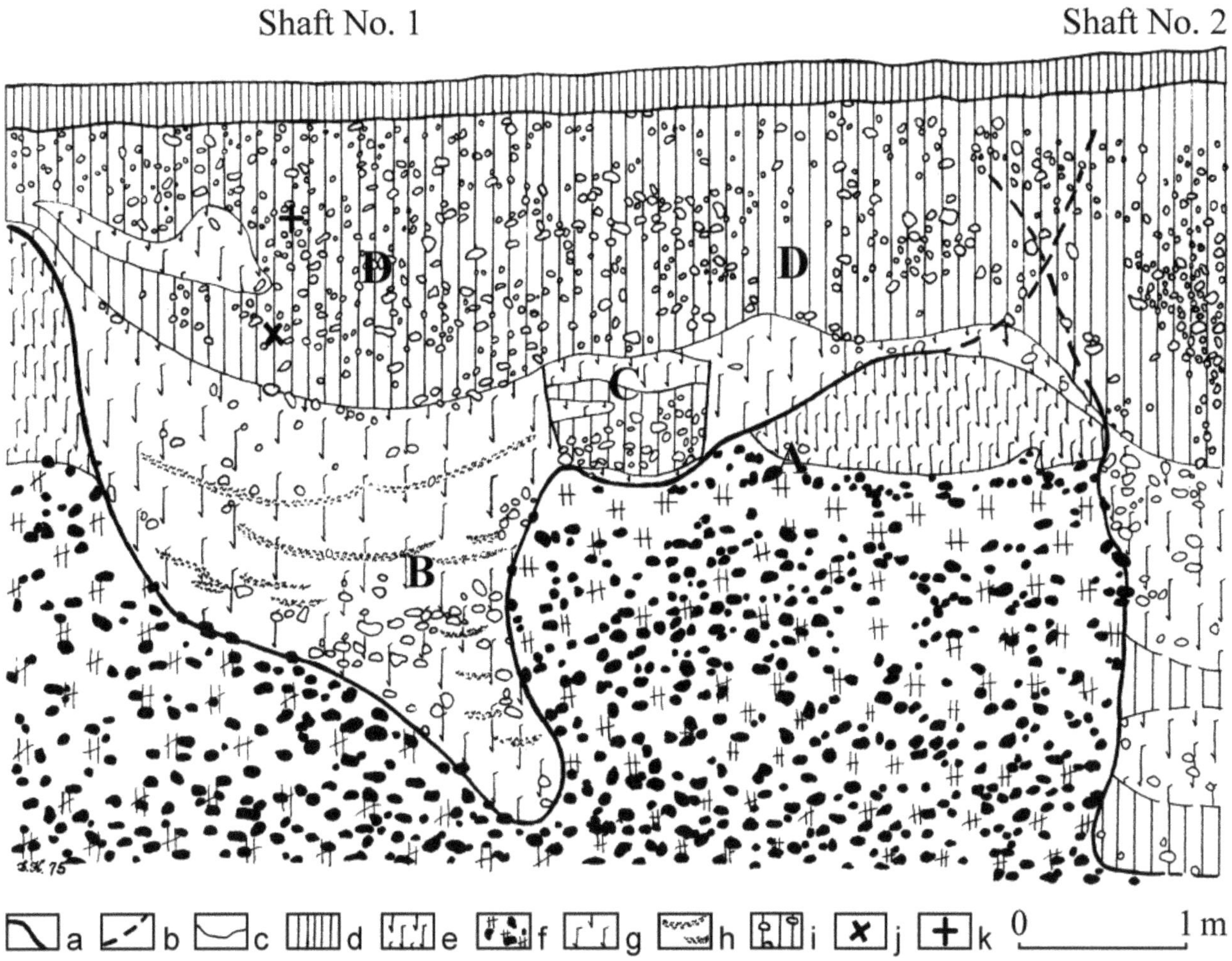

Figure 2. Sąspów, Cracow dist. Cross-section of Shaft No. 1 and fragment of Shaft No. 2. a – shaft walls; b – reconstructed shaft walls; c – limits of layers; d – soil and sub-soil; e – loess in primary position; f – karstic clay with flint nodules – exploited deposit; g – loess till from filling-in of shafts; h – grey-bluish loess loam; i – karstic clay with loess and chipped and natural flint material from filling-in of shafts; j – location of charcoal sample collected for ^{14}C dating 5325 +/- 90 BP (GrN-7052 C); k - location of charcoal sample collected for ^{14}C dating 5295 +/- 60 BP (Bln-1461). Markers A – D have been described in the article.

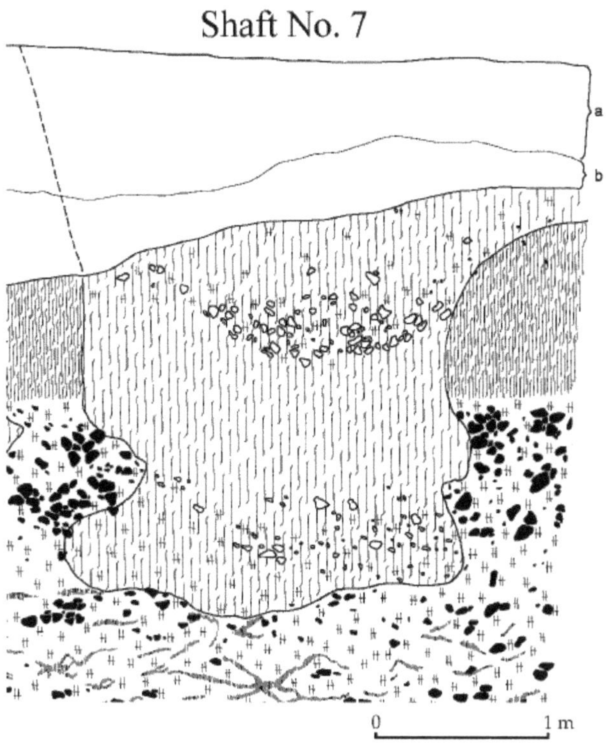

Figure 3. Sąspów, Cracow dist. Cross-section of Shaft No. 7. Graphic markers same as in Fig. 2. Also: a – layer removed by buldozer during road construction, not examined; b – layer mixed mechanically during road construction, not examined.

2. The mining field and shafts

The first shafts were discovered during salvage excavations in 1970. Investigations showed that the mine field was densely dotted with them (Figure 1). All in all, 14 shafts were excavated over several seasons. Researchers also uncovered many often extensive flint floors, remains of chipping floors and flattened waste heaps around the shafts.

The mining field at Sąspów revealed a concentration of shafts of different sizes. On the surface, the shafts were oval in shape and measured from 3.5 to 8m lengthwise. Their depths ranged between 2 and 5.2m, with the most shallow shafts located in the lower parts of the valley slope. The largest shaft no 1 was 4.3m deep, elliptical in shape on the surface, with a diameter of 8m lengthwise and 5m widthwise (Figure 1 and 2). While it was being exploited, waste material was thrown into the already partly filled in older shaft 2. In shaft 1 a platform was left to facilitate transport of waste material in the direction of shaft 2 (Figure 2A). After flint exploitation ceased, shaft 1 gradually filled up with loess flowing down from the surface to over half way up from the bottom (Figure 2B). The process probably lasted several years, as evidenced by traces of stagnant water in the loess filling of the shaft. After this interval, there were attempts to resume exploitation of flint. Probably using a wooden spade-like tool, a straight-walled probe hole was dug in the shaft filling to determine where to start excavating for flint (Figure 2C). The probe hole was made to check whether the platform which had been left behind earlier was worth exploiting. However, because the diggers found only karstic clay without flint nodules at the bottom of the probe hole, exploitation of shaft no 1 was not resumed. Nevertheless, their assumptions had been correct. The platform would have yielded much raw material (Figure 2A). The miners who dug the probe hole must have known about the unexcavated platform, as they were probably the same group which had earlier exploited the flint from that shaft.

Exploitation of shaft 1 was not resumed, but next to it another shaft, No. 4, was dug and the waste heap from the new shaft filled the remaining space in shaft No. 1 (Figure 1 and 2D). Such a conclusion is confirmed by the cross-section of the waste heap of shaft 4, located in the filling-in of shaft No. 1. Shaft 4 was smaller than shaft 1. At Sąspów there were both large shafts, such as Nos 1 and 8 (see Figure 6 in Lech 2008) and small ones, such as shaft 7 (Figure 3). In the case of shafts 1 and 7, it is possible to estimate the volume of extracted material. The amounts vary from about 20 cubic meters for shaft 7 to about 80 cubic meters for shaft 1. This shows that demand for flint varied and that the size of the Danubian groups which mined the flint also differed over time. It should be added that shaft No. 7 is older than No. 1, probably by about 250 years.

In the 3 to 5m shafts several jobs needed to be done. Apart from the miners extracting the flint nodules, someone had

No.	Feature	Laboratory Number	conv. bp	Dating in calibrated years	
				62%	95%
1.	Shaft No. 3	GrN-7692	5700±135 BP	4710-4440 BC	4900-4250 BC
2.	Shaft No. 7	GrN-7993	5575±75 BP	4490-4340 BC	4590-4310 BC
3.	Workshop 3/1970	Bln-1462	5325±60 BP	4240-4050 BC	4270-4030 BC
4.	Shaft No.1	GrN-7052C	5325±90 BP	4260-4040 BC	4340-3970 BC
5.	Shaft No. 1	Bln-1461	5295±60 BP	4180-4040 BC	4270-3980 BC
6.	Shaft No. 6 and workshops 2/1970	BM-1128	5046±102 BP	3960-3760 BC	4050-3640 BC
7.	Shaft No. 11	BM-2941	4620±180 BP	3650-3050 BC	3800-2850 BC

Figure 4. Radiocarbon dating of archaeological features from the Sąspów mine.

to transport the clay-covered flint to the surface, then clean the nodules by removing the clay and selecting those pieces which could be worked. The size of shaft 1 at the surface (Figure 1) indicates that quite a large number of miners were involved in the digging and in the extraction of the flint. Probably, during the initial stage there were five to ten people. Once karstic clay with flint nodules was reached, division of labour took place. Two to four miners could work at the bottom of the shaft while the same number would help them on the surface. Exploitation of the older shaft 7 needed the cooperation of two to four miners.

One may ask how laborious was exploitation of flint from the Sąspów shafts? In the 1960s, a Soviet archaeologist from Leningrad, S.A. Semenov (1898-1978), calculated that excavating one cubic meter of light sand sediment using tools available in the Neolithic would take one man four to five hours, and extracting one cubic meter of chalk takes seven to eight hours (Semenov 1968, 21-22). It can be assumed that at Sąspów extracting one cubic meter of deposit took about five to six hours, in which case exploitation of the flint from shaft No. 1 would take 400-480 hours and from shaft No. 7 about 100-120 hours. We do not know how many hours a day people worked but the data suggests that the exploitation of the shafts was labourious and needed at least several days of work, even with several miners working together.

3. Chronology and culture history

The Sąspów mine was exploited by Danubian communities, beginning with the LBK, most intensively by the Pleszów and Malice groups of the Lengyel-Polgár Cultural Complex from Little Poland – 5th millennium BC in calibrated radiocarbon years (Figure 4). The end of exploitation can be associated with late Lengyel-Polgár communities (see Whittle 1996, 146-193; Milisauskas and Kruk 2002, 193-197).

Owing to the discovery of several vessel fragments, we know that the miners who exploited the flint here belonged to the Danubian communities. The same is indicated by the morphology of the flint artefacts, radiocarbon dating (Figure 4) and the presence of flint from Sąspów in settle-

	Shaft 11	Shaft 12	Shaft 13
Number of analysed specimens:	**1628**	**1296**	**1343**
First morphological group:	**26.53%**	**9.80%**	**15.86%**
natural nodules	2.82%	1.16%	1.56%
pre-core forms and cores	23.71%	8.64%	14.30%
Second morphological group:			
blades and blade fragments	**0.43%**	**1.39%**	**0.45%**
Third morphological group:	**72.97%**	**88.81%**	**83.69%**
flakes and waste	44.78%	58.26%	41.32%
smaller natural flint pieces	28.19%	30.55%	42.37%
Fourth morphological group:			
morphological tools	**0.06%**	**0.00%**	**0.00%**
Total	**99.99%**	**100.00%**	**100.00%**

Figure 5. General structure of flint materials from shafts Nos. 11, 12 and 13.

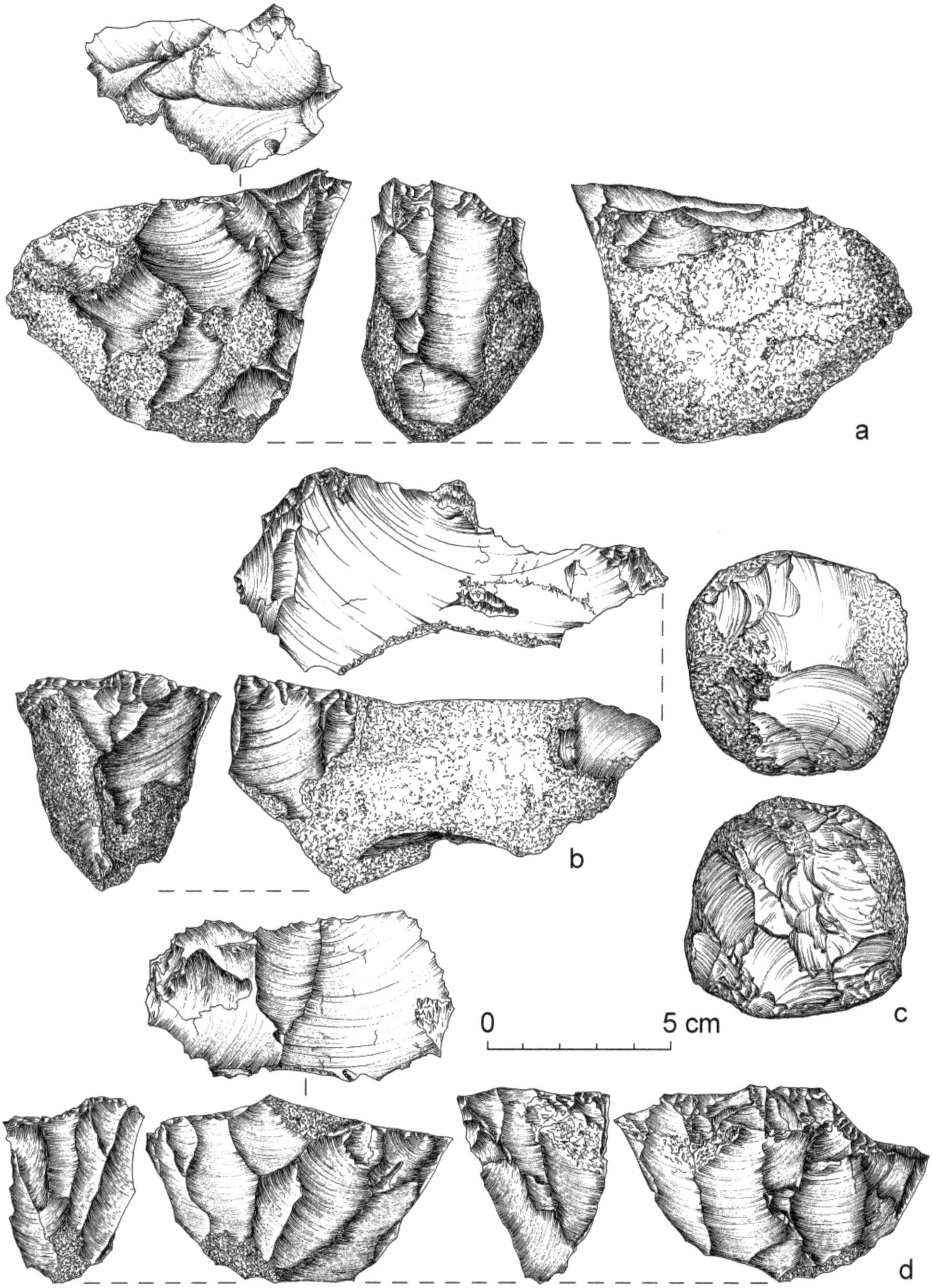

Figure 6. Sąspów, Cracow dist. Selection of flint specimens from the filling in of shafts Nos. 1 and 2. a, b, d – precore and early core forms; c – flint hammer stone. Drawn by H. Lech.

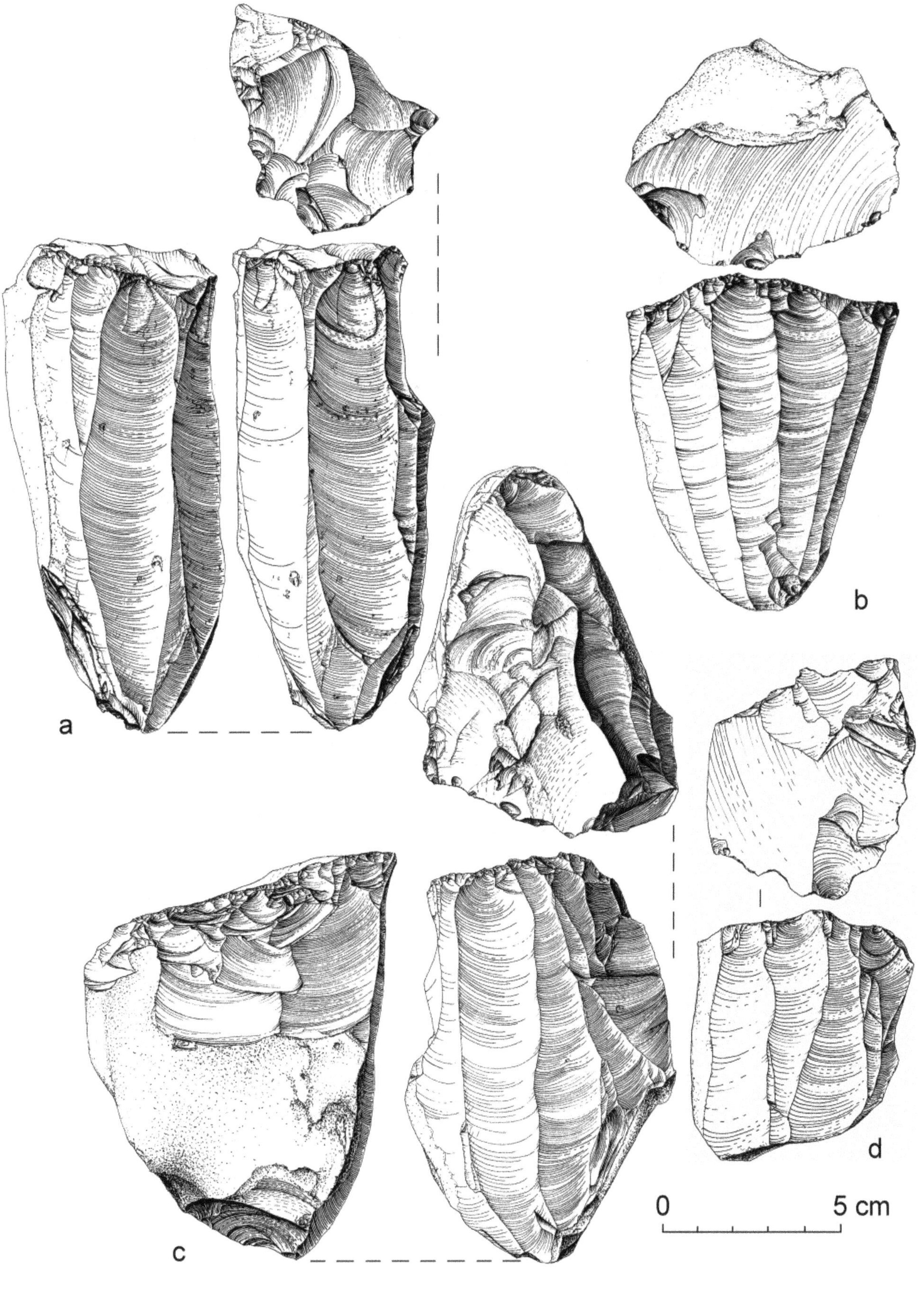

Figure 7. Sąspów, Cracow dist. Selection of cores for blades from workshop No. 3/1960. a – core No. 213; b – core No. 225; c – core No. 155; d – core No. 242. For detailed description of the cores see: A. Dzieduszycka-Machnikowa and J. Lech (1976, 203-209). Drawn by J. Krzepkowska.

ments of Danubian communities at various distances from the deposit (e.g., Lech 1997, 624-631; 2008, 287-288).

4. Working of flint

Flint working in Danubian communities in the Cracow region concentrated mainly on production of blades and their further processing. Production of flake blanks was of secondary importance. In Sąspów, pre-cores and early cores, as well as blade blanks were prepared (Figure 6 and 7). The importance of flakes in the production process is difficult to determine. They were used, to a certain extent, to make retouched tools, but most were probably simply production waste. No evidence was found that flint axe blades were produced. Flint axes were unknown among Danubian communities.

Analysis of the material indicates that initial selection of nodules for further working and primary treatment was done on the waste heaps by the shafts, thus confirming earlier ethnoarchaeological observations (Holmes 1919, 177-178). This provenance is indicated by the morphological structure of sample flint material obtained from shafts 11 and 13 during rescue excavations in 1994 (Figure 5). This was material which remained after selection of flint nodules which were taken away for further working elsewhere.

At the shafts, nodules were cleaned by removing the clay and examined for quality. Sometimes, one or several flakes with cortex were struck off the nodule, using a flint or stone hard hammer. The places where this was done are referred to as initial preparation workshops. Selected nodules were next taken to advanced preparation workshops, situated further from the shafts, sometimes next to the miners' camps. Here the final product was made. Two types of advanced preparation workshops were determined:

a) workshops preparing the pre-core and initial core forms (e.g., chipping floor 1/1971; see Figure 8 and 9);

and

b) workshops for advanced preparation and exploitation of cores (for blades produced from cores) – assemblages 1/1960 and 3/1960; see Figure 8 and 9).

The material from the workshops where pre-cores and early cores were made contains a large number of nodules left in their natural state, and failed pre-cores. There are much fewer early cores. The amounts of flakes and waste are very high, while the number of blades and blade fragments is very small. In the case of chipping floor 1/1971 the share of small natural flint pieces was very large (Figure 9), probably because flint nodules from which weathering clay had not yet been removed were brought here.

The material from workshops where blades were produced contains many exploited or damaged cores and the share of blades and blade fragments is considerable. The blades are those which were left after selection, therefore many have various defects (Dzieduszycka-Machnikowa and Lech 1976, 39-113; Lech 1982, 54-58).

At all the workshops and in other flint inventories on the mining field, there were many natural fragments of flint and very rare examples of prepared (morphological) tools (Figure 5 and 9). When needed, pieces left over from flint working were used. The size of the final product taken from the mine depended on the demand for flint and the number of miners, since transport of the flint was human-powered. It seems that the weight of the final product taken away from the mine was somewhere between 20 and 100kg. Analysis of material from two flint workshops where blades were produced suggests that, from each of them, about 3500-4000 blades were taken away to the settlements (1/1960 and 3/1960). From another, probably older workshop preparing pre-core and initial core forms (1/1971), several tens of pre-cores and early cores were carried away.

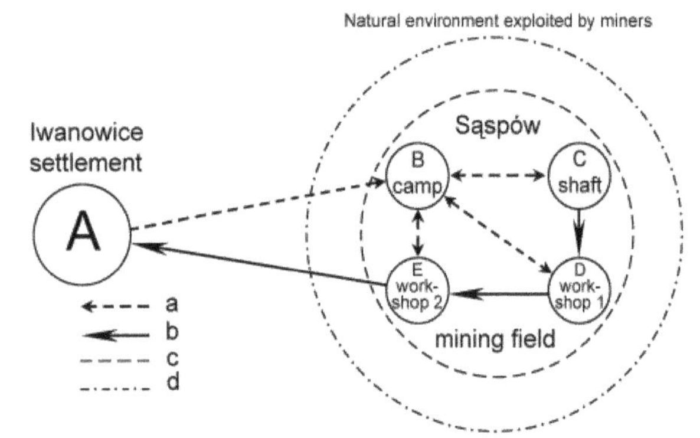

Figure 8. Sąspów, Cracow dist. Model for the organization of labour within the area of the mine, based on analysis of structures and assemblages of Lengyel-Polgár communities (later Danubians). a – main direction in which miners moved; b – direction of transport of mined flint nodules and semi products made from nodules (pre-core and early core forms, blade blank); c – mining field; d – zone of exploitation of natural environment surrounding the mine. Workshop 1 = initial preparation workshop; workshop 2 = workshop for advanced preparation including blade production from cores.

5. The Sąspów flint mine, settlement and distribution of raw material

In the Neolithic, the heights of the Polish Jura near Cracow were a specific region whose natural conditions were not conducive to 'permanent occupation for the purpose of the cultivation and herding', with the exception of the eastern rim of the 'Jura proper' (Kruk 1973, 15-16). The author of this article reached the same conclusions in his

	Workshop assemblages		
	1/1960	**3/1960**	**1/1971**
Number of analysed specimens:	**12757**	**10344**	**2748**
First morphological group:	**4.20%**	**4.34%**	**7.06%**
natural nodules	0.16	0.42%	4.00%
pre-core forms and cores	4.04%	3.92%	3.06%
Second morphological group:			
blades and blade fragments	**26.58%**	**28.33%**	**1.02%**
Third morphological group:	**69.15%**	**66.94%**	**91.59%**
flakes and waste	57.64%	46.58%	32.60%
smaller natural flint pieces	11.51%	20.36%	58.99%
Fourth morphological group:			
morphological tools	**0.07%**	**0.38%**	**0.33%**
Total	**100.00%**	**99.99%**	**100.00%**

Figure 9. General structure of workshop assemblages from the mining field.

studies (Lech 1971, 128-130; 1981b, 189-196; 2006, 390-418). The Ojców Jura and neighbouring areas were included in a programme of systematic archaeological field survey (Lech, Rook and Stępniowski 1984). From this we know that the mining field at Sąspów lies outside the area of permanent settlement of Danubian communities. Since exploitation of flint was a process which took at least several days, miners had to set up camps on or near the mining field. The remains of fireplaces on the site of the mining field may indicate traces of camps, but most camps were probably situated nearer water. There were many springs in the Sąspów Valley, near the mine. Camps were also set up in caves and rock shelters in Sąspów Valley, not far from the mining field, as was determined during excavations of the rock shelter 'Beside the mine' and the cave 'Below the Church' (Lech 1982, 56-64).

Studies of flint material from settlements of Danubian communities lying at various distances from the deposits confirm that Jurassic-Cracow flint arrived there in three forms:

a) natural nodules, sometimes with scars after removal of a single flake;

b) advanced pre-cores and early cores,

c) blade blanks and probably also flakes.

In some cases, one variant of Jurassic-Cracow flint, from one mining field dominates, in others, material from various deposits (mines), lying in the southern part of the Polish Jura, was used.

Settlements of Danubian farmers associated with the exploitation of Jurassic-Cracow flint were discovered in lower-lying areas, to the east and south-east of Sąspów. Their location was decided by the requirements of agriculture. The large LBK settlement in Olszanica was made up of long houses (Milisauskas 1986, 2002, 176-177). At the edge of the settlement an area of economic activity was found, with large amounts of flint waste from flint working. The flint had been mined, but did not come

Figure 10. Jerzmanowice, Cracow dist. Surroundings of 'Bat' cave (Jaskinia Nietoperzowa). Place where depot of pre-cores was found (see Figure 11).

from Sąspów. Olszanica was a settlement of producers, introducing Jurassic-Cracow flint into the system of long-distance, indirect exchange. Flint and several other siliceous rocks were objects of indirect interregional exchange between LBK settlements in the Vistula and Oder river basins and to the South (Lech 1989, 1990, 1997, 623-632; 2003, 22-27; see de Grooth 1994). The Sąspów mining field was only one source of Jurassic-Cracow flint for LBK communities, but flint from there was found, among others, in the north, in an LBK settlement at Boguszewo – 450km from Sąspów and at the famous settlement in Bylany, today in the Czech Republic – 300km west of Sąspów (Lech 1989, 116; Małecka-Kukawka 2001, 32-37; Werra 2010). Much more popular was another variant of Jurassic-Cracow flint, known from the settlement at Olszanica, but a depot of blades of Sąspów flint discovered in an LBK settlement at Borovce on the West-Slovakian Lowland, 300km south of the mine, confi rms the importance of the Sąspów raw material. Settlements at Boguszewo, Bylany and Borovce were typical settlements of users. Jurassic-Cracow fl int, mainly the Olszanica variant, is also known from LBK settlements in Lower Austria and Hungary (Figure 6 in Lech 2003). It was the most popular raw material among LBK communities in the eastern part of Central Europe.

Figure 11. Jerzmanowice, Cracow dist. The depot of pre-cores (see Figure 10).

LBK communities began fl int mining in Sąspów, but all the shafts dated so far were exploited by later Danubians (Figure 4). A settlement from a later period, associated with exploitation of the Sąspów fl int, was discovered in Iwanowice, 15km east of Sąspów. Pits belonging to a late Pleszów group (Modlnica phase) revealed tools, blade blanks and cores from Sąspów flint (Dzieduszycka-Machnikowa and Lech 1976, 150 and Plate XVIII, XIX). Shaft No. 1 from Sąspów and the workshops 1/1960 and 3/1960 mentioned earlier (Figure 9), producing blade blanks, can be associated with this phase of development of the Danubian communities. This allows us to propose a model showing how work was organized at the Sąspów mine, based on analysis of structures and assemblages (Figure 8).

The routes taken by miners carrying the mined flint and the use of caves are confi rmed by the discovery of a depot of pre-cores found in 2008 about 50cm beneath the surface (Figure 10), near the Nietoperzowa Cave, several kilometers from the Sąspów mine. The 11 buried pre-cores (Figure 11) of poor quality look like specimens eliminated from a transport of raw material. The fact that they were buried is a manifestation of some unknown beliefs of Danubian miners, evidenced by several similar discoveries in the region. Most of the settlements of later Danubian communities exploiting Jurassic-Cracow fl int were located within a 30km radius of mines in the Polish Jura. Apart from Iwanowice, there is also the settlement in Pleszów, on a terrace of the Vistula, which at this point is still a small river. Flint processing workshops were found in two of the settlement's development phases. The grave of a specialist fl int knapper was also discovered. The Vistula river was probably an important communication route to the north. Among later Danubians, Jurassic-Cracow fl int was still an object of widespread interregional indirect exchange but its role was smaller than among the LBK communities (Lech 1987, 243-247). The Sąspów mine fi eld never formed part of the immediate zone of economic activity of the Lengyel-Polgár settlements (Figure 8).

References

Bąbel, J. 2008. The Krzemionki fl int mines latest underground research 2001-2004, in P. Allard, F. Bostyn, F. Giligny and J. Lech (eds.), *Flint Mining in Prehistoric Europe. Interpreting the archaeological records,* 97-107. Oxford, BAR Publishing. British Archaeological Reports International Series 1891.

Burnez-Lanotte, L. (ed.) 2003. *Production and Management of Lithic Materials in the European Linearban-dkeramik.* Oxford, BAR Publishing. British Archaeological Reports International Series 1200.

Childe, V. G. 1925. *The Dawn of European Civilization.* London, Kegan Paul, Trench, Trubner.

Childe, V. G. 1929. *The Danube in Prehistory.* Oxford, Clarendon Press.

Dzieduszycka-Machnikowa, A. and Lech, J. 1976. *Neolityczne zespoły pracowniane z kopalni krzemienia w Sąspowie.* Wrocław, Zakład Narodowy imienia Ossolińskich.

Grooth, M. E. Th. de. 1994. *Studies on Neolithic flint exploitation. Socio-economic interpretations of the flint assemblages of Langweiler 8, Beek, Elsloo, Rijckholt, Hienheim and Meindling.* Maastricht, Scorpio.

Holmes, W. H. 1919. *Handbook of Aboriginal American Antiquities. Part I. Introductory. The Lithic Industries.* Washington, Government Printing Office.

Kruk, J. 1973. *Studia osadnicze nad neolitem wyżyn lessowych.* Wrocław, Zakład Narodowy imienia Ossolińskich.

Lech, J. 1971. Z badań nad kopalnią krzemienia na stanowisku I w Sąspowie, pow. Olkusz. In J. K. Kozłowski (ed.), *Études sur industries de la pierre taillée du néo-enéolithique,* 115-133. Kraków, Polskie Towarzystwo Archeologiczne.

Lech, J. 1975. Neolithic fl int mine and workshops at Saspów, near Cracow. In F. H. G. Engelen (ed.), *Tweede Internationale Symposium over Vuursteen. 8-11 Mei 1975 - Maastricht,* 70-71. Maastricht, Nederlandse Geologische Vereniging. Staringia 3.

Lech, J. 1980. PL 15 Sąspów I, Jerzmanowice, Wojew. Kraków, in G. Weisgerber, R. Slotta and J. Weiner (eds.), *5000 Jahre Feuersteinbergbau. Die Suche nach dem Stahl der Steinzeit,* 616-619. Bochum, Deutsches Bergbau-Museum. Veröffentlichungen aus dem Deutschen Bergbau-Museum Bochum 22. Also in second edition in 1981 and in third edition in 1999.

Lech, J. 1981a. Flint mining among the early farming communities of Poland. In F. H. G. Engelen (ed.), *Derde Internationale Symposium over Vuursteen. 24-27 Mei 1979 - Maastricht,* 39-45. Maastricht, Nederlandse Geologische Vereniging. Staringia 6.

Lech. J. 1981b. *Górnictwo krzemienia społeczności wczesno rolniczych na Wyżynie Krakowskiej: koniec VI tysiąclecia – 1 połowa IV tysiąclecia p.n.e.* Wrocław, Zakład Narodowy imienia Ossolińskich.

Lech, J. 1982. Flint mining among the early farming commu-nities of central Europe. Part II – The basis of research into fl int workshops. *Przegląd Archeologiczny* 30 (1983), 47-80.

Lech, J. 1987. Danubian raw material distributrion pattern in eastern central Europe, in G. de G. Sieveking and M. H. Newcomer (eds.), *The human uses of fl int and chert,* 241-248. Cambridge, Cambridge University Press.

Lech, J. 1989. Danubian raw material exchange network: a case study from Bylany. In J. Rulf (ed.), *Bylany Seminar 1987.* Collected Papers, 111-120. Praha, The Archaeological Institute of the Czechoslovak Academy of Sciences.

Lech, J. 1990. The organization of siliceous rock supplies to the Danubian early farming communities (LBK): central European examples, in D. Cahen and M. Otte (eds.), *Rubane & Cardial,* Liège, Etudes et Recherches Archéologiques du l'Université de Liège 39.

Lech, J. 1997. Remarks on prehistoric fl int mining and fl int supply in European archaeology, in A. Ramos-Millán and M. A. Bustillo (eds.), *Siliceous rocks and culture,* 611-637. Granada, Editorial Universidad de Granada.

Lech, J. 2001. Neolityczna kopalnia krzemienia na stanowisku I w Sąspowie, pow. Kraków i jej badania ratownicze w latach 1994 i 1996, in J. Lech and J. Partyka (eds.), *Z archeologii Ukrainy i Jury Ojcowskiej,* 349-372. Ojców, Ojcowski Park Narodowy.

Lech, J. 2003. Mining and siliceous rock supply to the Danubian early farming communities (LBK) in eastern central Europe: a second approach. In L. Burnez-Lanotte (ed.), *Production and management of lithic materials in the European Linearbankeramik,* 19-30. Oxford, BAR Publishing. British Archeological Reports International Series 1200.

Lech, J. 2006. Wczesny i środkowy neolit Jury Ojcowskiej, in J. Lech and J. Partyka (eds.), *Jura Ojcowska w pradziejach i w początkach państwa polskiego,* 387-438. Ojców, Ojcowski Park Narodowy.

Lech, J. 2008. Mining and distribution of fl int from Little Poland in the Lengyel, Polgár and related communities in the Middle/Late Neolithic – a brief outline, in Z. Sulgostowska and A. J. Tomaszewski (eds.), *Man – Millennia – Environment. Studies in Honour of Romuald Schild,* 281-292. Warsaw, Institute of Archaeology and Ethnology, Polish Academy of Sciences.

Lech, J., Rook, E. and Stępniowski, F.M. 1984. Archeologiczne badania poszukiwawcze i weryfi kacyjne w dorzeczu Prądnika. *Sprawozdania Archeologiczne* 36, 213-266.

Małecka-Kukawka, J. 2001. *Między formą a funkcją. Traseologia neolitycznych zabytków krzemiennych z ziemi chełmińskiej.* Toruń, Uniwersytet Mikołaja Kopernika.

Milisauskas, S. 1986. *Early Neolithic Settlement and Society at Olsznica.* Ann Arbor, The University of Michigan. The Museum of Anthropology.

Milisauskas, S. 2002. Early Neolithic. The First Farmers in Europe, 7000-5500/5000 BC. In S. Milisauskas (ed.), *European Prehistory. A Survey,* 143-192. New York, Kluwer Acdemic/Plenum Publishers.

Milisauskas, S. and Kruk, J. 2002. Middle Neolithic, continuity, diversity, innovations, and greater complexity, 5500/5000-3500/3000 BC. In S. Milisauskas (ed.), *European Prehistory. A Survey,* 193-246. New York, Kluwer Acdemic/Plenum Publishers.

Ruttkay, E. 1970. Das neolithische Hornsteinbergwerk von Mauer (Wien 23). *Mitteilungen der anthropologischen Gesellschaft in Wien* 100, 70-115.

Semenov, S. A. 1968. *Razvitie tekhniki v kamennom veke.* Leningrad, Izdatelstvo 'Nauka', Leningradskoe Otdelenie.

Werra, D. i.p. Longhouses and long-distance contacts in the Linearbandkeramik communities on the north-east borders of the ecumene: *"á parois doubles"* in Chełmno Land (Poland). *Anthropologica et præhistorica* 121.

Whittle, A. 1996. *Europe in the Neolithic. The creation of new worlds.* Cambridge, University Press.

Antlers in flint mining: technological opportunism and symbolism

Andrzej BOGUSZEWSKI and Ludomir R. LOZNY

Abstract

Terrain's morphology and geology determine the form and method of subterranean extraction, which in turn affects the choice of tools used. In this paper we look into the question of whether technological opportunism and optimization were the only decisive factors in choosing tools, and suggest that tradition may have influenced the Neolithic miners' tool usage.

Keywords

Antler tools. Flint mining. Symbolism. Neolithic Period.

1. Introduction

The use of antler, horn and bone tools in the extraction of minerals has been observed since the beginning of mining activity. In the Upper Paleolithic chert mine at Nazlet Khater 4 (Upper Egypt), dated to 35000 to 30000 BC antlers of gazelles and antelopes (*Alcelaphus buselaphus*) were employed along with stone implements (Vermeersch *et al.* 1997). Antler tools, probably used for extraction of rhyolite (Valde-Nowak 1991), were also found in layer VIII of the Oblazowa Cave in the Polish Carpathian Mountains and dated to the Gravettien. Antler and bone tools were probably used in the mining of the chocolate flint recovered from the Late Mesolithic burial at Janislawice, in Central Poland (Cyrek 1995). Evidence suggesting the extraction of this type of flint during the Late Mesolithic was also recorded at the mine at Tomaszow, located 200km south from the Janislawice burial (Schild *et al.* 1985).

The quantity of antler and bone mining tools increased during the Neolithic Period, when the extraction of silica-saturated (igneous) rocks reached its peak. Neolithic miners were equipped with antler, bone, rock, flint, and probably wooden tools, as recorded in Mur-de-Barrez, Blackpatch, Jablines, Szentagal or Krzemionki (Pull 1932, Bostyn *et al.* 1992; Biró 1995; Bąbel 2008).

2. Antler mining tools: manufacture, use, and symbolism

A comparison of frequencies of tools made of the above-mentioned materials in different flint mines suggests that antler tools were common and present a spectrum of morphological (typological) diversity. While stone and flint-made tools are very typical for mines (this fact allowed the identification of the so-called 'culture campignienne'), antler tools found in mines are not different from those found in settlements. Undoubtedly, the natural shape of antler determines the tool's function. However, to challenge the pragmatically understood correlation between shape and function, we suggest that the antler tool morphology may have also been influenced by tradition and possibly beliefs of prehistoric miners, something noticeably difficult to trace archaeologically. Lack of methodologies and data impedes analyses of symbolic meaning of mines and research regarding miner's behavior. Using selected data we would like to suggest an interpretation that may not relate to strictly technical or economic conditions of subterranean exploitation.

We limit our analysis to antler mining tools. Taking a closer look at this category of tools we noticed that about 70% of identifiable antler found on Neolithic settlements was locally collected (Billamboz 1977), and in flint mines the percentage is even higher. In shaft 1 at Grimes Graves, tools made of collected antler constituted 84% (Clarke 1915), at Loewenbourg 87% (Böckner 1980) and in shaft 2 Grimes Graves 98% (Clason 1979). These numbers suggest that seasonally collected and stored antler was the tool-makers preferred material. It is also possible that the exploitation of mines was seasonal and took place in winter and early spring, which is typical for societies following agricultural calendar. Taphonomic analysis of some tools also suggests that antler used to make tools had lain in forests for some time before it was collected, which is evidenced by the

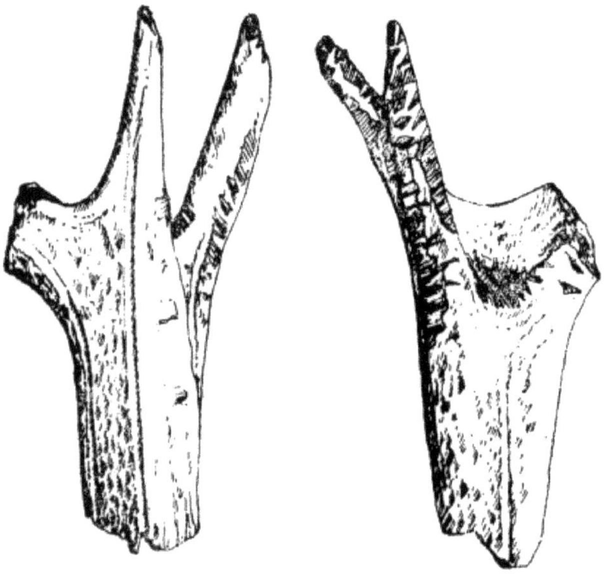

Figure. 1. Gnawing marks on antler. Krasne Sielo (after Boguszewski 1984a).

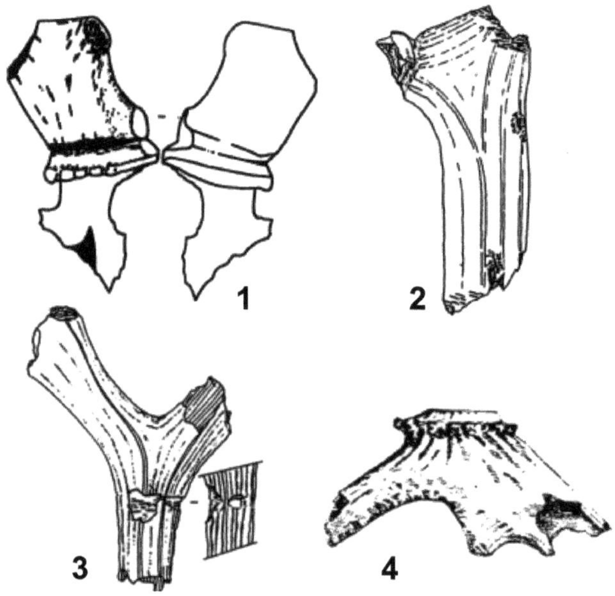

Figure. 3. Antler debris of the manufacturing of mining tools. 1: Villemaur, 2,3 : Serbonnes (France, all after Sidéra 1995); and 4: Krasne Sielo, Bielarus (after Boguszewski 1984a).

gnawing marks made by rodents, foxes, wolfs and deer (Figure 1) (Boguszewski 1984a).

The right-side antler was another clear preference: it makes up 70% of antler tools in Bretteville-le-Rabet mine (Desloges 1986), its ration is similar in other mines (Figure 2). Shaft 4 at Breteville is a notable exception—here left-side antler dominates. Does this discrepancy suggest individual preferences of right- or left-handed users? Whatever the interpretation, the data suggest two patterns in systematic and premeditated acquisition of antler for tool-making: incidental seasonal collection or organized and large-scale collection of antler, and selection of the left-side antler.

Spacious mines with deep shafts were used for many months, and archaeologists found hundreds of antler tools there. Is seems impossible to make such a number of tools prior to their use and to stock them up in the mine. G. de G. Sieveking (1978) suggested that 800 antler «picks » were needed to exploit one shaft at Grimes Graves, not counting other tool types. More than 570 antler picks have been found in four mines at Grimes Graves, most made of collected antler. If we assume a preference for the right-sided antler, it seems that antler from at least 120 deer was needed to exploit one shaft (Holgate 1995). The collection and manufacturing of such a number of tools in limited time were determined by the overall organization of mining activities. Even if it is an exceptional case, it illustrates the need for organization of the mining process including careful planning of the logistics and underground activities.

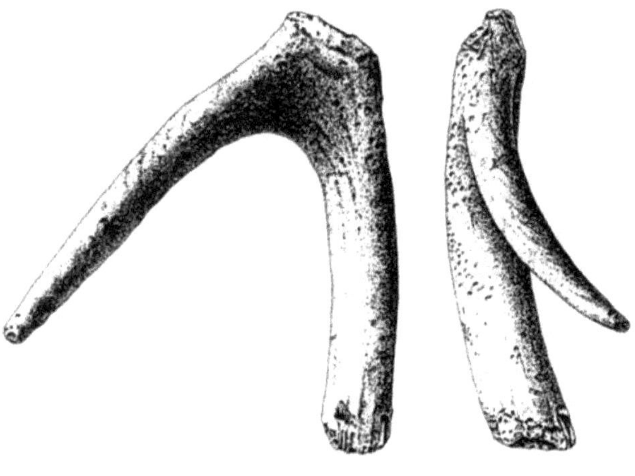

Figure. 2. Right-side "pic" from Bretteville-le Rabet, France (after Desloges 1986).

Antler is heavy and difficult to transport (some beams weight up to 8kg). None of the settlements located in close proximity to flint mines show evidence of antler storage. Where then were this tools produced—away from mines or locally, in the mines' immediate vicinity? Certain data suggest that tools may have been manufactured near the mines. Evidence of tool-making activities in form of antler debris, suggesting either the production of tools or their modifications, has been recorded in several mines, namely at Krasne Sielo (Bielarus), Loewenburg (Switzerland), St. Mihiel, Serbonnes, Villemaur « Le Grand Bois Margot » (France), Spiennes (Belgium) and others, and these finds corroborate this (Figure 3).

Antler preparation/softening required a large supply of water. Soaked and saturated antler was cut, incised, smoothed, etc, in order to make a desired tool (MacGregor *et al.* 1985). Antler dries out and hardens quickly. But it has to be soggy during the tool production, and therefore a tool-making workshop must have been close to a source of

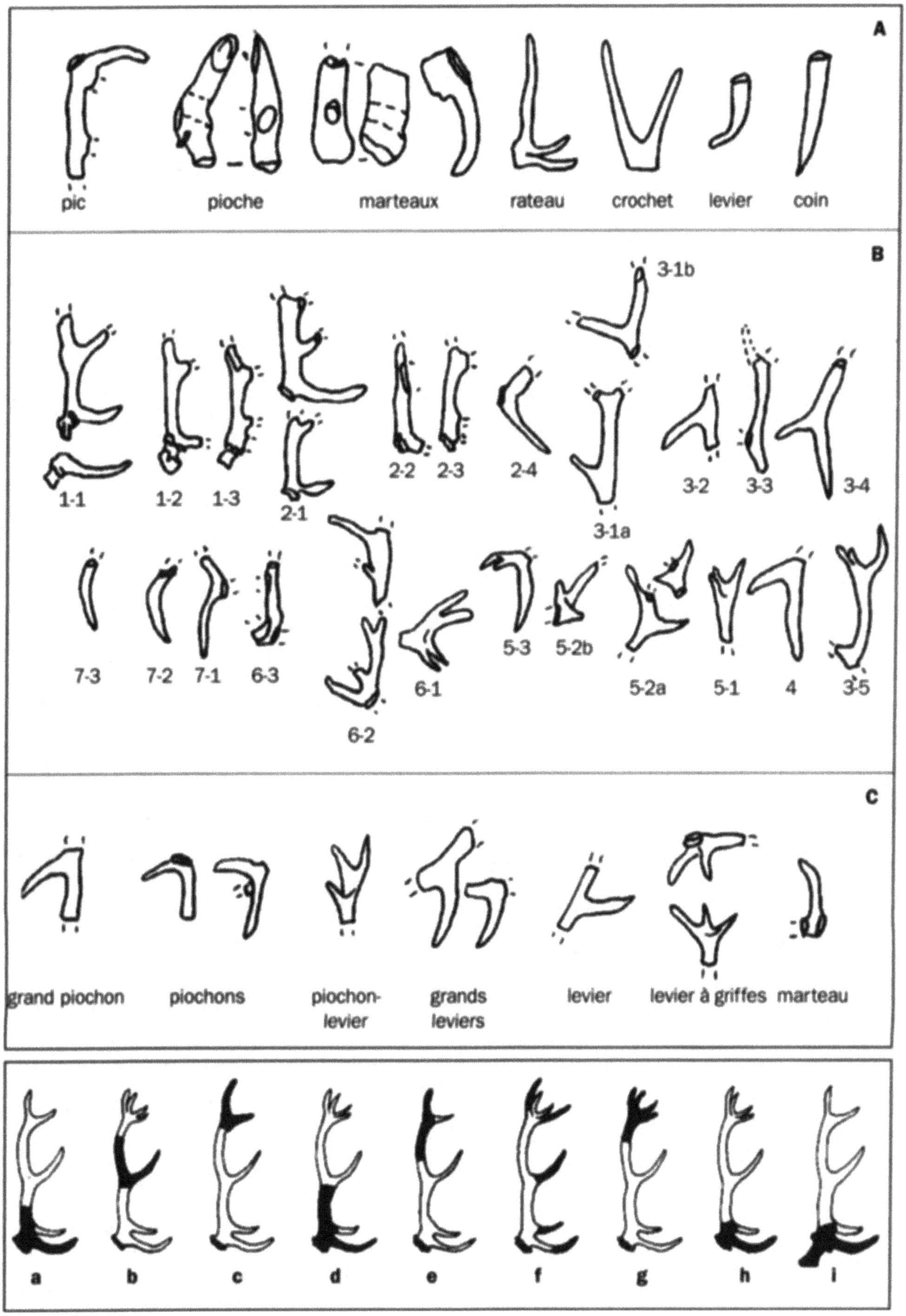

Figure. 4. Examples of different typologies of antler mining tools. (A): Soulier 1971; (B): Böckner 1980; (C): Desloges 1986. Below: parts of the red deer antler employed to produced differents types of tools. (after Bostyn et al. 1992).

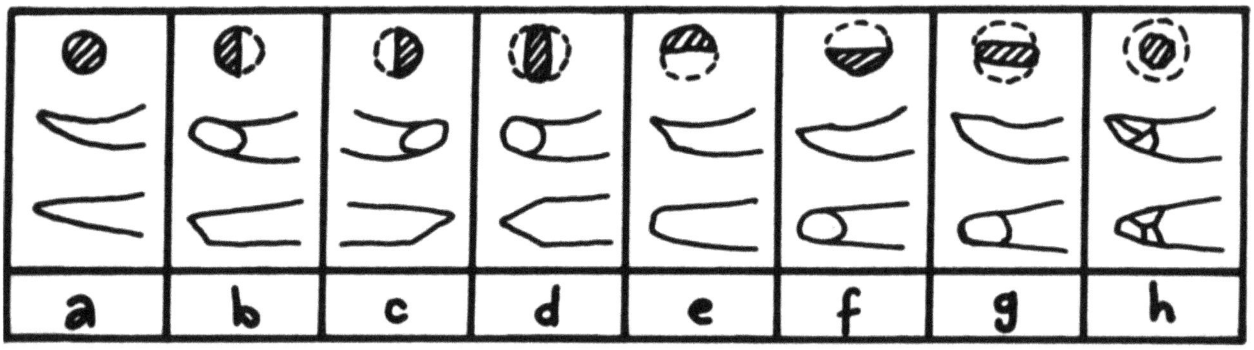

Figure. 5. Forms of points of antler minnig tools : natural tip (a), intentionaly shaped points (b-h). (after Boguszewski 1991a).

water. It had to be a source distinct from the one used for consumption because the tool production process contaminates water with debris, blood, etc. Although difficult to detect archaeologically, such production places may have existed in close proximity to mines (e.g. Zalewski 1988). It is also possible that tool-makers used special containers to soak antler. However, archaeological remains of such containers might be difficult to recover.

More than a dozen of antler mining tool-types are known. The morphological inspections of tools suggest a strong correlation between the anticipated tool function and its production (Figure 4). Certain tools found in different mines around Europe resist functional classification. Nevertheless, their frequencies suggest that certain shapes were preferred, facilitating identification or their function. Often parts not directly involved in flint extraction were prepared without care and frequently irregular (made with the use of flint axe), while the working edge area was carefully prepared. Some mining antler tools can be distinguished from "domestic" antler tools by the existence of bifacial or unifacial preparation of the working edge, elongated cuts, etc. This preparation makes the edge more effective and prevents quick abrasion, or maybe there were the marks of the reparations measurements. Antler prepared in such a way is rarely found in settlements. Since bone is more brittle than antler, bone-made tools are hardly ever found in mines but very common in settlements (for instance, perforators). Only sporadically 'chisels' were made of long bones and 'shovels' made of scapulas of different animals (proximal femur fragment and ox scapula) and - very rarely- human.

There are two categories of antler mining tools: those used directly to crush rocks, and the others used as handles for other tools. The tools of the second category are not common and the record suggests only sporadic use in some mines (for instance Krzemionki). The reasons for such discrepancy are not certain and might be attributed to the chronology of mines, the fact that miners followed different traditions, mining techniques, the type of sediment exploited, opportunistic behavior, or, perhaps, the place of mining in the overall economic system of societies.

Migal (1990) suggested that tools were selected according to their use (function) and the type of rock in which the flint was imbedded. It seems that the latter characteristic was not always decisive. For instance, no antler tools are known from Defensola 'A' mine (Di Lernia *et al.* 1995), where flint appears in hard sediment, while at Sümeg they dominate (e.g. Fülöp 1975; Bácskay 1980, 1995). The reason for such discrepancies may have been related to different tradition of flint extraction and techniques used. It is highly unlikely that miners did not have access to antler.

Several observations regarding tool use were made during experimental extraction of flint. In experiments conducted at the Krzemionki mine in Poland, A. Boguszewski used two antler 'chisels' made of points (one with a natural tip and the other with an elongated cut) and a 'pick' made of a point and beam fragment (Boguszewski 1991). The gallery was 15m below surface, where two rock formations converged: a not weathered and soft limestone containing large flint concretions was overlaid by hard and slightly weathered 20cm thick limestone. The use of classic antler 'pick' was not very productive in either formation and the tool's edge was quickly damaged. A better result was achieved with the use of antler 'wedge-chisel' 20cm long, made of carefully cut point with well-smoothed end. The experiment lasted for two hours and the quantity of flint obtained with the use of wedge-chisel exceeded the amount of flint obtained with the pick. It seems that a flint-made 'pick' was more effective in harder limestone deposits. Its working edge fit the cracks and worked well as a wedge or chisel. The results of this experiment suggest that the classic antler 'pick' (point + beam fragment) is productive in the extraction of flint imbedded in soft limestone or chalk, wedge-chisel antler tools is effective in the extraction of flint imbedded in various rocks, while flint-made picks can be used to extract flint imbedded in hard rocks.

3. Preparation of antler tools

Most antler tools, except handles and hammers, were used to crush rocks in order to extract flint. It is therefore important to briefly discuss the tool preparation process. It relates predominantly to the preparation of points. Obviously it was the most frequently damaged part of the tool and its damage related to the rock type and intensity of use. In some cases the wear was ignored and heavily used tools continued to be used, while in other cases they were repaired, evidenced by traces of repair in form of sharpening, thinning ends, etc. In some cases the tip was remodeled only after serious damage, so that the tool could be used again. Such approach to repair of antler tools may have been linked to local traditions or specific conditions of the time (Figure 5).

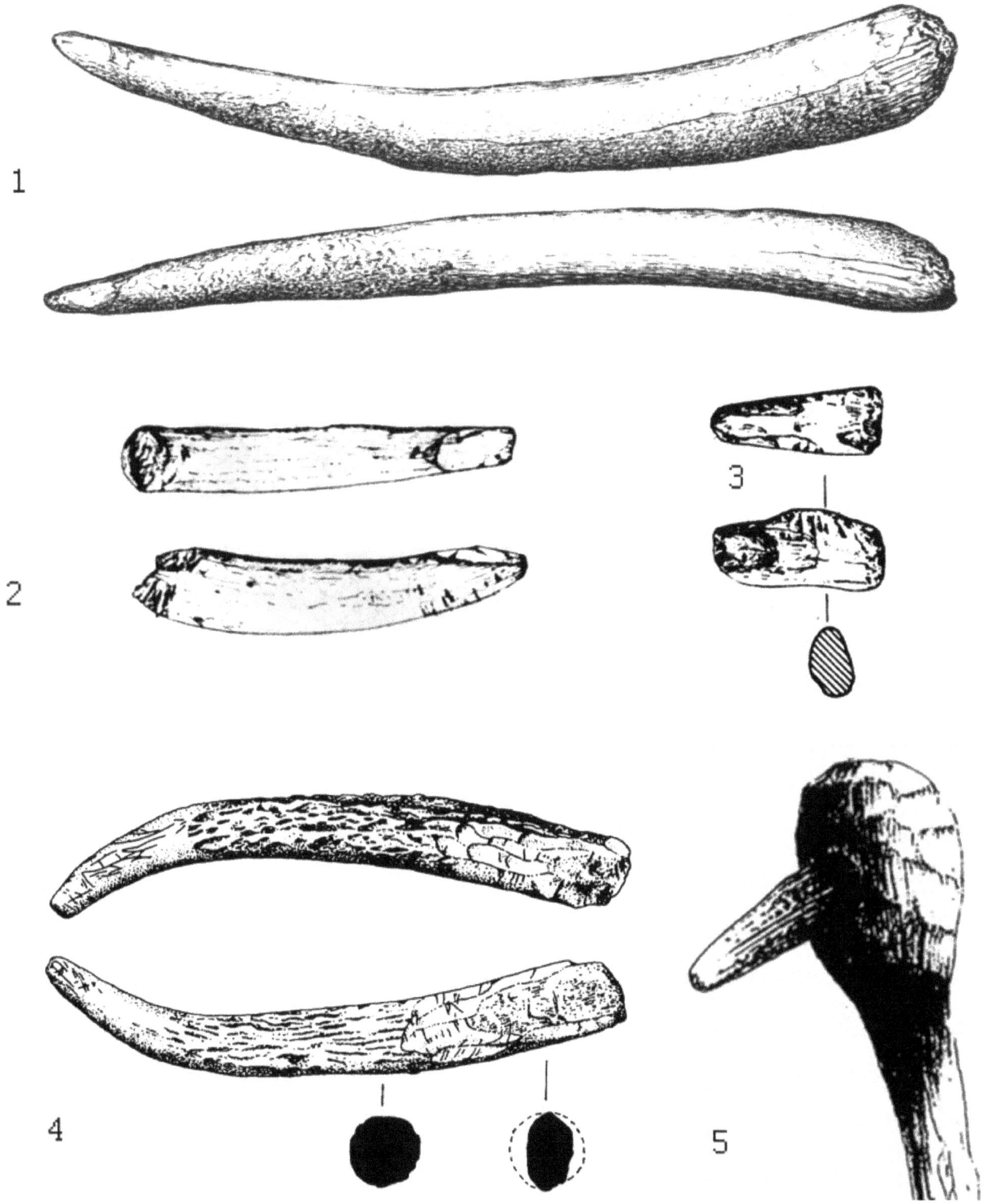

Figure 6. Antler tines. 1: tine working without handle (Polany-Kolonie, Poland; after Schild and Królik 1995); 2,3,4: tines working in a handle (2, 3: Krzemionki, Poland; 4: Ecton (cooper mine), England; after Boguszewski 1984, Barnatt and Thomas 1998).

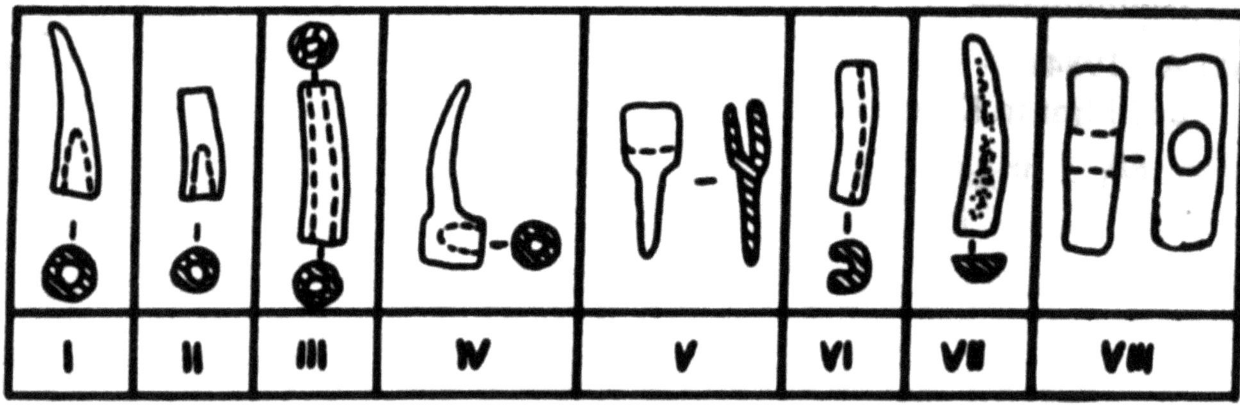

Figure. 7. Differents types of antler handles. (after Boguszewski 1991a).

4. Antler handles and antler tools in wooden handles

Both antler handles and antler tools in wooden handles are present in flint mines. Interestingly, distal ends of antler were used as tools hafted in wooden handles, while usable broken antler fragments still suitable for mounting in a handle were discarded in mines. It does not seem that their size and morphology may have been the reason for discarding the broken fragments. Is it then possible that a broken tool was a 'taboo'? Perhaps such tools could not been reutilized and must have been left in the mine. Antler tools and their fragments discarded in mines could attest the existence of specific rituals and customs related to underground extraction. There are clearly two sets of evidence that should be taken into consideration: 1) antler fragments prepared to be used as tools with wooden handles and 2) fragments, broken during mining but still usable if mounted in handles, which for some reasons have not been reutilized (Figure 6).

Although there are many antler handles found in mines, Borkowski's (1995) analysis of the relationship among the type of rock, shaft depth, and tools used suggests that tools with handles were not used in Krzemionki. Therefore, their presence in other mines requires explanation, as well as a more detailed examination of antler tool manufacture and maintenance and reasons behind miners' choice (Figure 7).

5. The angle between the working edge and the handle

The angle between the working edge (active part) and the handle (passive part) has to do with the relationship between the tool's tip that strikes the rock and the part held by the miner (Figure 8). The physics of this action is simple: the angle of strike and force used determine the result, i.e. the quantity of crushed rock. Therefore, the tool's shape is critical for productivity of mining activities. Following this assumption, we conclude that the morphology of antler tools suggest their (optimal) use. The Neolithic miners established their tool-kit according to planned activities. A closer examination of antler mining tools reveals a variety of angles between working edge and handle (Boguszewski 1991). It might further suggest different methods of use. It seems evident that despite morphological similarities tools were used differently.

6. Antler hammers

The basic rule in extraction of flint imbedded in rocks is that the tool (chisel) transmitting the kinetic energy from the human muscle through the hammer is harder than the hammer itself (commonly known as indirect percussion, 'percussion posée' in French (see e.g. Tixier *et al.* 1980; Bertouille 1989; Inizan *et al.*. 1995). This rule is followed in works requiring precision. Logically, hammers can not be of the same material as chisels. Analyzing hammers found in flint mines we may indicate the type of materials used to make chisels. This type of analysis also contributes to a better understanding of the mining tools and techniques to extract flint.

7. Conclusions

Godoy (1985) analyzed human behavior related to mining activities and suggested three phases of the mining process that should be considered (i.e. exploration, development and production) in order to understand such actions. His model was tested by Erskens *et al.* (2009) using evidence from the Nazca Region, Peru, and Knapp (1998) and Knapp and Pigott (1997), whose studies also addressed the economic aspect of mining. In our approach we limit our discussion only to the first phase (exploration) and suggest other, not strictly economic, explanations for certain behaviors in flint extraction.

As briefly discussed, the process of making antler mining tools requires time and energy investments. But antler tools were frequently damaged and many discarded fragments have been found in mines along with other debris of flint extraction. Interestingly, numerous discarded antler tools

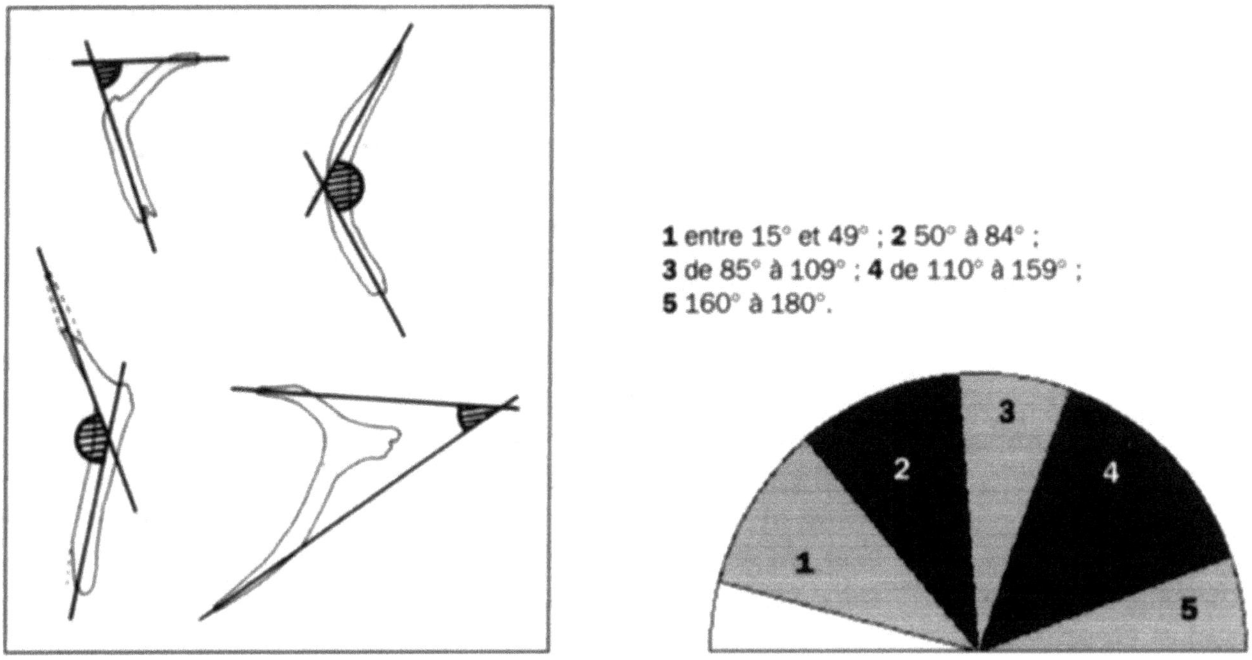

Figure. 8. Relationship between the active and passive part of the mining tool). (after Boguszewski 1991a).

found in abandoned mines were still in usable conditions. Apparently, despite their laborious production process, usable tools were discarded in abandoned shafts. Such evidence suggests that besides technological aspects of flint mining, which seems to be well recorded, there also was a ritualistic facet. Similarly to rituals associated with hunting and farming, there were rituals in mining. Perhaps discarding of usable tools in places of their use may also be linked to a ritualistic behavior. Such finds are common and worth further investigation in order to better understand such 'wasteful behavior' in flint mining. Not all miners behaved in such a way. At Aalborg and Chow in Denmark traces of the use of antler tools are well visible but none was discarded there (Becker 1951). Perhaps those miners favored pragmatism over superstitious ways and saved tools? However, in those cases considered throughout this paper, usable tools were discarded in mines. While not ruling out the possibility of accidental discards, we suggest that the discarding of usable tools may have been associated with ritualistic behavior.

References

Barnatt, J. and Thomas, G. H. 1998. Prehistoric mining at Ecton, Staffordshire: a dated antler tool and its context. *The Bulletin of the Peak District Mibes Historical Society* 13 (5), 72-78.

Bąbel, J. T. 2008. The Krzemionki flint mines. Latest underground research 2001-2004, in P. Allard, F. Bostyn, F. Giligny and J. Lech (eds.), *Flint mining in prehistoric Europe*, 97-109. Oxford, British Archaeological Reports International Series 1891.

Bácskay, E. 1980. Zum Stand der Erforschung prähistorischer Feuersteinbergbäu in Ungarn, in G. Weisgerber, R. Slotta and J. Weiner (eds.). *5000 Jahre Feuersteinbergbau. Die Suche nach dem Stahl der Steinzeit*, 179-182. Bochum, Veröffentlichungen aus dem Deutchen Bergbau-Museum 22.

Bácskay, E. 1995. Flint mine of Sümeg-Mogyorosdomb. Catalogue of flint mines : Hungary. *Archaelogia Polona* 33, 383-395.

Becker, C. J. 1951. Late Neolithic Flint Mines at Aalborg. *Acta AchaeologicA XXII* : 135-152.

Bertouille, H. 1989. *Théories physiques et mathématiques de la taille des outils*. Paris, Cahiers du Quaternaire 15, Centre National de la Recherche Scientifique.

Billamboz, A. 1977. L'industrie en bois de cerf en Franche-Comte au Néolithique et au début de l'Âge du Bronze. *Gallia Préhistoire* XX (1), 91-176.

Biró, K. 1995. Szentagal-Tüzköveshegy, Veszprem county. Catalogue of flint mines : Hungary. *Archaelogia Polona* 33, 402-408.

Boguszewski, A. 1984. Antler tools from the Neolithic Period and Early Bronze Age flint mine at Krzemionki. *Wiadomosci Archeologiczne* XLIX (2), 197-231.

Boguszewski, A. 1984a. Horn tools from the early Bronze Age flint mine in Krasne Sielo on the Roś river. *Wiadomosci Archeologiczne* XLIX (2), 233-242.

Boguszewski, A. 1991. Experimental use of antler tools in flint mine in Krzemionki. In *Archéologie Expérimentale*, t. 2, La terre : l'os et la pierre, la maison et les champs. Actes du Colloque International 'Expérimentation en archéologie: Bilan et perspectives', 46-48. Paris, Ed. Errance.

Boguszewski, A. 1991a. Horn mining tools typology, the proposition of the method and terminology. In *VI International Flint Symposium. Abstracts. October 1991*, 131-135. Madrid, Instituto Tecnológico GeoMinero de España.

Bostyn, F., Lanchon, Y. Boguszewski, A., Frugier, C., Jérémie, S., Laporte, L., Vacher, S. and Valero, C. 1992. Jablines "Le Haut Château". Une minière de silex au Néolithique, in F. Bostyn, and Y. Lanchon (eds.), 114 -116. París, DAF.

Böckner, G. 1980. Geweihgezâhe neolitischer Silexabbaulanger am Beispiel Löwenbourg-Neumühlefeld III. Ein Betraig zur Methode, in G. Weisgerber, R. Slotta and J. Weiner (eds.). *5000 Jahre Feuersteinbergbau. Die Suche nach dem Stahl der Steinzeit*, 48-66. Bochum, Veröffentlichungen aus dem Deutchen Bergbau-Museum 22.

Borkowski, W. 1995. *Krzemionki Mining Complex. Deposit management system*, 101-124. Warsaw, Studia nad gospodarką surowcami krzemiennymi w pradziejach, 2. Panstwowe Muzeum Archeologiczne.

Clarke, W.G. 1915. *Report on the excavations at Grime's Graves, Weeting, Norfolk, March-May, 1914*, 142-146. London, H.K. Lewis.

Clason, A.T. 1979. The Flintminers as a Farmer, Hunter and Antler Collector. III Symposium International du Silex. *Staringia* 6, 13-125.

Cyrek, K. 1995. On the distribution of chocolate flint in the Late Mesolithic of the Vistula basin, in P. Allard, F. Bostyn, F. Giligny and J. Lech (eds.), *Flint mining in prehistoric Europe*, 99-109. Oxford, British Archaeological Reports International Series 1891.

Desloges, J. 1986. Fouilles de mine à silex sur le site néolithique de Breteville-le-Rabet (Calvados). *Revue Archéologique de l'Ouest. Supplémént* 1, 73-101.

Di Lernia, S., Fiorentino, G., Galiberti, A. and Basili, R. 1995. The Early Neolithic mine of Defensola «A» (I 18): flint exploitation in Gargano area. *Archeologia Polona* 33, 119-132.

Erskens, W.E., Vaughn, K. J. and Grados, M. L. 2009. Pre-Inca mining and in the Southern Nasca region, Peru. *Antiquity* 83, 738-750.

Fülöp, J. 1975. Relics of prahistoric flint mining in Hungary. In Tweede Internat. Symosium over Vuursteen, Maastricht. *Staringia* 3, 72-77.

Godoy, R. 1985. Mining: Anthropological Perspectives. *Annual Review of Anthropology* 14, 199-217.

Holgate, R. 1995. Neolithic flint mining in Britain. *Archaelogia Polona* 33, 133-161.

Inizan, M.L., Reduron-Bachinger, M., Roche, H. and Tixier, J.1995. Technologie de la pierre taillée, vol 4, Préhistoire de la pierre taillée. Meudon, Cercle de Recherche et d'Études Préhistoriques, CNR.

Knapp, A.B. 1998. Social approaches to the archaeology and anthropology of mining, in A.B.

Knapp, V.C. Pigott and E. W. Herbert (eds.), *Social approaches to an industrial past: the archaeology and anthropology of mining*, 1-23. London, Routlege.

Knapp, A. B. and Pigott, V. C. 1997. The archaeology and anthropology of mining: social approaches to an industrial past. *Current Anthropology* 38, 300-304.

MacGregor, A. 1985. Bone, Antler, Ivory and Horn. The Technology of skeletal materials since Roman Period. London and Sydney, Barnes and Noble.

Migal, W. 1990. Organization and planning of flint mining operations in Krzemionki, Poland. In *Le silex de sa genèse à l'outil. Actes du Ve Colloque International sur le Silex*, 197-199. París, Cahiers du Quaternaire 17, CNR.

Pull, J. 1932. *The flintminers of Blackpatch*. London, Williams and Norgate.

Schild, R., Królik, H. and Marczak, M. 1985. Kopalnia krzemienia czekoladowego w Tomaszowie. Wrocław, Ossolineum.

Schild, R. And Królik, H. 1995. Polany-Kolonie II. Catalogue of flint mines: Poland. *Archeologia Polona* 33, 486.

Sidéra, I. 1995. Relations minières/habitat : un problème de méthode le potentiel des artefacts osseux. In J. Pelegrin y A. Richard (eds.), *Les Mines de silex au Néolithique en Europe: avancées récentes*. Actes de la table-ronde internationale de Vesoul (1991), 115-134. París, Comite Des Travaux Historiques et Scientifiques.

Sieveking, G.de G. 1978. Grime's Graves and european flint mining. In H. E. W. Crawford (ed)., *Subterranean Britain: Aspect of underground archaeology*, 87-90. London, Palgrave Macmillan.

Soulier, P. 1971. *L'Extraction du silex en Europe occidentale*. Maîtrise, Université de Paris I.

Tixier, J., M.-L. Inizan, H. Roche 1980. *Préhistoire de la pierre taillée*, vol 1, Terminologie et technologie. Valbonne, Cercle de Recherche et d'Études Préhistoriques.

Valde-Nowak, P. 1991. Studies in Pleistocene settlement in Polish Carpathians. *Antiquity* 65/248, 593-606.

Vermeersch, P.M., Paulissen, E., Stokes, S. And Van Peer, P. 1997. Middle Palaeolithic chert mining in Egypt. In A. Ramos-Millan and M.A. Bustillo (eds.), *Siliceous rocks and Culture*, VI Flint International Symposium – Madrid 1991, 173-194. Granada, Universidad de Granada.

Zalewski, M. 1988. Wstępne wyniki badań przeprowadzonych w rejonie leja krasowego w Krzemionkach. *Sprawozdania Archeologiczne* XL, 107-112.

Problem of the flint tools from the Sąspów mine site in the light of use-wear analysis

Jolanta MAŁECKA-KUKAWKA

Abstract

Retouched tools of a household character are rarely found in flint mine workshops. The 2008 report of flint material recovered in the course of the 1962 excavation of the flint mines in Sąspów singled out some few hundred tools, suggesting them to be of a household character. Exhibiting various types of retouch, these tools were said to be functional. This paper presents the results of a use-wear analysis of 91 of this most characteristic tool forms. The analysis has demonstrated that 80 (i.e. the majority) of the specimens analyzed under a microscope are pseudotools, containing no use-wear traces. Retouch observed on edges in many cases was performed in modern times, during and after the excavations. Only a few of these pseudo-retouches were performed in prehistoric times in the course of flint exploitation. However, clear traces of usage were identified on 11 specimens, in most cases these were employed for scraping weathered clay off of mined flint concretions. Single tools are of a household character.

Keywords

Use-wear analysis. Flint tools. Functional tools. Neolithic mine site in Sąspów. Jurassic Krakow flint.

1. Introduction

One of the most critical questions for the scholars of archaeology today is the relationship between motives and actions of people in the past and the material record of their lives uncovered during the excavation. Answering this question would be impossible without establishing suitable research procedures, incorporating current state of knowledge about a specific subject, appropriate methods of analysis and conclusions. The source critique is an important component of such study procedures, a stage that also incorporates various levels of analysis—evaluation of the quality of excavation based on the produced documentation and validity of formed conclusions concerning chronology, culture affiliation or function of uncovered layout of sources (their homogeneity). I would call this stage t*he critique of context of discovery*.

To an archaeologist the past is revealed through material products and remains uncovered at sites. But, in what way do modern events subsequent to the discovery of a site, followed by the excavation, transportation, washing, recording, storage effect physical characteristics of uncovered archaeological sources, thus considerably influencing quality/validity of formed conclusions about prehistoric processes? For the scholars of flint material this critical question, which I call *the critique of grounds context* and illustrate in this article, is particularly important.

2. Of the history of excavations of the flint mine in Sąspów

The history of flint material from the flint mine in Sąspów is long and abundant in all sorts of events. The events I have in mind are not the ones from distant past which produced the layout and character of sources found by archaeologists, but the contemporary ones—following the discovery of the site and its excavation.

In 1976 Anna Dzieduszycka-Machnikowa and Jacek Lech published a monograph *The Neolithic workshop assemblages from the flint mine of Sąspów*, which analyzed the flint material recovered during the first of three excavation seasons of 1960, 1962, and 1971-72. The study of materials from the year 1962 was written in 2008, 46 years after the

Figure 2. Examples of pseudo-tools from flint mine in Sąspów.

excavation. This difficult task was undertaken by a scholar from Jagiellonian University in Krakow Agnieszka Klimek, resulting in a monograph entitled *Uporządkowanie i wstępna klasyfikacja morfologiczna z elementami analizy nakopalnianych materiałów krzemiennych z badań mgr Anny Dzieduszyckiej-Machnikowej na stanowisku I w Sąspowie, pow. Kraków, w 1962 r.*

In the introduction the author wrote:

> 'Documentation of excavation conducted in year 1962 is very modest (...) The only information concerning the linkage of studied material with archaeological features are card records kept in wooden boxes. Unfortunately some of them were lost to small rodents and dampness, as between October 1962 and July 1969 the boxes were stored in a small wooden shack erected next to one of the warehouses of the Archaeological Workshop of ZAM IHKM PAN in Igołomia, the so-called "Stary Barak" (Old Shed). At the end of July 1969, the entire stored material from Sąspów site I was transferred to Lower Rotunda and Preparatory. In 1971 materials from excavation season 1962 were further transferred to the attic of the Palace.
>
> The wooden boxes used for storage of objects from 1962 excavations were recycled packaging and in vast majority unfit for transportation and storage of flint (...). Flint material was placed in boxes without any additional packaging (not even paper wrapping) usually filling boxes up to the rim. Transportation of boxes filled in this manner along bumpy roads of Sąspów caused formation of accidental retouch on flint, similar to the kind produced by washing them three times (first in Archaeological Museum in Krakow directly after excavation, and than in years 1969 and 2008 in Igołomia). Some of the boxes fell apart already in 1969. In such cases the material was collected and placed in stronger boxes emptied during study of material from assemblage of pits 1 and 3/1960. When material became mixed both card records were placed in the boxes. More boxes fell apart in following years in the attic of the Palace. When material spilled out of the boxes stored on piles more accidental retouch could have been formed on edges of some objects'.

Agnieszka Klimek was obviously fully aware of the affects of all sorts of after-excavation events on the archaeological material, and thus her investigation fulfilled the requirement of the thorough critique of sources (i. e. *grounds context critique*).

3. Problem of flint tools in materials from the flint mine in Sąspów

The study conducted by Agnieszka Klimek treated 16,982 flint specimens, whose total weight reached almost 2 tons. In the study we find a comment:

> '(...) the author has noted presence of additional forms, which she was willing to accept to be bulky, irregular mining tool forms, whose random shape and carelessness of manufacture makes their attribution disputable in some cases. It is the case for the forms described by the author as "side scrapers" (10 examples) and "mining end scrapers" (9 examples), characterized by irregular retouch covering considerable parts of the edges, thick flakes with irregular large niche-tooth retouch, usually accepted to be the attributes of "side scrapers" (they were supposedly used for removal of eluvial clay of the surface of excavated flint).
>
> Considering the case of the mentioned "tools", the way in which they were acquired as well as stored should be taken into account, as well as the character of the deposit itself. It cannot be excluded, that the "retouch" observed by the author is of accidental character, and was formed as a result of interactional chipping of material both during the piling on the surface of Sąspów mine, and during transport, washing and storage after excavation. Yet in some cases, it seems that the "retouch" is too regular and "logical" to eliminate its intentionality and tool-like character. This matter no doubt requires further study with the use of traceological analysis in case of functional tools, which is not the purpose of this study (...)'.

In response to her suggestion and for reasons mentioned above, 91 tools of the most distinctive character, selected by Klimek, were send to the Traceological Laboratory in Toruń for microscopic use-wear analysis.

4. Use-wear analysis results

The Institute of Archaeology of the Nicolaus Copernicus University in Toruń has been performing microscopic analyses for 15 years. Currently it employs Zeiss Axiotech optical microscope, with an objective magnification 10x, 20x and 50x, as well as a binocular Nikon microscope with a fluently altered magnification from 2 to 12.6x. So far, this kind of analysis has been performed primarily on various flint objects from settlement sites (see Małecka-Kukawka 2001). Materials from the Sąspów flint mines constitute the first collection of this kind to be subjected to microscopic analysis.

This type of analysis presents a number of challenges. Firstly, the usefulness of traceological examination is not explicit. Secondly, the analysis of materials of considerable size under a microscope can be problematic, considering the limitation of microscope construction. Nevertheless, practice has shown that microscopic observations, including those defining the tools function, can be made using a Nikon binocular microscope. The big problem was production of good quality microphotographs. As a rule they are produced with Zeiss Axiotech microscope. Unfortunately, most products from Sąspów mine were too large to fit between the stage and the objective of the microscope.

The results of the microscopic analysis of selected Sąspów materials are presented in Figure 1.

Figure 3. Example of a pseudo-tool formed in modern times (Figure 1: 30).

5. Conclusions

The microscopic analysis of specimens from Sąspów showed that the absolute majority of the selected products, defined as tools, consciously created and probably used by humans in prehistory, are pseudo-tools. Pseudo-retouch was formed accidentally. In some cases it was done in the past, during work involved in the extraction and initial processing of flint. However, in most cases (80 of 91) the pseudo-retouch is modern (Figure 2).

Of all the material analyzed, 11 specimens have evident (visible under a microscope) traces of intentional use. These specimens are functional tools, used by humans to perform some actions – they have an edge, edges or faces, defined by us as 'working'. Most of them were identified as side scrapers for strongly abrasive material (possibly for scraping eluvial clay off of flint), they have shining polish along smoothed edges and sometimes legible striations.

Undoubtedly particularly noteworthy are the functional tools, not directly related to mining and processing of flint. Although their number is small, their presence demonstrates that, in addition to the extraction and initial processing of flint, a variety of activities related to everyday life was performed in the workshops. Therefore, traceological analysis of selected

Figure 4. Pseudo-tool with modern retouch recognized as an initial bifacial tool (Figure 1: 47).

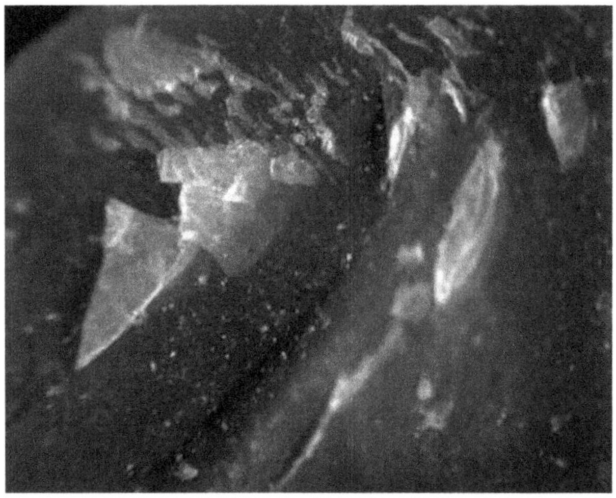

Figure 5. Microphotography of modern damages on flint recognized to be a morphological tool (Figure 1: 33).

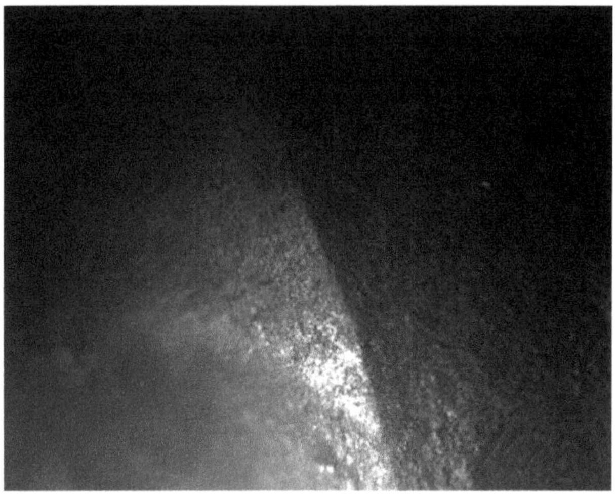

Figure 6a. Side scraper for abrasive material (clay?), negative sides are polished on the arrises (Figure 1: 45).

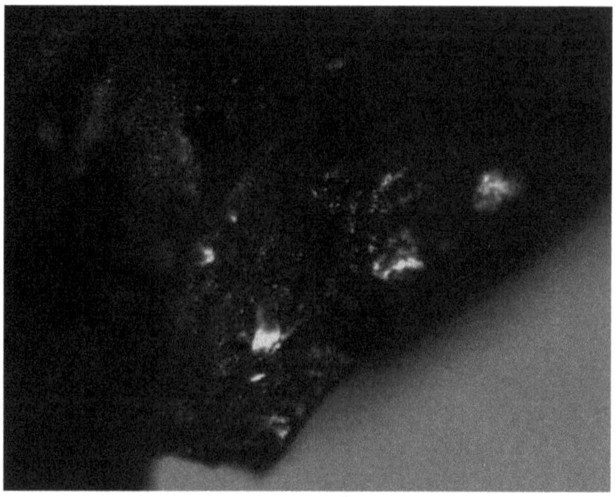

Figure 6b. Microphotography of polish inside negatives on an abrasive material side scraper (clay?) (Figure 1: 45).

materials from mines and accompanying workshop is a worthwhile enterprise, providing us with additional clues about quotidian human activities not directly related to flint mining.

The conducted analysis of selected materials from the workshops of Jurassic Krakow flint mines in Sąspów showed that the applied microscopic observation can potentially guide us away from drawing false conclusions or distorting the real nature of analyzed material.

References

Dzieduszycka-Machnikowa, A. and Lech, J. 1976. *Neolityczne zespoły pracowniane z kopalni krzemienia w Sąspowie* [*The Neolithic workshop assemblages from the flint mine of Sąspów*]. Polskie Badania Archeologiczne 19, Wrocław-Warszawa-Kraków-Gdańsk, Zakład Narodowy imienia Ossolińskich.

Klimek A. 2008. *Uporządkowanie i wstępna klasyfikacja morfologiczna z elementami analizy nakopalnianych materiałów krzemiennych z badań mgr Anny Dzieduszyckiej-Machnikowej na stanowisku I w Sąspowie, pow. Kraków, w 1962 r.* [*Alignment and preliminary morphological classification, with elements of analysis, of flint material of mine's workshop from excavation of Anna Dzieduszycka-Machnikowa on site I in Sąspów, Krakow district, year 1962*]. Unpublished.

Małecka-Kukawka, J. 2001. *Między formą a funkcją. Traseologia neolitycznych zabytków krzemiennych z ziemi chełmińskiej* [*Between form and function. Traceological analysis of the Neolithic flint assemblages from Chełmno Land*]. Uniwersytetu Mikołaja Kopernika, Toruń. Wydawnictwo.

Figure 1. Morphological structure of the flint assemblages from Sąspów (tools after analysis by A. Klimek) and results of the microwear analysis.

No.	inventory No.	classification categories (after A. Klimek)	classification categories (after J. Małecka-Kukawka)	Size, length x width (in mm)	function	comments
1	24	selected tool	flake	113x29	no traces of use	modern retouch on the edge
2	24	selected tool	debris	123x33	no traces of use	intentional retouch – partially modern
3	24	selected tool	debris	67x15	no traces of use	
4	26	supposed tools	flake	37x9	no traces of use	
5	18	initial form of a bifacial tool	debris	72x20	no traces of use	
6	26	side scraper	flake	63x23	no traces of use	pseudo-retouch on the edge – partially modern
7	26	Corbiac burin	blade with a fractured apex part	54x24x9	no traces of use	modern retouch
8	27	selected object	flake	107x50	no traces of use	spontaneous niche – prehistoric, within the niche another, smaller one – modern
9	27	as above	flake	47x17	no traces of use	modern retouch
10	27	as above	single striking platform microlithic blade core	length 29	no traces of use	
11	27	as above	flake	121x32	no traces of use	modern grinds/crush outs
12	13	as above	technical flake with a pseudo end scraper front	80x33	no traces of use	retouch is partially prehistoric, partially modern, no use-wear traces
13	13	as above	flake	76x29	no traces of use	
14	26	as above	debris	73x25	no traces of use	cracked in high temperature, part of the edge's damage is modern
15	26	as above	flake	60x15	no traces of use	
16	26	as above	flake	104x28	no traces of use	modern retouch
17	26	as above	residual core	61	no traces of use	
18	26	side scraper	side scraper	109x28	side scraper for wood – on a fragment of the left edge	part of the edge damaged in modern
19	22	selected tools	flake side scraper	42x18	no traces of use	fine edge neighbouring retouch – prehistoric, smoothed edges and arrises, dip, notch-like scale negatives – modern, areata postdepositional polish on the ventral side
20	22	as above	debris	48x14	no traces of use	
21	22	as above	debris	65x23	no traces of use	large flake negative on the ventral side – modern
22	22	as above	residual core?	61	no traces of use	
23	22	as above	flake	112x25	no traces of use	modern pseudo-retouch on one edge
24	8	initial form of a bifacial tool?	debris	66x24	no traces of use	modern pseudo-retouch
25	26	truncated blade-trapeze	blade's mesial fragment	31x19x4	no traces of use	pseudo-retouch on a perpendicular edge – modern

26	26	?	whole blade	58x20x7	no traces of use	modern pseudo-retouch
27	29	side scraper	debris	85x27	no traces of use	cracked in high temperature, small scales fell off during transportation? modern retouch
28	13	bifacial tool	debris	83x26	no traces of use	modern pseudo-retouch
29	63	selected object	flake	84x19	no traces of use	modern pseudo-retouch
30	63	as above	flake	69x16	no traces of use	modern pseudo-retouch (Figure 3)
31	48	selected object	debris	79x29	no traces of use	modern pseudo-retouch
32	48	as above	initial splinter core?	33x12	no traces of use	prehistoric retouch
33	35	as above	flake	54x14	no traces of use	modern pseudo-retouch, jagged, irregular edge, arrises not smoothed (Figure 5)
34	35	as above	debris, heavily cracked in high temperature	74x23	no traces of use	pseudo-retouch – modern niche
35	35	as above	flake	109x19	no traces of use	modern pseudo-retouch, more matt then the face, arrises not smoothed
36	64	as above	flake cracked in high temperature	130x23	no traces of use	pseudo-retouch, possibly prehistoric
37	66 ?	as above	flake	152x45	no traces of use	one edge is crashed (jagged) but not as a result of use
38	28	supposed sickle blade inset	whole blade	40x14x5	no traces of use	Shiny face, visible with a naked eye – crystalline inclusions in silica, no traces of use
39	28	as above	debris	52x17	no traces of use	
40	18	selected tools	flake	71x26	side scraper for hard wood/bone, antler	
41	22	as above	flake	94x20	no traces of use	on the edges there are visible scale negatives, formed partially in prehistory, arrises smoothed down, there are visible grate marks, some of which are overlapped by modern damage
42	22	as above	debris	34x15	no traces of use	prehistoric pseudo-retouch
43	22	as above	debris	37x11	a fragment of a whittling knife	
44	22	as above	debris	46x19	no traces of use	single scale negatives on the edge – prehistoric, on them overlapping modern damages
45	18	retouched blade	apex part of a technical blade	63x26x17	side scraper for abrasive material (clay ?)	visible polish on parts of working edges, insides of scale negatives lack of traces of use, but the negative sides are polished on arrises, on the faces there are visible chaotic flat patches of polish, possible it could be a trace of contact with stone during use (Figure 6a, 6b)

46	23	selected object	retouched flake	101x28	side scraper for abrasive material	as above
47	28	initial form of a bifacial tool	debris (core fragment)	80x26	no traces of use	modern pseudo-retouch (Figure 4)
48	32	selected tools	flake	79x22	no traces of use	pseudo-retouch, partially modern
49	29	selected object	debris	68x26	no traces of use	nothing
50	40		debris	41x13	no traces of use	one edge with a prehistoric retouch
51	40		debris with a single retouched edge	67x35	side scraper for abrasive material	comments as position No. 45
52	54		core tablet	68x15	no traces of use	pseudo-retouch, partially modern
53	38	selected object	apex part of a blade	55x31x8	no traces of use	pseudo-retouch on a fragment of the edge
54	38	as above	debris	51x20	no traces of use	pseudo-retouch and glossy postde-positional damage
55	55		retouched flake	31x9	no traces of use	prehistoric accidental retouch, overlapped by modern negatives
56	55		apex part of a blade	21x21x5	no traces of use	scale negatives on the edge, seem to be of utilitarian character and are a result of a low angle of edges sharpening, some negatives are of a different colour and texture then the patina covered face of the blade, probably modern
57	55		debris	125x36	no traces of use	pseudo-retouch on a fragment of an edge, partially prehistoric, partially modern, in some places visibly scales are almost loose
58	34	selected tools	flake	131x43	no traces of use	accidental retouch on one edge, on the other a modern one, areata postdepositional glossy polish on the face
59	26	supposed sickle blade insert	blade's mesial part	34x26x10	no traces of use	
60	26	as above	as above	37x23x7	no traces of use	
61	26	as above	blade with a fractured butt	42x17x9	no traces of use	
62	26	as above	whole blade	62x20x15`	no traces of use	Pseudo-retouch on one edge
63	26	as above	apex part of a blade	22x13x2	no traces of use	on the edges there are visible patch like, shiny polish areas, irregularly scattered along the edges, on the versal face also spots of polish, in a loose formation, most probably it is the result of undefined natural factors
64	17	3 tools, core, refitting	flake preform	62x19	no traces of use	
65	17	as above	flake preform	82x28	no traces of use	
66	17	as above	flake	63x19	no traces of use	
67	17	as above	debris	58x19	no traces of use	
68	17	as above	flake with nega-tives of single percussions	53x26	no traces of use	
69	17	as above	flake	73x27	no traces of use	on the face patch like and linear postdepositional polish areas, from contact with stones

70	36		flake	52x17	no traces of use	
71	36		flake	57x2	no traces of use	pseudo-retouch, areata postdepositional traces resulting of contact with rocks
72	36		debris	98x32	no traces of use	pseudo-retouch, partially modern
73	36		debris	90x40	no traces of use	modern pseudo-retouch
74	36		debris	68x23	no traces of use	pseudo-retouch, could be mistaken with intentional during macroscopic analysis, under the microscope lack of visible use-wear traces, the edge is rugged, chaotically jagged, in various places on the edge and faces there are visible patches of shiny polish
75	36		flake, with a partially parallel course of edges	80x22	side scraper for hard material (hard wood, bones, antler)	blunted and rounded working edge, polish and striations on the edge, traces visible on arrises
76	26	supposed tools	flake preform	66x25	no traces of use	pseudo-retouch – partially prehistoric, partially modern
77	50		debris, with a bifacial retouch of an edge	64x24	no traces of use	
78	37	selected tools	flake	46x9	no traces of use	
79	37	as above	debris	29x8	no traces of use	accidental retouch
80	37	as above	butt part of a blade	44x28x8	side scraper for abrasive material (pigments, ochre?)	on the working edge and neighbouring faces visible rust-red coloured patches, use-wear traces are visible on their face, which suggests that the patches are not the result of a postdepositional processes
81	37	as above	retouched blade	67x22x14	side scraper for abrasive material (clay?)	visible mirror like, flat polish areas on parts of working edges, insides of scale negatives lack of traces of use, but the negative sides are polished on arrises, on the faces there are visible chaotic flat patches of polish, possible these could be traces of contact with stone, or maybe of friction against organic setting during use
82	67		flake	61x25	no traces of use	pseudo-retouch, partially modern
83	26		blade with a fractured apex part	49x16x7	no traces of use	pseudo-retouch, partially modern
84	26		whole blade	76x24x11	probably a side scraper for wood	working edge largely ruined by younger, prehistoric and modern damages
85	26	traces of wear in medial part, above a notch, the apex could have been a shaft, knife-like tool	whole blade	71x20x9	possibly, used as a side scraper	traces are only just visible, niche and few other scale negatives are later, possibly modern

86	26	tool-like retouch, polish on both sides – associated with setting, no specific traces of use	debris	46x8	no traces of use	spontaneous niche retouch, in general the specimen has slightly smoothed down edges, these may seem use-wear traces alike, it is the effect of postdepositional processes
87	26	perforator, visible traces of polish on the tip of the apex	debris	43x17	no traces of use	generally the specimen's edges are lightly smoothed and polished, can seem like traces of use, caused by postdepositional processes
88	26	knife-like use-wear traces near edges, worked as a knife	blade with a fractured apex part	52x18x5	no traces of use	the specimen has thin delicate edges, fine scale negatives spread chaotically along the edges were not formed as a result of work
89	26	retouched truncated blade, knife-like use-wear traces near edges	blade with a fractured apex part	43x19x7	no traces of use	the specimen's edges are lightly smoothed and polished, can seem like traces of use, caused by postdepositional processes, in few places negatives caused by later, also modern damages
90	26	grinds near the apex, worked as a perforator – probably	apex part of a blade	49x25x6	no traces of use	accidental retouch on the edge, sharply ending tip suggests probable use in a function of a perforator, no unambiguous traces
91	26	as above	whole blade	85x24x14	no traces of use	accidental retouch on the edge, sharply ending tip suggests probable use in a function of a perforator, no unambiguous traces

Pedreira do Aires and Monte das Pedras: two Neolithic flint 'mines' in the Lisbon Peninsula

Marco António ANDRADE and Henrique MATIAS

Abstract

This paper describes the Neolithic sites of Pedreira do Aires and Monte das Pedras, both located in the Lisbon Peninsula, Portugal. The archaeological assemblages recovered from these sites - mostly lithic materials such as debitage debris, both 'tested' and shaped flint blocks, preparation flakes and core rejuvenation elements - and the geological context (Upper Cenomanian limestone rich in flint nodules), suggest that both should be interpreted as flint extraction localities (although not necessarily mines, the flint being recovered in a secondary position from detritic deposits) and occasional workshops oriented towards bladelet production. These sites can therefore be understood as small, seasonal campsites located within the sphere of influence of a larger settlement and part of a complex spatial use and resource exploitation network that lasted from the 5th to the 3rd millennia BC (the Neolithic and Chalcolithic of the region). Models of flint procurement must take into account that different types of site existed: settlements, funerary sites and resource procurement sites. The relationship with contextually similar sites highlights the strategies used in the exploitation of flint at this time in the Lisbon Peninsula.

Keywords

Flint exploitation. Flint procurement. Workshops. Neolithic. Chalcolithic. Lisbon Peninsula.

1. Geographic, geological and archaeological contexts of Pedreira do Aires and Monte das Pedras

The Neolithic sites of Pedreira do Aires and Monte das Pedras are located in the lower Lisbon Peninsula, in the area of influence of the Tagus estuary (Figure 1). They reflect different implantation models. The first site is located at the bottom of a gentle slope, on the bank of a stream, without good visibility over the surrounding area. The second is located at the top of a large platform bounded by the deep valleys of two confluent streams; visibility over the surrounding area is very good, and a visual relationship with synchronic settlements and funerary areas existed.

Geologically, both sites are located on a strip of Upper Cenomanian limestone (Cretaceous) separating the basalts of the 'Lisbon volcanic complex' from Albian-Lower and Middle Cenomanian limestone (Figure 2). Flint nodules are present in the exposed outcrops of this limestone strip. They are also found in a secondary position in adjacent debris levels and along the banks of streams. In the immediate vicinity of Pedreira do Aires and Monte das Pedras several sites of similar age existed, possibly part of the same settlement network.

The archaeological context of Pedreira do Aires includes several megalithic monuments (dolmens of Trigache 1 to 4, Pedras Grandes and Batalhas), several poorly defined Neolithic and Chalcolithic settlements (Castelo da Amoreira, Quinta do Castelo Nascente, Gaitadas and Casal do Murtal) and other possible flint 'mines'/workshops (Casal Novo and the Pedernais Cave). The archaeological context of Monte das Pedras includes several megalithic monuments of different typology (the rock-cut tombs of Carenque 1 to 4 and Baútas, the dolmens of Pego Longo, Monte Abraão, Estria, Pedra dos Mouros and Carrascal, and the *tholoi* of Agualva and Pedreira do Campo), several Neolithic and Chalcolithic settlements (Serra das Éguas, Espargueira, Baútas and Tojal de Vila Chã Norte) and another possible flint 'mine'/workshop (Moinhos da Funcheira).

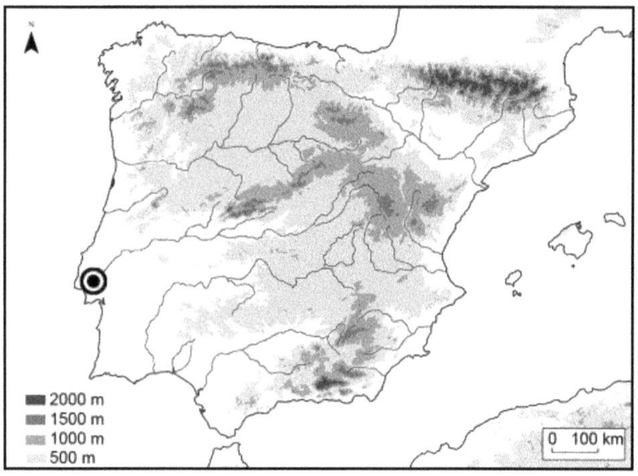

Figure 1. Location of Pedreira do Aires and Monte das Pedras in the geographical context of the Iberian Peninsula.

The above indicates that organized communities with well limited, functionally specialized areas of occupation (e.g., for settlement, burial or resource procurement) existed between the Late Neolithic (identified by dented-rim and carenated bowls found in settlements and by engraved schist plaques at funerary sites) and the Late Chalcolithic (recognized by the presence of bell-beaker pottery at settlements and funerary sites). The Early Neolithic of the area is poorly known and is currently represented only by a few shards of pottery decorated with incised or impressed motifs and a few geometric microliths found at the settlements of Espargueira and Baútas. Nonetheless, an important settlement from this time, Zibreira, is located 3km north of Monte das Pedras.

2. Context of archaeological data recovery and of the studied assemblages

The Pedreira do Aires and Monte das Pedras sites find themselves in an advanced state of destruction, which seriously undermines their potential for archaeological interpretation. Recent excavations have detected no *in situ*-preserved contexts.

The analysis of the recovered assemblages and their geological contexts show both sites were dedicated to a specific activity during the Neolithic and (possibly) the Chalcolithic: the exploitation, extraction and procurement of siliceous raw materials, with the occasional production of knapped artifacts. Lithic materials are abundant at both sites, corresponding to about 98% of all archaeological finds and characterized by the presence of:

1) exploitation and debitage debris (preparation flakes reflecting the first stages of prismatic debitage).

2) shaped flint blocks and their respective cortex removal flakes.

3) core pre-forms (flint blocks from which the cortex has been totally or partially removed).

4) tool pre-forms (unfinished tools).

5) prismatic cores abandoned in the full debitage stage (not exhausted).

6) core preparation and rejuvenation elements (cortical and partly cortical flakes, flanks, tablets).

7) finished tools (few in number compared to the number of debitage debris and rejuvenation elements).

Unworked and 'tested' flint nodules and cobbles (with less than three flakes extracted) along with cortex removal flakes are abundant at both sites. This abundance prevented the systematic sampling of these items; they are therefore not considered in the statistical analysis (Figure 3). The discarding of such blocks after experimentation is possibly due to their poor quality (presence of geodes and cleavages). This problem also affected later phases of production; the majority of prismatic cores show scars of hinged removals (possibly the reason for their being abandoned).

The recovered tools include retouched blades, bladelets and flakes, as well as scrapers, notches, denticulates and perforators. An arrowhead pre-form was also found at Pedreira do Aires. The presence of these artifacts could be the result of *in loco* knapping (confirming the vocation of both sites as workshops, even if only occasionally). Alternatively, they may simply represent the discarding of tools made elsewhere. At Monte das Pedras, cortical or partly cortical flakes were used (albeit not exclusively) as the blanks for tools (mostly retouched flakes, scrapers and denticulates), suggesting a rigorous economy of raw material use.

Bladelet production is attested to at both sites by the presence of prismatic bladelet cores and core rejuvenation elements with bladelet extraction negatives, as well as by tool blanks (Figure 4). These were mostly extracted using pressure flaking with heat pre-treatment of the cores intended to facilitate the knapping process, confirming the Neolithic chronology of both sites. The presence of crested bladelets and cortical/partly cortical bladelets indicates that the first stages of the production of these artifacts took place at the extraction sites. However, several exhausted bladelet cores indicate that the subsequent stages of the reduction sequence was also undertaken in these places.

The colour of most Pedreira do Aires and Monte das Pedras flints (blocks, cores and artifacts) varies between light/dark grey and greyish green, typical of Upper Cenomanian geological contexts. Flints of different colour (white, pinkish red, reddish brown and yellowish green) are also present, although none is inconsistent with the geological source (Almeida, Araújo and Aubry 2003). Moreover, at

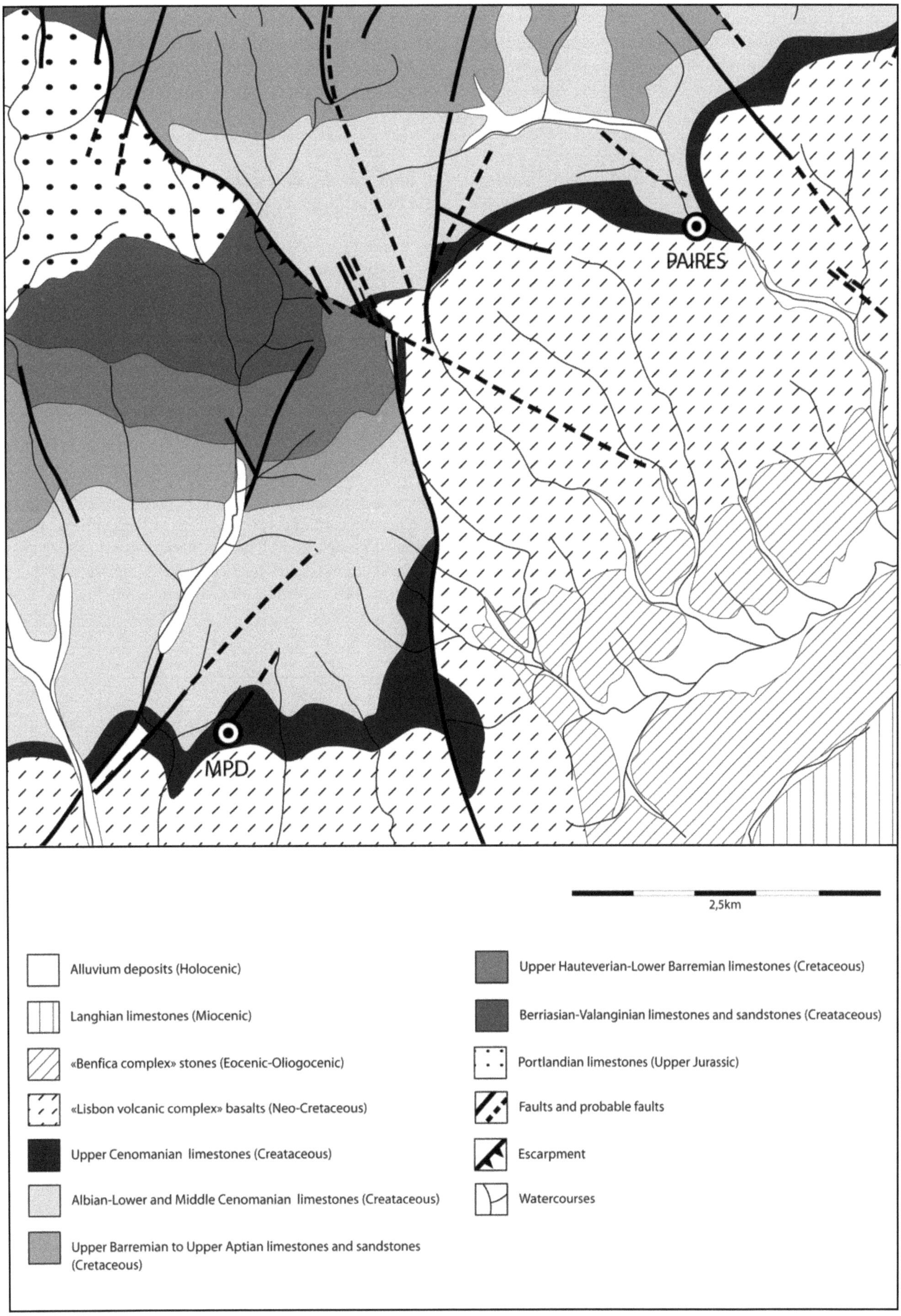

Figure 2. Geological context of Pedreira do Aires (PAIRES) and Monte das Pedras (MPD).

both sites flint gradually darkens from the surface to the core, a consequence of weathering. Therefore, differently coloured flints may not only have come from the same source, they may even have belonged to the same original blocks.

Tools related to the extraction of raw materials and the production of artifacts have also been recovered. These include a hammerstone/anvil made from a quartzite cobble from Pedreira do Aires that features a characteristic dome-shaped depression created by percussion (used in bipolar core debitage), as well percussion marks on its edges.

Pottery is represented by a few, mostly uncharacteristic shards. The only element with a clear chronological attribution is the fragment of a dented-rim bowl recovered at Monte das Pedras, which is typical of the Late Neolithic of Portuguese Estremadura (*c.* 3300-2900 BC).

The above data suggest that both sites functioned as flint procurement localities and occasional workshops over an extended, although not necessarily continuous, period of time. This period may have begun in the later part of the Early Neolithic, as indicated by the presence of lithic artifacts characteristic of this period (small sized blades, bladelets and perforators made from flakes), and continued until at least the Late Neolithic (as indicated by the presence of large blades and dented-rim bowls) - a period of some 1600 years (4500-2900 BC). Chalcolithic use can be hypothesized given nearby settlement evidence, but at present this is not supported by diagnostic finds.

3. Flint procurement at Pedreira do Aires and Monte das Pedras: primary position vs secondary position

As mentioned above, both sites are located on a strip of Upper Cenomanian limestone where flint can be found in a primary position embedded in the rock matrix, as well as in secondary positions on a slope and in the alluvial deposits of the valley bottoms (Figure 5). In Portugal, there are countless geological formations with rocks appropriate for knapping (Portuguese Geological Chart, *Instituto Geológico e Mineiro*) although most are found in a secondary position in debris deposits (Almeida, Araújo and Aubry 2003).

Simple macroscopic observation of an artifact's cortex can indicate the kind of source that was exploited. Most of the knapped blocks, as well as the blanks, show a rolled cortex with characteristic impact marks made during fluvial transport. The preferred source for collecting flint was therefore secondary deposits on river terraces. Some of the

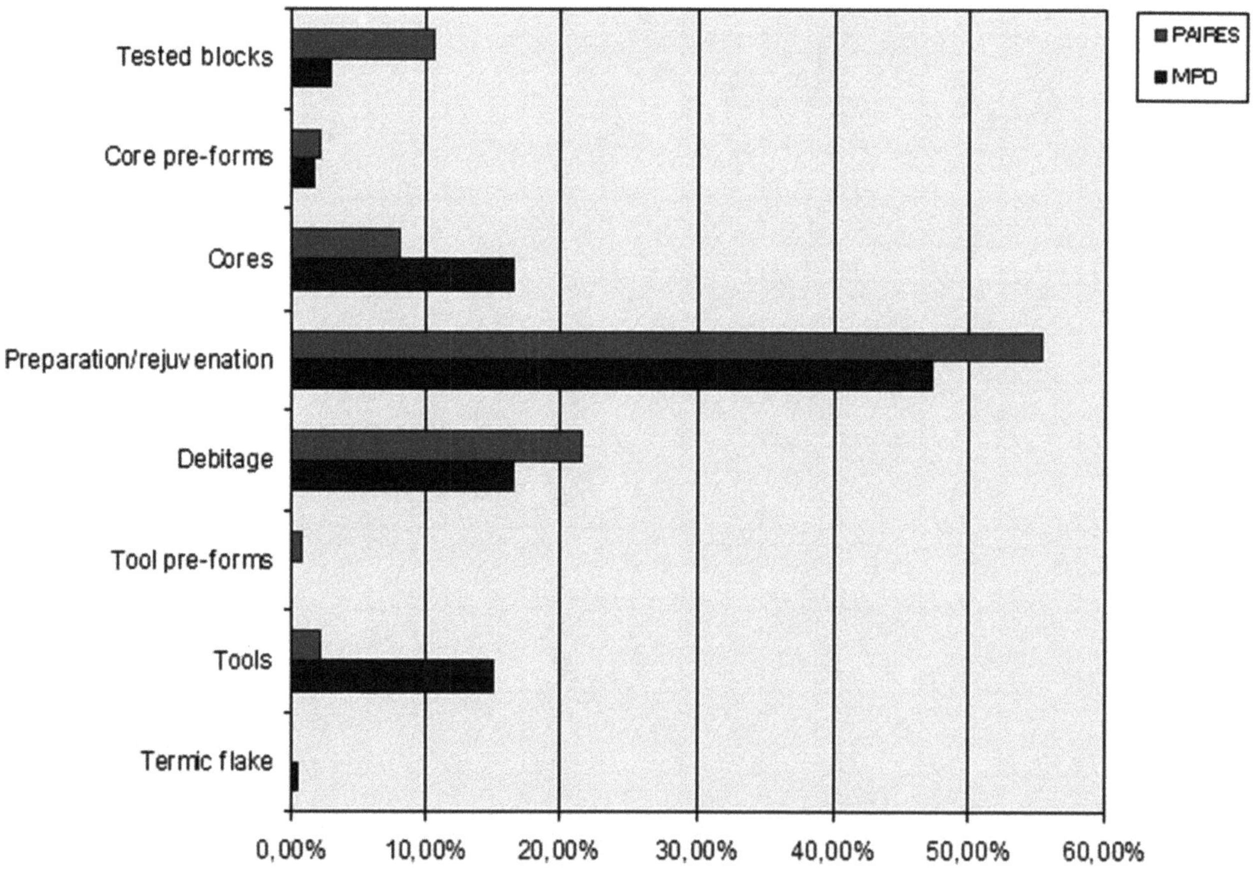

Figure 3. Comparison of the lithic industries of Pedreira do Aires (PAIRES) and Monte das Pedras (MPD), organized by typology. Unworked and tested flint blocks and cortical flakes - abundant at both sites but collected apart - are not represented in this count.

lithic artifacts have a rolled neo-cortex, perhaps indicating that flints from slope deposits formed a backup source. No evidence was found of the knapping of flints from the primary limestone formation.

This procurement pattern is related to the different knapping qualities of the flint from different types of source. This region is highly affected by tectonics (Figure 2), a consequence of the orogeny of both the Sintra mountain range and the Lisbon volcanic complex. This orogeny cut and folded the adjacent formations, including the flint-bearing Cenomanian limestone. As a result, the *in situ* flints show abundant cleavages as well as different sized geodes - obstacles to knapping. Debris with fewer imperfections can be found in the slope deposits originated by the erosion of the outcrop. In small terraces and along stream banks, however, flint cobbles of various sizes with a quite different interior can be found. As a result of mechanical transport and attendant shock fracturing, most of the internal flaws have been eliminated. Even though some cleavages and geodes are still found in this material (the cobbles were only transported a few kilometres at most, therefore not all flaws would have been eliminated) it is possible to find raw material perfectly adequate for knapping. Sorting out the best material would have required testing, i.e., striking blocks once or twice at their point of collection. This explains the large number of cobbles in the studied assemblages with only 1-3 flake extractions.

4. Conclusions. Pedreira do Aires and Monte das Pedras: flint procurement strategies and lithic artifact production in the Lisbon Peninsula

No clear instance of Neolithic and Chalcolithic flint mining such as recorded in other areas of Iberia, e.g., Murcia (Jimenez Lorente 1983; Jimenez Lorente, Ayala Juan and Navarro Hervás 1999), Casa Montero (Consuegra Rodríguez, Gallego Garcia and Castañeda Clemente 2004), La Venta (Ramos Millán *et al.* 1993) or Granada (Martínez Fernandez *et al.* 2006), has ever been found in the Lisbon Peninsula. Although the superficial excavation of debris deposits seems to have occurred, no gallery excavation has been identified in the limestone formations of Portuguese Estremadura (Almeida, Araújo and Aubry 2003). Recent surveys attempting to identify the flint sources of this region have shown flint nodules to be found in secondary positions: detritic deposits detached from limestone formations and alluvial deposits (naturally transported from the original source). Such easily accessed deposits naturally represent preferred procurement sources.

The large number of Neolithic and Chalcolithic settlements in the Lisbon Peninsula, and therefore the need for raw materials for tool manufacture, coupled with the occurrence of siliceous rocks in the local geology, ought to correlate with the existence of many sites such as those discussed here. However, perhaps due to lack of research, this cannot yet be said to be the case.

The only possible prehistoric flint mine *sensu stricto* mentioned in the literature is that of Campolide (Choffat 1889, 1907). However: 1) there is no unambiguous evidence of gallery excavation (mining was inferred by reference to other instances of flint exploitation in prehistoric times and on the basis of the recovery of Neolithic hammerstones); 2) flint blocks in a secondary position, which are easier to access than those in the hard Cenomanian limestone, exist in the vicinity; and 3) the supposed galleries follow the natural dip of the geological layers and could simply have been natural caves with prehistoric deposits. The Neolithic and Chalcolithic sites of the Lisbon Peninsula that have been generically characterized as mines and/or workshops suggest four different flint procurement models:

1) small-scale exploitation sites where lithic artifacts were occasionally made, such as Pedreira do Aires, Monte das Pedras and Casal Novo (Andrade and Cardoso 2004; Andrade, i.p.).

2) extensive exploitation areas with specialized lithic production, such as Casas de Baixo and Arruda de Pisões (Zilhão 1994; Forenbaher 1999, 2006).

3) settlements located near flint sources and directed towards the exploitation and production of lithic artifacts, such as Vila Pouca and Santana, related to the putative Campolide flint mines (Forenbaher 1999).

4) small camp-sites installed near a larger settlement with intensive lithic production, such as Barotas and Monte do Castelo, related to the fortified settlement of Leceia (Cardoso and Costa 1992; Cardoso and Norton 1997, 1998).

When taking chronology into account, these four models can be grouped into just two basic models:

1) occasional exploitation sites, used in the framework of seasonal group movements to satisfy immediate needs, and related to the advent and affirmation of the first farming communities (Neolithic).

2) permanent exploitation sites specialized in artifact production, determined by permanent raw-material procurement needs, and related to the consolidation of stable farming communities (Chalcolithic).

The archaeological context of the sites discussed here suggests two possible interpretations (similar scenarios have been proposed for the workshops of Murcia, after Jiménez Lorente, Ayala Juan and Navarro Hervás 1999).

1) that they are exploitation sites located in the resource procurement area of one or more settlements and where the shaping of flint blocks and occasional manufacture of blanks and tools was undertaken.

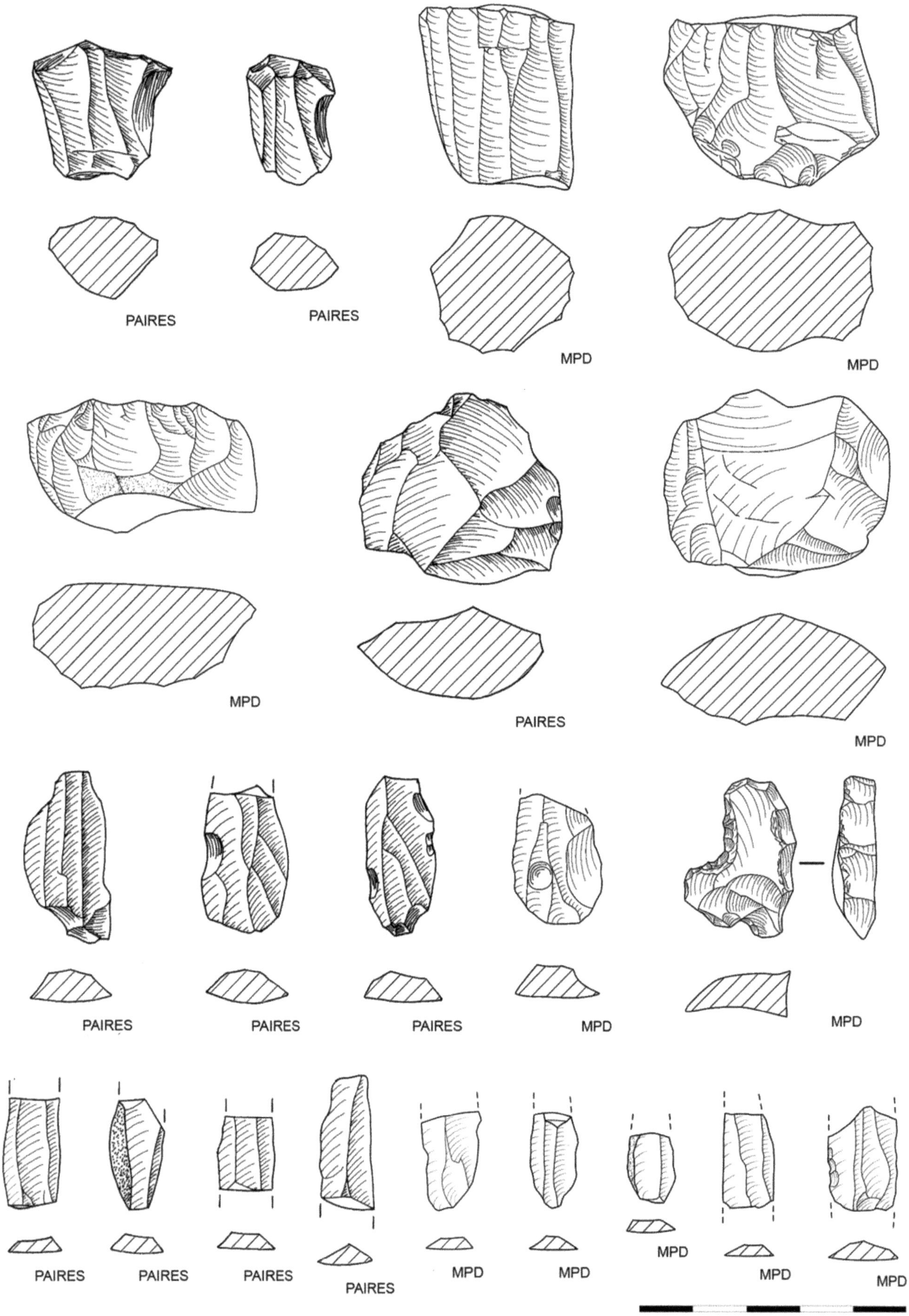

Figure 4. Prismatic bladelet and flake cores, discoidal cores, core rejuvenation elements one of which used as tool after introducing a notch, bladelets and small blades recovered at Pedreira do Aires (PAIRES) and Monte das Pedras (MPD). Scale 2:3.

2) that they are exploitation sites where flint blocks were shaped into core pre-forms, with subsequent reduction taking place at settlements located nearby.

Therefore, Pedreira do Aires and Monte das Pedras may have been workshops and/or craftsman campsites located near raw material sources, with the recovered assemblages reflecting not only workshop activities but also the discarding of daily use tools (retouched bladelets and flakes, scrapers, notches, denticulates and perforators). Certainly, both the production and consumption of lithic artifacts (as defined by Jiménez Lorente, Ayala Juan and Navarro Hervás 1999) are documented. The present authors favour this scenario, but given the available data a specialized site interpretation cannot be rejected.

A preliminary analysis of the lithic industry of Early Neolithic and Late Neolithic sites near Pedreira do Aires and Monte das Pedras (Zibreira and Vale de Lobos) indicates that cortical elements (mostly re-used as tools, such as retouched flakes or scrapers) are poorly represented. This suggests that flint was introduced into settlements as core pre-forms, which is consistent with the interpretation of the two flint knapping areas discussed here as procure-

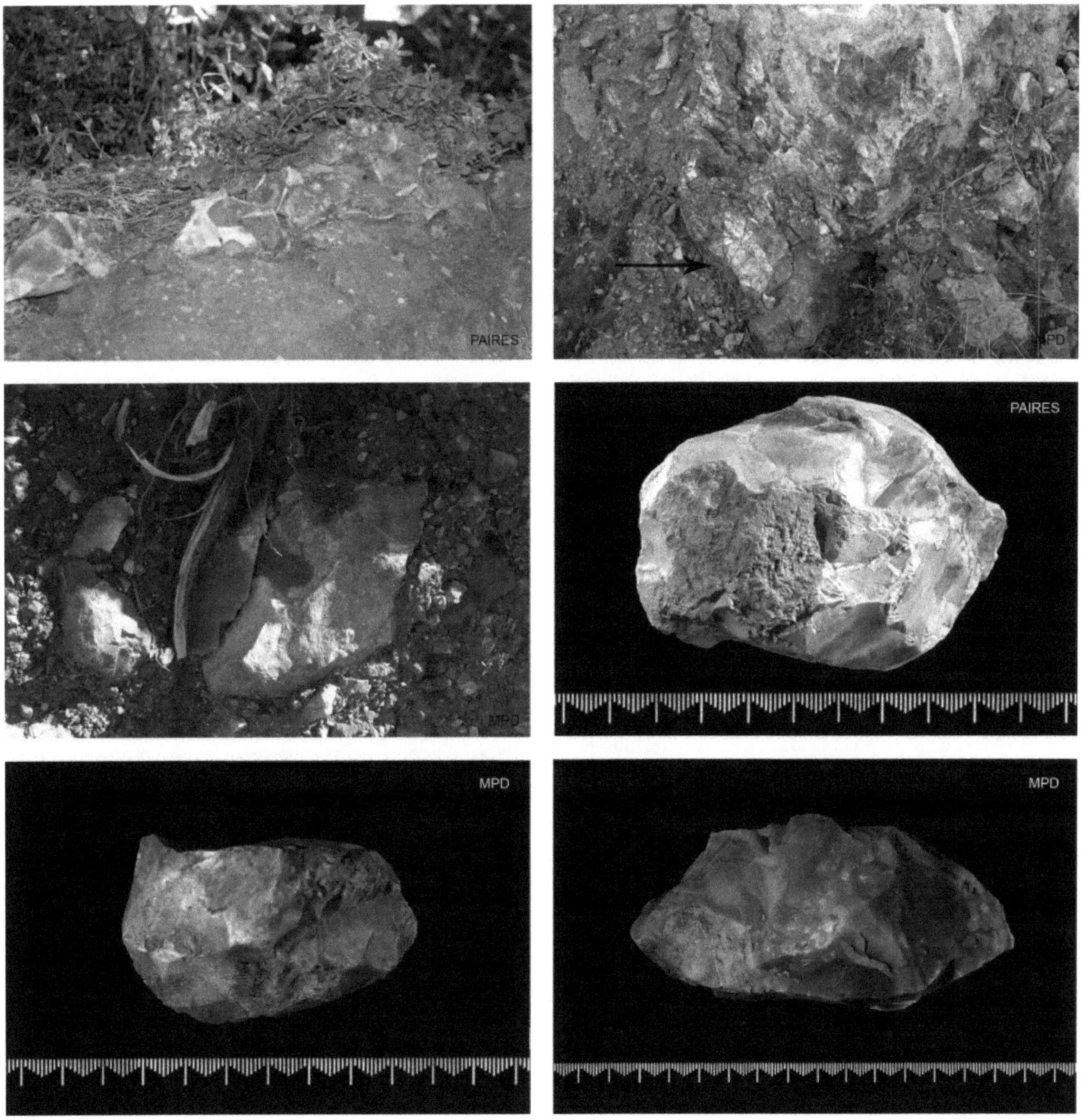

Figure 5. Flint nodules in the limestone outcrops at Pedreira do Aires (PAIRES, photograph 1) and Monte das Pedras (MPD, photograph 2); flint cobble from debris at Monte das Pedras (MPD, photograph 3); tested flint cobbles recovered at Pedreira do Aires (PAIRES, photograph 4) and Monte das Pedras (MPD, photographs 5 and 6).

ment sites for such communities. Thus, following Chapman (1990), these sites are not indicative of intra-site functional differentiation but rather inter-site functional variation between the components of a single settlement system.

Acknowledgments

The authors thank Thierry Aubry for his input in the interpretation of the Campolide mines.

References

Almeida, F., Araújo, A. C.; Aubry, Th. 2003. Paleotecnologia lítica: dos objectos aos comportamentos, in J. E. Mateus and M. Moreno-García (eds.), *Paleoecologia Humana e Arqueociências: Um programa multidisciplinar para a arqueologia sob a tutela da cultura*, 299-349. Lisboa, Instituto Português de Arqueologia.

Andrade, M. A. i.p. O sítio pré-histórico de Monte das Pedras (Mina, Amadora): identificação e caracterização de uma possível oficina de talhe neolítica. *Revista Portuguesa de Aqueologia.*

Andrade, M. A. and Cardoso, M. S. 2004. O sítio pré-histórico da Pedreira do Aires (Ramada, Odivelas): notícia da sua identificação. *Revista Portuguesa de Arqueologia* 7 (1), 137-163.

Cardoso, J. L. and Costa, J. B. 1992. Estação pré-histórica de Barotas (Oeiras). *Setúbal Arqueológica* 9/10, 229-245.

Cardoso, J. L. and Norton, J. 1997-98. A oficina de talhe do sílex do Monte do Castelo (Leceia, Oeiras). *Estudos Arqueológicos de Oeiras* 7, 35-45.

Carrión Méndez, F., García González, D. and Lozano Rodríguez, J. A. 2006. Métodos y técnicas para la identificación de las fuentes de materias primas líticas durante la Prehistoria reciente, in G. Martínez Fernández, A. Morgado Rodríguez and J. A. Afonso Marrero (cords.), *Sociedades prehistoricas, recursos abióticos y territorio. Actas de la III Reunión de Trabajo sobre el Aprovisionamiento de Recursos Abióticos en la Prehistoria,* 45-61. Granada, Fundación Ibn Al-Jatib.

Chapman, R. 1990. E*merging Complexity: the Later Prehistory of South-East Spain, Iberia and West Mediterranean.* Cambridge, University Press.

Choffat, P. 1889. *Étude géologique du tunnel du Rocio. Contribution à la connaissance du sous-sol de Lisbonne.* Lisboa, Comissão dos Trabalhos Geológicos de Portugal.

Choffat, P. 1907. Exploitation souterraine de silex à Campolide aux temps néolithiques. *O Archeólogo Português* 12, 338-342.

Consuegra Rodríguez, S., Gallego García, M. M. and Castañeda Clemente, N. 2004. Minería neolítica de sílex de Casa Montero (Vicálvaro, Madrid). *Trabajos de Prehistoria* 61(2), 127-140.

Forenbaher, S. 1999. *Production and Exchange of Bifacial Flaked Stone Artifacts during the Portuguese Chalcolithic.* Oxford, BAR Publishing. British Archaeological Reports In-ternational Series 756.

Forenbaher, S. 2006. Lithic production at Casas de Baixo and the prismatic blades of the Portuguese Neolithic. In N. F. Bicho (ed.), *From the Mediterranean Basin to the Portuguese Atlantic Shore: Papers in Honor of Anthony Marks. Actas do IV Congresso de Arqueologia Peninsular*, 231-244. Faro, Centro de Estudos do Património/Departamento de História, Arqueologia e Património, Universidade do Algarve.

Jimenez Lorente, S. 1983. Introduccion a la problematica de los talleres de silex al aire libre en la provincia de Murcia. *Cronica del XVI Congreso Nacional de Arqueologia*, 53-63. Zaragoza.

Jimenez Lorente, S., Ayala Juan, M. M. and Navarro Hervás, F. 1999. Nuevos talleres de sílex al aire libre en Murcia. *XXIV Congreso Nacional de Arqueologia*, 83-93. Cartagena.

Martínez Fernandez, G., Morgado Rodríguez, A., Afonso Marrero, J. A., Cámara Serrano, J. A. and Cultrone. G. 2006. Explotación de rocas silíceas y producción líticas especializadas en el Subbético central de Granada (IV-III mil. Cal B.C.), in G. Martínez Fernandez, A. Morgado Rodríguez and J. A. Afonso Marrero (cords.), *Sociedades prehistoricas, recursos abióticos y territorio. Actas de la III Reunión de Trabajo sobre el Aprovisionamiento de Recursos Abióticos en la Prehistoria*, 293-313. Granada, Fundación Ibn Al-Jatib.

Ramos Millán, A., Peña González, B., Osuna Vargas, M., Tapia Espinosa, A. and Aznar Pérez, J. C. 1993. La mina de silex de la Venta. Investigaciones arqueológicas de 1990-91. *Anuario Arqueologico de Andalucía* 2, 212-224.

Vallespi Perez, E. 1968. Tallers de silex al aire libre en el pais vasco meridional. *Estudios de Arqueologia Alavesa* 3, 7-27.

Vallespi, E., Ramos Muñoz, J., Espejo, M. and Cancalejo, P. 1988. Talleres liticos andaluces del Calcolitico y Bronce. *Revista de Arqueología* 90, 14-24.

Zilhão, J. 1994. A ofi cina de talhe neo-calcolítica de Casas de Baixo (Caxarias, Vila Nova de Ourém). *Trabalhos de Arqueologia da EAM* 2, 35-45.

The gathering, stocking and knapping of flint during the Chalcolithic at Casal Barril (Portugal)

Ana Catarina SOUSA and Victor S. GONÇALVES

Abstract

Casal Barril, a Chalcolithic site near Ericeira (Portuguese Estremadura), was discovered in 2006 during the construction of the A21 motorway. All data seem to indicate that Casal Barril was almost contemporaneous with the establishment of the first copper metallurgical communities in the region (Beta-260629: 2860-2490; Beta-260628: 2630-2470 both Cal BC 2 sigma). However, it is impossible to know by radiocarbon dating if its flint sources were exploited in earlier times. Casal Barril belongs to a network of flint exploitations, debitage sites and movement at the regional and even larger scale.

Keywords

Portuguese Chalcolithic. Flint extraction. 3rd Millennium BCE.

1. A preliminary approach... from random find to first case study

The Casal Barril site was discovered in March 2006 during the archaeological supervision of the construction of the A21 motorway (Malveira – Mafra – Ericeira). Archaeological excavations took place between the 4th and 20th May of that same year. 'Mine-type' or 'flint workshop' sites are often discovered during large infrastructure projects, such as the making of motorways. Indeed, the great majority of Neolithic mines known to date were discovered during such projects. For example, the Casa Montero site in Spain was identified during the construction of the M50 motorway (Consuegra *et al.* 2004), and the Neolithic mine of Ri in northwestern France was discovered during the course of work on the A88 motorway. On a more regional level, the only primary context known for the extraction of flint, the Campolide site in Lisbon, was identified during the construction of a railway in 1888 (Choffat 1907).

The archaeological work carried out at Casal Barril has supplied a large body of information concerning the extraction, deposition and circulation of flint in the Portuguese Estremadura region during the 3rd millennium. Studies in this area are still ongoing. A preliminary inventory of the materials recovered has been made, and a typological and technological study of the assemblage will take place in the near future. No palaeobotanical nor geoarchaeological study of the site has yet been undertaken. The geological characterisation of the site and of its raw materials (i.e., those used by the inhabitants of contemporaneous settlements) will form one of the basic stages in the characterisation of the strategies used to procure mineral resources at Casal Barril and in the surrounding area. This present presentation is, therefore, of a preliminary nature. An exhaustive monograph is planned for the future.

2. The data under study

2.1. Location of the site

Casal Barril is situated near the town of Ericeira, in Mafra County, in the District of Lisbon (UTM coordinates 29S 464977E 4313948; military geographic coordinates 38°58'13,173"N, 09°24'15,938"W). The site lies in a valley with an absolute altitude of 60m, on the right bank of the Ribeira da Fonte Boa, a small stream that flows into the sea at Praia do Matadouro. The site lies in Cretaceous strata (orbitolinid limestone), a substrate found along the main watercourses of the region (Ribeira da Fonte Boa, Ribeira de Cheleiros to the south, and Ribeira do Cuco to the north). This rocky substrate occurs in association with patches of Cenomanian rocks and the deposition areas of ancient beaches. Pleistocene fluvial terraces are also visible. The presence of these Pleistocene patches led some researchers (Ferreira 1984) to propose, somewhat indiscriminately, a Palaeolithic chronology for this area.

2.2. Stratigraphic contexts

The stratigraphic interpretation of the site is strongly affected by changes caused by the extraction of soil. Before the building of the motorway, the area was profoundly affected by vine cultivation, although the valley was left relatively intact. During work on the motorway, the valley where the site is located was subjected to the removal of some of its uppermost sediments.

Although systematic surface surveying and archaeological supervision of the motorway construction work was carried out over a 50m-wide area, no other evidence of flint exploitation was identified. Nevertheless, other archaeological remains abound in the area, some of which were also investigated during the course of the above construction work. In the valley where the site lies, numerous limestone outcrops of the *lapias* type were identified, while above, limestone banks abound.

In section, the Casal Barril flint deposit shows continuity beyond the area of the motorway. Thus, it is plausible that the site may extend beyond the valley, a possibility that can only be checked with extensive excavation work; no other surface traces are visible. The archaeological remains of Casal Barril were therefore found in a restricted area. An excavation grid of approximately 20 square meters was set up where the archaeological deposits were found, with remains concentrated on an area of around 9 square meters. The flint material of the deposit was surrounded by limestone outcrops of the *lapias* type. Given the nature of the site, the small size of the area and the large amount of materials present, an excavation by artificial units within natural layers was adopted. This involved 12 artificial units and the making of both photographic and graphic records (scale 1:10). All the materials found were registered according to their stratigraphic unit, layer and grid reference.

As mentioned above, the removal of the upper strata at the site by mechanical means has hindered a complete knowledge of its stratigraphy being gained. However, a pit with a very irregular profile (US 7) has been identified with a maximum depth of 1.22m and a width of 2 x 2m at the top. This pit was filled with a sediment of a very dark colour (Munsell, 10YR 5/3, brown) abounding in charcoals (US 4). At the base of the pit a clay sediment of a near orange colour (Munsell 10YR 5/6, yellowish brown) (US 2) was seen. The latter sedimentary deposits differ in terms of the materials they incorporate. In US 4, besides the abundant evidence of lithic industry, a few ceramic shards, a shell (*Glycimeris* sp.) and the tooth of a mammal were recovered, while in US 2 more boulders of un-worked raw material are seen, as well as a large number of pebbles. Archaeological materials were found exclusively in these two units, both deposited over a marly layer (US 6).

2.3. Material culture and functional specialisation of the site

The classification and study of Casal Barril is ongoing. Although the area excavated was small, an assemblage of some 67,000 lithic pieces was recovered. All were organised in a preliminary fashion and quantified by sample. Area K.6 (see photograph in Figure 11) was selected for detailed study and was found to contain 7,274 items, of which 6,456 were of flint (Figure 1). Besides lithic material, occasional artefacts and ecofacts were registered, although these corresponded to a mere 0.04% of the total. Casal Barril is therefore a highly specialised site, focused on the exploitation, storage and debitage of flint.

2.3.1. Raw materials

The specialisation of the site is clearly demonstrated by the almost dominant presence of flint, corresponding largely to a single type. Only a few examples of exogenous flint have been found. The local flint has a brown colour, and shows a number of inclusions no doubt responsible for the fractures identified. Two types of raw material nodule are found: flint *rognons*, with a fine cortex layer, and large flint boulders, with a thicker cortex and a number of inclusions.

Quartzite is also present in the form of pebbles. The great majority have not been knapped: of the 310 items recovered only three are flakes, one being the remains of knapping and one a nucleus. Thus, the numerous pebbles found seem to be in their primary context.

2.3.2. Preliminary reconstitution of the flint working process

Since the recovered assemblage was so voluminous, counts was made at the level of working process (Figure 2):

1. Nodules of raw-material composed of unprocessed boulders.

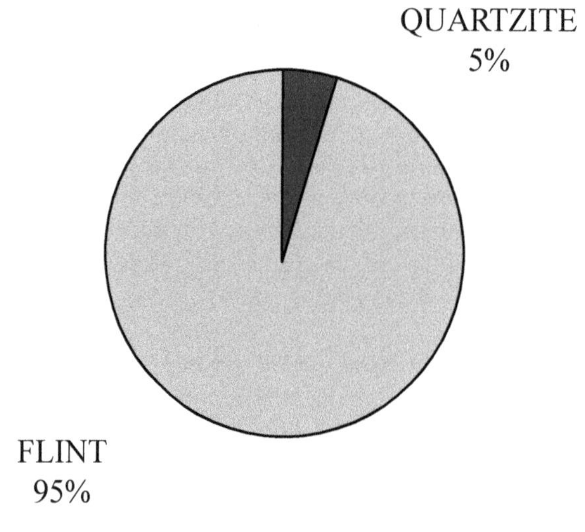

Figure 1. Raw materials. Area K.6.

2. Nuclei of different stages, according to different debitage schemes (prismatic and centripetal).

3. Debitage by-products (flakes predominate but some elongated products, blades and bladelets are also present).

4. Tools: retouched blades and scrapers, and denticulated and retouched tools (tools for general use).

5. Residual materials resulting from debitage.

6. Flint fragments showing natural (non-anthropic) fractures (unworked raw material, rejected material).

In absolute terms, flint fragments that might correspond to fractured raw-materials predominate. Residual materials make up the second largest group, followed by the products (mostly flakes). In numerical terms, nodules represent 1.5% of the total, although by relative weight they in fact correspond to just 0.05% (Figure 3). The least numerous group consists of nuclei; these must have been the main product circulated.

With respect to the vertical distribution of the various types of flint materials, two main concentrations were noted, in layers 2 and 10, possibly representing several, short episodes of deposition. The nodules were concentrated at the base of the stratigraphy, especially the larger ones.

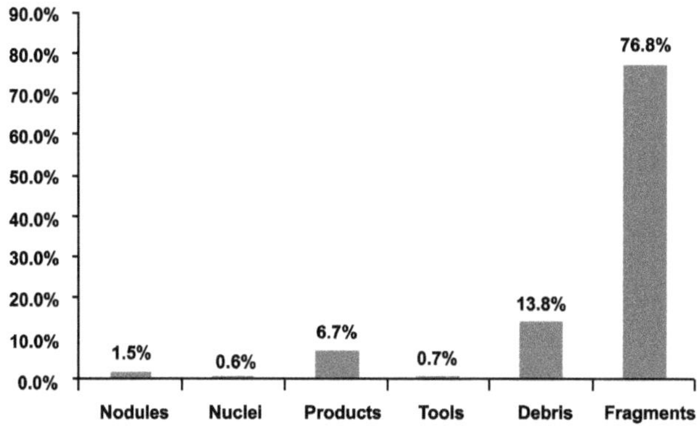

Figure 2. Chaîne operatoire in area K.6 (number of objects and artefacts: 6456).

The distribution of the materials by stratigraphic unit also presents certain peculiarities. US 2 included 4,501 items (73%) and US 4 1,656 (27%). US.4 has a pit that shows intense signs of combustion, and has a larger number of debitage products and tools than US 2. In contrast, US 2 has a larger number of flint cores and nodules (Figure 4).

Layer	Nodules		Nuclei		Products		Tools		Debris		Fragments		Total	
	Nº	%	Nº	%	Nº	%	Nº	%	Nº	%	Nº	%	Nº	%
1	1	1%	7	18%	32	7%	4	8%	256	29%	156	3%	456	7%
2	15	16%	7	18%	162	**38%**	13	**27%**	264	**30%**	966	**19%**	1427	**22%**
3	19	**20%**	7	18%	83	19%	5	10%	73	8%	618	12%	805	12%
4	4	4%	0	0%	27	6%	4	8%	64	7%	278	6%	377	6%
5	6	6%	8	**21%**	59	14%	5	10%	78	9%	379	8%	535	8%
6	9	10%	4	11%	32	7%	8	17%	43	5%	272	5%	368	6%
7	0	0%	3	8%	7	2%	1	2%	13	1%	127	3%	151	2%
8	1	1%	1	3%	10	2%	1	2%	36	4%	193	4%	242	4%
9	7	7%	0	0%	7	2%	0	0%	8	1%	592	12%	614	9%
10	17	18%	0	0%	6	1%	5	10%	23	3%	939	19%	990	15%
11	15	16%	1	3%	5	1%	2	4%	31	3%	446	9%	500	8%
	94	100%	38	100%	430	100%	48	100%	889	100%	4966	100%	6465	100%

Figure 3. Flint materials in K.6 – (# 6456).

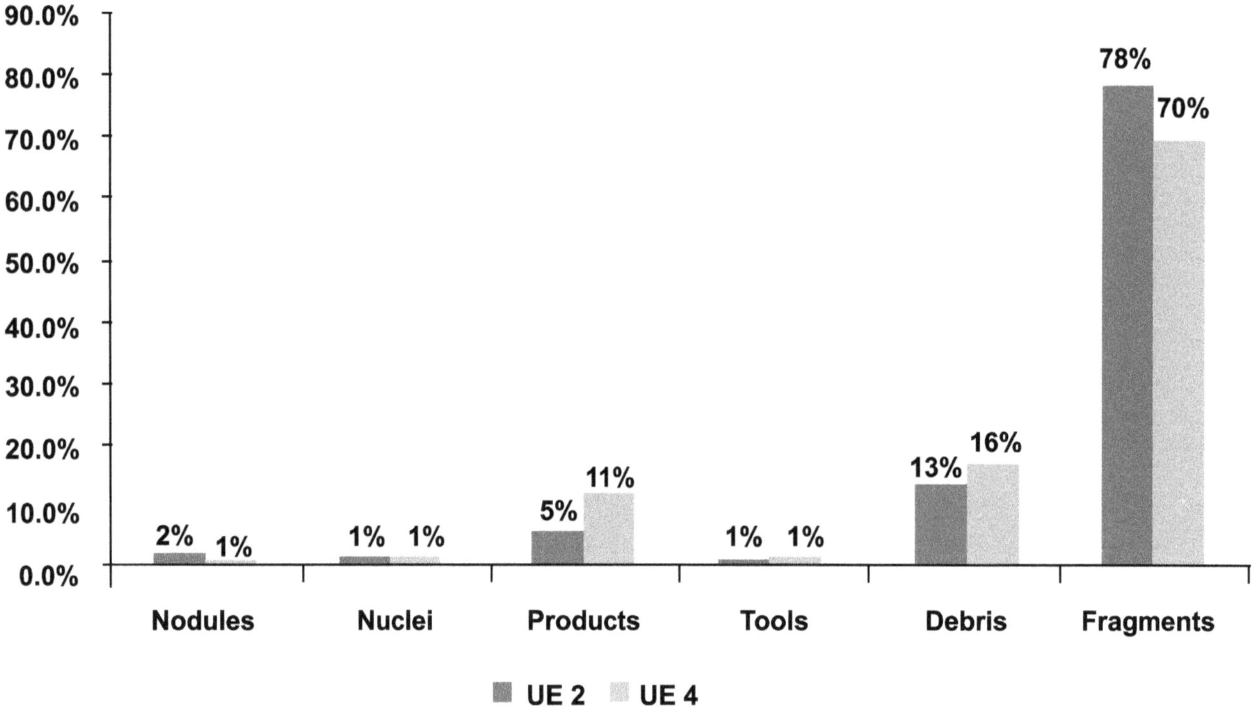

Figure 4. Quantitative comparison of the chaîne operatoire in stratification units 2 and 4. Area K.6 (# 6456).

2.3.3. Equipment carried by knappers

Besides evidence of knapping at the site, other evidence regarding the communities that exploited the flint at Casal Barril has been found. Ceramic shards (three plain rims) and a polished stone axe of sub-rectangular section have been recovered. Except for a polished stone axe and three quartzite pounders, little evidence of the heavy tools used in the exploitation of the flint has been found. A few artefacts of exogenous flint, bladelets and retouched flakes have also been found.

Alongside these artefacts, 18 charcoal samples were recovered, although these have not been analysed anthracologically. US 4 shows signs of combustion, including the presence of charcoal, coloured sediment and (sometimes)

Lab Reference	Sample	Context	$\delta^{13}C$ (0/00)	Radiocarbon date (BP)	Cal BC (1σ)	Cal BC (2σ)
Beta-260629	Charcoal	CBR.J5.P4.385.Z4.L2	-25.110/00	4090 +- 40	2840-2570	2860-2490
Beta-260628	Charcoal	CBR.J5.P4.384.Z7.L4	-24.8/0.00	4020 +- 40	2580-2480	2630-2470

Figure 5. Radiocarbon dates – Casal Barril.

Habitat	Distance (m) measured from Casal Barril	Early Chalcolithic	MiddlE Chalcolithic	Late Chalcolithic (Beaker)
Quinta dos Loureiros	578	○	●	●
Casal Cordeiro 5	936	○	○	●
Casal Romeirão	2098	○	○	●
Mil Regos	1780	●?	●?	○

● present / ○ absent

Figure 6. Settlements around Casal Barril.

the presence of thermoclasts. However, there is no sign of any thermal treatment of the local flint material; such evidence has only been found for artefacts made from exogenous flint.

Faunal remains are practically absent, except for a shell of the *Glycimeris sp.* and the tooth of a mammal.

2.4. Absolute chronology

Radiocarbon dating was performed on charcoal from level 2 (Beta-260629) and level 4 (Beta-260628). Since the other 16 samples come from the same levels, their analysis was deemed unnecessary. Figure 5 shows results obtained.

Although differences are seen in terms of the upper date interval, these are not statistically significant, implying rapid sedimentation at the site.

3. Flint extraction, knapping and distribution: preliminary interpretations

3.1. Casal Barril and local Chalcolithic settlements

Casal Barril is situated in a suburban area where traces of Neolithic and Chalcolithic occupation have only recently been discovered. The area has thick sediments covering the 4th and 3rd millennia BCE, and the type of land use (urban areas alternating with areas of dense vegetation cover) renders surface surveying difficult. The settlements recently identified in the area correspond entirely to random findings made during the course of motorway construction.

To date, four settlements have been identified in the area surrounding Casal Barril (Figure 6).

An element common to all four is their strategy of implantation; all are open settlements without any type of defence, and all are situated at the top of inter-fluvial areas that lie on Cretaceous limestone substrata. Their chronologies also imply a certain unity. Except for Mil Regos where the material culture allows no clear culture classification to be made, Beaker ceramics are found at all these sites.

Only Casal Cordeiro 5 has been subjected to archaeological excavation (Sousa 2008), and it can be clearly designated as a permanent Chalcolithic settlement. Its stratigraphy is well preserved and three areas of occupation can be distinguished. Domestic structures with floors and hearths have been identified in two places. The absolute dates for Casal Cordeiro 5 have been obtained exclusively from shell samples; the bone samples analysed contained no collagen. The finding of other organic materials (charcoal and mammalian remains) suggests better radiocarbon analyses could be made in the near future. Leaving aside certain reservations regarding the interpretation of dates, comparisons of data supplied by the material culture place Casal Cordeiro 5 in the same chronological spectrum as Casal Barril.

Although Casal Cordeiro 5 is very close to Casal Barril, very little evidence of lithic industry was found there. Therefore, the knappers seem not to have used Casal Cordeiro 5 for the collection of flint, even though flint was available at this site.

As for the majority of Chalcolithic settlements in the Portuguese region of Estremadura, few remains survive of the early stages of the flint processing operation: nodules, preparation tools and unworked raw materials are in the minority while tools predominate. Macroscopic observations reveal that a diversity of flint types were used, although it is difficult to find any parallels with flint from Casal Barril. Lithographic studies of the various contemporary assemblages at Casal Barril are planned, especially of the fortified settlement of Penedo do Lexim about 12km away.

3.2. Provisioning, knapping and circulation of flint within and beyond the region of Estremadura

Cretaceous and Jurassic formations make up the geological substrate of Estremadura, the region where Casal Barril is located. Veins of flint abound in these formations. The quantity, accessibility, variety and quality of this flint is unparalleled in Portugal (Forenbaher 1999, 31). Estremadura has been long referred to as the main source of flint on a 'national level'; the abundance of sources of flint contrasts with a notable lack in the surrounding regions (especially the Alentejo region). Extra-regional routes of exchange therefore grew up around Estremaduran flint, which was exchanged for amphibolite from the northern part of the Alentejo region, and for copper from the southwest Iberian pyrite belt.

Although this theoretical model of extra-regional exchange appears in much of the literature referring to the Neolithic and Chalcolithic of Estremadura, the archaeological basis for it is very limited. No systematic geo-archaeological surveying of the areas of procurement has been performed, and excavations are needed to characterise the specialised sites (mines or workshops) along with work to characterise the lithic industries of Estremadura and the Alentejo.

In southern Portugal, the only mine-type of flint exploitation is that seen at Campolide, where galleries for flint extraction were identified at the end of the 19th century (Choffat 1907). The latter, very old publication makes the curious statement: 'pourquoi les hommes néolithiques de Campolide exploitaient-ils le silex par galeries souterraines, ce qui, à cette époque, presentait des difficultés incomparablemente plus grands qu'actuellement?' (Choffat 1907, 341). A clear functional model for the exploitation of these flint mines is yet to be proposed.

The other known contexts are open sites; usually these have been characterised only by surface surveying and co-

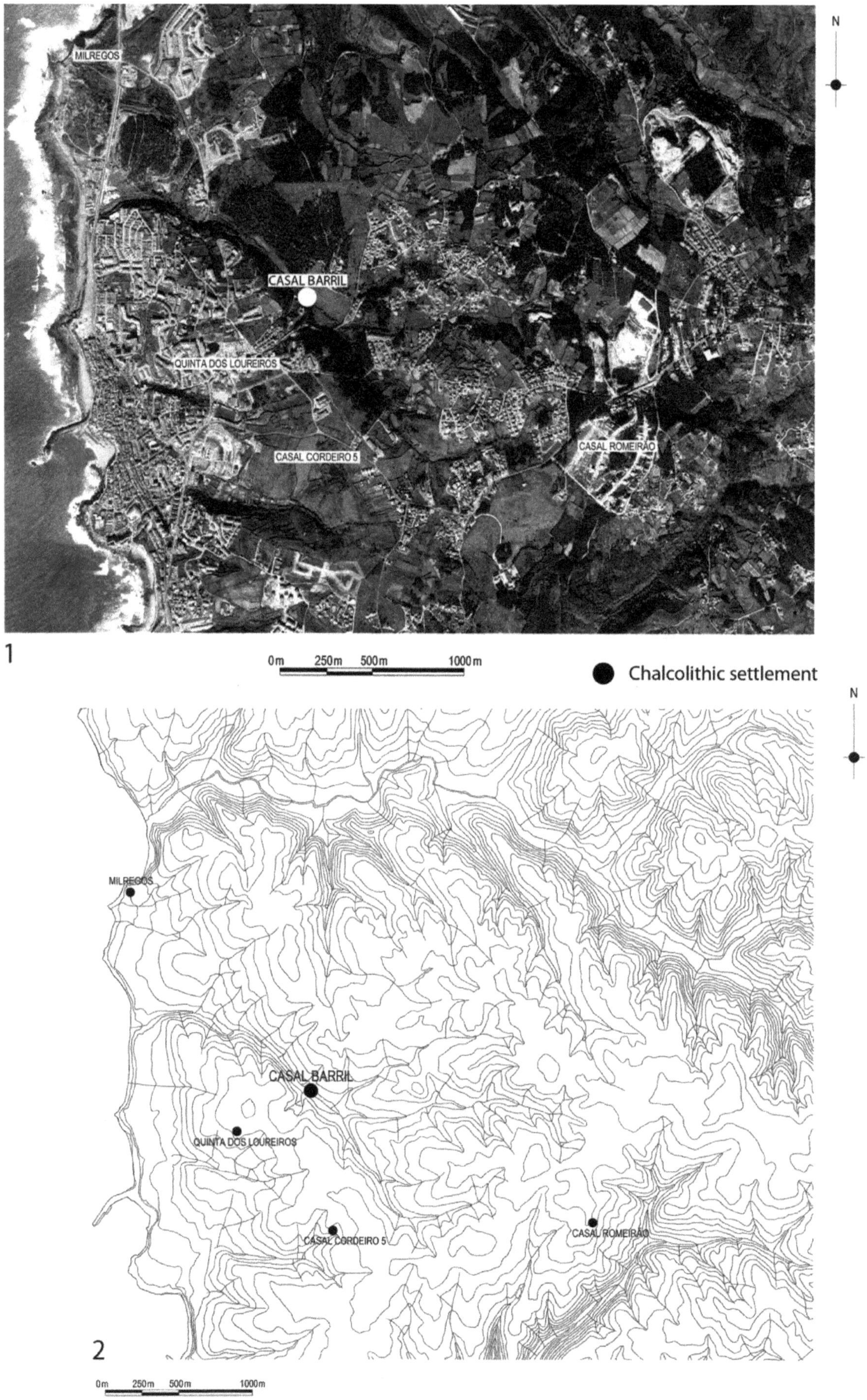

Figure 7. 1 and 2: The location and Chalcolithic entourage of Casal Barril, near Ericeira (Mafra, Portugal).

Figure 8. 1: the site when discovered, before excavation; 2: flint nodules; 3: layer 4; 4: the pit and the bedrock after the excavations.

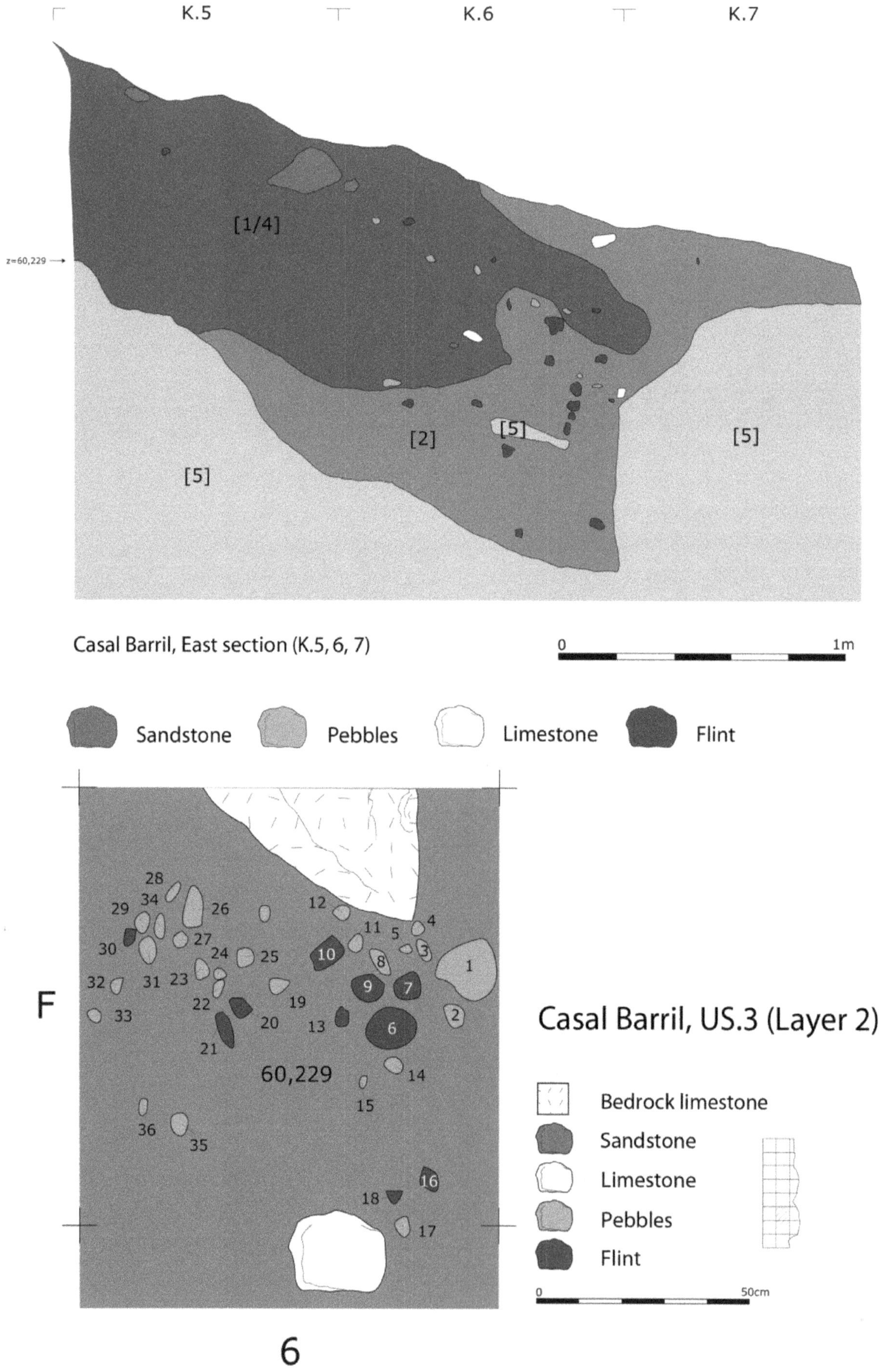

Figure 9. Above: Casal Barril, East section (K5, 6, 7). Below: Casal Barril, US.3 (Layer 2).

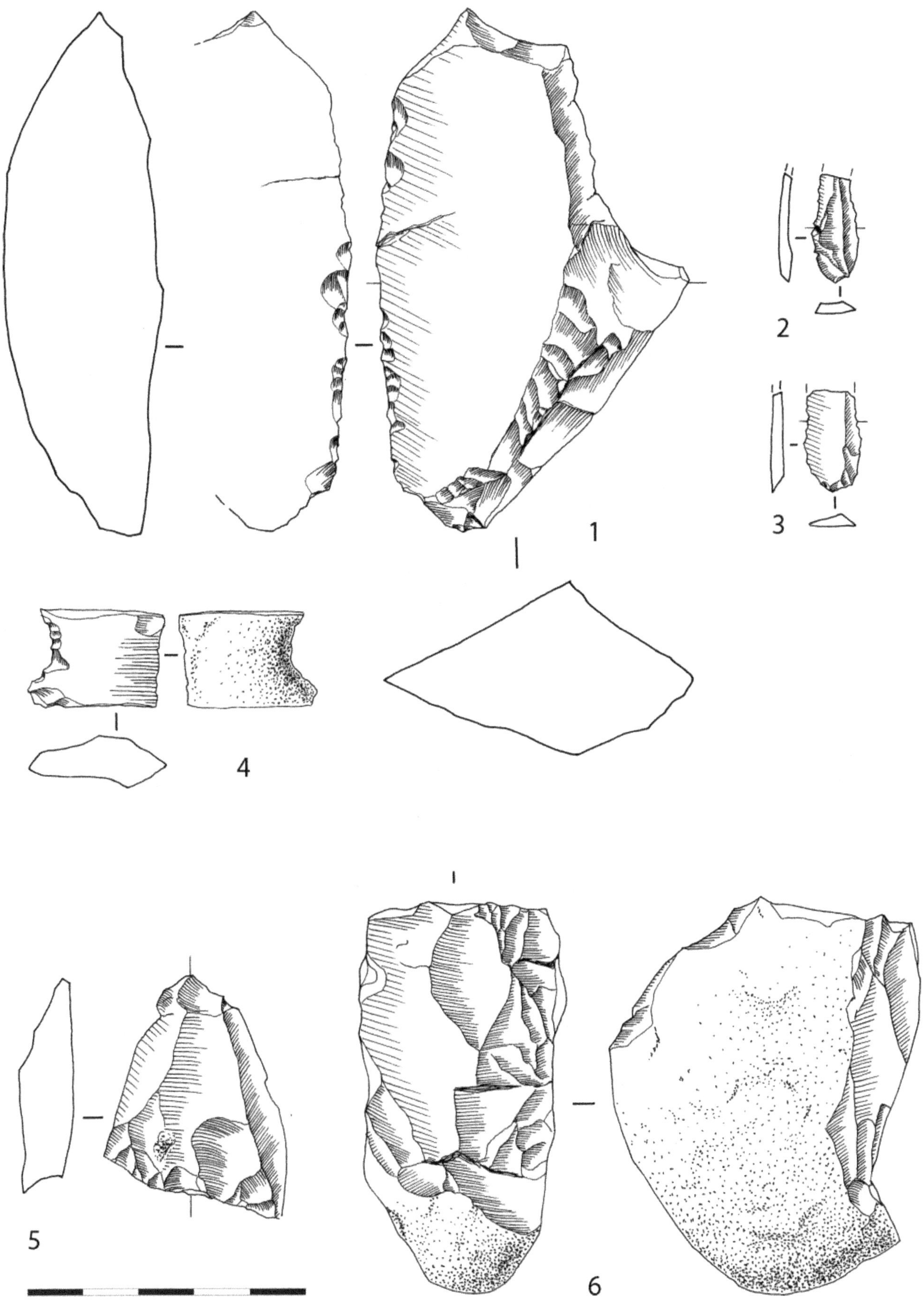

Figure 10. 1: scraper CBR-457; 2, 3: bladelets CBR 301, 417; 4: denticulated CBR-345; 5: flake CBR-435; 6: nucleus CBR-297.

Figure 11. 1: flint nodule CBR-523, probably crushed for quality inspection; 2: nucleus CBR-456; 3: scraper CBR-457.

Figure 12. 1, 2, 3: used polished stone axe CBR-397; 4: ceramic shard CBR-520.

llection. Among these, Barotas (Cardoso and Costa 1992) and Monte do Castelo (Cardoso and Norton 1997-1998), in the area of influence of the fortified settlement of Liceia, deserve special mention. The mapping of the surface at Pedreira do Carrascal took place in the 19th century, adjacent to the dolmens of Belas (Ribeiro 1880), while the Santana site in Lisbon was mapped in the early 20th century (Forenbaher 1999). Recently, surface sites have also been identified at Cortegaça (Sousa 1998) and Pedreira do Aires (Andrade and Cardoso 2004).

Of all the flint exploitation sites known in Estremadura, the results of only two excavations have been published: Olival do Passal (Zilhão 1995) and Casas de Baixo (Zilhão 1994; Forenbaher 1999). Both are situated in the area of the Rio Maior, one of the main sources of flint raw material in the northern region of Estremadura, although these sites were discovered during the course of a research project on the Upper Palaeolithic and were initially classified as being of Solutrean chronology. Although the area excavated was restricted (Casas de Baixo – 3.5 square meters, Olival do Passal – 5 trenches), the information retrieved is of importance in understanding the strategies of flint exploitation, knapping and circulation in Estremadura. These sites specialised in the first stages of debitage of leaf-shaped products, widely represented in the lithic industry of the end of the 4th and all of the 3rd millennium BCE. As at Casal Barril, Casas de Baixo is also home to an irregular pit with a depth of 1.14m (Forenbaher 1999, 49).

The only absolute date available for lithic production contexts in Estremadura is that obtained for Casas de Baixo. However, it would seem to be too old when compared to the dates attributed to the large leaf-shaped products at Liceia (Soares and Cardoso 1995). Thus, Casas de Baixo cannot be associated with Casal Barril with any certainty, although in typological terms they seem to be contemporaneous. The exploitation strategy followed at the two sites, however, are clearly distinct. While Casas de Baixo is dominated by the early stages of the work process required to make leaf-shaped products, at Casal Barril the initial stages of debitage predominate, with the preparation of cores. The lithic industries of contemporaneous settlements, such as at Casal Cordeiro 5, seem to show the debitage process followed during the 3rd millennium BCE was complex. This process was not entirely undertaken within settlements. Rather, leaf-shape pre-forms, cores and probably blades entered circulation for finishing in other places.

3.3. Mine, workshop or stocking place? The functional characterisation of Casal Barril

The interpretation of Casal Barril is ongoing with geoarchaeological and technological studies of the site still to be performed. The lack of similar sites and the limited number of lithic assemblages in contemporaneous contexts make the interpretation of the data obtained difficult.

In Estremadura, a region that abounds in flint, mines with shafts and galleries for flint extraction are unknown. This unexplained situation finds parallels in other regions, such as at Grand-Pressigny (Villes 2004, 312), where flint knapping and circulation was also important. Further research is required, although the pits identified at Casal Barril and Casas de Baixo may represent forms of systematic exploitation of flint outcrops occurring close to the surface. At least in terms of pit filling, one might interpret Casal Barril in a manner similar to Casa Montero, (Consuegra *et al.* 2004, 134), where the excavated pit was gradually filled with knapping remains and rejected flint boulders. However, at Casal Barril the concentration of lithic materials is not found at the base of the pit.

Faced with such an abundance of lithic materials, and a location at the bottom of a valley, it might be that the flint extraction site associated with Casal Barril was located in an adjacent area. A similar situation may have existed at the La Venta flint mine, where two areas of exploitation have been identified at a distance of 200m from one another. The slopes of these areas have been designated "workshop areas" (Ramos Millán 1997).

The analysis of the lithic industry of settlements contemporaneous with Casal Barril seems to indicate that several types of flint were in circulation in Estremadura, each originating from distinct supply areas. The data available for the lithic industry of Estremadura in the 3rd millennium BCE suggests the existence of parallel chains of production for leaf-shaped products and elongated products (blades).

Casal Barril belongs, therefore, to a network of exploitation, debitage and circulation of flint, with yet-to-be-identified intensification and diffusion routes transcending the domestic scale of exploitation. Only from a standpoint of stratigraphic contexts, with absolute chronologies in place, and by comparing data with lithic assemblages from domestic and funerary contexts, can further progress be made.

Finally, this work would be incomplete without some words on the absolute chronology of Casal Barril. The dates proposed so far (see Figure 5) confirm one another. Rapid sedimentation at the site might imply an uninterrupted sequence of use, or two successive phases close together in time. The first interpretation is suggested by level 2 (Camada 2), which has more pebbles and nodules of raw-material and rather fewer artefacts. The second is suggested by level 4 (Camada 4), the contents of which fill the pit in US 7, which has the highest percentage of artefacts, including ceramics. However, in both contexts the same type of flint was used. The absolute chronology of the settlements in the region is important in connecting their activities. For Penedo do Lexim, 12km from Casal Barril, three dates very similar to that obtained for Casal Barril have been suggested: Beta-186854 (2,866-2,482 cal BC 2 sigma), Beta-175775 (2,862-2,489 cal BC 2 sigma), Beta-175774 (2,870-2,498 cal BC 2 sigma) (see Gonçalves and Sousa

2006, 242). Further, one of the dates obtained for Casal Cordeiro 5, located just 936m from Casal Barril, falls into the same period: Sac-2159, 2883-2614 cal BC 2 sigma (Sousa 2008). In fact, all the available data seem to indicate that Casal Barril was almost contemporaneous with the establishment of the fi rst metallurgical communities in the region. It is impossible to know, however, whether its fl int was used in earlier times. The present work highlights the complexity of the fl int industry in the 3rd millennium BCE. in Estremadura and southern Portugal.

Acknowledgements

The authors thank Álvaro de Figueiredo and Adrian Burton for their help with the English version of the manuscript.

References

Andrade, M. and Cardoso, M. 2004. O sítio pré-histórico da Pedreira do Aires (Ramada, Odivelas): notícia da sua identificação. *Revista Portuguesa de Arqueologia.* 7 (1), 137-163.

Cardoso, J. L. and Costa, J. B. 1992. Estação pré-histórica de Barotas (Oeiras). Setúbal Arqueológica. *Setúbal* 9-10, 229-245.

Cardoso, J. L. and Norton, M. F. 1997-98. A oficina de talhe do sílex do Monte do Castelo (Leceia). *Estudos Arqueológicos de Oeiras* 7, 35-46.

Cardoso, J. L., Norton, J. and Carreira, J. L. 1996. Ocupação calcolítica do Monte do Castelo (Leceia, Oeiras). *Estudos Arqueológicos de Oeiras* 6, 287-299.

Choffat, P. 1907. Exploitation souterraine du silex à Campolide aux temps pré-historiques. *O Archeologo Português* 12, 338-342.

Consuegra Rodríguez, S., Gallego García, M. M. and Castañeda Clemente, N. 2004. Mineria neolítica de sílex de Casa Montero (Vicálvaro, Madrid). *Trabajos de Prehistoria.* 61 (2), 127-140.

Ferreira, O. da V. 1984. A "Pebble Culture" ou "Pebble Industry". In *Lucerna - Homenagem a D. Domingos de Pinho Brandão.* Porto, Centro de Estudios Humanísticos.

Forenbaher, S. 1999. *Production and Exchange of Bifacial Flaked Stone Artifacts during the Portuguese Chal-colithic.* Oxford, BAR Publishing, British Archaeological Reports International Series 756.

Gonçalves, V. S. and Sousa, A. C. 2006. Algumas breves refl exões sobre quatro datas ^{14}C para o Castro da Rotura e o 3° milénio nas Penínsulas de Lisboa e Setúbal. *O Arqueólogo Português.* S.4 (24), 233-266.

Ramos-Millán, A. 1997. La Venta. A prehistoric flint mine in a tribal society (Iberian Southeast). In Schild, R.; Sulgostowska, Z. - *Man and Flint. Proceedings of the VII International Flint Symposium,* 117-121. Warszawa, Institut of Archeology and Ethnology Polish Academy of Sciences.

Ribeiro, C. 1880. *Notícia de algumas estações e monumentos pré-históricas do Concelho da Figueira,* vol II, Monumentos megaliticos das vizinhanças de Belas. Memória apresentada à Academia Real das Sciências de Lisboa.

Sousa, A. C. 1998. *O Neolítico final e o Calcolítico na área da Ribeira de Cheleiros.* Trabalhos de Arqueologia 11. Lisboa, Instituto Português de Arqueologia.

Sousa, A. C. 2008. Arqueologia na A21 : uma análise preliminar dos trabalhos arqueológicos 2004-2007. *Boletim Cultural,* 411-497.

Villes, A. 2004. Quel est, aujourd'hui, le "vrai visage" du Grand-Pressigny? In J. Evin (ed)., *Un siècle de construction du discours scientifique en Préhistoire.* XXVI Congrés préhistorique de France. Avignon, 21-25 Septembre 2004, 311-330. Paris, Societé préhistorique française.

Zilhão, J. 1994. A oficina de talhe neo-eneolítica de Casas de Baixo (Caxarias, Vila Nova de Ourém). *Trabalhos de Arqueologia da EAM* 2, 35-45.

Zilhão, J. 1995 '[1997]. O *Paleolítico Superior da Estremadura Portuguesa.* PhD thesis, Universidad de Lisboa, Edições Colibri.

Evidence of flint mining in the Treviño syncline (Basque-Cantabrian Basin, western Pyrenees, Spain)

Antonio TARRIÑO, Alfonso BENITO-CALVO, Pedro J. LOBO, Iosu JUNGUITU and David LARREINA

Abstract

This paper examines the exploitation of lithic resources in the Araico mountains of the Basque-Cantabrian Basin, Spain. The arrangement of silicified strata in this area allows flint to be exploited producing in the terrain essentially trenches and dumps. Archaeological surveys have recovered a great deal of lithic evidence of mining activity, including hundreds of hammer-stones and tens of maces and cores etc., as well as remains related to primary extraction labours, such as dumps associated with the trenches. These trenches were analysed from both geological and archaeological standpoints and the findings related to the lithic remains present. Light Detection and Ranging (LiDAR) technology was shown to be a powerful tool throughout the investigation. The trenches were confirmed to be the product of flint exploitation that made use of specific tools employed in primary extraction and posterior treatment *in situ*. The outstanding quality of the flint from the study area led to its demand in different periods - from the Pleistocene to the Holocene and even to historic times; flint from the area has been found at archaeological sites several tens of kilometres away.

Keywords

Flint mining. Geomorphological cartography. Prehistory. DEM LiDAR. Miranda-Treviño syncline. Mining mace. Mining trenches. Dumps.

1. Introduction

Archaeological background

The first surveys in the study area were undertaken by Deogracias Estavillo, who studied the middle and lower basin of the Ayuda valley, specifically the area around the Araico mountains (Figure 2).

Estavillo's discovery of about 50 ophite hammerstones, the majority complete and with a central groove to allow handling, is remarkable. These hammers, used to exploit flint veins, provide valuable evidence of the intense mining activity that occurred in the region. Estavillo also recorded more than 1000 cores (Estavillo 1957) and a number of structures associated with outcrops that he described as 'pits' (Estavillo 1955, 1957, 1975). Based on flint knapping technology, Estavillo framed his findings within 'a basic Campignoide facies that chronologically showed a strong Upper Palaeolithic tradition along with some pieces from the Metal Ages' (Estavillo, 1957).

Juan Maluquer de Motes related the Araico industries with others previously reviewed by Salvador Vilaseca in the Tarraconense Priorate (Vilaseca 1936, 1973). Maluquer, however, framed them more accurately within the local Neolithic, showing them to be associated with flint workshops and unfortified villages similar to those unearthed in the La Rioja and Navarra regions (Maluquer de Motes 1957, 1966).

Enrique Vallespí made an important contribution to the study of this phenomenon of open air settlements in Álava and the surrounding territories by being the first to report the problems associated with these sites: chronology, reasons for the settlement or activities developed in this type of sites. He analyses a large number of them in the province of Álava and groups them into different geographical

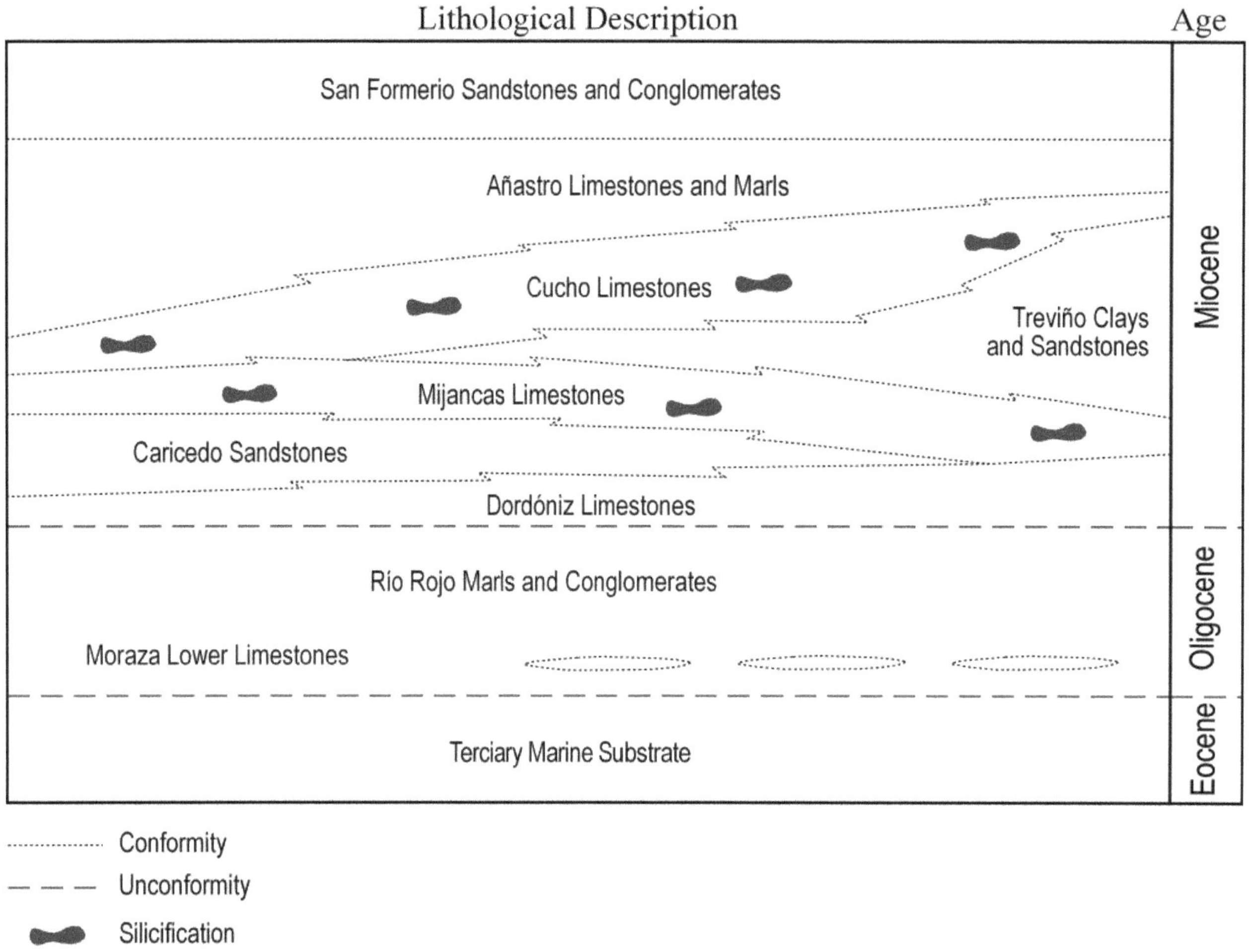

Figure 1. Geological profile of the Araico mountain range and Cucho elevations.

areas: Mountains of Vitoria, basin of the Ayuda river, etc. Based on his previous experience in the region called *Bajo Aragón* (Vallespí 1959, 1961) he distinguishes two groups: flint workshops or quarry settlements and inhabitation settlements. Culturally he puts its beginning in the Neolithic times with a massive development during the Chalcolithic, extrapolating these characteristics to the whole Ebro Valley (Vallespí 1967, 1968, 1972).

In addition to these pioneering works, part of our group had a good experience in field survey projects initiated in the 1980s (Ferreira *et al.* 1984 ; Ortiz *et al.* 1990), and linked to the introduction into Spain of the spatial archaeological methodologies proposed by I. Hodder, C. Orton and D.L. Clarke (Hodder and Orton 1976; Clarke 1978).

The meeting *I Coloquio sobre Distribución y Relaciones entre los Asentamientos*, held in Teruel in 1984, which was attending by Hodder (Hodder 1984; Mederos 1997), helped to consolidate the practice of systematic surveying in Álava, focusing on recent Prehistory. Among the most intensively examined areas there were the Rojo and Ayuda valleys, and some of the surrounding higher areas, such as the Araico-Cucho and Portilla-Moraza mountain ranges. During the due course of this work, some of the sites discovered by Estavillo were revisited and new evidence of mining was found (Ferreira *et al.* 1984; Ortiz *et al.* 1990).

In the last years the use of new GIS and Remote Sensing technologies has allowed to detect and even to chart linear structures associated to mining works, which are just outlined on the ground as small erosions following the silicified layers.

Geographical and geological contexts

The study area lies in the middle-lower basin of the River Ayuda and its tributary, the River Rojo, toward the east. This area involves several villages belonging to thee municipalities of Berantevilla and Treviño.

The Araico mountains, a monoclinal structure that reaches 900m of maximun altitude on its main peak (El Cerro), separates the basins of the rivers Ayuda and Rojo whose talweg ranges between 450-550m form the sea level. This structure continues to the north of the Ayuda, between the villages of Treviño, Cucho, Busto and Golernio in the Cucho Mountains (Figure 2).

Geologically the region is located to the south of the Basque-Cantabrian basin, in the Southern Pyrenees syncline

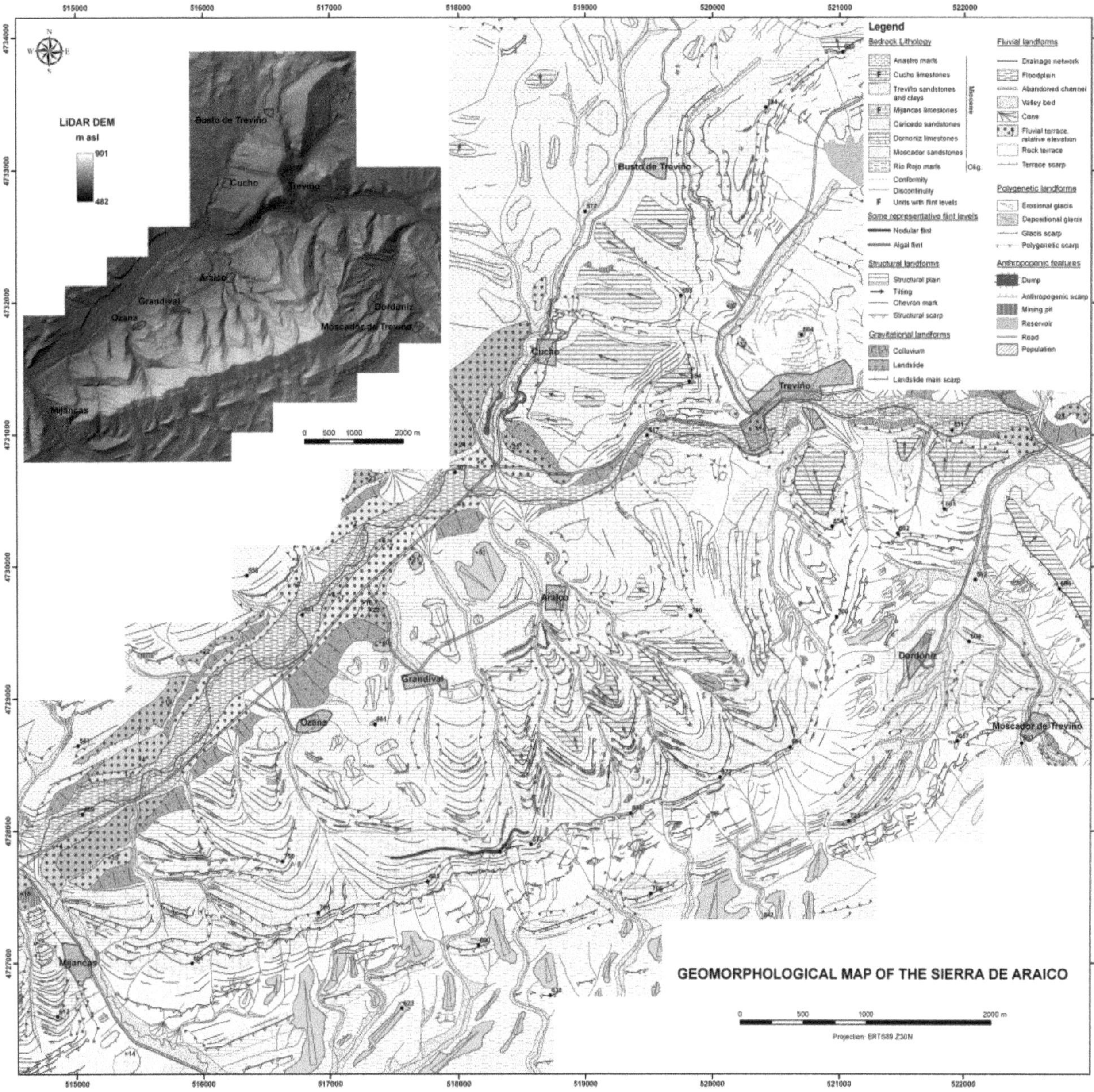

Figure 2. Geomorphological map of Araico mountain range (Benito-Calvo et al. 2010).

(Miranda-Treviño depression), where Miocene continental materials (Aquitanian-Burdigalian) unfolding a platter syncline about 60km in length and 20km wide (oriented approximately E-W). These materials contain significant silicifications of lacustrine-palustrine origin.

2. Analysis and Results

Use of light detection and ranging technology to locate and analyse mining exploitations

Locating and estimating the size of archaeological sites amid mountainous and wooded environments entails substantial difficulties (Kvamme 2005). The remote sensing technologies usually used (orthophotos, stereographic images, satellite images, etc.) only partially solve these problems, which must be tackled by the rather arduous task of field-walking, and even despite these efforts, large areas of considerable archaeological interest remain out of the studies. In the Province of Álava, 47% of which is plant-covered, this 'archeological desert' is easily noticeable within mountanious areas (Badaya, Arkamo, Cantabria, etc.). Light Detection and Ranging (LiDAR) technology can, however, substantially mitigates these problems.

LiDAR: Techonological principles and application

LiDAR technology is based on the emission of laser pulses that bounce off a target (terrain, object, etc.) and the timing of their return to the emission point. This technique allows exceptional high precision measurements to be taken and is used in many disciplines, from the exploration of the universe to ophthalmological surgery.

The development of a Digital Elevations Model (DEM) requires the use of airborne scanning LiDAR system (ALS). This system emits a fan laser beams with a frequency ranging from 30,000-50,000 pulses per second, by a side scanning system, covering an ample terrain surface. An inertial unit controls the acceleration and angular speed on 3 axes of coordinates. GPS equipment registers the exact trajectory of the beam. The GPS timing is also recorded, in addition to the beam deflection angle, the time taken for the beam to return to its, origin and its intensity on return. From these variables the 3D coordinates of signal bounce points can be determined (Maune 2007).

LiDAR technology affords advantages for locating archaeological sites. Each laser beam hits against a more or less circular area ranging from 10cm to 1.5m in diameter. If this 'footprint' contains solid elements, e.g., a rocky surface, each beam is reflected and sends to the receiving scanner a single signal. However, when the beam strike a porous element, e.g., the leaves of a tree, the reflection of the beam will be different depending on the exact impact points. If the power beam is sufficient, part of it will be reflected by vegetation and part will penetrate through the crown cover reflecting back from the ground, creating two replies for the same pulse. The receiving scanner records the differencies between the reply time and the intensity of the reflected beams, allowing the differential measurement of ground and plant canopy parameters. These data are reflected by the generated computer database allowing the production of a high accuracy DEM in areas with dense plant canopy, as well as the identifications of anthropical structures covered by it.

LiDAR in the Basque Country: DEM of the Araico Mountains. Methodology.

The Basque Country has a good digital archive of aerial pictures, orthophotos and topographical maps; which enable a quite accurate knowledge of the terrain for the present day and also from a diachronic point of view, since we have aerial images from the 1930s. All this information is completed with the public access database courtesy of the *Servicio de Cartografía del Gobierno Vasco, Dpto. de Medio Ambiente y Urbanismo* (Cartographic Service of the Basque Country Government. Enviromment and Urban Planning Department). LiDAR images were obtained for the territory thanks to the collaboration of this entity with the *SGIKER. Servicio general de Cartografía y Sistemas de Información Geográfica de la Universidad del País Vasco* (Cartographic Service of the University of the Basque Country). Due to the novelty of LiDAR technology, we did not have a single software package designed *ex profeso* for reading and operating the archives. Therefore, this process has been carried out by using two different platforms.

Most of the work was undertaken using gvSIG software produced by the *Consejería de Infraestructuras y Transportes de la Generalitat Valenciana* (Regional Council of Infraestructures and Transport of the Valencia Government) and developed by the company IVER and the Jaume I University (Castellón de la Plana), in collaboration with several public and private entities.

The reading of LiDAR formats (.LAS and .BIN) was possible thanks to the DielmoOpenLiDAR extension from the company DIELMO 3D SL which interprets the data as point vectorial information, allowing the acces and operation of the database as well as the symbolic representation through altitude, intensity or point classification criteria (Conselleria de Infraestructuras y Transportes 2008).

This application associated to the SEXTANTE module (Equipo SEXTANTE 2008) developed by the University of Extremadura and the Regional Government of Extremadura, allows rasterizing, softening and filling of DEM's empty data cells generated previously.

All the above mentioned applications are distributed as open code software with a GNU GPL license.

Due to the complications of the gvSIG application for operating with large information volumes, commercial software (ArcGis® Desktop 9.3 and 3D Analyst™ software by ESRI™) was used to select the data by category. Only data corresponding to the beams refleted from the ground were used; removing the interference caused by vegetation and other elements not corresponding to the geological surface.

The last phase of the work consisted of shading, editing and printing the final images for being used both during fieldwork and during the production of high-accuracy maps of the research area. The resulting Digital Elevation Model (DEM) was used for the exact localization of archaeological, geological and geomorphological elements. A minimum of 2 points per square meters were used, allowing a DEM raster with 1m side pixels and a precisions of up to 15-20cm.

Geological study

Geology of the silicifications

The geological studies of these terrains are scant and oldfashioned, most of them unpublished and oriented towards the petroleum research performed during the 1950s and 1960s (1956; 1961). Among these studies it is worth mentioning the work of Riba (1976) that analyses the syntectonic evolution of the Treviño-Miranda syncline and the formation of a clearly asymmetric structure due to the northward displacement of the depocentres when the filling of the tertiary depression took place. The axis of this syncline shows an arched shape and its surroundings involve several diapirs (e.g., the Salinas de Añana, Maestu, Ocio, Peñacerrada diapirs and the hidden diapir of Treviño). Evidence of prehistoric flint mining linked to silicifications has been identified in the area (Estavillo 1955, 1957, 1975; Ortiz *et al.* 1990).

These silicifications are very well represented within the geological record, but they went unnoticed by researchers, and were first mentioned by Ramirez del Pozo (1979) in his geological report on the province of Álava. These silicifications are included within the carbonated layers that outcrop from the syncline. They can surpass 1000m in thickness - at Mijancas a thickness of 1100m has been recorded- consist mainly of lacustrine marlstones and limestones (IGME 1978). Next to suchs rocks there are also compacted limestone, dolomite rocks, dolomitic limestone and calcareous dolomites with different degrees of cementation. Some layers have small amounts of clay minerals and some others have large amounts of organic materials usually associated with silicifications. Authors as Siever (1962) suggest that the decomposition of such organic material can cause the formation of flint during early diagenesis (Figure 1).

Regardless of the mineralogical nature of the host rock, initially four microfacies have been distinguished:

- *Micrites with fenestral porosity*. They appear within the limestones and calcareous dolomites including flint nodules with *liesegang* rings. Their formation environment is typically lacustrine, showing the occasional presence of ostracods. The exterior appearance of the flint is characterized by dark colours. Grumous structures are also noticeable by the grouping of peloids referred to as 'clotted micrite' by Esteban and Klappa (1983).

- *Bioclastical micrites with fenestral porosity and abundant ostracods and gastropods*. This microfacies appears in several silcrete horizons at the bottom of the columns. The silicifications from these microfacies have light colours (light brown). Their enviromment of deposit is also lacustrine.

- *Brechoid silcretes*. This denomination covers a large number of layers with silicifications (silcretes) that appear in the column. Pedotubules are noticeable in some of them. There also typical vadoses cement formed at the subaereal-freatic interphase together with eminently brechoids facies which are typical in palustrine formation environments. They tend to be associated to carbonose materials that are responsible of its blackish colouring. Sometimes multitude of little vegetal remains silicificated appear. It can be observed that there are sequences of fibrose quartz filling ancient porous that are interpreted as cementations. Within these sequences appear alternances of quarzine and chalcedonite. Arribas and Bustillo (1985) finds the same textures and although it cannot be discarded that they are caused by the reemplacements of esparitic cements that fill a primary porosity. The fact of the absence of esparite remains points out the most probably existence of primary cementation in vadose environments. They present very dark colours, even black.

- *Micrites with algae lamination.* These are banded micrites with certain content of organic material that stress the parallel lamination. This is a 2m deep level that can be use as guide level due to its characteristic banded aspect. More or less located at the centre of the packet, there is a single layer of banded flint ranging from 5 to 10cm. It is a very singular flint whose presence has been identified largely in archaeological sites. The formation environment is also lacustrine.

All of these silicifications can be interpreted as early diagenetic silicifications in which the silica could came from the alteration of surrounding silicificated minerals (clay minerals constituent of the lacustrine facies where flint is embedded), during humid stages, in a similar way as it happens in numerous silcretes of the lower basin of the Tajo river (Bustillo 1976).

Regarding the silicification process, after the analyses of the lithology and petrography of the flint, these are similar examples to the ones described by Arribas and Bustillo (1985) for lacustrine-palustrine carbonates from the Paleogene of the lower Tajo river basin. Two silicifications models are interpreted: i) silicification of freatic vadose environments related to vertical variations of the phreatic level and ii) silicifications under water layers produced when the deepness of the water diminishes and therefore the concentration of silica increases (Tarriño 2006). In some cases silcretes with a double silicification have been detected.

The lacustrine sedimentation detected in Miocene of the Miranda-Treviño syncline follows a model of carbonate shallow-lake type facies. Using the classification of carbonated shallow lakes proposed by Freytet (1973), the examined sediments correspond to the ones of a lake with a well developed palustrine strip. Every small fluctuation in these lakes can cause the exposition of great areas of calcareous mud to the effects of aerial and phreatic diagenesis and pedogenesis. This type of sedimentation is associated to a great abundance of organic material.

Geomorphological study

Araico mountain range constitutes a structural relief developed in the southern flank of a syncline with WNW-ESE direction that is incised by the Ayuda valley from NNE to SSW. Toward East, this relief is limited by a breached anticline of NW-SE direction (Moscador-Treviño) which forms a semicircular depression (Moscador-Dordóniz) (Figure 2) (Benito-Calvo *et al.* 2010).

The northern hillside of the Araico mountains presents a general slope of 8-12° and it is characterized by chevron marks and cuestas developed in lacustrine limestones dipping between 19-30°. These landforms are incised by V-shaped valleys (Figure 2), characterized by narrow alluvial-colluvial bed valleys flowing towards the main valley of the Ayuda river, ultimately forming torrential cones. On the other hand, the southern hillside is defined by a stepper slope (19°), with valleys incised in a more gentle degree draining into Rojo river, tributary of the Ayuda.

The gravitational landforms are restricted to colluvium -mainly located on the lower part of the valleys- as well as some landslides. The most significant landslide is situated near the syncline axis, associated with a clay formation interbedded within limestones (Figure 2). This translational slide that provokes the diversion of the stream has occurred in the right margin of a narrow valley.

Asymmetrical fluvial terraces and the floodplain are developed in the lower part of the Ayuda valley (Figure 2). The fluvial terraces have a thickness of ~2-4m, composed by sub-rounded and rounded gravels with sands and clays matrix. The relative elevations of the terraces are +40, +30-36, +18-26, +16, +7-9 and +2m above the Ayuda river, framing the fluvial sequence between Middle Pleistocene and Holocene (Santisteban and Schulte 2007) (Benito-Calvo *et al.* 2008). Alternatively, erosive plains associated with rock terraces without significant deposits were described. LiDAR DEM has allowed the identification of elongated depressions within the floodplain that have been related with abandoned channels.

Polygenic landforms corresponding to glacis are also represented in the Araico mountains (Figure 2). These landforms have been identified in relative positions between +160m and +14m. The glacis of higher positions -poorly preserved- probably corresponds to the pediments associated with Middle Miocene-Pliocene planation surfaces described in other zones of the Basque-Cantabrian Range (Benito-Calvo *et al.* 2007). The glacis located in lower positions is related to Quaternary evolution of the valleys and may contain occasional thin covers of sediments (subangular clasts and clays).

Anthropogenic features are systematically associated with populations, roads, hydraulic and farming activities, etc., nevertheless the oldest human features in the earth surface frequently respond to mining activities. The application of LiDAR DEM technology (see below) has allowed us mapping the most representative mining evidences; trenches and dumps (Figure 3). The Araico mountains are plenty of trenches and dumps, although the highest concentration is located on the northern flank of the mountains to the south of the village of Araico. These mining trenches are usually characterized by elongated morphologies that follow the limestone layers and appear associated with dumps.

As a result of the combined analysis of DEM LIDAR and the geomorphological cartography three main conclusions have been drawn:

- Firstly, it has been stated a high correlation between structures that follows the direction of the strata and the most suitable quarry areas.

- Parallel to the previous observation, it has revealed that most of the structures perpendicular to the stratification correspond to erosive and slope processes.

- Finally, the majority of the scarps that follow the contour lines are due to agricultural terracing systems acconditioned during historical times.

Archaeological remains, the mining maces

The present research focuses in a specific item among the assemblage patiently recovered from surface in Álava by Deogracias Estavillo: Mining maces, indispensable tool for the above mentioned extensive mining.

Estavillo (1975) described around fifty of these maces but after our analysis some of them have been reclassified as hammer-stones, loom-weights or undefined implements, leaving the total number of maces in 30. Also noteworthy is that in this paper we name 'mace' to the instrument used for the primary extraction of flint nodules from surrounding rocks in contrast with 'hammer-stone' that is used referring to flint knapping works. Nonetheless this assessment, it is likely that the biggest exemplars of hammer-stones were employed as well for extraction of nodules albeit we establish (see below) a criteria for discriminate them.

The specimens analysed have oblong shape, the smallest ones usually tending to be spherical (Figure 4). They ranged in length from 62mm to 200mm with an average measure between 100 to 150mm. The lightness exemplar weights less than half a kilo (0.46kg) in stark contrast with the heaviest (5.66kg), being the most of them in an intermediate position ranging from 0.70 to 2kg. All the maces show a groove of variable deepness -from barely 3-4mm to 50mm- surfacing them usually in their middle part. This groove is interpreted as a mechanism of handling (Eiroa *et al.* 1999) and can appear surrounding the entire piece, ¾ of it or only in two opposite sides showing a polished surface stronger as more intense is the profoundness. On the orthogonal sides to the grooves there are other marks consisting of circular close-pitted bruising and occasionally end-flaking coupled. These traces have been described unequivocally as symptoms of hammering hard rocks (Pickin and Timberlake 1988). Regarding the raw material, all the maces are made out of ophite. It is necessary to note that there is no such rock available in Treviño, althrough it is available in the near Peñacerrada diapir.

On the other hand, the hammer-stones have spherical shape with symptoms of pitting disposed erratically around its entire surface. As a rule, if they are manufactured utilizing flint they have bigger size (around 150mm of diameter) than if they use river cobbles. There are no traces of handing mechanism or end-flaking.

Apart from the specimens of the Estavillo collection here examined, several other examples of maces have been recovered from Treviño (Ortiz *et al.* 1990; Tarriño 2006) and the occasional appearance of another exemplar picked up while fieldwork seasons were taking place is not infrequent.

3. Discussion and Conclussion

Application of LiDAR technology and geomorphological mapping

Within the present study framework, the analysis and specifically the information treatment of LiDAR has opened a genuine revolution regarding the methodology of photointerpretation and fieldwork. Nonetheless, the high accuracy Digital Elevation Model (DEM) –previously obtained through the cloud points LiDAR- has permitted to demarcate in an exceedingly effective way a substantial number of susceptible flint mining points that, in some other way, would have been imperceptible hidden by the plant canopy in some cases or by an insufficient detail of the DEM interpolated from other cartographical resources (topographic maps).

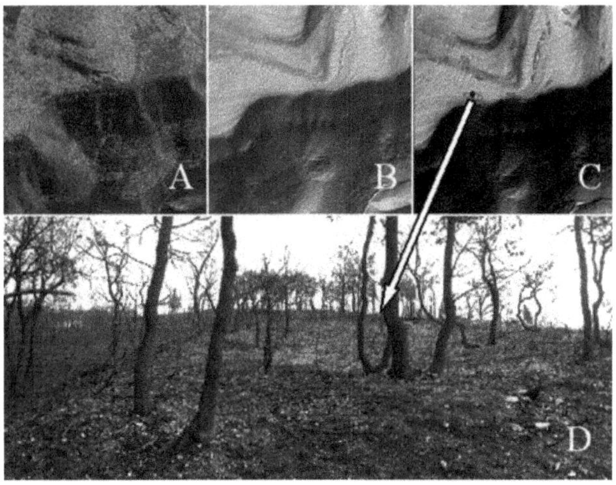

Figure 3. Outline of the methodological procedure applied to LiDAR information analysis focused on a specific zone of the research area (Alto de San Miguel surroundings). A) Digital Surface Model (DSM), including vegetation and anthropogenic features, generated through the cloud points LiDAR B) Digital Terrain Model (DTM), obtained interpolating ground points previously selected and classified. C) Photo interpretation and location of potential interest areas. D) Image of a pit used for flint exploitation previously detected with DTM LiDAR and lately checked on the field.

Additionally, the mentioned areas have been characterised in a great scale detail attending to its geological and geomorphological context, relating them with the relief and processes where they are framed. Therefore, although fieldwalking seasons are indispensable in any case, there must not pass unnoticed the primordial role on the preli-

Figure 4. Image showing the best preserved maces analysed. The exemplar of the right was handled for its exhibition in the Archaeology Museum of Álava. At the centre of the picture, middle row, there is a hammer-stone made of flint, being noticeable the absence of groove and the spherical shape. Miguel surroundings). A) Digital Surface Model (DSM), including vegetation and anthropogenic features, generated through the cloud points LiDAR B) Digital Terrain Model (DTM), obtained interpolating ground points previously selected and classified. C) Photo interpretation and location of potential interest areas. D) Image of a pit used for flint exploitation previously detected with DTM LiDAR and lately checked on the field.

minary stage of potential areas of interest as well on the previous selection of the points to check on the field.

Finally, it is remarkable the utilities derived from the utilisation of geomorphological mapping applied to the pursued aims. As a matter of fact, uniquely by means of this method it has been possible to discriminate the morphogenetic context each of the detected features that are inscribed in. Therefore, this phase has revealed as crucial to classify the mentioned features in two main outfits; part of them corresponds to anthropic processes whilst the rest is associated to natural landform evolution processes.

Microespatial perspective: Treviño flint mining exploitation

From a microespatial point of view, the homoclinal disposition of the Araico mountains permits that its colluvial deposits could have been exploited from the Pleistocene. From Neolithic onwards, the existence of silicifications included within poorly cemented host rocks has allowed to carry out relatively easy simple mining exploitations: trenches with or without dumps that can reach several tens of meters of development and pits apparently shallow. The anthropical character of the mentioned structures is reinforced by its disposition on the lines of the layers outlined by the silicifications (two of those have been mapped) corresponding to nodular flint and algae stripped flint (Figure 2) and by the existence of tens of maces and hammerstones, as well as thousand of waste flakes, nucleus and chipping products.

Macroespatial perspective: Treviño flint diffusion

Recent researches concerning the provenance of lithic raw materials reveal that the different types of Treviño flint have been largely distributed within archaeological sites during the Western Pyrenees Prehistory (Figure 5). Some of them are framed in the Pleistocene, e.g., Altamira (around 130km away in straight line from Treviño) (Tarriño *et al.* in press), Las Caldas (260km) (Corchon *et al.* 2009), Antoliñako Koba (75km) (Tarriño 2006), Bolinkoba (50km) (Tarriño and Aguirre 1997), Urratxa-III (45km) (Tarriño 1997), Labeko Koba (50km) (Tarriño 2000), Brassempouy (210km) (Tarriño *et al.* 2007), Isturitz (140km) (Tarriño and Normand 2006) or Mugarduia (50km), the last one currently in study, whereas other are settled in the Holocene, e.g., Mendandia (15km) (Tarriño 2005), Kanpanoste Goikoa (25km) (Tarriño 1998), Las Yurdinas-II (10km) (Tarriño 2003), Atxoste (25km) or Los Cascajos (50km), the last ones currently in study.

In these cases it is noticeable a significant transportation of Treviño flint to further distances that can exceed considerably a hundred of kilometers. This shows up the important displacements related to the ways of life of prehistoric populations, which currently are under discussion with the head archaeologists of the involved archaeological sites.

Flint mining evidences: the human traces

Based on the lithic assemblage found on surface during the surveying of the Araico mountains, Estavillo (Estavillo 1975) classified the settlements into two different groups: 'villages' and 'workshops'. The first ones were characterised by the presence of fireplaces, pottery and finished lithic objects like knives, arrow heads, axes, hand-mills, etc. whereas the workshops showed absence of those three elements being replaced by maces, picks, hammer-stones, flint cores, flakes and chipping floors.

It is our opinion that the workshops correspond to the evidences of mining activities (trenches, pits, dumps, etc.) previously described. Migal (1997) outlines four methods of extracting flint depending on the complexity and technological degree employed. He mentions small exploitations -similar to those located at Treviño- as the simplest method, pointing out that they correspond to seasonal activities, but he also introduce a more specialized type of mine 'deeper niche units dug into the most dangerous layers of eroded rock. Here we see a tendency for extensive exploitation utilizing natural fissures in the rock mass, with the miners selecting directions of work offering the easiest conditions.'

For the accomplishment of such mining operations heavy tools like maces are crucial. The described grooved maces are a well known type of mining tool in Europe (Piel-Desruisseaux- 1989), also it is often recorded in metal mining contexts sometimes with variants in morphology or handling (Pickin and Timberlake 1988; Pickin and Worthington 1989; Guilbert 1994). These implements were made out of a specific rock –ophite- harder than the host rock and appropriate to attack it. This practice of utilizing rocks foreign to the site for mining activities has been reported before.

The appearance of massive chipping floors, flakes, cores, hammer-stones, profiling-hammers and other evidences of knapping close to the maces and picks is perfectly coherent with the mining industry, 'the objective of such onsite flint working was to reduce the mass of the blanks and semi-finished tools for transport' (Lech 2008).

The exploitation of the above exposed flint mining resources, should be understood as the first step of a chaine-operatoire that, in the case of Treviño, reveals a complex lithic production that involves far distance contacts.

In this context of human progress the management of resources of Treviño flint fits perfectly since, on the one hand this material has outstanding characteristics and quality and his long distance distribution has already been demonstrated elsewhere (e.g., Tarriño 2006).

Chronology

There are few data about the chronological period in which such activities took place; Estavillo (Estavillo 1975) pro-

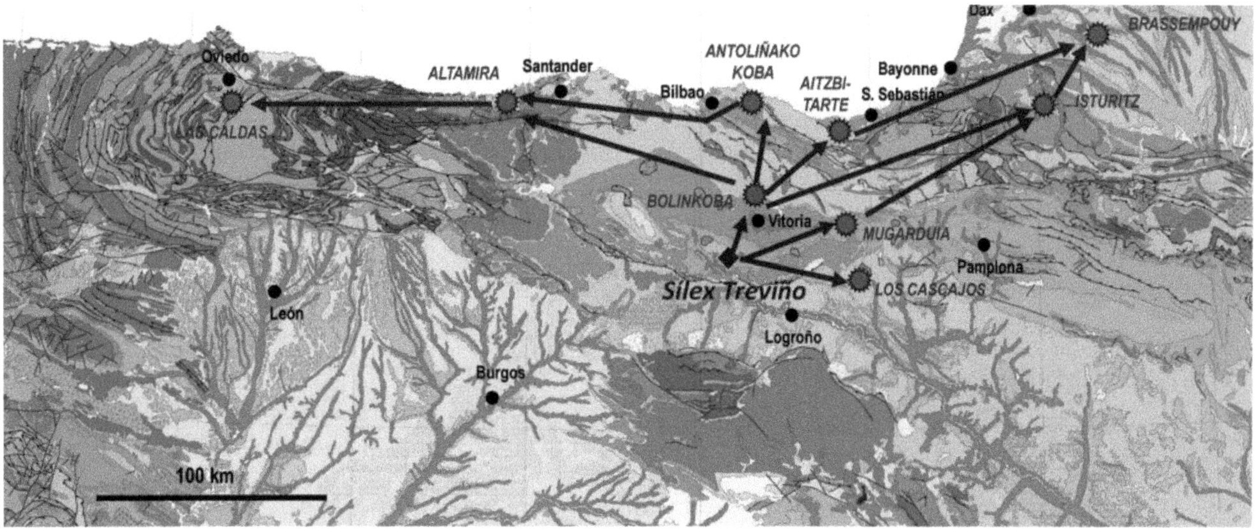

Figure 5. Diffusion of Treviño flint towards Pleistocene and Holocene archaeological sites located on the Cantabrian corniche and Western Pyrenees.

posed three moments: Bronze Age, Iron Age and Roman times. From our point of view this chronological frame is rather short since there are indications pointing out a longer development from Palaeolithic to Contemporary history.

During the last decade, the advance in the characterisation and identification of Treviño flint (Tarriño 2006) has permitted the identification of this raw material in cave settlements along the entire Franco-Cantabrian region, ascertaining its use at least from Aurignacian period (Brassempouy, Isturitz, etc.) (Tarriño and Normand 2006) (Tarriño *et al.* 2007).

However, the tools employed for primary extraction collected from surface (hammerstones, maces, etc.) as well as the very exploitation system of trenches and dumps seem to relate more intensively the Treviño flint mining activities with periods of the Recent Prehistory. According to these facts, the systematic surveying of the Rojo river valley during the 1980s revealed large concentrations of post-Palaeolithic (Neolithic, Chalcolithic and Bronze Age) settlements next to the southern slope of the Araico mountains (Ortiz *et al.* 1990), immediately under the maximum concentration point of mining activities (Alto de San Miguel).

It is obvious notwithstanding, that the chronological sequence of Treviño flint exploitation only can be satisfactorily resolved by means of test-pitting or archaeological excavation on the located points of extraction.

Acknowledgements

This work has been funding by the HAR2008-05797 project and Ramon y Cajal Programme, RYC-2007-01626 both of the *Ministerio de Ciencia e Innovación* (Science & Innovation Department, Spanish Government) also with contribution of the European Union (FSE); PhD fellowship funded by the *Dirección de Política Científica del Departamento de Educación, Universidades e Investigación del Gobierno Vasco* (Direction of Scientific Policy of Education, Universities and Research Department of the Basque Country Government). Cartography courtesy of the *Servicio General de Cartografía y Sistemas de Información Geográfica de la Universidad del País Vasco (SGIKER)* (Service of Cartography and GIS of the Basque Country University (SGIKER). LiDAR images courtesy of the *Servicio General de Cartografía del Gobierno Vasco* (Cartographic Service of the Basque Country Government). Thanks as well to the *Museo de Arqueología de Álava* (Museum of Archaeology of Álava) for their help and attention.

References

Arribas, M. and Bustillo, M. A. 1985. Modelos de silicificación en los carbonatos lacustres-palustres del Paleógeno del borde NE de la Cuenca del Tajo. *Boletín Geológico y Minero* 96, 325-343.

Benito-Calvo, A. and Pérez-González, A. 2007. Erosion surfaces and Neogene landscape evolution in the NE Duero Basin (north-central Spain). *Geomorphology* 88, 226-241.

Benito-Calvo, A., Pérez-González, A. and Parés, J.M. 2008. Quantitative reconstruction of late Cenozoic landscapes: a case study in the Sierra de Atapuerca (Burgos, Spain). *Earth Surface Processes and Landforms* 33, 196-208.

Benito-Calvo, A., Tarriño, A., Lobo, P.J., Junguitu, I. and Larreina, D. 2010. Geomorphology and prehistoric flint mining evidence in the Sierra de Araico (Basque-

Cantabrian Basin) Burgos-Alava, Spain. *Journal of Maps* v2010, 584-590. 10.4113/jom.2010.1147

Bustillo, M. A. 1976. Estudio petrológico de las rocas silíceas miocenas de la Cuenca del Tajo. *Estudios Geológicos* 32, 451-497.

Clarke, D.L. 1978. *Analytical Archeology*. 2ª Ed. London. Methuen.

Conselleria e Infraestructuras y Transportes and DIELMO 3D SL. 2008. *GvSIG 1.1.2. Extensión Dielmo Open LiDAR 0.1. Manual del Usuario 1.0.* Valencia. Generalitat Valenciana.

Corchón, M. S., Tarriño, A. and Martínez-Quintana, J. 2009. Mobilité, térritoires et relations culturelles au début du Magdalénien moyen cantabrique: nouvelles perspectives, in F. Djindjian, J. Kozlowski and N. Bicho, *Proceedings of the XV World Congress UISPP (Lisbon, 4-9 September 2006). Le concept de territoires dans le Paléolithique supérieur européen.* Oxford, BAR Publishing. British Archaeological Reports International Series 1938.

Devereaux, B. J., Amable, G. S., Crow, P. and Cliff, A. D. 2005. The potential of airborne LiDAR for detection of ar-chaeological features under woodland canopies. *Antiquity* 79, 648-660.

Eiroa, J. J., Bachiller, J. A., Castro, L. and Lomba, J. 1999. *Nociones de tecnología y tipología en Prehistoria.* Barce-lona. Ariel.

Equipo SEXTANTE. 2008. *Conceptos Generales de SEXTANTE sobre gvSIG*. Plasencia. Universidad de Extremadura.

Estavillo, D. 1955. Las industrias líticas de Araico (Condado de Treviño). *Zephyrus* 6, 172-177.

Estavillo, D. 1957. El Neolítico de facies Campiñiense de Araico-Treviño. *IV Congreso Internacional de Ciencias Prehistóricas y Protohistóricas*. Madrid, 1954, 433-443.

Estavillo, D. 1975. Contribución a la prehistoria del País Vasco - Los hallazgos líticos de la zona de Araico; un po-blado protohistórico en Portilla la alta y otros datos de ar-queología Treviñesa. *Estudios de Arqueología Alavesa* 8, 11-47.

Estavillo, D. 1980. *El Condado de Treviño*. Vitoria-Gas-teiz, Institución Sancho el Sabio.

Esteban, M. and Klappa, C. F. 1983. Subaerial exposure environments, in P. A. Scholle, D. G. Bebout and C. H. Moore, *Carbonate Depositional Environments, American Association Petroleum Geologists Memoir* 33, 1-54.

Ferreira, A., Tarriño, J. M., Lobo, P. J., Ortiz, L. and Vi-vanco, J. J. 1984. Avance al estudio de los asentamientos al aire libre en la cuenca del Río Rojo (Álava). *Arqueología Espacial* 3, 119-136.

Freytet, P. 1973. Petrography and paleo-environment of continental carbonate deposits with particular reference to the Upper Cretaceous and Lower Eocene of Languedoc. *Sedimentary Geology* 10, 25-60.

Guilbert, G. 1994. Hammer-stones from the copper-mining site at Ecton, Staffordshire. *Bulletin of the Peak District Mines Historical Society*. 12 (3), 26-27.

Hodder, I. 1984. New generations of spatial analysis in Archaeology. *Arqueología Espacial* 1, 7-24.

Hodder, I. and Orton, C. 1976. Spatial Analysis in Archaeology. *New Studies in Archaeology* 1. Cambridge. Cambridge University Press.

IGME 1978. *Mapa Geológico de España, 1/50.000. Hoja núm. 138 (Treviño)*. Madrid. Ministerio de Industria y Energía.

Kvamme, K. L. 2005. Terrestrial Remote Sensing in Archaeology, in H. D. G. Maschner and C. Chippindale (eds.), *Handbook of Archaeological Methods*, 423-477. Lanham, Altamira Press.

Lech, J. 2008. In the realm of the Devil. *Focus on Archaeology Academia* 2 (18), 16-19.

Maluquer de Motes, J. 1957. Las comunidades prehistóricas alavesas y sus problemas. *Boletín de la Institución Sancho el Sabio* Año I Tomo 1. (1-2), 51-64.

Maluquer de Motes, J . 1966. Consideraciones sobre el problema de la formación de los vascos. IV Symposium de Prehistoria Peninsular (Pamplona 1965). *Pyrenae* 2 (192-93), 115-129.

Maune, D. F. 2007. *Digital elevation Technologies and Applications: The DEM Users*. Bethesda, ASPRS.

Mederos Martín, A. 1997. Una revolución aparente. El impulso de los coloquios sobre Arqueología Espacial en la renovación disciplinar española de los años ochenta. *Zephyrus* 50, 305-321.

Migal, W. 1997. Selected aspects of specialization in mining and flint knapping, in R. Schild and Z. Sulgostowska (eds.), *Proceedings of the VIIIth International Flint Symposium Warszawa-Ostrowice, September 1995*, 99-101. Warszawa, Institute of Archaeology and Ethnology, Polish Academy of Sciences.

Ortíz, L., Vivanco, J. J., Ferreira, A., Lobo, P. J., Muñoz, M. D., Pinillos, R., Tarriño, A. and Tarriño, J. M. 1990. *El hábitat en la prehistoria del valle del Río Rojo (Álava).* Cuadernos de Sección de Eusko-Ikaskuntza (Prehistoria-Arqueología) 3. Donostia, Eusko Ikaskuntza.

Piel-Desruisseaux, J. L. 1989. *Instrumental Prehistórico, forma, fabricación, utilización.* Barcelona, Masson.

Pickin, J. and Timberlake, S. 1988. Stone hammers and fire setting: a preliminary experiment at Cwmystwyth mine, Dyfed. *Bulletin of the Peak District Mines Historical Society* 10 (3), 165-167.

Pickin, J. and Worthington, T. 1989. Prehistoric mining hammers from Bradda Head, Isle of Man. *Bulletin of the Peak District Mines Historical Society* 10 (5), 274-275.

Ramírez del Pozo, J. 1973. *Síntesis geológica de la provincia de Álava.* Vitoria-Gasteiz, Caja de Ahorros Municipal de la ciudad de Vitoria.

Riba, O. 1956. *La cuenca terciaria de Miranda-Treviño.* Unpublished report, CIEPSA.

Riba, O. 1961. *Informe sobre el terciario de Treviño.* Unpublished report, CIEPSA.

Riba, O. 1976. Tectogenèse et sédimentation: deux modèles de discordances syntectoniques pyrénéennes. *Bulletin du Bureau de Recherches Géologiques et Minières* 4, 85-103.

Santisteban, J. I. and Schulte, L. 2007. Fluvial networks of the Iberian Peninsula: a chronological framework. *Quaternary Science Reviews* 26, 2738-2757.

Siever, D. F. 1962. Silica Solubility, 0ºC-200ºC, and the diagenesis of siliceous sediments. *Journal of Geology* 70, 127-150.

Tarriño, A. 1997. Análisis de restos líticos silíceos del yacimiento arqueológico de Urratxa III (Orozko, Bizkaia), in M. Muñoz Salvatierra y E. Berganza (eds.), *El yacimiento de la cueva de Urratxa III (Orozko, Bizkaia)*, 93-121. Bilbao, Universidad de Deusto.

Tarriño, A. 1998. Análisis de restos líticos silíceos del yacimiento arqueológico de Kanpanoste Goikoa (Vírgala, Álava). In A. Alday (dir.), *El depósito arqueológico de Kanpanoste Goikoa (Vírgala, Álava).* Memoria de Yacimientos Alaveses 5, 105-108. Vitoria-Gasteiz. Diputación Foral de Álava.

Tarriño, A. 2000. Estudio de la procedencia de los sílex recuperados en el yacimiento de *Labeko Koba* (Arrasate, Gipuzkoa), in A. Arrizabalaga and J. Altuna (eds.), Labeko Koba (País Vasco). *Hienas y humanos en los albores del Paleolítico superior*, Munibe 52 (Antropología-Arkeologia), 345-354. San Sebastian.

Tarriño, A. 2003. Estudio de la procedencia de los efectivos líticos del yacimiento arqueológico de las Yurdinas-II (Peñacerrada-Urizaharra, Álava). In J. Fernández Eraso (dir.), *Las Yurdinas-II: un depósito funerario entre finales del IV y comienzos del III milenio BC.* Memorias de Yacimientos Alaveses 8, 93-99.

Tarriño, A. 2005. Fuentes de aprovisionamiento de los silex del yacimiento de Mendandia (Sáseta, Condado de Treviño –Burgos-). In A. Alday (ed.), *El campamento prehistórico de Mendandia: Ocupaciones mesolíticas y neolíticas entre el 8500 y el 6400 BP,* 465-486. San Sebastián, Fundación J.M. Barandiarán.

Tarriño, A. 2006. *El silex en la cuenca vasco-cantábrica y pirineo navarro: caracterización y su aprovechamiento en la prehistoria.* Museo Nacional y Centro de Investigación de Altamira 21. Madrid, Ministerio de Cultura.

Tarriño, A. and Aguirre, M. 1997. Datos preliminares sobre fuentes de aprovisionamiento de rocas silíceas en algunos yacimientos paleolíticos y postpaleolíticos del sector oriental de la Cuenca Vasco-cantábrica. *Veleia* 14, 101-106.

Tarriño, A. and Normand, Ch. 2006. Procedencia de los restos líticos en el Auriñaciense Antiguo (C 4b1) de Isturitz (Pyrénées-Atlantiques, Francia). *Espacio, Tiempo y Forma. Serie I, Prehistoria y Arqueología.* 15, 135 – 146.

Tarriño, A., Bon, F. and Normand, Ch. 2007. Disponibilidad del sílex como materia prima en la Prehistoria del Pirineo occidental, in N. Cazals, J. González Urquijo and X. Terradas (coords.), *Fronteras naturales y fronteras culturales en los Pirineos Prehistóricos.* 103 – 123. Santander. Monografías del Instituto Internacional de Investigaciones Prehistóricas de Cantabria.

Tarriño, A., Cava, A. and Barandiarán, I. i.p. Materias Primas líticas de las industrias del Solutrense Cantábrico: el ejemplo de Altamira (Cantabria, España). *Actas del Congreso: Le Solutréen ... 40 ans après Smith'66 (Preuilly-sur-Claise, Touraine, Francia, 2007).*

Vallespí Pérez, E. 1959. Bases arqueológicas para el estudio de los talleres de sílex del Bajo Aragón. *Caesaraugusta* 13-14, 7-20.

Vallespí Pérez, E. 1961. Síntesis del estado actual del conocimiento de las industrias macrolíticas post-paleolíticas del cuadrante NE de España. *VI Congreso Nacional de Arqueología* (Oviedo, 1959), 64-71.

Vallespí Pérez, E. 1967. Talleres de sílex al aire libre en Álava. *Ampurias* XXIX, 231-234.

Vallespí Pérez, E. 1968. Talleres de sílex al aire libre en el País Vasco Meridional. *Estudios de Arqueología Alavesa* 3, 7-27.

Vallespí Pérez, E. 1972. Conjuntos líticos de superficie del Museo Arqueológico de Álava *Estudios de Arqueología Alavesa* 5, 7-79.

Vilaseca, S. 1936. *Les estacions-tallers del Priorat i extensions.* Reus, Llibreria Nacional i Estrangera.

Vilaseca, S. 1973. Las industrias de cantos rodados del Cabo de Salou. *Ampurias* 35, 1-108.

Verri, G., Barkai, R., Bordenau, C., Gopher, A., Hass, A., Kaufman, A., Kubik, P., Montanari, E., Paul, M., Ronen, A., Weiner, S. and Boaretto, E. 2004. Flint mining in prehistory recorded by in situ-produced cosmogenic ^{10}Be. *Proceedings of the National Academy of Sciences of the United States of America* 21, 7880-7883.

The impact of geological factors on flint minig and large blade production in the Betic Cordillera (Spain) in the 4th–3rd mill. BC

Antonio MORGADO and José A. LOZANO

Abstract

Siliceous rocks are abundant in the Betic Cordillera (Andalusia, Spain). These resources were used by human groups in the Late Neolithic and the Copper Age for manufacturing large flint blades (4th-3rd mil. BC), but their exploitation has not been studied systematically. The present work describes the geological background to flint exploitation in this region. The geological features of the areas with flint outcrops determined how readily raw materials could be found, which in turn influenced the mining techniques used. One group of exploitations (those of the Campo de Gibraltar Complex), set against a background of conglomerates and tectonic *mélanges*, involved the use of open air quarrying techniques. A second group, characterized by the continuity of the flint levels of the Milanos Formation (Middle Subbetic), required pit mining often be practised for the sought-after flint to be extracted.

Keywords

Flint mining. Large blade production. Copper Age. Betic Cordillera. Spain.

1. Introduction

Archaeological evidence indicates that a technical change in mid-sized and large flint blade production took place in southern Spain in the Late Neolithic (*c.* 3700 BC). The knapping methods and techniques used in this period are very different to those of the Early Neolithic (Martínez and Morgado 2005; Pelegrin and Morgado 2007) (Figure 1). Large blade craftsmanship remained a major human activity until well into the Copper Age, when such blades disappeared from the archaeological record alongside the dissemination of Beaker culture pottery (*c.* 2400/2300 BC).

The technical features of the blades from this period underline the need for highly selected raw materials, both in terms of quality and block size. The best flint outcrops in southern Spain were exploited intensely throughout the Neolithic, and a series of innovations in the blade production process made this region stand out from the rest of Europe (Morgado *et al.* 2008; Morgado *et al.* 2009). A number of flint mining sites associated with blade production have been located throughout the Betic Cordillera (Fernández Ruiz and Márquez Romero 1985; Vallespí *et al.* 1988; Ramos Muñoz 1997; Morgado 2002). However, there has been a dearth of petrological analyses of the raw materials they contain, and no archaeological survey has been conducted to document the mining activities associated with these sites.

The aim of the present study was to establish the geological characteristics of the flint outcrops known to date in southern Spain. This should help in the conceptualisation of the geological factors involved in the flint mining processes followed in the Betic Cordillera. This study is part of a wider research line on flint technology in southern Spain.

2. Geological context

The mining sites included in this analysis are located in the Betic Cordillera, in the south and southeast of the Iberian Peninsula. Three major geological units exist in the Betic Cordillera, established on the basis of lithological,

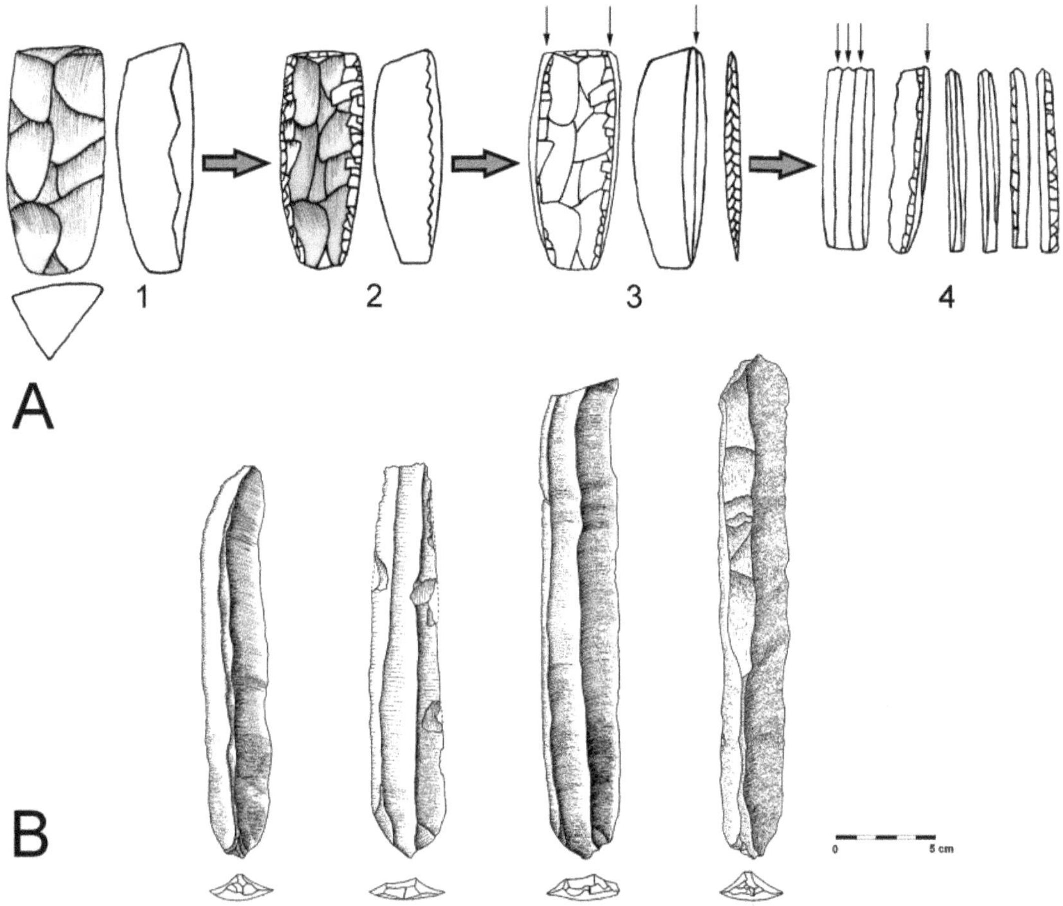

Figure 1. A. Method of blade making (Late Neolithic-Copper Age, southern Spain). 1 and 2: Preparation of core. 3: First extractions of crested blades. 4: Debitage. B. Large Blades from Cueva de Los Molinos (Granada Museum, Spain).

stratigraphic, petrological and structural factors (see Vera 2004 for a review): the External Betic Zone, the Campo de Gibraltar Complex, and the Internal Betic Zone. Each unit has its own subdivisions (Figure 2).

Two major tectono-stratigraphic settings make up the External Betic Zone: the Prebetic and Subbetic. The Prebetic area is located towards the north of the Betic Cordillera and consists mainly of quasi-autochthonous units with locally allochthonous subdivisions. The Subbetic area is composed of intensively deformed allochthonous units. The materials of the External Betic Zone were deposited at the southern Iberian palaeomargin during the Mesozoic and Cenozoic as a result of episodes of intracontinental fracturing, convergence and collision (García Hernández *et al.* 1980, 1989; Vera 1988, 2004). The Prebetic unit is composed of Triassic to Miocenic sedimentary rocks deposited at the southeastern edge of the southern Iberian palaeomargin. Sediments from shallow marine waters with short intervals of continental episodes prevail in the Prebetic area. Consequently, a profusion of detritic materials is found in this area. No flint mining sites associated with blade technology have to date been found in the Prebetic area.

The Subbetic unit covers the southernmost External Betic Zone and is composed of a sequence of sedimentary materials ranging from the Triassic to the Middle Miocene. In addition, pelagic facies are abundant starting in the Upper Lias. Volcanic and subvolcanic rocks have been also documented. The Subbetic unit has been divided into three paleogeographic subdomains: the External, Middle and Internal Subbetic subdomains. The External Subbetic domain expands throughout the northernmost section of the Subbetic area and is composed of Jurassic limestones and condensed facies depositions in boundary environments. Most of the Middle Subbetic subdomain is part of a subsiding sedimentary environment. Jurassic and Cretaceous marly facies are abundant. There are also interspersed volcanic rocks from the Middle and Upper Jurassic. The Internal Subbetic subdomain is located in the southernmost sector of the Subbetic unit, and possesses calcareous facies that were deposited in ocean-boundary environments during the Middle and Upper Jurassic, a period during which this subdivision was not actively subsiding. An additional subdomain with specific stratigraphic, paleogeographic and tectonic features - the Penibetic unit - emerges in the western sectors of the Betic Cordillera.

The Campo de Gibraltar Complex is located over the remaining units of the Betic Cordillera. It has a variable tectonic setting and is composed of a set of allochthonous units detached from their original substrate. It is unaffected by Alpine metamorphism. These units make up the Meso-Cenozoic cover that expands from the Middle-Upper Jurassic to the Lower Miocene. This cover is composed of deep marine facies made from clay and, to a lesser extent, marls. Siliciclastic turbidites (flysch facies) are abundant in the Campo de Gibraltar Complex, especially for the Lower Cretaceous and the period between the Upper Oligocene and Lower Miocene.

The Internal Betic Zone has traditionally been divided into three tectonically superimposed complexes: the Malaguide, Nevado-Filabride, and Alpujarride complexes (from top to bottom). They also incorporate a frontal unit (frontal units of the Internal Betic Zone) in contact with the Campo de Gibraltar Complex and the External Betic Zone. These frontal units are at the front of the Internal Betic Zone in the Central and Western sectors of the Betic Cordillera (between the Malaguide Complex and the External Betic Zone). External and internal frontal units can be distinguished according to their tectonic setting, stratigraphic features and facies (Martín-Algarra *et al.* 2004). The Malaguide Complex at the top is composed of a slightly metamorphosed or unmetamorphosed Palaeozoic basement of detritic material (Sanz de Galdeano 1997). Red Permic-Triasic detritic sediments, and a thin marine sedimentary layer of Liasic dolomites and limestones overlay the Palaeozoic basement. The remaining Jurassic and Cretaceous materials are mostly limestones or marly limestones with pelagic influences. According to Sanz de Galdeano (1997) three superimposed groups of units exist in the Alpujarride Complex: phyllites and quartzites, gypsum and basic igneous intrusions, and thick sections of dolomite or dolomitic marbles and limestones with metapelitic intercalations (from bottom to top). Bodies of peridotites are located at the base of some of the units. Most of the shale results from the Palaeozoic sediments under the influence of the Alpine orogeny, while the phyllites result from Lower Triassic sediments. Finally, the carbonates composing the cover have been dated to the Middle and Upper Triassic. The Nevado-Filabride Complex is the deepest tectonic unit and is composed exclusively of metamorphic rocks. Its stratigraphic sequences are similar to those of the Alpujarride Complex, although with a more ancient Palaeozoic basement (Precambrian) and a partly carbonated Triassic/post-Triassic cover.

3. Mining sites associated with the production of large flint blades

Flint mining sites from the Late Neolithic and Copper Age have been documented in the central and western sections of the Betic Cordillera between the Provinces of Granada and Malaga (Figure 2). Taking the geological units described above as a reference, these mining sites can be grouped into three distinct areas: the Middle Subbetic area, the

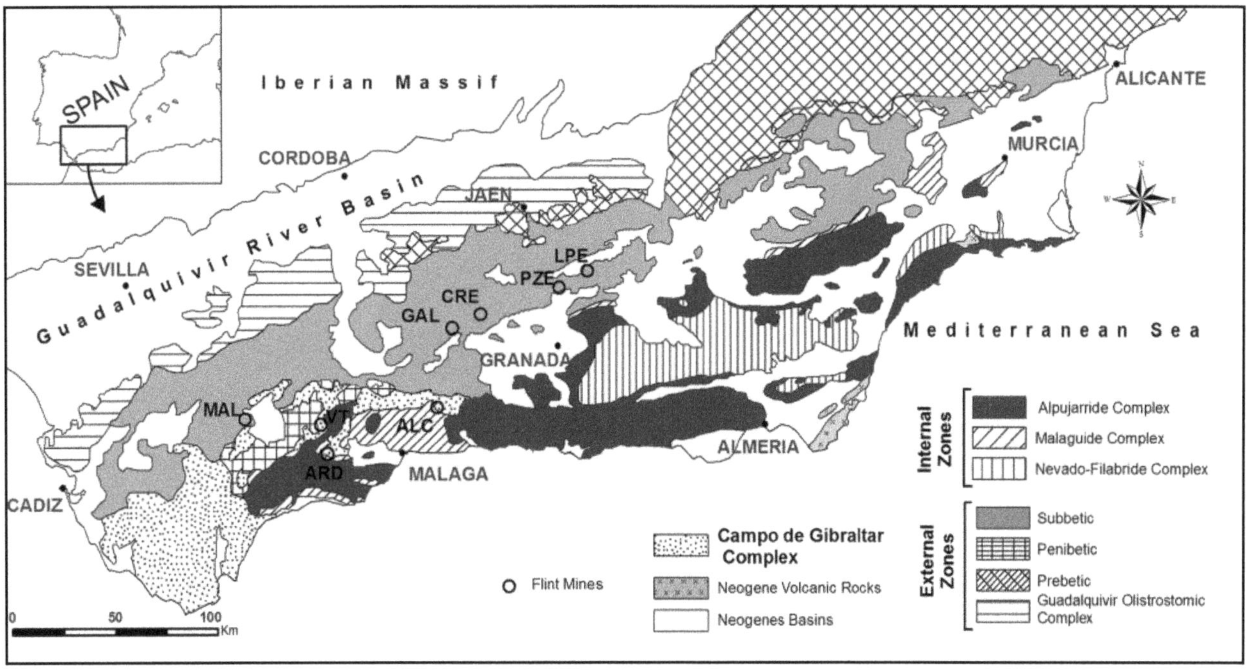

Figure 2. Geological map of the studied flint mines, southern Iberian Peninsula.

Campo de Gibraltar Complex, and the Malaver unit. The Middle Subbetic unit and Campo de Gibraltar Complex are part of the Neonumidian flysch.

a. Mining sites in the Middle Subbetic Zone

Scattered references since the early 20th century mention the existence of flint quarries dated to late prehistoric times in the Middle Subbetic Zone. Early findings were made in the vicinity of the towns of Piñar and Iznalloz (Obermaier 1934). However, no thorough analysis of these sites has ever been conducted.

The best known site is located in the Gallumbares valley (GAL-Loja, Granada). Since the 16th century, military engineers referred to this valley as Spain's largest quarry for gun-flint production (Morgado and Roncal 2009). The site has been studied by the Prehistory Department at the University of Granada (Morgado *et al.* 2001). Additional sites have been documented at Cerro del Reloj (CRE, Montefrio, Granada), Cortijo del Zegrí/Onitar (CZE-Iznalloz) and Loma de Los Pedernales/El Cuarterón (LPE-Iznalloz, Granada).

The flint found at these sites (GAL, CRE, CZE and LPE) comes from the calcicastites level (calcarenites and calcilimolites), which belongs to the Milanos Formation (Middle Subbetic; Molina and Vera 1996a, 1996b) dated to the late Jurassic (Kimmeridgian-Berriasian) (Molina *et al.* 2008). Levels with calcilimolites and calcarenites present structures of internal order including parallel lamination, hummocky cross-stratification and ripples. These structures are associated with storms affecting the bottom of a shallow ocean located away from the continent (Molina and Vera 1996a, 1996b).

The flint found at the Milanos Formation is of variable colour, with tones ranging from grey-white to grey-blue or even very dark grey. Microfacies show wakestone-packstones of peletoidal type with rounded quartz grains, iron oxide, sponge spicules and other bioclasts. Oolitic flint (grainstone) is also present.

b. Mining sites of the Neonumidian Formation in the Campo de Gibraltar Complex

These flint mining sites were first discovered in the 1980s as a result of several archaeological surveys. The nucleus of the mining resources is located in the Valle del Turón (Málaga). Additional mining sites are located at Ardite/El Garrotal (ARD) and Cerro Alcolea (ALC) (Ramos Muñoz *et al.* 1986; Espejo and Cantalejo 1989-90).

Our team has recently conducted a petrological analysis of the mining sites at the Valle del Turón (Lozano *et al.* i.p.). Eight sites have been identified in this area: La Galeota (GAL), Castillo del Turón (CTR), Canchal Herriza the Ram (CHC), Reconco (REC), Puerto de los Martínez (PLM), Espíldora (ESP), Cortijo el Pilar (PIL) and El Chorrito (ECH). Macroscopically, the flint from these sites is of a grey-black tone, semi-translucent, fine-grained and of homogeneous texture. According to the microfacies of the samples studied, this flint originated diagenetically from hemipelagites derived from limestone, mudstones and wackestones. Radiolarians, abundant sponge spicules, and, to a lesser extent, filaments and foraminifera dated to the Early and Middle Jurassic are present. Local dolomitization is observed frequently in the presence of idiomorphic and rhombohedral dolomite microcrystals ($<10\mu$). Phycosiphon and chondrite ichnotaxa have been identified by ichnological analysis, a new means of analysing archaeological flint that allows the non-destructive characterisation of samples (Rodríguez-Tovar *et al.* i.p.). Other mining sites of the Campo de Gibraltar Complex can be found throughout the Province of Malaga, such as Ardite/El Garrotal (Alozaina, Málaga) (Fernandez and Marquez 1985; Marquez *et al.* 2004) and Cerro Alcolea (ALC) in the upper Velez river valley (Ramos Muñoz 1997).

Together, all the above sites are part of the Neonumidian Formation (Martín-Algarra 1987; Martín Algarra *et al.* 2004), which is made up predominantly of clays and sandstones with some *mélange*-like features from the Miocene (Aquitanian-Burdigalian). Mesozoic and Tertiary materials are incorporated in varying proportions. In addition, flint blocks are found among the Mesozoic materials identified above. The Neonumidian lies primarily along the front of the internal areas. Therefore, the flint embraced by the *mélange* comes from the edge of the Internal Betic Zone (i.e., frontal units) (Lozano *et al.* i.p.). Thus, all mining sites in the Valle del Turón (LGA, CTR, CHC, REC, PLM, ESP, PIL and ECH), in addition to those of Ardite/El Garrotal site (ARD), are located near the outcrops in the External Frontal sections of the Internal Betic Zone (i.e., the Pereila type unit; Martín Algarra 1987). The Pereila unit has limestones and marly layers with Lower Jurassic flint, which provided the raw material exploited by Man.

The Cerro Alcolea site (LAC, Periana) is composed of thick conglomerate strata. The microfacies of these flints provide evidence of its connection with the External Frontal units of the Internal Betic Zone.

c. Malaver Formation mining site (Campo de Gibraltar Complex)

The mining site at the Malaver Formation (Ronda, Málaga) was among the first to be subjected to archaeological analysis (Vallespí and Cabrero 1980-81; Espin Canovas 1989-90; Aguayo and Moreno Jimenez 1998). Some scholars have regarded this site as 'the most important flint workshop in Andalusia' (Aguayo and Moreno Jimenez 1998, 111). However, the nature of its raw materials and their geological genesis have not yet been extensively studied.

The Malaver Formation emerges in the northwest section of the city of Ronda, between the municipalities of Montecorto (Málaga) and El Gastor (Cádiz) and is part of the

Campo de Gibraltar Complex. The unit is located near the Subbetic-Penibetic boundary overlying the Montecorto Unit, which in turn backthrusts over the Campo de Gibraltar Complex. The Malaver Formation is a chaotic clay unit also composed of clay blocks (Gutiérrez-Mas *et al.* 1991). The diversity of flint types found in this source provides an indication of the geological complexity of the Malaver Formation. This formation originated in the Lower Miocene and is composed of megabreccias packages, massive conglomerates hundreds of meters thick, giant blocks (e.g., Malaver mountain) from the Middle Subbetic (e.g., Sierra del Pinar), and scaly clays from the Campo de Gibraltar Complex. All these materials rest on top of the Permo-Triassic layer of the Montecorto unit (Gutiérrez-Mas *et al.* 1991).

The flint exploited by Man came from the conglomerates of the Malaver Formation. These conglomerates originated due to the dismantling of the sedimentary series of the Middle Subbetic. The Middle Subbetic series (e.g., Sierra del Pinar type) possesses marly silts and oolitic limestones along with flint from the Middle and Upper Lias. Martín-Algarra (1987) has also documented marls and radiolaritic marls intercalated with micritic limestones with flint, and filaments from the Dogger-Malm boundary. It should be noted that the raw materials for large blade production was primarily large oolitic flint nodules.

4. Mining types

The geological features of flint-rich areas determined how readily raw materials could be found, which in turn affected the mining technique of choice. In this respect, the tectosedimentary features of the different areas required mining systems different to those used at flint sources in the Middle Subbetic area and the Campo de Gibraltar Complex.

a. Mining sites in the Middle Subbetic (External Betic areas and Central Betic Cordillera).

The flint extracted at the four mining sites currently known in this area belongs to the Milanos Formation and has been dated to the Upper Jurassic (Malm: Kimmeridgian-Tithonian) (Vera and Molina 1996) (sites: Los Gallumbares, Cortijo del Zegrí/Onitar, Cerro del Reloj, and Loma de Los Pedernales/El Cuarterón). The Milanos Formation has marl and limestone levels that originated in a pelagic platform. The extension of the Milanos Formation strata, the integrity of the flint levels, and the abundance of flint nodules in this area made this geological unit highly attractive to human communities of recent prehistory. Flint sources were readily located and exploited by shafts (Figure 3).

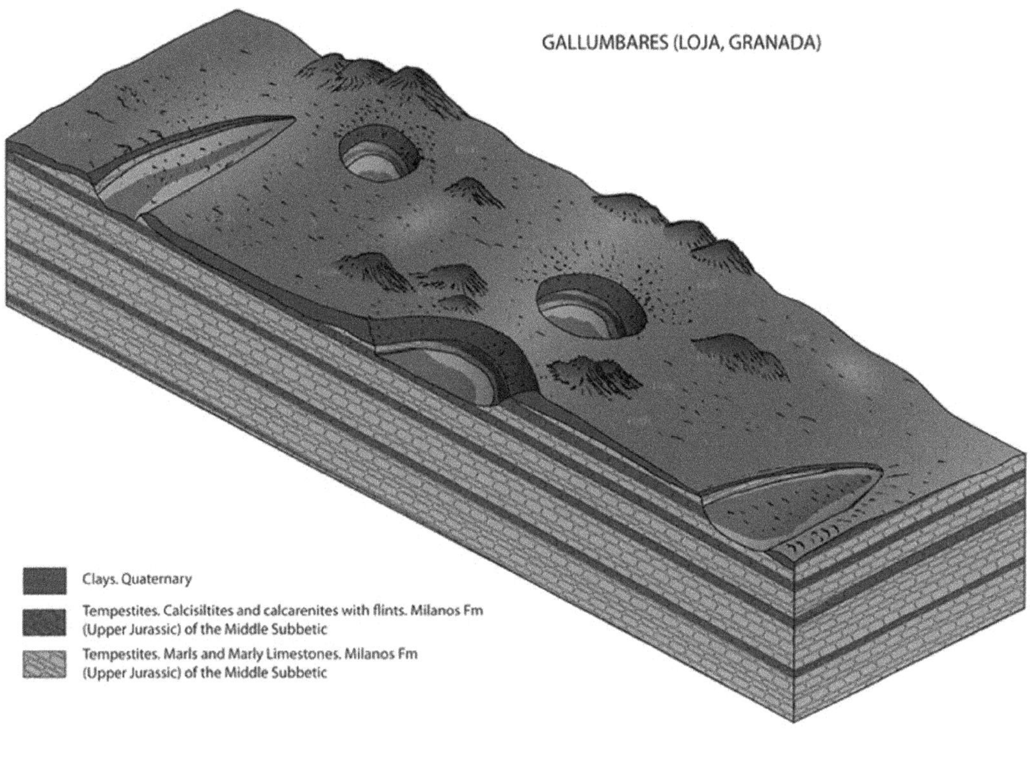

Figure 3. Model of flint exploitation, Milanos Formation (Middle Subbetic).

b. Mining in the Malaver Formation (Campo de Gibraltar Complex, western Betic Cordillera)

The mining activities here focused on extracting large flint blocks from the conglomerates and megabreccias. Conglomerates were taken directly from the Malaver mountain's cliffs via quarrying activities.

c. Mining sites of the Campo de Gibraltar Complex (Figure 4 and 5)

The Campo de Gibraltar Complex is paleogeographically composed of sedimentary Cenozoic materials deposited over the ancient oceanic crust. Large flint nodules have been extracted from conglomerates and *mélanges* in the mining sites in this area (Cerro Alcolea, Ardite/El Garrotal and the Valle del Turón: La Galeota, Castillo del Turón, Espíldora, Cachal H. Carnero, Chorrito, Los Reconcos and El Pilar). The flint found in the conglomerates (primarily *mélanges*) resulted from the erosion or dismantling of the frontal units of the Internal Betic Zone. The absence of continuous flint levels determined the need for quarrying activities and trenching for exploitation to be possible.

5. Conclusions

The geological characteristics of the flint outcrops in the south of the Iberian Peninsula influenced the strategies required for their exploitation. From the record of the fifteen mentioned exploitations linked to flint blade production, it can be concluded that the background to one group of exploitations involved conglomerates and tectonic *mélanges*. The impossibility of predicting the size and quality of their flint nodules predetermined the use of open air quarrying techniques. This behaviour was based on opening large areas to locate and select the best nodules for the production of large blades. This extensive quarrying was also favoured by the steep geomorphology of the Malaver and the Campo de Gibraltar exploitations. The background of a second group is characterized by the continuity of flint levels based on the alternation of the marl and limestone of the Milanos Formation (Middle Subbetic). In addition,

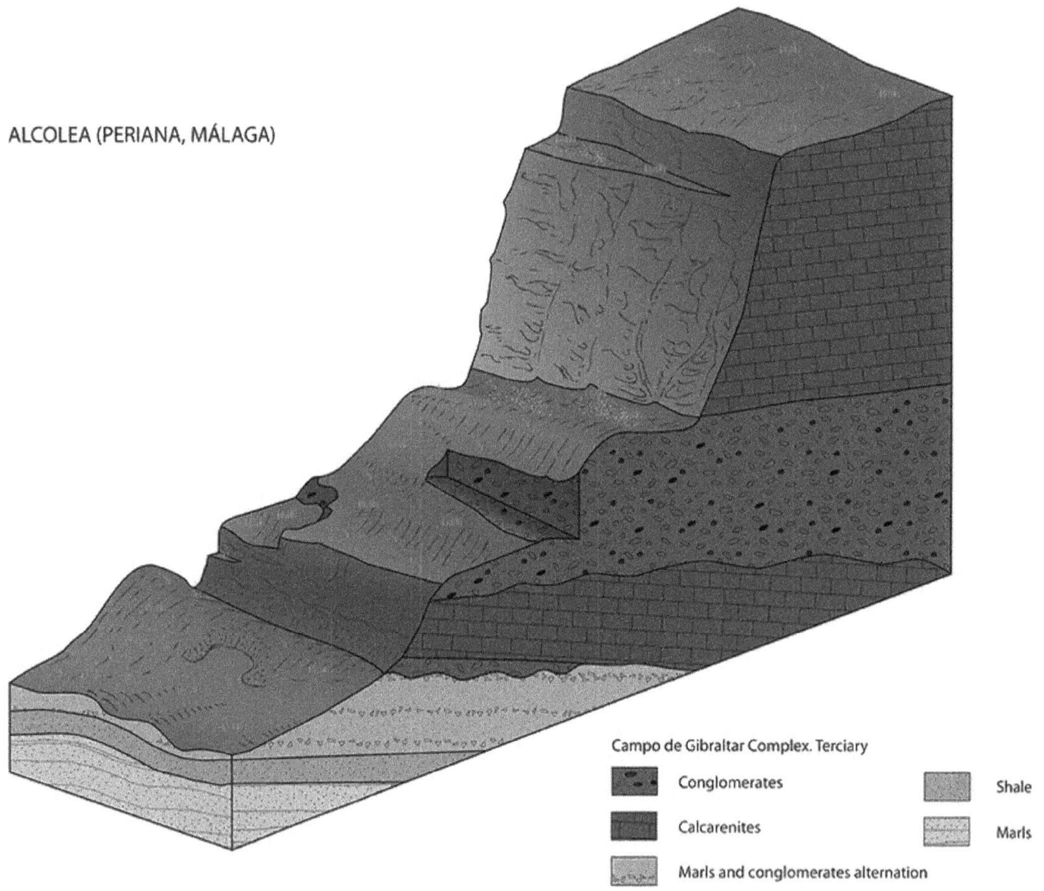

Figure 4. Model of flint exploitation, Campo de Gibraltar Complex (Alcolea conglomerate).

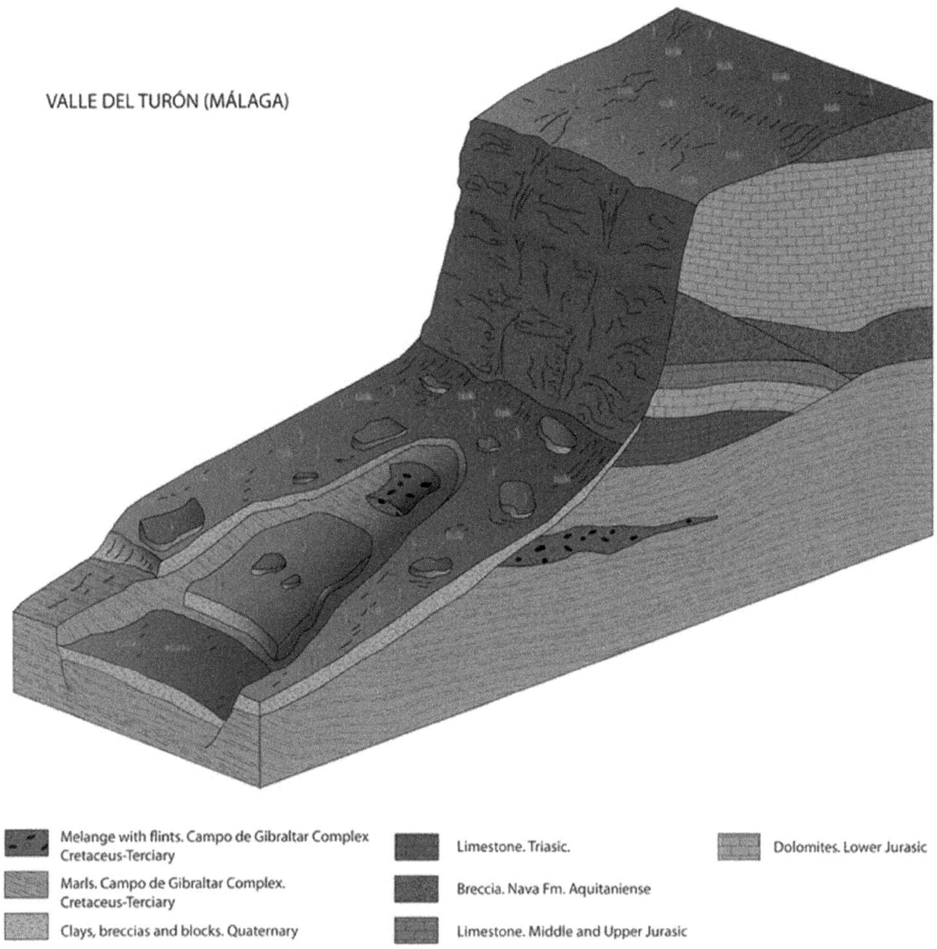

Figure 5. Model of flint exploitation, Campo de Gibraltar Complex (Valle del Turón mélanges).

the Milanos Formation has a smoother geomorphology. This facilitated the location of large nodules of flint by pit mining.

Acknowledgements

This research was supported by projects HAR2009-07283, PO6-HUM-01658 and the HUM-274 group.

References

Aguayo, P. and Moreno Jiménez, F. 1998. El complejo arqueológico de Malaver-Lagarín y su significación en el suministro de rocas silíceas en el mediodía peninsular, in J. Bernabeu, T. Orozco and X. Terradas (eds.), *Los recursos abióticos en la Prehistoria. Caracterización, aprovisionamiento e intercambio*, 111–125. Valencia, Universitat de València.

Espejo, M. M. and Cantalejo, P. 1989-90. La Galeota, un taller de sílex calcolítico (Ardales, Málaga). Colección Martín Domínguez. *Mainake* XI-XII, 21–40.

Espín Cánovas, C. 1989-90. La colección E. Rudilla Cayuela. Contribución al estudio del taller de sílex calcolítico del Pico Malaver (Montecorto, Málaga). *Mainake* XI-XII, 41–56.

Fernández Ruiz, J. and Márquez Romero, J. E. 1985. El taller de Ardite, Coín (Málaga). *Cuadernos de Prehistoria de la Universidad de Granada 10*, 103–129.

García-Hernández, M., López-Garrido, A. C., Rivas, P., Sanz De Galdeano, C. and Vera, J. A. 1980. Mesozoic paleoeogeographic evolution of the External Zones of the Betic Cordillera. *Geologie en Mijnbouw* 59, 155–168.

García-Hernández, M., López-Garrido, A. C., Martín-Algarra, A., Molina, J. M., Ruiz Ortiz, P. A. and Vera, J. A. 1989. Las discontinuidades mayores del Jurásico de las Zonas Externas de las Cordilleras Béticas: análisis e interpretación de los ciclos sedimentarios. *Cuadernos de Geología Ibérica* 13, 35–52.

Gutiérrez-Mas, J. M., Martín-Algarra, A., Moral-Cardona, J. and Domínguez-Bella, S. 1991. *Introducción a la Geología de la provincia de Cádiz*. Cádiz, Universidad de Cádiz.

Lozano, J. A., Morgado, A., Puga, E. and Martín-Algarra, A. 2010. Explotaciones del sílex tipo "Turón" (Málaga, España): localización y caracterización petrológica y geoquímica. *Geogaceta* 48, 163-166.

Márquez Romero, J. E., Fernández Ruiz, J., Conejo Pedrosa, M. T. and González Martín, J. 2004. Proyecto de investigación del río Grande: prospecciones arqueológicas y electrofísicas en el tramo III (Arroyo del Lugar al Arroyo de Los Valles). Campaña 2001. *Anuario Arqueológico de Andalucía* 2001 (2), 39–48.

Martín-Algarra, A. 1987. *Evolución geológica alpina del contacto entre las Zonas Internas y las Zonas Externas de la Cordillera Bética*. Unpublished PhD thesis, University of Granada.

Martín-Algarra, A. (coord.), Alonso-Chaves, F. M., Andreo, B., Azañón, J. M., Balanyá, J. C., Booth-Rea, G., Crespo-Blanc, A., Delgado, F., Díaz de Federico, A., Estévez, A., Galindo-Zaldívar, J., García-Casco, A., García-Dueñas, V., Garrido, C. J., Gervilla, F., González-Lodeiro, F., Jabaloy, A., López-Garrido, A. C., Martín-Algarra, A., Martín-Martín, M., Nieto, J. M., O'Dogherty, L., Orozco, M., Puga, E., Rodríguez-Calero, R., Ruiz-Cruz, M. D., Sánchez-Gómez, M., Sánchez-Navas, A., Sanz de Galdeano, C., Soto, J. I., Torres-Roldán, R. L. and Vera, J. A. 2004. Zonas Internas Béticas. In Vera, J. A. (ed.), *Geología de España*, 395–437. Madrid, SGE-IGME.

Martínez, G. and Morgado, A. 2005. Los contextos de elaboración de hojas prismáticas de sílex en Andalucía oriental durante el Neolítico reciente. Aspectos técnicos, modelos de trabajo y estructuración social. *III Congreso del Neolítico en la Península Ibérica*, 359-368. Santander, Universidad de Cantabria.

Molina, J. M. and Vera, J. A. 1996a. La Formación Milanos en el Subbético Medio (Jurásico superior): definición y descripción. *Geogaceta* 20 (1), 39–42.

Molina, J. M. and Vera, J. A. 1996b. Tempestitas en el Subbético medio (Fm. Milanos, Jurásico Superior). Sus características y facies relacionadas. *Geogaceta* 20 (1), 56–59.

Molina, J. M., Martín-Algarra, A., Nieto, L. N., Rey, F. J., Ruiz-Ortiz, P. A. and Vera, J. A. 2008. Stratigraphy of the jurassic in the subbetic (Southern Spain): A sintesis. *Abstracts of the International Geological Congress*, Oslo 2008 (August 6–14th), online.

Morgado, A. 2002. *Transformación social y producción de hojas de sílex durante la Prehistoria Reciente de Andalucía oriental. La estrategia de la complejidad.* PhD thesis, Granada, University of Granada (electronic edition, 2008).

Morgado, A. and Roncal, E. 2009. *Los últimos talladores del sílex. Estudio histórico-arqueológico sobre la explotación del sílex en las tierras de Loja y la producción militar de piedras de chispa del reino de Granada durante los siglos XVIII y XIX.* Granada, Fundación Ibn-al Jatib de estudios y cooperación cultural.

Morgado, A., Martínez, G., Roncal, E. and Martín, J. J. 2001. Prospección arqueológica en relación con la explotación de rocas silíceas en el sector occidental de la región de Los Montes (Granada). *Anuario Arqueológico de Andalucía* 1997 (2), 77–85.

Morgado, A., Pelegrin, J., Aubry, T. and Lozano, J. A. 2009. La producción especializada de grandes láminas del Sur y Oeste de la Península Ibérica, in J. F. Gibaja, X. Terradas, A. Palomo and X. Clop (eds.), *Les grans fulles de sílex. Europa al fi nal de la Prehistòria*, 89–97. Barcelona, Museu d'Arqueologia de Catalunya.

Morgado, A., Pelegrin, J., Martínez, G. and Afonso, J. A. 2008. La production des grandes lames dans la Péninsule Ibérique (IVe et IIIe millénaires), in M.-H. Dias-Meirinho, V. Léa, K. Gernigon, P. Fouéré, F. Briois and M. Bailly (eds.), *Les industries lithiques taillées des IVe et IIIe millé-naires en Europe occidentale*, 309–330. Oxford, BAR Publishing. British Archaeological Report International Series 1884.

Obermaier, H. 1934. Estudios prehistóricos en la provincia de Granada. A*nuario del Cuerpo Facultativo de Archiveros, Bibliotecarios y Arqueólogos* 1, 255–292.

Pelegrin, J. and Morgado, A. 2007. Primeras experimentaciones sobre la producción laminar del Neolítico Reciente-Edad del Cobre del sur de la Península Ibérica, in M. L. Ramos Sáinz, J. E. González Urquijo and J. Baena Preysler (eds.), *Arqueología Experimental en la Península Ibérica. Investigación, didáctica y patrimonio*, 131–139. Santander, Asociación Española de Arqueología Experimental.

Ramos Muñoz, J. 1997. *Tecnología lítica de los talleres de cantera de la Axarquía de Málaga. Aproximación al estudio de las formaciones económicas de la Prehistoria Reciente.* Colección Monografías nº 10, Málaga, Diputación Provincial de Málaga.

Ramos Muñoz, J., Espejo Herrerías, M. and Cantalejo Duarte, P. 1986. *Taller calcolítico del Castillo del Turón. Ardales (Málaga).* Málaga, Ayuntamiento de la villa de Ardales.

Rodríguez-Tovar, F., Morgado, A. and Lozano, J. A. 2010. Using Ichnofossils to Characterize Chert Tools: A Preliminary Study from Southern Iberia. *Geoarchaeology* 25 (4), 514-526.

Sanz de Galdeano, C. 1997. *La Zona Interna Bético-Rifeña*. Monografías Tierras del Sur. Granada, Universidad de Granada.

Vallespí, J. E. and Cabrero, R. 1980-81. Calcolítico y Bronce Pleno en El Moral de Montecorto, Ronda (Colección Pérez Aguilar). *Mainake* II-III, 48–75.

Vallespí, J. E., Ramos Muñoz, J., Martín Córdoba, E., Espejo, M. M. and Cantalejo, P. 1988. Talleres líticos andaluces del Calcolítico y Bronce. *Revista de Arqueología* 90, 14–24.

Vera, J. A. 1988. Evolución de los sistemas de depósito en el margen continental ibérico de las Cordilleras Béticas. *Revista de la Sociedad Geológica de España* 1, 373–392.

Vera, J. A. 2004. Cordillera Bética y Baleares. In Vera, J. A. (ed.), *Geología de España*, 345–464. Madrid, SGE-IGME.

Prehistoric flint exploitation in Loma de Enmedio-Realillo (Tarifa coast, Cádiz, Spain)

Salvador DOMÍNGUEZ-BELLA, José RAMOS MUÑOZ and Javier MARTÍNEZ

To our colleague and beloved friend Lothar Bergman, in memoriam.

Abstract

In the present paper we analyze the dynamic of territorial occupation by the first class society in the Atlantic band of Cádiz in the 3rd Millennium BC. We consider environmental and geological factors, organization of settlements, and examine the relationship between the economic bases (production mode and economic and social organization) characterizing this society and the currently available archaeological data. While providing a detailed description of Loma de Enmedio-Realillo site's geographical and geological features, particularly its flint resources, we evaluate its technology, raw materials and the objects distributed in nearby villages. Drawing on the available data, we conclude that this territory of the lower Guadalquivir valley can be characterized as the location sporadically occupied by hunter-gatherer groups and, during the Late Prehistory, of the systematic flint exploitation by hierarchical societies.

Keywords

Flint mining. Atlantic Band of Cadiz. Archaeomineralogy. Raw materials.

1. Introduction

In contrast to the abundant studies of Phoenician and Protohistoric periods and the widely-acknowledged great historic interest in Gadir and Tartessos, the scholarship dealing with the prehistoric societies of the Atlantic Zone of Cadiz during 3rd and 2nd millennium BC had been very scant.

In the 90s, an interdisciplinary group of scientists from the University of Cadiz, headed by Jose Ramos, received an authorization of Junta de Andalucía to carry out a project titled 'The prehistoric occupation of the countryside and coast in the Atlantic Zone of Cadiz'. The study began with a series of archaeological survey programs in San Fernando, Chiclana de la Frontera, Conil de la Frontera, Medina Sidonia, Barbate and Tarifa, which enabled us to locate and document more than 100 settlements corresponding to Paleolithic hunter-gathered societies, Neolithic tribal communities and first class societies of the 3rd and 2nd millennium BC. Rescue excavations were carried out in the settlements of El Estanquillo (Ramos 1993), La Mesa (Ramos *et al.* 1999), El Retamar (Ramos and Lazarich 2002) and La Esparragosa (Pérez *et al.* 2005; Ramos *et al.* 2008).

We have departed from a hypothesis that in the 3rd and 2nd millennium BC the territory of Cadiz Bay and its surrounding area was the centre of agricultural production and mining, undergoing the process of hierarchization and transformation into the first class society.

We presume that during this period, and since the Copper Age, the political organization of Atlantic Zone had a nuclear centre in the Valencina de la Concepción and Gandul area, in the province of Sevilla (Arteaga *et al.* 1995; Arteaga and Cruz-Auñón 1995; Arteaga 2002).

The archaeological data confirm the importance of cereal agriculture, and the possible use of dryland farming. The territories of Cadiz Bay and their immediate vicinity were used as agricultural and pasture lands, controlled from a local centre. One of the important characteristics of this society was the use of underground storage. We also evaluate the changes in the social division of labour brought about by the appearance of craftsmen (working pottery, textile, and lithic production).

Using this general explanatory framework, we attempt to characterize lithic raw material exploitation by the prehistoric societies (Palaeolithic hunter-gatherer-fisher, Neolithic tribal communitarian and early stratified societies from the Late Prehistory), in the Gibraltar Strait area, drawing on the existing analysis of the production, exchange and consumption dynamic of lithics by different societies the South of the Iberian Peninsula and the Tingitan Peninsula (Domínguez-Bella 1999, 2002; Domínguez-Bella *et al.* 2002a, 2002b, 2004; Ramos *et al.* 1998, 2004, 2006, 2008a).

2. Natural environment and resources

The Atlantic Zone of Cádiz is located in the farmost western end of the Betic Cordillera. The Low Guadalquivir Valley lies to the northwest of these mountains, between the coast and the countryside (Gutierrez *et al.* 1991), and is characterized by low or almost flat relieves, never exceeding 100m altitude in the coastal area.

The adjacent countryside has individual hills, with few elevations never exceeding 200 meters. As for its geology, Cadiz Bay is composed of groups of quaternary materials (burden, sands, slimes and clays), post orogenic materials from the Miocene-Pliocene, especially calcarenites, and the quaternary filler of the fluvial beds of the Guadalete, Iro and Salado rivers (Gutierrez *et al.* 1991).

In the vicinity, other materials from different ages and lithologies can be found. These include materials belonging to the Middle Subbetic, clays and marls from the Triassic (facies Keuper), frequently containing dolerites. They represent material sources from the Jurassic and Cretaceous periods and occupy different areas of Medina-Sidonia and Chiclana de la Frontera.

To the east, the predominant material of the geological units of the Campo de Gibraltar is sandstone from the Aljibe Unit, with marl insertions from Miocene Age in the neighbouring district of La Janda.

The interaction of the geologic structure and natural processes has produced the conditions influencing the location of prehistoric settlements and the possibilities of their economic exploitation. In general, the interior area adjacent to Cádiz Bay is a region with great diversity of soils and a great agricultural potential (Guerra *et al.* 1963).

Marl soils of the Middle Subbetic offered the inhabitants of the surroundings of Medina-Sidonia and Chiclana de la Frontera an excellent resource for agricultural production, since they were optimally suited for cultivation of legumes, cereals and pasture. This region also has significant resources for the development of lithic industry based especially in dolerite exploitation (Domínguez-Bella *et al.* 2002).

Since the emergence of agriculture, soils have been a product of social action upon the environment and have been subjected to the physical-chemical effects of the vegetation throughout history. Therefore, the development of human activities, in many cases, is associated with intensification of the erosive effects caused by the human transformation of the natural environment.

The countryside and coast of Cádiz have suffered tectonic processes and have undergone a substantial sedimentary evolution that has transformed the coast (Zazo 1989; Gracia *et al.* 1999; Arteaga and Cruz-Auñón 1995; Arteaga and Schulz 2008; Arteaga *et al.* 2001; Alonso *et al.* 2008), which begin with the last transgressive maximum Flandrian and generate interesting processes of sedimentary evolution, as well as alluvial phenomena of rivers. These stages contrast with more arid periods, represented by dunes of eolic origin.

3. Geographical and geological setting

Loma de Enmedio site is located in the vicinity of the Cueva del Moro, approximately 500m west of it, in the Sierra de la Plata, near the Cortijada del Realillo de Bolonia, 1000m from the coast line of the Atlantic Ocean, near Tarifa, Cádiz, Spain (Figure 1).

This site was discovered by Lothar Bergmann, who in 1998 kindly shared with us the location of this find. Considering it position in the so-called Atlantic Band of Cadiz, the site is peculiar in terms of its geographical and geological features. It area is limited to a 200m zone, located between the Cortijo de la Gloria and the Cortijada del Realillo in Bolonia, Tarifa, Cádiz (Figure 1).

Geologically it is located in Tertiary materials, near the sandstones from the Aljibe Unit (Gutiérrez *et al.* 1991), in a fracture zone marked by the so-called fault of Cabo de la Plata. In this area, great outcrops of Aljibe sandstone appear, with an Aquitanian age, alternating layers of yellow sandstone with brown clay, all with large dips to the NW. They contrast with the so-called Almarchal Units, consisting of clays with blocks layered with rigid materials such as carbonated sandstones and some packages of isolated banded flint (Figure 1). This outcrop of lithic materials appears crusted in the bedrock of Aljibe sandstones

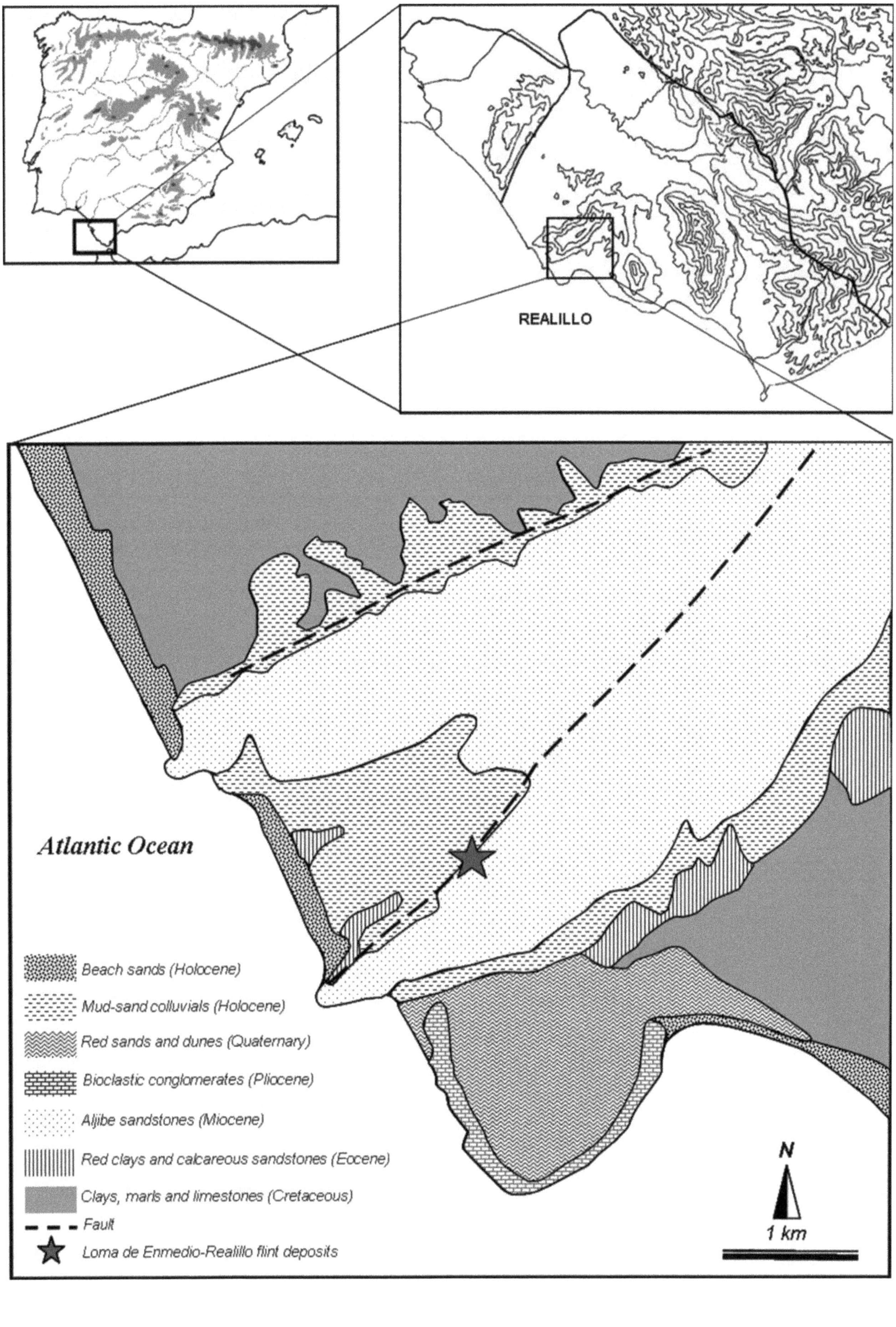

Figure 1. Geological map of the Loma de Enmedio-Realillo site area.

Figure 2. General view of Loma de Enmedio-Realillo site. At the background, Aljibe sandstones hill face.

and carbonated levels, all into Upper Cretaceous-Lower Miocene clays (Figure 2). Here we find block and layers of flint, reaching 25cm in thickness (Figure 3), which have been macroscopically classified as 'porous flint and banded flint' (Ramos *et al.* 2002). Immediately on top of the mentioned outcrop, we find the evidence of lithic industry (traces of flaking and splitting). In the zones where soil coverage appears, this is of brown forrestial land type.

4. Lithic resources in the area

The siliceous materials exploitation in the North slope of the Gibraltar Strait is a topic studied by numerous scholars for the past 15 years (Ramos *et al.* 1998, 2004, 2008b; Domínguez-Bella *et al.* 2002a; Giles *et al.* 1989, 2003; Gutiérrez *et al.* 1994).The lithic study of siliceous and basic rocks resources in this territory helps us understand the economic activities in the settlements (Dominguez-Bella 1999; Dominguez-Bella 2002a; Pérez *et al.* 1998; Ramos *et al.* 1998, 2004), and provides us with means to understand the processes of production, distribution and consumption, i. e., the economic structure of this society.

We analyse the work artefacts in the context of raw material procurement, and their characteristics as means of production. The study of these artefacts facilitate our un-

Figure 3. General view of a carbonated sandstone with flint strata. Detail view of the banded flint.

derstanding of farming practices, which have obvious environmental impact (Pérez 1997, 1998).

The geological context of the region explains the presence of specific raw materials (Pérez *et al.* 1998). On the one hand, we observe the materials from the Middle Subbetic, especially clays and gypsums from the Triassic (Keuper facies) found in the environments of Medina Sidonia and Chiclana de la Frontera (Domínguez-Bella 1999), where the presence of dolerites are frequent (Morata 1993). And, on the other, we find the units from the Campo de Gibraltar constituted mainly by sandstones from El Aljibe.

We have found numerous artefacts elaborated from igneous rocks (dolerites) and metamorphic rocks (amphibolites, quartzites, micaschist). Sillimanite adzes have also been identified, as well as axes made of sedimentary rocks: detrital (lutites, sandstones, and conglomerates) and carbonated (calcareous and calcarenites).

Siliceous resources in this area have a great variety of registers, such as pebbles, mainly from the Betic Cordillera, that can be recovered all the way into the Cádiz Bay as a result of fluvial transport, mainly of the Guadalete river. We have documented sources of siliceous material in the countryside, such as Loma de Enmedio-El Realillo site.

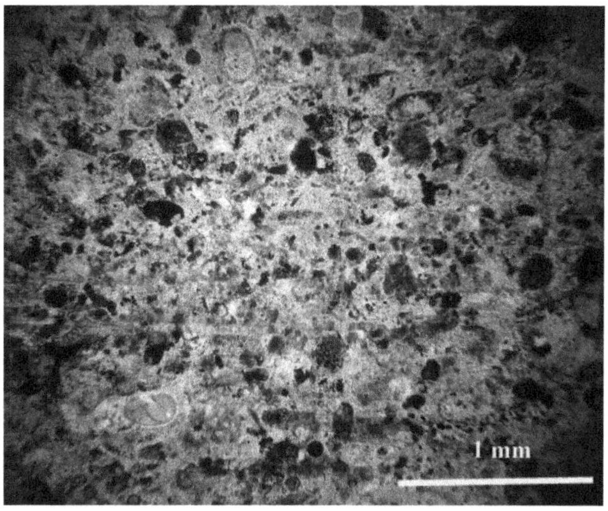

Figure 4. Optical microscopy view (unpolarized light) of the banded flint, with fossil remains, ooids, etc.

5. Results. Archaeological products and chronology

Flint knapping took place on the silex outcrop, in the quarrying area itself and its vicinity.

The petrological study shows that this is highly porous silica, with visible banding even on a macroscopic scale. It

Figure 5. Laminar products with porous banded flint from Los Charcones (III millennium B.C. site).

Lithic products	
Types	Number
Cores	44
Blades	114
Debris	44
Retouched Artefacts (Scrapers, Notches & Denticulates)	16
Total	218

Figure 6. Table with types of the knapped lithic products, found in the Loma de Enmedio-Realillo site sampled work area.

has a significant amount of fossils (Figure 4). An optical microscopic study of thin fragments analysed through x-ray diffraction has been conducted, demonstrating that the material is composed predominantly of quartz and traces of calcite. Similar petrological and compositional features have been identified in flint from archaeological sites in the Atlantic Band of Cadiz, as well as the site of Los Charcones (Figure 5).

We have made an initial classification of the knapped lithic products (Figure 6), found in the sampled work area.

We have documented 218 lithic knapped products in this site. Forty four of them are cores, showing well defined traces of work processes. They are unipolar, bipolar, centripetal, polyhedral cores and for blades. The objective of flint processing appears to have been the production of flakes and blades. The sample of blades consists of pieces representing the beginning of the knapping process, size and internal knapping, levallois configuration and crested blades. Lithic production areas contain evidence of debris. Among the retouched products, we have documented scrapers, notches and denticulates (Figure 6).

We have observed a superficial use of raw materials in the areas populated by human groups associated with a Middle Palaeolithic hunter-gatherer technology. Work processes are aimed at the production of centripetal cores and flakes of Levallois technique (Ramos *et al.* 2008a). They are explained by the mobile character of these human groups in the territory of the Atlantic band of Cádiz (Ramos *et al.* 2008a). Traces of the subsequent occupation of the site point to a more intensively use.

The site is, therefore, constitutes a direct production area, sourced through the direct extraction from the outcrops incrusted in the mother rock of Miocene clay; a clear example of the so-called 'flint workshops'. The knapping work typology has a clear affinity to the technologies employed in the Late Prehistory, by societies dated to the 3rd-2nd millenniums BC.

6. Socio-economical synthesis

The zone of production of Loma de Enmedio-Realillo is directly connected with the exploitation by human groups of the nearby nuclear settlement of Los Charcones, in the Laguna de La Janda area. It was a procurement area where lithic products were initially centralized and subsequently redistributed to other minor settlements.

The chronological frame is well defined in the 3rd-2nd millenniums BC and relates to the historical processes of production, distribution and consumption of the initial class societies (Arteaga 1992; Nocete 2001) in the Atlantic band of Cadiz (Ramos *et al.* 2008b).

The archaeological relation of both sites shows extraction and redistribution processes from nuclear centres in the initial class societies. The analysis of the distribution processes provides scholars with valuable data about the social control of the territory and the social and productive relations between areas of procurement and consumption.

These economic processes are directly related to the nature of settlements and their social organization. So, the presence of manufactured artefacts made of raw materials originating outside the study area, in the Atlantic Zone of Cadiz, in the 3rd millennium BC, point to a phenomenon of product redistribution, consisting in the actual movement of goods, towards a centre (Arteaga 1992; Nocete 1989, 1994, 2001; Pérez 1998), which structured the nuclear organization of the settlements.

The control of surplus agricultural products in the settlements such as La Mesa and Los Charcones explains the discovery of artefacts manufactured from exogenous raw material. Their presence could be associated with the phenomena of redistribution of goods and artefacts legitimating territorial and social control.

All this evidence leads us to the hypothesis that we are dealing with the centralized socio-economic organization of some settlements.

In addition to the exogenous materials, the settlement of Los Charcones contains a significant amount of raw materials from Loma de Enmedio-El Realillo, demonstrating that the raw materials of local origin were frequently used for tool manufacturing.

7. Conclusions

Loma de Enmedio-Realillo site is an interesting example of *in situ* open-air flint extraction and production area, located by the outcrop itself. It is chronologically situated in the Recent Prehistory and exploited by 3rd millennium BC groups (Ramos *et al.* 2002).

In this workshop, the flint knapping products are represented by different types of cores: single, prismatic, and especially polyhedral cores. Blade production from cores apparently predominates here and is characteristic of the Recent Prehistory of the region.

This site contains a large number of documented blades and retouched products. These include notches and denticulate. In the course of the inventory of retouched products we have also encountered a number of good quality cores as well as retouched artefacts of scraper type, reminiscent of the characteristic products of the Mousterian techno-complex.

All this evidence leads us to conclude that groups of Mousterian techno-complex repetedly visited this area, generating some retouched products. The flint outcrop was intensively exploited in later periods of the Late Prehistory, during the 3rd millennium BC. Since then, new and more complex political structures were established. Territorial centralization and the development of agricultural and pastoral practices lead to the formation of an initial social hierarchization, paralleled by an even bigger environmental transformation.

Acknowledgements

To Purificación García Díaz and Ana Durante for their editorial assistance. This work was supported by the investigation group P.A.I. HUM-440 of the Junta de Andalucía government and the I+D+I Spanish national research DGI project. HAR2008-06477-C03-02/HIST. 'La implantación de las especies domésticas en la Europa Atlántica: Sociedades Neolíticas y uso de materias primas líticas en la región del Estrecho de Gibraltar'.

References

Alonso Villalobos, C., Gracia, J. and Benavente, J. 2008. Evolución histórica de la línea de costa en el sector meridional de la Bahía de Cádiz. *Revista Atlántica-Mediterránea de Prehistoria y Arqueología Social* 11, 13-37.

Arteaga, O. 1992. Tribalización, jerarquización y estado en el territorio de El Argar. *Spal* 1, 179-208.

Arteaga, O. 2002. Las teorías explicativas de los 'cambios culturales' durante la Prehistoria en Andalucía: nuevas alternativas de investigación. In *Actas del III Congreso de Historia de Andalucía. Prehistoria*, 247-311. Córdoba, Publicaciones de la Obra Social y Cultural de CajaSur.

Arteaga, O. and Cruz-Auñón, R. 1995. Acerca de un campo de silos y un foso de cierre prehistóricos ubicados en 'La Estacada Larga' (Valencina de la Concepción, Sevilla). *Anuario Arqueológico de Andalucía* 1995-III, 600-607.

Arteaga, O. and Hoffmann, G. 1999. Dialéctica del proceso natural y sociohistórico en las costas mediterráneas de Andalucía. *Revista Atlántica-Mediterránea de Prehistoria y Arqueología Social* 2, 13-121.

Arteaga, O., Kölling, A., Kölling, M., Roos, A. M., Schulz, H. and Schulz, H. D. 2001. El Puerto de Gadir. Investigación geoarqueológica en el casco antiguo de Cádiz. *Revista Atlántico-Mediterránea de Prehistoria y Arqueología Social* 4, 345-415.

Arteaga, O., Schulz, H. and Roos, A. M. 1995. El problema del 'Lacus Ligustinus'. Investigaciones geoarqueológicas en torno a las marismas del Bajo Guadalquivir. *In Tartessos 25 años después 1968-1993*, 99-135. Jerez de la Frontera, Ayuntamiento de Jerez de la Frontera.

Arteaga, O. and Schulz, H. (Eds.). 2008. Geoarqueología y proceso histórico en la Bahía de Cádiz. *Revista Atlántica-Mediterránea de Prehistoria y Arqueología Social* 10.

Domínguez-Bella, S. 1999. Los recursos líticos de las sociedades prehistóricas. Aplicación de las técnicas geoarqueológicas y arqueométricas. El caso de La Mesa y otros ejemplos de la Banda Atlántica de Cádiz, in J. Ramos, M. Montañés, M. Pérez, V. Castañeda, N. Herrero, M.E. García and I. Cáceres (eds.), *Excavaciones arqueológicas en La Mesa (Chiclana de la Frontera, Cádiz). Campaña de 1998. Aproximación al estudio del proceso histórico de su ocupación,* 165-182. Chiclana de la Frontera, Cádiz, Ayuntamiento de Chiclana de la Frontera, Fundación Vi-prem and Universidad de Cádiz.

Domínguez-Bella, S. 2002. Geología del Arco de Gibraltar. El Sur de la Península Ibérica y el Norte de África, como fuentes potenciales de materias primas minerales en la Prehistoria, in M. Tilmatine, J. Ramos and V. Castañeda (eds.), *Libro de Actas de las 1ª Jornadas de Estudios Históricos y Lingüísticos. El Norte de África y el Sur de la Península Ibérica*, 219-232. Cádiz, Universidad de Cádiz.

Domínguez-Bella, S., Pérez, M., Ramos, J., Morata, D. and Castañeda, V. 2002a. Raw materials, source areas and technological relationships between minerals, rocks and prehistoric non-flint stone tools from the Atlantic zone, Cadiz province, SSW Spain, in E. Jerem and K. T. Biró (eds.), *Archaeometry 98*, 723-728. Oxford, BAR Publishing. British Archaeological Reports International Series 1043 (2).

Domínguez-Bella, S., Sánchez Aragón, M., Baños Pozo, C. and Jurado Fresnadillo, G. 2002b. La industria lítica del yacimiento de El Retamar y los materiales líticos de sus estructuras arqueológicas. Estudio arqueométrico e inferencias arqueológicas, in J. Ramos and M. Lazarich (eds.), *El asentamiento de El Retamar (Puerto Real, Cádiz). Contribución al estudio de la formación social tribal y a los inicios de la economía de producción en la Bahía de Cádiz, 393-402. Cádiz,* Servicio Publicaciones Universidad de Cádiz.

Domínguez-Bella, S., Calado, D., Cardoso, J. L., Clop, X. and Tarriño, A. 2004 . Raw materials in the Neolithic-Aeneolithic of the Iberian Peninsula. *Slovak Geological Magazine* 1-2, 17-42.

Giles, F., Santiago, A., Gutiérrez, J. M., Mata, E. and Aguilera, L. 1989. El poblamiento paleolítico en el valle del río Guadalete, in F. Díaz del Olmo and J. Rodríguez (eds.), *El Cuaternario en Andalucía Occidental. Trabajos de Paleolítico y Cuaternario.* AEQUA, Monografías 1, 43-57. Sevilla.

Giles, F., Santiago, A., Aguilera, L., Gutiérrez, J.M. and Finlayson, J. C. 2003. Paleolítico Inferior y Medio en la Sierra de Cádiz, evidencias de grupos de cazadores-recolectores del Pleistoceno Medio y Superior. *Almajar* 1, 8-35.

Gracia, J., Rodríguez, J., Benavente, J., Cáceres, I. and López, F. 1999. Tectónica cuaternaria en la Bahía de Cádiz, in L. Pallí and C. Roque (eds.). *Avances en el estudio del Cuaternario español. X Reunión Nacional sobre Cuaternario,* 67-74. Gerona

Guerra, A., *et al.* 1963. *Estudio agrobiológico de la provincia de Cádiz.* Cádiz. Excma. Diputación Provincial de Cádiz. Cádiz.

Gutiérrez Mas, J. M., Martín Algarra, A., Domínguez-Bella, S. and Moral, J. P. 1991. *Introducción a la geología de la provincia de Cádiz.* Cádiz, Servicio de Publicaciones Universidad de Cádiz.

Gutiérrez Mas, J.M., Santiago, A., Giles, F., Gracia, J. and Mata, E. 1994. Áreas de transformación de recursos líticos en glacis de la Depresión de Arcos de la Frontera (Cádiz). In J.F. Jordá Pardo (ed.), *Geoarqueología*, 305-316. Madrid. I.T.G.E.

Morata, D. 1993. *Petrología y Geoquímica de las ofitas de las Zonas Externas de las Cordilleras Béticas.* Unpublished PhD thesis, Universidad de Granada.

Nocete, F. 1989. *El espacio de la coerción. La transición al estado en las campiñas del Alto Guadalquivir (España). 3000-1500 a.C.* British Archaeological Reports Internacional Series 492. Oxford, BAR Publishing.

Nocete, F. 1994. *La formación del estado en las campiñas del Alto Guadalquivir (3000-1500 a.n.e.). Análisis de un proceso de transición.* Monográfi ca Arte y Arqueología. Granada, Universidad de Granada.

Nocete, F. 2001. *Tercer milenio antes de nuestra era. Relaciones centro / periferia en el Valle del Guadalquivir.* Barcelona, Bellaterra Arqueología.

Pérez, M. 1997. *La producción de instrumentos líticos pulimentados en la Prehistoria Reciente de la Banda Atlántica de Cádiz. Cádiz*, Universidad de Cádiz.

Pérez, M. 1998. La producción de instrumentos de trabajo pulimentados en el territorio de la banda atlántica de Cádiz. *Revista Atlántica-Mediterránea de Prehistoria y Arqueología Social* 1, 97-124.

Pérez, M., Domínguez-Bella, S., Morata, D. and Ramos, J. 1998. La industria lítica pulimentada en la Prehistoria Reciente de la Banda Atlántica de Cádiz. Estudio de áreas fuente y relaciones entre litología y yacimientos. *Cuaternario y Geomorfología* 12 (3-4), 57-67.

Pérez, M., Ramos, J., Vijande, E. and Castañeda, V. 2005. Informe preliminar de la excavación arqueológica de urgencia en el asentamiento prehistórico de La Esparragosa (Chiclana de la Frontera, Cádiz), *Anuario Arqueológico de Andalucía* 2002-III, 93-103.

Ramos, J. 1993. *El hábitat prehistórico de 'El Estanquillo'. San Fernando.* San Fernando, Fundación Municipal de Cultura. Ayuntamiento de San Fernando.

Ramos, J., Domínguez-Bella, S., Morata D., Pérez M., Montañéz, M., Castañeda, V., Herrero, N. and García, M. E. 1998. Aplicación de las técnicas geoarqueológicas en el estudio del proceso histórico entre el V y III milenios a.n.e. en la comarca de La Janda (Cádiz). *Trabajos de Prehistoria* 55 (2), 163-176.

Ramos, J., Montañéz, M., Pérez, M., Castañeda, V., Herrero, N., García, M. E. and Cáceres, I. (eds.). 1999. *Excavaciones arqueológicas en La Mesa (Chiclana de la Frontera, Cádiz). Campaña de 1998. Aproximación al estudio del proceso histórico de su ocupación.* Chiclana de la Frontera, Ayuntamiento de Chiclana, Fundación Vipren y Universidad de Cádiz.

Ramos, J., Castañeda, V., Domínguez-Bella, S., Montañéz, M., Aragón, A., Moncalyo, F., Castañeda, A., Vijande, E. and Expósito, J. A. 2002. Informe de la campaña de prospecciones superfi ciales desarrollada en los términos de Barbate y Tarifa. Valoración en el proyecto de investigación: La ocupación prehistórica de la campiña litoral y banda atlántica de Cádiz. *Anuario Arqueológico de Andalucía* 1999-II, 9-20.

Ramos, J. and Lazarich, M. (eds.). 2002. *El asentamiento de 'El Retamar' (Puerto Real, Cádiz). Contribución al estudio de la formación social tribal y a los inicios de la economía de producción en la Bahía de Cádiz.* Cádiz, Universidad de Cádiz y Ayuntamiento de Puerto Real.

Ramos, J., Domínguez-Bella, S., Pérez, M., Montañéz, M., Castañeda, V., Herrero, N. and García, E. 2004. Producción, distribución y consumo de productos líticos en el marco de la formación económico social clasista inicial en la banda atlántica de Cádiz. In *III Simposio de Prehistoria Cueva de Nerja,* 386-392. Málaga, Fundación Cueva de Nerja.

Ramos, J., Domínguez-Bella, S. and Pérez, M. 2006. Producción, distribución y consumo de productos líticos en la banda atlántica de Cádiz por sociedades clasistas iniciales, in G. Martínez, A. Morgado and J.A. Afonso (coords.), *III Reunión de trabajo sobre aprovisionamiento de recursos abióticos en la Prehistoria,* 327-348. Granada, Universidad de Granada.

Ramos, J., Domínguez-Bella, S., Pérez, M., Vijande, E. and Cantillo, J. J. 2008a. Materias primas y análisis técnico de los productos líticos tallados por sociedades tribales comunitarias y clasistas iniciales. In J. Ramos (coord.), *La ocupación prehistórica de la Campiña Litoral y Banda Atlántica de Cádiz.* Arqueología Monografías, 167-175. Sevilla, Junta de Andalucía.

Ramos, J., Pérez, M., Clemente, I., García, V., Ruiz, B., Gil, M. J., Vijande, E., Soriguer, M., Hernando, J. and Zabala, C. 2008b. La Esparragosa (Chiclana de la Frontera). Un asentamiento con campo de silos en la campiña de Cádiz, del IV° milenio a.n.e., in M. Hernández, A. A. Soler and J. A. López (coord.), *IV Congreso del Neolítico Peninsular,* 385-392. Alicante, Museo Arqueológico de Alicante-MARQ.

Ramos, J., (coord.) 2008. Memoria del proyecto de investigación: La ocupación prehistórica de la campiña litoral y Banda Atlántica de Cádiz aproximación al estudio de las sociedades cazadoras-recolectoras, tribales-comunitarias y clasistas iniciales. Arqueología Monografías. Sevilla, Junta de Andalucía.

Zazo, C. 1989. Los depósitos marinos cuaternarios en el Golfo de Cádiz. In F. Díaz del Olmo and J. Rodríguez Vidal (eds.). *El Cuaternario en Andalucía Occidental.* AEQUA, Monografías 1, 113-122. Sevilla.

Searchers and miners: first signs of Flint exploitation in Madrid's region

Javier BAENA PREYSLER, Sergio BARÉZ, Alfredo PÉREZ-GONZÁLEZ, Marta ROCA, Ana LÁZARO, Raúl MÁRQUEZ, Inmaculada RUS, Carmen MANZANO, Felipe CUARTERO, Irene ORTIZ, Pedro RODRÍGUEZ, Teresa PÉREZ, Iván GONZÁLEZ, José POLO, Daniel RUBIO, Manuel AlCARAZ and Ana ESCOBAR

Abstract

Since the 19th century Madrid region has been well known for its Lower and Middle Palaeolithic sites. Fluvial terraces from the main rivers of the region have allowed us to detect a great number of sites. During the past few years, heritage conservation initiatives undertaken by the administration of Madrid's regional government have also facilitated the discovery of new sites, which in the future will undoubtedly improve our knowledge about the activities of Mousterian groups inhabiting this region. In this paper, we present a preliminary report of the site of El Cañaveral (Madrid), whose excavation revealed a large concentration of flint knapping artifacts and debris, as well as information on the sources of flint procurement and its first exploitation steps.

Keywords

Iberia. Madrid. Mousterian. Flint procurement. Numerical dating.

1. Background

Madrid is undoubtedly one of the most thoroughly studied Spanish regions of Palaeolithic occupation. However, it is also a host of great number of scholarly misunderstandings. The earliest human occupation of this region occurred 450,000 years ago, in the Middle Pleistocene, during the Mindel-Riss interglacial (OIS 13-12). This period is characterized by the presence of lithic industry on fluvial terrace levels, as exemplified by the finding from Pinedo site (Querol and Santonja 1979) in Aranjuez (Madrid), Transfesa or San Isidro archaeological sites (Baena *et al.* 2000).

It is somewhat paradoxical that one of the largest gaps in the scholarship on this region is the absence of a reliable regional chronological sequence of the Palaeolithic occupation, due to the scarcity of geo-chronological data. Therefore, the subsequent section is inevitably based on general assumptions and traditional interpretations of the known archaeological record.

2. The dynamic of the Lower to Middle Palaeolithic transition in Madrid

On a regional scale, Acheulean chronocultural assemblages are traditionally defined by the presence of handaxes. However, this criterion should be applied with certain caution. Our actual knowledge of supply and catchment areas suggest that testing pieces, trials, mistakes, or even initial knapping stages may well be taken for bifacial items. On the other hand, the traditional elements used to differentiate Lower Palaeolithic (Acheulean) from Middle Palaeolithic (Mousterian), have been revised thanks to the recent archaeological discoveries and new geochronological data (Gamble 1999).

The Acheulean classification cannot be applied to some collections lacking handaxes, a large number of flaking products (flake debitage) or retouched elements from programmed schemes. The presence of configured material departing from trifacial schemes (Böeda 1994; Baena and Baquedano 2004; Cuartero 2004, i.p.) and the absence of discoid and Levallois débitage characterized by the pre-

Figure 1. Location of El Cañaveral site in the Iberian Peninsula.

sence of multipolar debitage appear to be the best indicators for an Ancient Palaeolithic in our area. The typological repertory of Lower Acheulean in Madrid and façonage chains coexist with assemblages composed by configured tools made of flakes or fragments, turned into scrapers, retouched flakes, denticulates, notches, and knife-like tools among others, with a high frequency (even though variable) of Levallois products, which consist of recurrent centripetal and preferential lineal modalities (Böeda 1988).

During the development of Acheulean moments, the presence of Levallois knapping sequences, initially recurrent, were more frequent within lithic remains. In advanced stages, it is possible to confirm the presence of preferential modalities (uni and bipolar) and Quina retouches (Bourguignon 1996, 1997). Current data suggest that the traditional classification of Acheulean artifacts from Madrid is actually based on some really old collections along with others, typical of the Middle Palaeolithic, in which the presence of bifaces (MTA and MTB) led to their categorization as Acheulean.

It has been pointed out that the scholarship on the transition to the Middle Palaeolithic has been focused on the manner and proportion in which some techniques and typological characteristics appear and gave no serious consideration to the environmental exploitation models or the socioeconomic structures of the population.

However, during these periods some important changes occurred. One of them is the substantial amplification of the raw material selection strategies, unrelated to the supply sources, which do not vary, given the abundance of resources in the territory. The technical changes required by the new operative schemes entailed the demand for better quality raw materials and an improved knowledge of the environment and its productive potential.

In our area, the most characteristic types of items are sidescrapers, denticulates, notches, backed knifes and points. New modes of reduction appear are documented, preferentially Levallois, unipolar, bipolar, Quina- (Boëda 1988, 1989; Turq 1992; Bourguignon 1996, 1997; amongst others), schemes that will be a constant during the subsequent Upper Pleistocene. Although we can detect a continuous pattern during the initial stages, the final moments are difficult to systematize due to the lack of dating information. The dates obtained recently have offered results quite similar to the Upper Unit of 'Soto e Hijos' (Manzanares), (39.000 ± 3500 and 32.000 ± 2500 BP), associated to a highly developed Mousterian, or the so-called

Figura 2. Contact between lower member and upper member of the Middle Unit in the M-45 profiles.

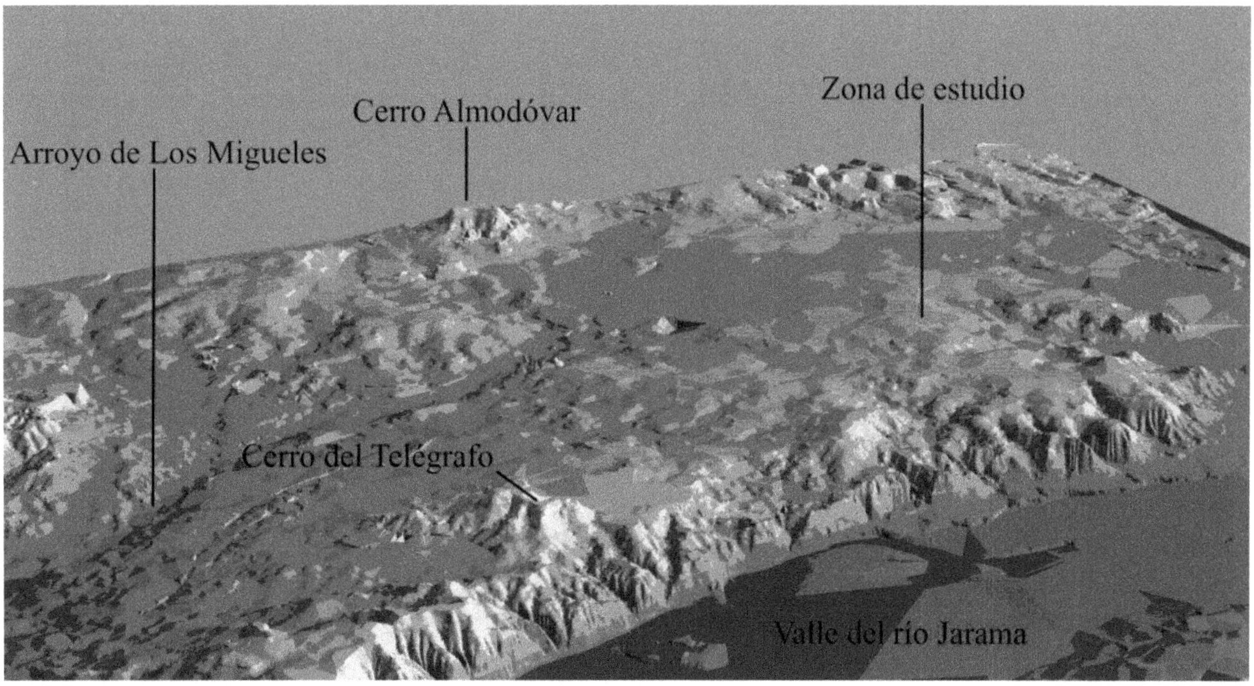

Figure 3. Digital terrain nmodel with the study area.

'12 de Octubre' archaeological site, which goes between 25 and 40 ky BP (López Recio *et al.* 2007). The existence in fluvial contexts of re-deposited materials corresponding to the erosion of the oldest terraces is problematic when assigning a chronology to the industry. Additionally, we have to be cautious when considering the possible continuity between both periods, since our territory could present a model of cultural change much slower than in other peninsular areas.

Numerous archaeological sites have been documented in old river basins of the main and secondary terraces (Rus 1987; Baena *et al.* 2000). Among them, it is worth mentioning La Parra (Pérez de Barradas 1922, 1923,1926; Obermaier and Barradas 1924; Enamorado 1989), López Cañamero (Pérez de Barradas and Wernert 1921a; Pérez de Barradas 1924a; Obermaier 1925; Sánchez 1985), and Atajillo del Sastre (Pérez de Barradas 1923, 1926, 1929; Obermaier and Barradas 1924; Sánchez 1985), Casa del Moreno (Pérez de Barradas and Wernert 1921a ; Pérez de Barradas 1922, 1923, 1924a, 1926, 1929, 1934), El Sotillo (Pérez de Barradas 1922, 1924b,1926, 1929; Pérez de Barradas and Wernert 1930, 1931; Martínez de Merlo 1984; Enamorado 1984b), El Almendro (Pérez de Barradas and Wernert 1919; Pérez de Barradas 1924a, 1926, 1929, 1934), La Torrecilla (Pérez de Barradas 1926, 1929; Cobo et al. 1979; Enamorado 1989), Parador del Sol (Pérez de Barradas 1922, 1924 a, 1926, 1929; Obemaier 1925), Fuente de la Bruja (Pérez de Barrada 1922, 1923, 1926, 1929; Hermández Pacheco 1956). More recently we have some examples in Las Fronteras- Pinto (Baena and Ibañez 1996), La Torrijana-Pinto (personal communication of Belen Márquez and Juan Gómez), El Negralejo (Antonio Geanini's personal communication), Canteras de Vallecas-Camino de Salmedina (Pérez de Barradas *et al.* 1921; Obermaier *et al.* 1921; Baena *et al.* 1997), Los Cenagales (Javier Pastor's personal communication), Pedazo del Muerto (López Covacho *et al.* 1996), la Gavia (Rus 1987; Vega *et al.* 1999), the archaeological sites of 12 de Octubre and the complex terraces of Villaverde-Butraque (López Recio *et al.* 2007), Fábrica de los Ladrillos-Preresa (Joaquin Panera's personal communication) and surface sites such as Ciudad Lineal, Chamartín or San Blas (Rus and Velasco 1993).

The data gathered in all these sites has been used in the attempts to systematize and classify the collections according to the French Mousterian facies (Rus 1987; Rus and Velasco, 1993), resulting in most cases in the assignment

Figure 4. Aerial view of the sondages in the Exclusion Area, next to Vías Colectoras (Cortesy of the Junta de Compensación de la UZP2.01 El Cañaveral).

Figure 5. Aerial view of the sondage area and the Parcela 32 excavation (Cortesy of the Junta de Compensación de la UZP2.01 El Cañaveral).

of the assemblages to the MTA, solely on the basis of the presence of bifaces. However, this procedure must be reconsidered, especially in the case of the oldest collections, where the control of the material's spatial and stratigraphic position was precarious or simply non-existent (Enamorado 1984a, 1984b, 1989; Sánchez 1985).

3. Pleistocene in Madrid during the first decade of XXI century

For the past couple of years our region has experienced a deep transformation that affected interfluvial areas situated between the Manzanares and Jarama basins. For several decades the construction activities have mainly affected the fluvial areas (sand and gravel exploitations), but now, the urban growth has started to affect environments far beyond river basins.

Recently, the development of Urban Enlargement Projects (called *Proyectos de Ampliación Urbanística,* PAU) and important highway constructions, such as M-45, have lead to the discovery of new Pleistocene archaeological sites, while at the same time producing new methodological challenges. Until now, we had never recorded large open areas containing abundant archaeological remains at risk. During these past few years, heritage policy promoted by the Regional Agency responsible for Historical Heritage (*Dirección General de Patrimonio de la Comunidad de Madrid*) has encouraged the development of important research projects inside the areas affected by construction activities and facilitated the recovery of new material enabling the scholars to enhance our knowledge of the prehistory of the region.

Drawing on the support of Cañaveral Compensation Board (Junta de Compensación de Cañaveral), as well as the funding from the research project 1664/00/2007 from the Dirección General de Patrimonio de la Comunidad de Madrid, 'Study and documentation of Pleistocene archaeological sites in Madrilenian Community: watersheds of Manzanares and Jarama's basins', beginning in 2007, we carried out fieldwork study of the archaeological sites of El Cañaveral (Vicálvaro, Madrid). The archeological project has involved the collaboration between private institutions, the University and Administration, which has guaranteed the management and documentation of one of the biggest and most complex cases of an archaeological site recorded in our region.

El Cañaveral is a home of several sites belonging to different chronocultural frames. One thing all of them have in common is the great abundance of knapping lithic remains associated with flint outcrops. Our project has focused on those spaces where the conservation conditions guaranteed the best interpretational possibilities.

4. Geoarchaeological aspects of Madrid region

The geological context of the site is the Madrid basin, a part of a bigger morpho-structural unit called Tajo's basin (Figure1). It has a triangular shape and is bounded by the Central System and Gredos mountains in the NW, the Iberian Range in the NE, and by the mountains of Toledo in the S (Pérez-González 1994).

During the Miocene the basin underwent an evolution caused by various factors, leading to three big sedimentary stages, which were associated with three large stratigraphic units: Lower Unit (Middle Rambliense-Aragoniense), Middle Unit (Middle Aragoniense- Lower Vallesiense) and Upper Unit (Upper Vallesiense-Turoliense) (Calvo *et al.* 1996).

Within the project area, the Tertiary stratigraphic series is represented by the Lower Unit, made up of massive gypsums, sharply folded gypsums with green clays intersections, and the Intermediate Unit is made up by clays, dolomites and flint levels. At the same time, this Intermediate Unit is divided in two components. The Lower component is made of 12 meters of sharp green folded clays with some centimetres interceptions of muddy clays and sandy clays. Above it is the Upper component, made up by sharply folded clays, multicolor clays with some local carbonates and flint nodules levels (Figure 2) (Bárez and Pérez-González 2007). The Intermediate Unit presents frequent deformations caused by a karstification processes of the Lower Unit and some collapse phenomenon as well as the resulting deformation by the 'reflect effect' in the Intermediate Unit. The study of this Tertiary stratigraphic series has been very important, because of it impact on the genesis and the archaeological preservation (Figure 2).

Considering its geomorphology, this area has characteristic features mainly controlled by lithologic and climatic factors, with two clearly distinct domains—Jarama Valley in the East, and the erosive-structural plains in the West. Here, the wide Jarama valley presents a high asymmetry in his transversal cross section, with a lot of stepped terraces in the left bank and a strong erosive character in the right one, with only the lowest terraces (+8m and +12-

Figure 6. Sedimentary packages to postpone the geoarchaeological analysis.

15m). The western boundary of the valley is characterized by the cliffs formed on the gypsums of the Intermediate Unit extending several kilometers and reaching 100 meters of slope.

On top of the cliffs, from the right bank of Jarama to the left cliffs of the Manzanares River, there is a plateau forming a part of the great dividing platform situated between these two rivers (Figure 3). The area of the site is situated in that geomorphologic unit. It is a structural-erosive origin relief conditioned by the presence of flint layers and carbonated clays from the Upper Member of Intermediate Unit, that lead to wide plains in the highest areas, emphasized in the limit with Jarama valley by the presence of gypsum cliffs, while in more depressed areas the Lower Member appear, mainly made up by clays, and frequently is covered by bottom valley sediments.

In the course of the works we were able to recover the evidence of the existence of wide quaternary sedimentary series in the whole prospected area. They have colluvium sediment origin, fluvial and wind with a lot of quartz-feldespat sands in their composition. The thicknesses vary and sometimes reach 9 meters. One of the causes of these various thicknesses is the high quaternary karstic activity development in the Lower Unit of Miocene, that cause Intermediate Unit deformation and the presence of subsident areas in the surface where all the sediments are adapted. We have systematically observed edaphic processes on the roof of sedimentary series, producing type BT argilic horizons and carbonates precipitation horizons (Ck) (fersialitic ground, typical from Mediterranean fields). Eventually, edaphic developments intercalates in the sequences also appeared.

From a Geological perspective, the archaeological sites located along the M-45 present a similar context. Geomorphologically, they are situated on a platform that defines the watershed between the Manzanares and Jarama rivers, where a lot of silica rocks come to the surface (flint, opal, etc). The unusual abundance of these rocks explains the presence of a high density of archaeological sites linked to supply/catchment patterns of raw materials.

The preservation of a great number of these archaeological sites has been made possible by the different sedimentary stages, eolian and colluvian types, which happened during the quaternary, lacking high energy and almost any transport capacity that covered and buried the anthropological activities protecting them from external erosive agents. The numerical dating series obtained by OSL in all areas have registered the activities of human groups in OIS 3 and latter.

5. The archaeological interventions in Pleistocene archaeological sites: El Cañaveral

The archaeological fieldwork carried out in El Cañaveral has been one of the first studies aimed at the excavation of Pleistocene sites of large dimensions and containing great amounts of archaeological materials, undertaken in the context of an salvage project. The excavation of such archaeological sites has its own inherent difficulties, requiring a specific protocol of action, which we proceed to explain below (Baena *et al.* 2008).

The first step consists in a superficial survey of the area, carried out in the sedimentary units whose characteristics suggest greater archaeological potential. We paid special attention to the areas which would undergo future urbanization and where the task of heritage preservation would be particularly urgent.

Drawing on the previous studies of this region, we established an *Archaeological Exclusion Area* to be assessed and studied through manual and mechanic excavation of test pits, which allowed us to proceed to the total or partial clearing of the area. Backhoe pit testing has been essential for identifying the areas of real archaeological interest.

Therefore, we decided to apply a method based on backhoe test pits (ditches of 20m in length) around the areas with known archaeological potential for further examination. The pits were dug until they reached Tertiary levels, allowing us to obtain a complete and broad view of the quaternary sequence (Figures 4 and 5).

Based on the evidence collected on the surface and on the sections of the test pits with possible sedimentary packages containing archaeological remains in good preservation conditions, we then adopted an excavation strategy that could account for the varying qualities of different kinds of sediments. In practice it meant the excavation of natural levels with the techniques commonly used for these chronological periods and sample collection. Occasionally, in the course of the development El Cañaveral construction project, we had to confront emergency circumstances forcing us to take the complete sedimentary packages in order to guarantee the safety of the geoarchaeological evidence. In these exceptional cases we have followed the rule of conserving the series and referenced blocks whenever possible, in order to postpone the geoarchaeological analysis (Figure 6).

6. The archaeological excavation areas of El Cañaveral

One of the main problems we had encountered in the course of the study was the persistent evidence of isolated prospecting activities carried out during the Neolithic as well as an abundant presence of pits and structures of the contemporary period in the entire geoarchaeological unit. The identification of lithic remains result of the contemporary mining activities was done based on sedimentological studies of the deposits, the raw material employed (of very bad quality due to its productive orientation), on the absence of some technical categories (cores and minor size flakes), as well as on the presence of metallic stigmas in

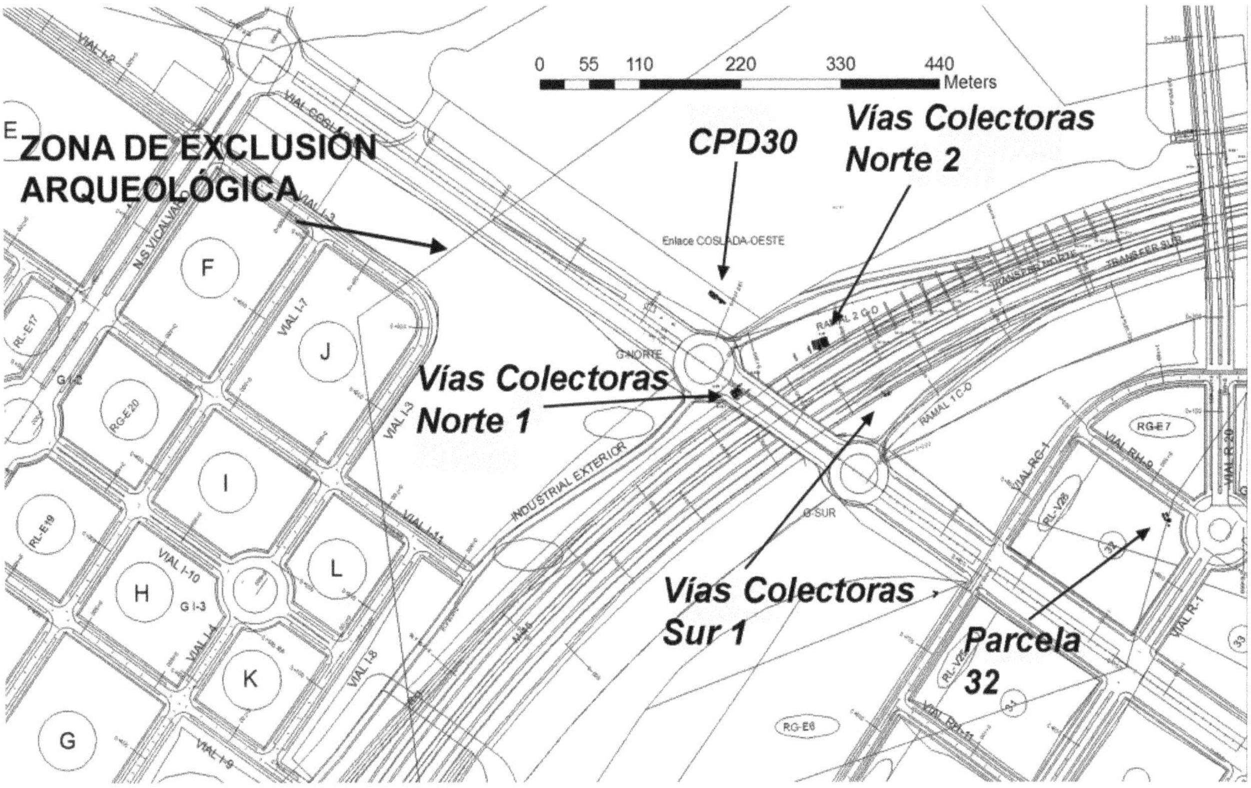

Figura 7. Building map of El Cañaveral with the main excavated areas.

the flake platform. Even more challenging has been the identification of the pits and ditches of Neolithic period, which involved the examination of the repertory and dating of the sedimentary contexts. In addition to this evidence, we have compiled substantial distribution data of excavated areas whose datings and technical features confirm the existence of an important Mousterian occupation in the zone (Figure 7).

1. North-South Vicalvaro Vial Excavation (Cañaveral 1)

In the North – South Vial Vicálvaro, five excavation areas were identified based on the evidence of the lithic industry concentrations visible on the surface. This initial data pointed to the delimited areas where flaking activities took place, corroborated by the finding of a large number of flint nodules carried there from its source (fluvial deposits and specially, coluvionar deposits). However, the presence of raw material was low in comparison to the rest of areas that were discovered later.

Therefore, these first test pits contained not only the evidence of lithic industry but also large concentrations of rough flint nodules, which appears to be the common characteristic of most of the excavated areas.

Vías colectoras

In the surrounding area of the M-45 highway we used a backhoe test pits for identifying sites and excavation areas of special archaeological interest detected during the survey. The result obtained from some test pits was negative, because only boulder lithic materials were located.

On the other hand, in a different set of test pits we located a cemented slimes level under the red clays stratum (paleosoil), containing well-preserved lithic fragments. These findings called for the organization of several rescue excavations.

1.1) Excavation of Estribo Norte de la Estructura 1 (Vías Colectoras Norte 1)

The archaeological intervention had two phases: a mechanical soil clearance and a further delimitation of an area for manual excavation (80 square meters). The sequence recognized presented three archaeological strata over which different soil episodes took place.

In level III, composed of prismatic brown clays, the lithic industry documented seems exceptionally fresh, with only some carbonate concretions. However, the orientations and inclination of the materials point to geological processes of vertical alteration, possibly linked to the formation of the different paleosoils.

It is necessary to emphasize the presence of big massive flint blocks which reveal percussion and knapping activity, being common in levels II and III, in the clays stratum that contact the Tertiary level. We used mechanical means to

Figure 8. Vías colectoras Norte 1. A. General view of the excavated area with the Flint blocks and the debitage clusters. B. Detail of an in situ refitted big flake with the Flint block.

remove the big flint nodules in order to proceed with the search for refitted materials. Once removed, we confirmed that, under them, there still appeared lithic materials. The question of the processes that led to the genesis of lithic industry deposits under big nodules is still an open one. Several reassemblages of lithic industry have been documented in field. One possible explanation is the presence of high intensity percussion processes, aided by the existence of natural fissures on the rock that allowed the fragmentation of the big blocks, whose smaller pieces later served in other debitage processes in the same area (Figure 8).

1.2) Area 3 excavation (Vias colectoras norte 2)

The area selected for excavation totalled 164 square meters and its sequence included at least one clay layer containing large assemblages of lithic industry represented by a great number of cores and hammer stones.

The distribution of the materials alternates between concentrations with predominantly horizontal layouts and sets of raw materials located along vertical slopes, probably resulting from the edafic processes that took place later. At the same time, it is likely that here the coluvionary deposits were the principal source of raw material and the basis over which the well-preserved traces of knapping activity were superimposed. New coluvionar-eolic episodes affected in a different way the distribution of various knapping zones with the partial dragging of some pieces and the intrusion of new boulder materials (Figure 9).

Taking this into consideration, we decided to increase the excavation area in order to locate those assemblages whose conditions of conservation were the best. During the excavation we recorded the distribution of oval or circular concentrations of fresh material corresponding to particular areas of debitage as well as hearth remains indicating a temporary occupation of this area. According to techno-typological criteria, the assemblage clearly belongs to the Mousterian period, characterized by a clear predominance of Levallois schemes, and contrasting with an occasional presence of blade schemes, similar to the ones detected inside the Mousterien in other European zones (Tuffreau 1990; Réveillon and Tuffreau 1994; Réveillon and Cliquet 1994; Cuartero *et al.* 2007). The occasional retouched pieces, two Mousterian points and two scrapers (as opposed to 14.000 pieces without retouch) point towards a Middle Palaeolithic chronological attribution, fitting the most classic Mousterian types.

1.3) Excavation of Vias Colectoras South

In this zone, a concentration of lithic industry was mostly composed of nodules, fragments, flakes and cores and some isolated retouched tools, found in their original position, evident from their conservation. Similarly to other test pits, the concentrations of lithic materials appear on deposits just on top of the tertiary formations, where we can see good quality flint blocks with a carbonated cortex, arranged in a natural way, and containing in many cases clear evidence of human manipulation.

Particularly interesting was the location of flint nodules in square D6, where we found large size blocks with large anthropic extractions and flake refittings. This sector contains traces of negative structures (small pits) with visible remains of charred wood (carbonaceous sediment) radially dispersed around some of the nodules. The study of these findings is still in progress, but its preliminary results point to the precence of a set of Neolithic mining episodes, thus cutting through the Mousterian sequence documented in the rest of the area. The numerical dates obtained are as follows: Beta-244210 5750 +/- 50 BP, 2 sigma calibration: Cal BC 4720 to 4470 (Cal BP 6660 to 6420); Beta-244211 6020 +/- 40 BP, 2 sigma calibration: Cal BC 5010 to 4800 (Cal BP 6960 to 6750) and Beta-244209 6490 +/- 50 BP, 2 sigma calibration: Cal BC 5530 to 5360 (Cal BP 7480 to 7310). They suggest the existence of mining activities in the surroundings of the close-by Neolithic flint mine of Casa Montero.

Near this excavated area, we documented some coluvionary levels containing a large volume of lithic remains and collected some samples for dating. Their typological and technological features confirm an Acheulian ascription, which prompted us to recover the maximum quantity of material in these levels.

2) C.P.D. 30

In this zone, neighboring the above-described VCN 2, the test pits showed the existence of important concentrations of very fresh lithic industry, in association with supports altered by atmospheric agents. The pits disclose a level with a great number of cores and some quartzite hammers. In the excavation layout we have also noted somewhat circular concentrations of fresh materials probably corresponding to knapping zones. Based on techno-typological parameters, there is still a clear predominance of Levallois schemes associated with some few blade schemes, similar to the ones recovered in other zones of El Cañaveral.

3) Parcela 32 (Cañaveral 3)

The fieldwork in Parcela 32 is still in progress. In this area we located at least three different levels with important concentrations of fresh material. The analyzed features indicated the existence of knapping areas similar to the ones documented in Vías Colectoras Norte. The documented material consists predominantly of flakes, blade flakes and cores, while the retouched material is very poorly represented. The initial study of these materials confirmed the existence of Levallois operative chaines of recurrent centripetal and unipolar modality.

At the same time, we have documented some intense blade exploitation schemes- unipolar modalities (Figure 10). As in previous cases, there is evidence of exploitation of high quality flint and opal sources, where sediment removal and quarrying activities seem to have taken place.

7. Cultural interpretation of the site and the variability of dense artefact contexts

The natural resources found in Madrid region were an attraction for its inhabitants since the earliest periods. The available data shows that these areas played an essential role as flint procurement zones, evidenced by the frequency of the deposits containing traces of knapping activity, that took place near primary sources as well as others in secondary position, similar to what has been named *quarrying facies* (Barandiarán and Vallespí 1982). The comparative study of the documented operative chains, both

Figure 9. General view of the Vías Colectoras Norte 2 (Área 3) excavation and detail of one of the debitage cluster.

Figure 10. General view of the Parcela 32 (Cañaveral 3) excavation. A, Profile with Flint outcrops showing sediment remotion. B, Detail of lithic distribution. C, In situ cores wit percussion stigmata.

in contexts of abundant lithic resources (interfluvial areas) and in contexts of consumption (close to riverbeds), has turned out to be an essential tool for gaining better understanding of the global strategies developed by the palaeolithic human groups (Geneste 1991).

In El Cañaveral we have mainly documented the first stages of the operative chain. Procurement, reduction and blank production (*entame and mise in forme*) aimed at the exploitation, together with steps of full production (*débitage*), took place within the same space. Traces of consumption activities (final stages of the operative chain) are very limited or non-existent, pointing to the short duration of human occupations.

As for the first stages of the operative chain, the strategies employed are characterized by the absence of extensive planning. Several models apparently predominate: the procurement of small blocks (up to 10-20cm long), frequently originating from coluvionary deposits; mid-size blocks (20 to 50cm long) coming from similar deposits or produced through fragmentation of big blocks; and the exploitation of big massive nodules (50 to 200cm long) through percussion (possibly throwing), taking advantage from open air outcrops whose exploitation, in some cases, seem to have been favored by the excavation of ditches (figures 10 and 11).

In the first case, the selection is based on the morphologies and dimensions near the volumetric model resulting from the Levallois-discoid *débitage*. In the second case, we observe the use of direct unifacial and unidirectional schemes (Martin Blanco and Djema 2005), or multipolar ones of discoid or Quina appearance (Figure 12), clearly intended for obtaining thick blanks. In both cases, the strategy consists of taking advantage of the block's fissures caused by meteorization, thus generating blanks of middle and small size, as well as fragments without clear knapping signs. The large flakes obtained from big and medium size blocks are equally close to the searched core morphology.

The technical and technological features of the El Cañaveral debitage can be grouped according to three major operative schemes (Figure 13 and 14):

a) Preferential and recurrent Levallois method (unipolar and bipolar modalities), most frequently employed, consists in direct percussion with hard stone hammer. With respect to the cores, we possess a huge number of elements, being dominant those which are initially exploited. Some pieces reveal a high technical and technological *débitage* skill, corresponding in many cases to a more intensified exploitation, characterized by a clearly 'wasteful' strategy, also called *gaspillage* strategy (Ba-

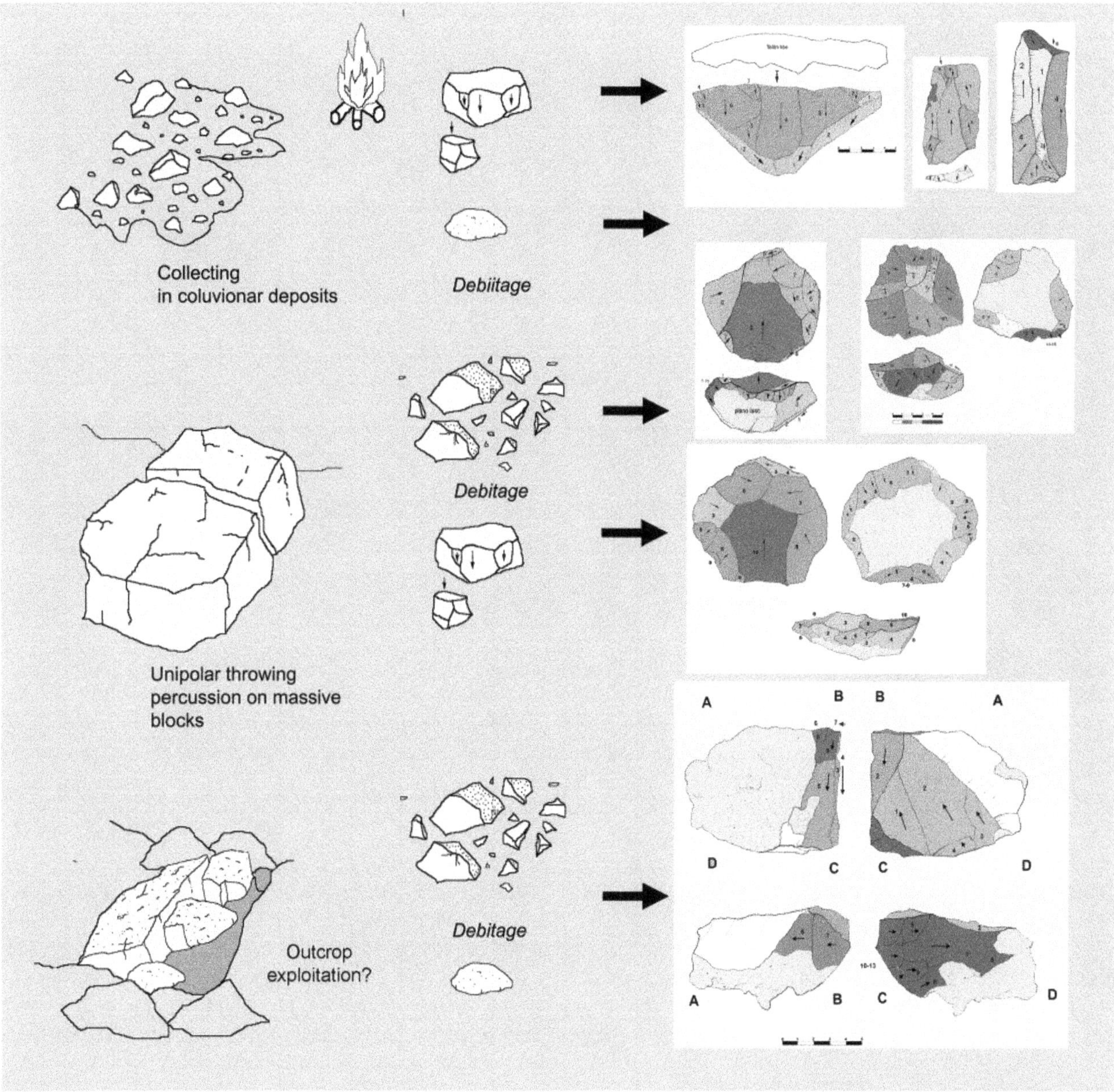

Figure 11. Blank support production strategies in Vías Colectoras. A. Unipolar schemas B. Multipolar/discoid (quina debitage appearance) schemas.

zile 2002) and evidenced by a great abundance of raw material in the zone.

b) At the same time, we have documented the presence of discoid and hierarchical discoid schemes, whose possible final products are pseudolevallois points, backed knives and quadrangular and triangular flakes. Although the nature of the relationship between these schemes and the previous ones (the role of recycling or ramification processes) (Bourguignon *et al.* 2004) is yet to be established, there are numerous cases demonstrating how different methods derive from each other and how different modalities are superimposed in the same core.

c) In other cases, we document unipolar blade schemes, resulting in cores of prismatic morphology, which could be linked to the initial phases of centripetal modalities of exploitation, similar to those described in French contexts (Révillion and Cliquet 1994; Klaric 2000).

Regarding the presence of structural categories, cortical flakes, cores and knapping debris, together with a scarce configuration of tools, are the common features of the entire analyzed areas.

The recently obtained magnetic susceptibility data from the minerals contained in the structure's sediments of Area 3 indicate the probable existence of combustion areas. This means that, if these combustion structures are indeed contemporary with the knapping activities, we may be dealing with temporary open air camps, where the main documented activity is the flint procurement. The possible confluence of both events in the same space would imply an exceptional finding within the peninsular context, i.e.,

Figure 12. Main catchment strategies documented in different areas at El Cañaveral.

the first evidence of an open air occupation exclusively associated with flint procurement and knapping activities (Baena 1994).

From a technological point of view, the common elements observed in the different excavated areas can be summarized as follows:

- If we take into account the amount of documented cores and compare it with other contexts (e.g., river areas), the morphological variation margin of the exploited blocks is very high. The high degree of variability in methods, modalities and morphologies can be explained by the abundance of resources, or the collective participation of the human group in the processes of procurement and production. Therefore, the features documented in El Cañaveral are coherent within contexts where the procurement of lithic resources and the knapping learning processes take place at the same time.

- The scale of the industries is big. However, there are clear differences between exploitation strategies at different stages (catchment, production of supports, full production). There is no evidence of microlithic production in any of the studied areas.

- The distribution of categories in these contexts is totally different from the one documented in the consumption environments. The absence of tools is particularly interesting especially if we compare it with cave contexts.

- The convergence of different knapping methods in the same item is frequent and could be explained by the recurrent occupation of the sites (recycling in palimpsests), or other recycling strategies that have yet to be analyzed. The above-mentioned overlapping, suggests that, as opposed to the ramification processes documented in other contexts, we may be facing here with reuse processes (as it is documented in the level J of Romaní, Vaquero, in this volume).

The scarcity of scholarship dealing with this kind of sites has complicated our task of making a comparative analysis of the behavioral variability within the Mousterian groups and those of the preceding stages. The data we have gathered so far in El Cañaveral suggest that a higher degree of variability is expressed in certain aspects of these kinds of occupations.

Different modalities of access, extraction and initial processing of the raw material found here are infrequent in lithic assemblages associated with sites of a higher occupational impact. These initial strategies, though relatively simple, reveal a diversity of flake/blank production methods, which are different from the ones developed in later stages.

The variability, that is evident in exploitation methods, is coexistent or interrelated in some of the other areas seemingly connected with the zones of higher material concentration, higher occupational impact, or higher visit rates. Many of the operative schemes showing different

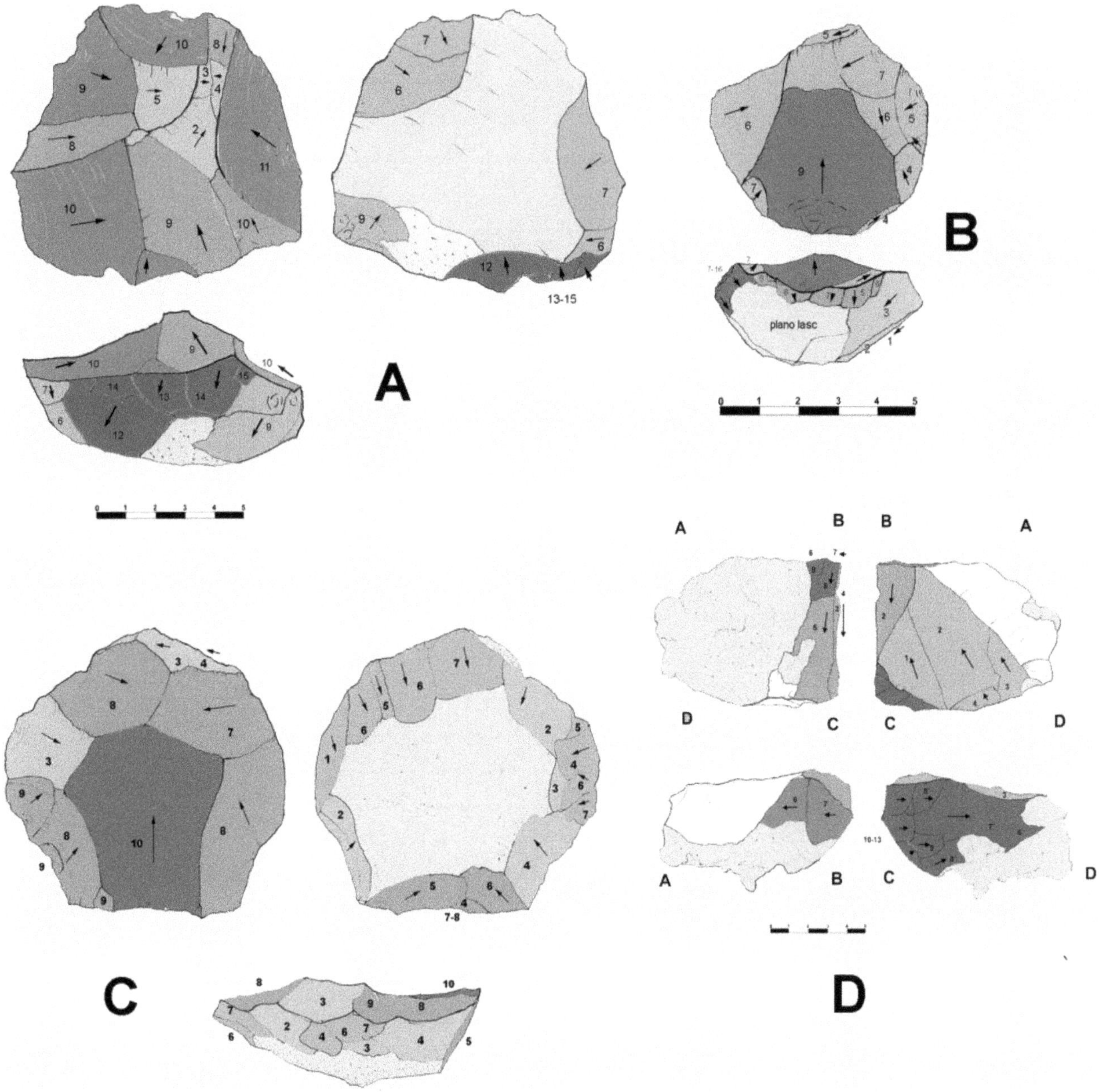

Figure 13. Diacritic schemas in cores. A, levallois recurrent centripetal core from Área 4 close to Vías Colectoras Norte 1. B y C, levallois (lineal and preferential) cores from Vías Colectoras Norte 2. D, blade core from Parcela 32.

degrees of hierarchization or alternation in the extractions call for a more detailed evaluation.

Additionally, in these places we see different degrees of technical skill, possibly linked to learning behavior. This prompts us to consider knapping activity as a group behavior and the procurement area an optimal place for the acquisition of technical and technological skills.

The virtual absence of variability in typological or raw material composition prods us to seriously consider some interesting nuances, such as the few pieces of foreign material (fundamentally quartz and quartzite), collected in riverbeds or terraces of the Jarama and Manzanares rivers, associated with the knapping tools. In the few cases that these materials are not linked to knapping elements, we identify cores and flakes that, as with the scarce retouched pieces made on flint, can be related with discarded products, rejected in the moment of arrival to the area, in the context of a tool kit renovation.

The characteristics of El Cañaveral have some commonalities with other sites such as Perales del Rio (Cobo *et al.* 1979; Gamazo and Merlo 1983; Baena 1992), Las Fronteras (Baena *et al.* 2002) or Los Berrocales (Iván Manzano's personal communication), where the existence of Mousterian occupations associated with the procurement and exploitation of lithic resources in areas distant from the fluvial basins has been confirmed. Similar cases have been recorded in other areas of the Iberian peninsula (Ebro's ba-

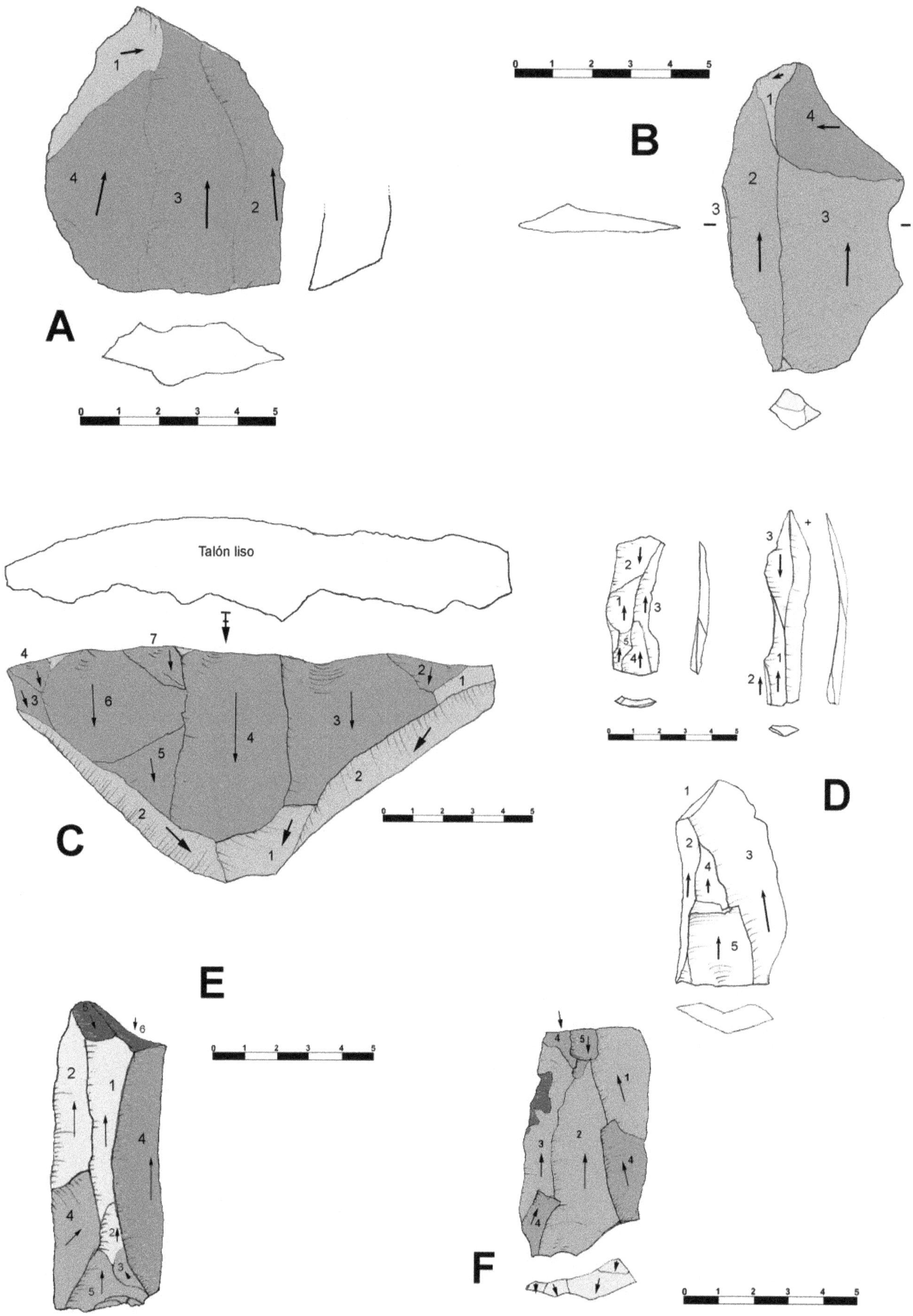

Figure 14. Diacritic schemas in flakes. A, B and C, unipolar schemas from Área 4 close to Vías Colectoras Norte 1, D, blade schemas from Parcela 32, E y F, unipolar or bipolar schemas from Vías Colectoras Norte 2.

sin: Montes 1988; Utrilla and Tilo 2001-2002; Casa de la Mina II in Ciudad Real: Sanguino *et al.* 1994; Valgrande in Salamanca: Santonja 1986; La Coca in Alicante: Fernández Péris 1998; Cerro de San Cristóbal in Toledo: López Recio and Baena 2001; Páramos del Duero: Díez Martín 2000), or in Central European contexts (Roebroeks 1988; Geneste 1988; Van Kolfschoten *et al.* 1993; Moncel and Perreve 1999; Slimak *et al.* 2005; among others). These studies demonstrate certain uniformity in the changes experienced by the groups of the end of the Mousterian in the context of increasing specialization.

Acknowledgements

This work has been funded by the project 'Documentary study on archaeological Pleistocene sites of the CAM' 1664/2007/00 of the Council of Culture and Tourism of the CAM. We are also would like to express our gratitude to the rest of the members of Arquex S.L., to D ª Ester Andreu (Arqueomedia) D. Iván Manzano Molina, D ª Laura Dapena Albiach, D. Dan Cabanes i Cruelles, D. Ángel Carrancho Alonso for their collaboration, as well as to the Dr. Nick Debenham and the Junta de Compensación ZP2.01 El Cañaveral.

References

Baena Preysler, J. 1992. *Talleres paleolíticos en el curso final del río Manzanares.* Patrimonio Arqueológico del Bajo Manzanares 1. Madrid, Universidad Autónoma de Madrid.

Baena Preysler, J. 1994. *El Paleolítico Inferior y Medio en la región sur de Madrid: Estudio de las facies de talleres.* Unpublished PhD thesis, Universidad Autónoma de Madrid.

Baena Preysler, J., Ibánez, F. 1996. *El yacimiento paleolítico de Las Fronteras. Actas de la Reunión de Arqueología Madrileña,* 181-186. Madrid.

Baena Preysler, J. and Baquedano, I. 2004. Avance de los trabajos arqueológicos realizados en el yacimiento Paleolítico de Tafesa, antiguo Transfesa (Villaverde-Madrid): principales rasgos tecnológicos del conjunto lítico. *Zona Arqueológica* 4, 31-47.

Baena Preysler, J., Conde, C. and Carrión, E. 1997. Estrategias de captación durante el Paleolítico Antiguo en la región sur de Madrid. *Rubricatum* 2, 69-76.

Baena Preysler, J., Conde, C., Carrión, E. and Pastor Muñoz, J. 2000. Paleolítico y Epipaleolítico. In, *La Arqueología madrileña en el final del siglo XX. Boletín de la Asociación de Amigos de la Arqueología* 39-40, 81-104.

Baena Preysler, J., Conde Ruiz, C., Gamazo, M., Sesé C. and Soto, E. 2002. Repertorio de yacimientos paleolíticos en el Manzanares y Jarama. In, *Bifaces y Elefantes. La investigación del Paleolítico Inferior en Madrid, 1, 460-491. Zona Arqueológica.* Madrid. Museo Arqueológico Regional.

Baena Preysler, J., Polo J., Bárez, S. Cuartero, F., Roca, M., Lázaro, A., Nebot,A., Pérez González, A., Pérez,T., Rus, I., Rubio, D., Martín-Puig, D., Manzano, C., González, I. and Márquez, R. 2008. Tecnología musteriense en la región madrileña: un discurso enfrentado entre valles y páramos de la Meseta sur. *Treballs d'Arqueologia* 14, 249-278. Universidad Autónoma de Barcelona. Barcelona.

Barandiarán, I. and Vallespí Pérez, E. 1982. Prehistoria de Navarra. *Trabajos de Arqueología Navarra* 3, 1-241.

Bárez, S. and Pérez-González, A. 2007. Patrones de aprovechamiento minero prehistórico del Sílex de Casa Montero (Vicálvaro, Madrid), in J. Lario and P. Silva (eds.), *XII Reunión Nacional de Cuaternario,* 91-92. Ávila. Publicaciones de la Escuela Técnica Superior de Ingenieros Industriales.

Bazile, F. 2002. Le premier Aurignacien en france méditerranéenne. Un bilan. *Espacio, Tiempo y Forma,* Série I, *Prehistoria y Arqueología,* 15, 215-236.

Boëda, E. 1988. Le concept Levallois et evaluation de son champ d'application. In M. OTTE (ed.), *L'Homme de Neandertal. La Technique* 4, 13-26. Lieja.

Boëda, E. 1991. La conception trifaciale d'un nouveau mode de taille paléolithique. In E. Bonifay and B. Vandermeersch (eds.): *Les premiers européens*, 251-263. Editions du Comité des Travaux Historiques et Scientifiques. Paris.

Bourguignon, L. 1996. La conception de débitage quina. *Quaternaria Nova* VI, 149-166.

Bourguignon, L. 1997. *Le Moustérien de Type Quina: Nouvelle Définition d'une entité technique.* Unpublished PhD thesis, University of París.

Bourguignon, L., Faivre, J. P. and Turq, A. 2004. Ramification des chaînes opératoires: une specificité du Mousterien?. *Paléo* 16, 37-49.

Calvo, J. P., Alonso Zarza, A. M., García del Cura, M. A., Ordoñez, S., Rodríguez Aranda, J. P. and Sanz Montero, M. E. 1996. Sedimentary evolution of lake systems through the Miocene of Madrid Basin, in P. F. Friend and J. C. Dabrio (eds.), *Paleoclimatic and paleohydrological constraint. Tertiary Basins of spain: The stratigraphic record of crustal kinematics*, 272-277. Cambridge, Cambridge University Press.

Cobo, A., Gamazo, M., Hoyos, M. and Soto, E. 1979. Los yacimientos paleolíticos de las terrazas del Manzanares. Estado actual de la cuestión. *I Jornadas de Estudios sobre la Provincia de Madrid,* 38-43. Diputación Provincial de Madrid. Madrid.

Cuartero, F. 2004. *Tecnología lítica del nivel IV de la Cova del Bolomor (Tavernes de Valldigna, València).* Unpublished PhD thesis, Universidad de Valencia.

Cuartero, F. i.p. Tecnología lítica en el nivel 4 de la Cova del bolomor: una economía de reciclado, *Saguntum.*

Cuartero, F., Martín Puig, D. and Baena, J. 2007. Propuesta experimental para el análisis tecno-económico de conjuntos líticos: el caso de la industria laminar del nivel XVII de El Esquilleu (Castrocillórigo, Cantabria), in M. L. Ramos, J. E. González and J. Baena (eds.). *Arqueología experimental en la Península ibérica; investigación, didáctica y Patrimonio.* Asociación Española de Arqueología Experimental, 113-120. Santander.

Díez Martín, F. 2000. El poblamiento Paleolítico en los Páramos del Duero. *Studia Archaeologica* 90, 521. Universidad de Valladolid.

Enamorado, J. 1984a. Las facies del Musteriense en el Valle del Manzanares según Pérez de Barradas: bases para una revisión. *Estudios de Prehistoria y Arqueología Madrileña* 3, 11-31.

Enamorado, J. 1984b. El esbaikiense: sus implicaciones en la definición de una facies del Musteriense en el Valle del Manzanares. *Trabajos de Prehistoria* 41 (1), 293-304.

Enamorado, J. 1989. La Torrecilla y La Parra: análisis de la industria lítica de dos yacimientos de época paleolítica en el Valle del Manzanares. Madrid. *Boletín del Museo Arqueológico Nacional* VII, 9-27.

Fernández Péris, J. 1998. La Coca (Aspe, Alicante). Área de Aprovisionamiento y talla del Paleolítico Medio. *Recerques del Museu d'Alcoi* 7, 9-46.

Gamazo, M. and Martínez de Merlo, A. 1983. El yacimiento Achelense de Perales del Río (campañas de excavación de 1980 y 1981). *Homenaje al Profesor M. Almagro Basch,* I. Ministerio de Cultura, 95-105. Madrid.

Gamble, C. 1999. *The Palaeolithic Societies of Europe. (Cambridge World Archaeology).* Cambridge, Cambridge University Press.

Geneste, J. M. 1988. Systèmes d'approvisionnement en matières premieres au Paléolithique moyen et au Paléolithique supérieur en Aquitaine. In J.K. Kozlowski (coord.). *L'homme de Neandertal. Vol 8:* La mutation, 61-70. Etudes et Recherches Archéologiques de l'Université de Liége, 35.

Geneste, J. M. 1991. Systèmes techniques de production lithique: variations techno-économiques dans les processus de réalistaion des outillages paléolithiques. *Techniqyes et Cultures* 17-18, 1-35.

Hernández Pacheco, E. 1956. Bos primigenius en la segunda terraza del Manzanares. *Boletín de la Real Sociedad Española de la Historia Natural* 54, 22-23.

Klaric, L. 2000. Note sur la présence de lames aménagé es por technique de Kostienki dans les couches gravettiennes du Blot (Cerzat, Haute Loire). *Bulletin de la Société Préhistorique Française* 97 (4), 625-636.

López Covacho, L., Ortiz del Cueto, J. R. and Rodríguez y Fuente, M. 1996. El yacimiento prehistórico de Pedazo del Muerto (Pinto, Madrid). *Actas de la Reunión de Arqueología Madrileña,* 213-214. Madrid.

López Recio, M., Baena Preysler, J. 2001. Captación de recursos líticos durante el Paleolítico Medio en la comarca de la Mancha Toledana: El Cerro del Molino de San Cristóbal (Camuñas). In *II Congreso de Arqueología de la Provincia de Toledo. La Mancha Occidental y la Mesa de Ocaña,* 11-28. Diputación Provincial de Toledo. Toledo.

López Recio, M., Silva, P. G., Alcalde, R., Cuartero, F., Expósito, A., Lázaro, A., González Hernández, F. M., Manzano, I., Martín, D., Moral, B., Morín, J., Sánchez, F., Simón, G. and Tapias, F. 2007. Los yacimientos paleolíticos del 12 de Octubre y Villaverde - Barrio del Butarque (Madrid) y su contextualización geológica en el Valle inferior del Manzanares, in J. Lario and P. G. Silva (eds.), *Contribuciones al Estudio del Periodo Cuaternario,* 115-116. Madrid-Ávila (AEQUA).

Martín Blanco, P. and Djema, H. 2005. Los sistemas operativos del complejo Musteriense. El problema de la variabilidad y sus implicaciones. *Monografías* 20, 315-332. Santander, Museo de Altamira.

Martínez de Merlo, A. 1984. El Paleolítico Superior en el Valle del Manzanares: el yacimiento de El Sotillo. In, *Boletín del Museo Arqueológico Nacional* II, 47-68. Madrid.

Moncel M.-H. and Perreve, V. 1999. Un atelier de taille moustérien : Le Clos du Charnier Lagorce (Ardèche, France). *Quelques remarques sur les modes de débitage L' Anthropologie* 103 (3), 471-484.

Montes, L. 1988. *El Musteriense de la cuenca del Ebro.* Zaragoza, Universidad de Zaragoza.

Obermaier, H. 1925. *El hombre fósil.* Comisión de Investigaciones Paleontológicas y Prehistóricas, Memoria 9.

Obermaier, H., Wernert, P. and Pérez de Barradas, J. 1921. El Cuaternario de las canteras de Vallecas (Madrid). Boletín del Instituto Geológico de España XLII, 305-332.

Obermaier, H. and Pérez de Barradas, J. 1924. Las diferentes facies del Musteriense español y especialmente de los yacimientos madrileños. *Revista de Bibliotecas, Archivos y Museos* I (2), 143-177.

Pérez de Barradas, J. 1922. Yacimientos paleolíticos del Valle del Manzanares (Madrid). Memoria acerca de los trabajos practicados en 1920-1921. *Junta Superior de Excavaciones y Antigüedades,* Memoria 42. Madrid.

Pérez de Barradas, J. 1923. Yacimientos paleolíticos de los valles del Manzanares y del Jarama (Madrid). Memoria acerca de los trabajos realizados en 1921-1922. *Junta Superior de Excavaciones y Antiguedades,* Memoria 50. Madrid.

Pérez de Barradas, J. 1924a. Yacimientos paleolíticos del Valle del Manzanares (Madrid). Memoria acerca de los trabajos realizados en 1922-1923. *Junta Superior de Excavaciones y Antiguedades,* Memoria 60. Madrid.

Pérez de Barradas, J. 1924b. Yacimientos paleolíticos del Valle del Manzanares (Madrid). Memoria acerca de los trabajos realizados en 1923-1924 . *Junta Superior de Excavaciones y Antiguedades,* Memoria 64. Madrid. .

Pérez de Barradas, J. 1926. *Estudios sobre el terreno cuaternario del Valle del Manzanares*, Madrid. Ayuntamiento de Madrid.

Pérez de Barradas, J. 1929. Los yacimientos prehistóricos de los alrededores de Madrid. *Boletín del Instituto Geológico y Minero* LI, 153-322.

Pérez de Barradas, J. 1934. El Achelense del Valle del Manzanares (Madrid). *Anuario del Cuerpo Facultativo de Archiveros, Bibliotecarios y Arqueólogos* I, 337-353. Madrid.

Pérez de Barradas, J. and Wernert, P. 1919. El nuevo yacimiento cuaternario en el Valle del Manzanares. *Boletín de la Sociedad Española de Excursionistas* XXVII, 238-269. Madrid.

Pérez de Barradas, J. and Wernert, P. 1921a. Yacimientos paleolíticos del Valle del Manzanares (Madrid). *Junta Superior de Excavaciones y Antigüedades,* Memoria 33, 146-166.

Pérez de Barradas, J. and Wernert, P. 1921b. Excursión geológica por el Valle inferior del Manzanares. *Boletín Sociedad Ibérica de Ciencias Naturales* XX(III), 1-21.

Pérez de Barradas, J. and Wernert, P. 1930. El yacimiento paleolítico de El Sotillo (Madrid). *Anuario de Prehistoria Madrileña* I, 54-95.

Pérez de Barradas, J. and Wernert, P. 1931. El yacimiento paleolítico de El Sotillo (Madrid). *Anuario de Prehistoria Madrileña* II, 61-106.

Pérez de Barradas, J., Obermaier, H. and Wernert, P. 1921. El Cuaternario de las canteras de Vallecas de Madrid. *Boletín del Instituto Geológico de España* XLII.

Pérez-González, A. 1994. La Depresión del Tajo. In M. Gutiérrez Elorza (ed.), *Geomorfología de España,* 389-436. Madrid, Rueda.

Querol, M. A. and Santonja, M. 1979. *El yacimiento Achelense de Pinedo (Toledo).* Excavaciones Arqueológicas en España 106.

Roebroeks, W. 1988. From finds scatters to early hominid behaviour & a study of Middle Palaeolithic riverside settlements at Maastricht-Belvedere (The Netherlands). *Analecta Praehistorica Leidensia* 21, 198.

Réveillon, S. and Tuffreau, A. 1994. Valeur et signification du débitage laminaire du gisemenet paléolitqhiue moyen du Secline (Nord), in S. Réveillon and A. Tuffreau (eds.), *Les industries laminaires au Paéolithque Moyen,* 19-43. Dossier de Documentation Archéologique 18. CNRS Editions. Paris.

Réveillon, S. and Cliquet, D. 1994. Technologie du débitage laminaire du gisement paléolithique Moyen de Saint-Germain des-Vaux (Port-Racine, secteur I) dans le contexte des industries du Paléolithique Moyen du Massif Armoricain, in S. Réveillon and A. Tuffreau (eds.), *Les industries laminaires au Paéolithque Moyen,* 45-62. Dossier de Documentation Archéologique 18. CNRS Editions. Paris.

Rus, I. 1987. El Paleolítico. *130 Años de Arqueología Madrileña,* 20-43. Madrid, Comunidad de Madrid.

Rus, I., Velasco, F. 1993. El poblamiento Prehistórico de Madrid. In *Historia de Madrid,* 67-86. Madrid. Ed. Complutense.

Sánchez, J. 1985. Los yacimientos paleolíticos de El Atajillo, Atajillo del Sastre y López Cañamero en el Valle del Manzanares. *Estudios de Prehistoria y Arqueología Madrileña* 4, 75-120.

Sanguino, J., Gómez de la Laguna, J. A., Jiménez, A. and Martín Blanco, P. 1994. Identificación de cadenas operativas líticas en el sitio arqueológico de 'Casa de la Mina II' (Argamasilla de Alba, C. Real): Consideraciones acerca de los yacimientos superficiales sin contexto estratigráfico. *Zephyrus: Revista de Prehistoria y Arqueología* 47, 15-40.

Santonja, M. 1986. Valgrande (Puebla de Yeltes, Salamanca): Área de talla y sitio de ocupación del Paleolítico Medio. In E. Baquedano Pérez, *Numantia II: Investigaciones Arqueológicas en Castilla y León,* 33-85. Valladolid, autor editor.

Slimak, L., Bressy, C., Guendon, J. L., Montoya, C., Ollivier, C., and Renault, S. 2005. Exploitation paléolithique de silex oligocènes en Haute Provence (France). Caractérisation des matières premières et processus d'acquisition. *Comptes Rendus Palevol* 4 (4), June 2005, 359-367.

Tuffreau, A. 1990. Le Paléolithique Moyen Récent dans le Nord de la France. In C. Farizy (ed.), *Paléolithique Moyen Recent et Paléolithique Superieur Ancien en Europe.* Actes du Colloque International de Nemours, 159-165. Memoires du Musée de Préhistoire d'Ille de France, APRAIF. Nemours.

Turq, A. 1992. Raw material and technological Studies of the Quina Mousterian in Perigord, in H. Dibble and P. Mellars, *The middle Paleolithic: Adaptation, Behavior, and Variability*, 75-85. University of Pennsylvania Press. Philadelphia.

Van Kolfschoten, T., Roebroeks, W. and Vandenberghe, J. 1993.The Middle and Late Pleistocene sedimentary and climatic sequence at Maastricht-Belvédère: the Type Locality of the Belvédère Interglacial. - *Mededelingen Rijks Geologische Dienst* 47, 81-91.

Utrilla, P. and Tilo, M. A. 2001-2002. Cabezo Marañán (Castelserás): un lugar de explotación del sílex durante el Paleolítico Medio. *Kalathos: Revista del Seminario de Arqueología y Etnología Turolense* 20-21, 85-99.

Vega, G., Raposo, L. and Santonja, M. 1999. Environments and settlement in the Middle Palaeolithic of the Iberian Peninsula, in W.Roebroeks and C. Gamble (eds.). *The Middle Palaeolithic Occupation of Europe,* 23-48. Leiden, University of Leiden.

Time for action. The chronology of mining events at Casa Montero (Madrid, Spain)

Pedro DÍAZ-DEL-RÍO and Susana CONSUEGRA

This day will last
A thousand years
If you want it to

Dawn is a feeling
The Moody Blues

Abstract

We present the radiocarbon dates for twelve charcoal samples covering the complete area of the mine field of Casa Montero (Madrid, Spain), a site with more than 4000 plotted shafts. The χ^2 test shows that eleven of them are statistically identical, with a 65% probability that all mining episodes occurred between 5337 and 5218 cal BC, a time span of approximately four generations. We test this probable hypothesis against other archaeological evidence and conclude that the comparatively large scale mining actions at Casa Montero would have necessarily required the mobilization of several small scale Early Neolithic groups into a succession of collective actions, probably performed in a seasonal manner. Neolithic flint mining in Europe was not a long term technical solution to a practical need, but an extraordinarily meaningful and timely -historically contingent- social activity. In order to understand mining actions in these terms, we would require a reevaluation of the statistical variability and meaning of series of radiocarbon dates already obtained at many other flint mines. When we do so, we might observe that, as in the case of Casa Montero, many of these radiocarbon dates actually represent sets of short term highly active 'generational' mining episodes separated in time.

Keywords

Early Neolithic. Flint mine. Central Iberia. Radiocarbon dates. Labor. Seasonality. Generational event.

1. Introduction

The general chronological framework of flint mining in Prehistoric Europe is as follows: there is scarce and occasionally ambiguous evidence for Pleistocene quarries (e.g., Baena *et al.* in this volume), some scanty Mesolithic mining evidence (e.g., Oliva in this volume), a few locations with sixth millennium BC Neolithic mines (Galiberti 2005; Díaz-del-Río *et al.* 2006; Schild 1995), and an important peak in the amount of evidence for mining activity for the late fifth through fourth millennia BC. Flint mining persists in certain regions of prehistoric Europe beyond those dates, with well-known cases such as the third millennium BC Grimes Graves (Longworth and Varndell 1996), while the social and economic transformations that took place during the Bronze Age modified the previous production, circulation and use of flint tools in much of the continent (e.g., Karimali 2005).

Beyond this impressionistic observation lie severe problems that have a direct effect on the possibility of establishing a comparative perspective on prehistoric flint mines. Among them is the lack of contextualized sets of radiocarbon dates for most known flint mines, particularly of dates from short-lived species, and when they do exist, the serious problems in establishing fine-grained sequences for each one of the individual sites, not to mention the difficulties of combining them at a regional or continental scale. One of the resulting problems of all this has been the limited opportunities for building the specific histories of mines in their local and regional contexts. Consequently, the assumption of long term procurement cycles for most mines, in many cases measured in millennia, has somehow restricted the possibility of developing historically informed interpretations. Not infrequently, the technical aspects of underground flint procurement have relegated mining to specialists, while the earliest monumental materialization of Neolithic societies has been used only as a secondary source for the interpretation of European prehistory.

This dynamic can only be subverted by increasing our interest in building historically contingent interpretations of mines. In order to do so we can begin by avoiding two assumptions. First, a set of radiocarbon dates does not define the length of a continuous exploitation of the mine, but the statistically determined chronological limits in which an unknown number of mining events actually occurred. This kind of reasoning ignores the probabilistic nature of radiocarbon dates, and circumvents the fact that all statistically identical dated events may (or may not) be dating the same mining episode.

The second aspect, in some way entangled with the previous point, is the acceptance of 'work-day per person' estimates as veracious reconstructions of past realities. These kinds of estimates are critical if we want to compare the labor deployed at different flint mines, or between different mining events. But, because the scale of these individual mining events is in most cases elusive to archaeologists, we not infrequently assume that these estimates do not just have a comparative purpose, but actually represent the scale and pace in which labor was effectively deployed. Thus, the most parsimonious interpretation of flint mines considers minimum cost in both persons and labor time. When we combine these figures with the long-term temporal range of most sites, we end up accepting that labor must have been deployed in small amounts, by small groups of people (male, of course –Gero 1991), for domestic or down-the-line exchange purposes, year after year, century after century, throughout millennia. This in fact was our initial hypothesis for Casa Montero.

Of course, all or parts of this interpretation may well have been the case. Archaeological interpretations are in most instances underdetermined: one can frequently build alter-

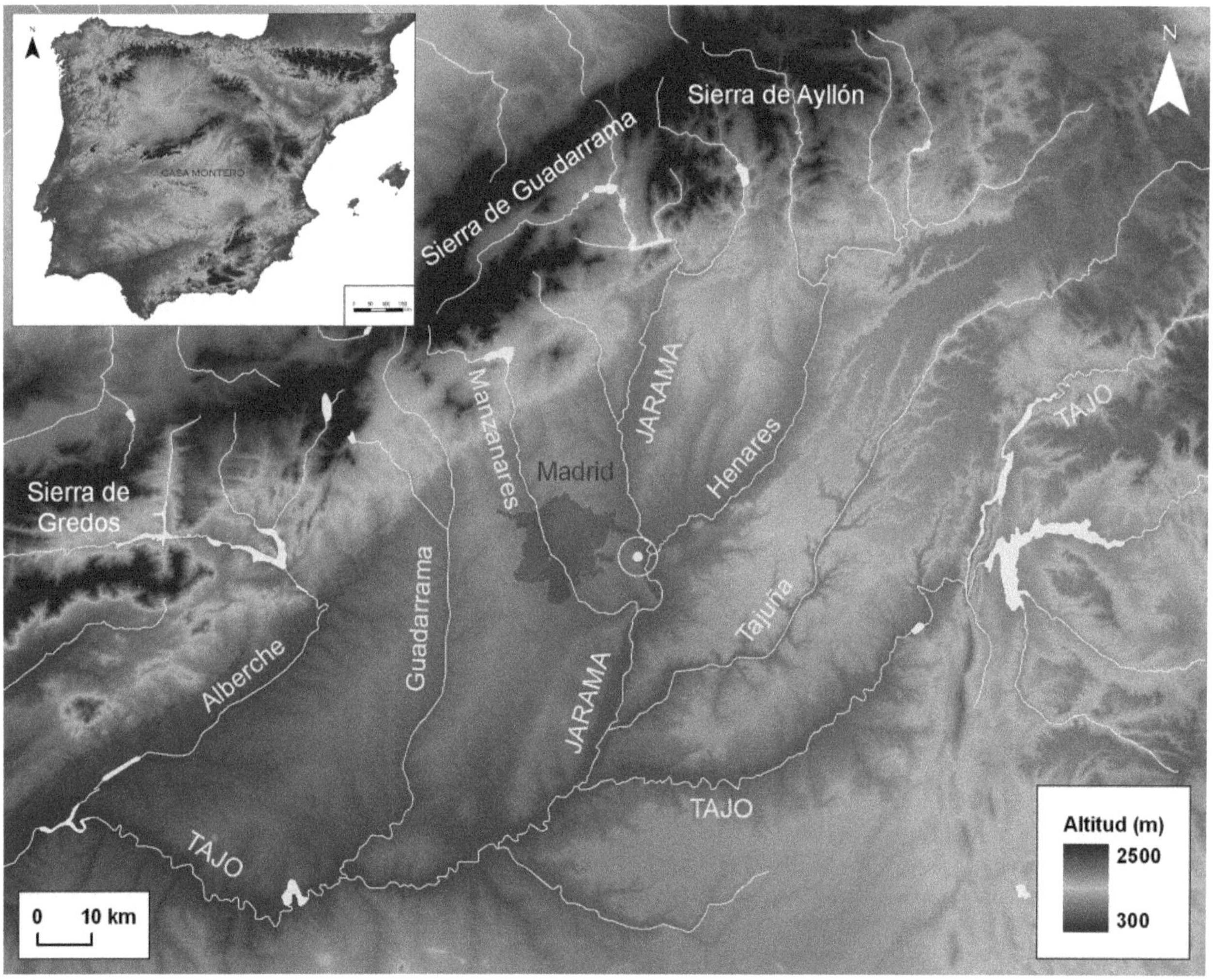

Figure 1. Location of the Early Neolithic flint mine of Casa Montero (Madrid).

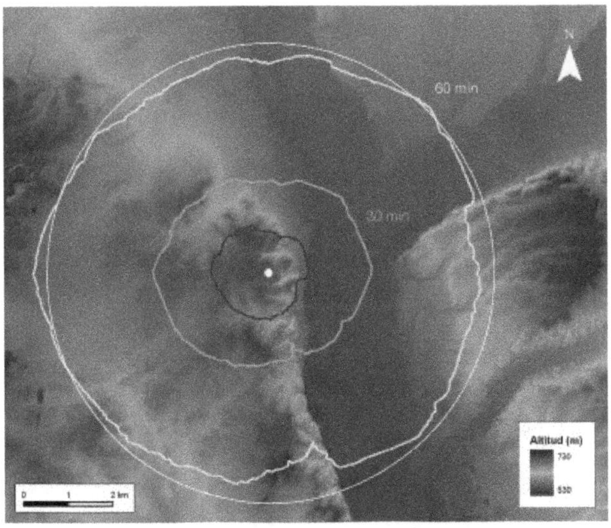

Figure 2. Area of intensive surface survey in the surroundings of the Early Neolithic flint mine of Casa Montero (Madrid).

native and reasonably parsimonious hypotheses with the same data sets. But our point is that this image of prehistoric flint mining frequently blurs the possibility of thinking about alternative social and political scenarios for each one of the subsequent mining events. To put it in other terms, it avoids the likelihood of diverse historically contingent mining circumstances. Yet, if we are able to build a solid and parsimonious alternative interpretation for at least one case study, then we may place ourselves in conditions to see and, most of all, think about Neolithic flint mining not as an homogeneous preindustrial activity, but as a varied set of historically dependent cultural practices.

2. Casa Montero

The Casa Montero flint mine is located on a bluff overlooking the confluence of two of the main rivers in the Madrid province, the Jarama and the Henares (Figure 1). The Neolithic mining complex occupies an extension of less than three hectares, almost completely mapped and partially excavated. Note that this is the correct extension of the site after the complete analysis of the existing evidence. We thus correct our previous publications where Casa Montero is said to have an extension of 6 to 8 hectares. The open area excavation revealed the existence of around four thousand cylindrical mining shafts, one meter mean in width and up to ten meters deep, none of which cut each other on the actual surface. Madrid's regional geology is known to have an abundant and ubiquitous presence of available flint. This fact makes the existence of mining activity in itself quite surprising, for both the region and time period of the events. During our regional Prehistory, flint was used at least from the Paleolithic up until the Bronze Age, when flint tools became extremely scarce. The peak of flint use seems to have been the Copper Age, when massive amounts of expedient tools were made out of local and, probably, secondary sources. The earliest Neolithic groups in central Iberia selected Casa Montero and not other primary or secondary deposits for flint procurement due to a combination of factors: geological (Bustillo *et al.* 2009), technological (Castañeda *et al.* 2008), locational and, most of all, social.

Our excavation design was specifically oriented towards the analysis of the potential variations in both the mining strategies and the chronology of mining events (Capote *et al.* 2008). Throughout the first two seasons we focused on defining the individual variability of shafts which later on allowed us to decide the best strategy for the third and final excavation campaign. For this last season, we designed and applied an aligned systematic sampling, digging all the shafts included in each one of the sixteen grid squares of ten by ten meters, regularly distributed throughout the site. The driving force behind this strategy was that comparing groups of shafts would be more enlightening than comparing randomly selected individual pits. And so it was.

The excavation was accompanied by a similar intensive surface sampling followed by a backhoe subsurface exploration in the two square kilometers west of the mining area, and an intensive surface survey in the 60 minute buffer zone (Figure 2). This allowed a more accurate image of the scale, temporality, and variability of flint mining techniques between Early Neolithic and Modern Era exploitations located in the vicinity of the excavated area. Sixty five tons of processed flint items were recovered during the excavation of 324 Neolithic shafts. The total amount of flint manipulated and discarded at the site during the Early Neolithic must have been more than 800 tons.

3. The radiocarbon dates

The excavations at Casa Montero have revealed few datable short-life organic remains. Animal bones are scanty, most of all rings recovered in various phases of production, from long bone shafts to final polished products (Figure 3) (Yravedra *et al.* 2008). We avoided using them as datable samples, mostly because of their scarcity in Neolithic Iberia and their potential museographic qualities.

In contrast, charcoal samples are comparatively abundant, the 3019 remains being the second biggest collection of identified charcoal in Early Neolithic Iberia. All the same, the deposits filling the mine pits were not rich in charcoal. Samples, mostly small scattered fragments of charred wood recovered by hand, were obtained from only 90 out of the 324 shafts excavated.

Charred remains were recovered from thirteen stratigraphic units filling five different shafts, all dug during the first field season. Remains were mainly *Quercus* ilex/coccifera. *Quercus* ilex (holm oak, carrasca in Spanish) is a large evergreen oak. *Quercus* coccifera (kermes oak, cos-

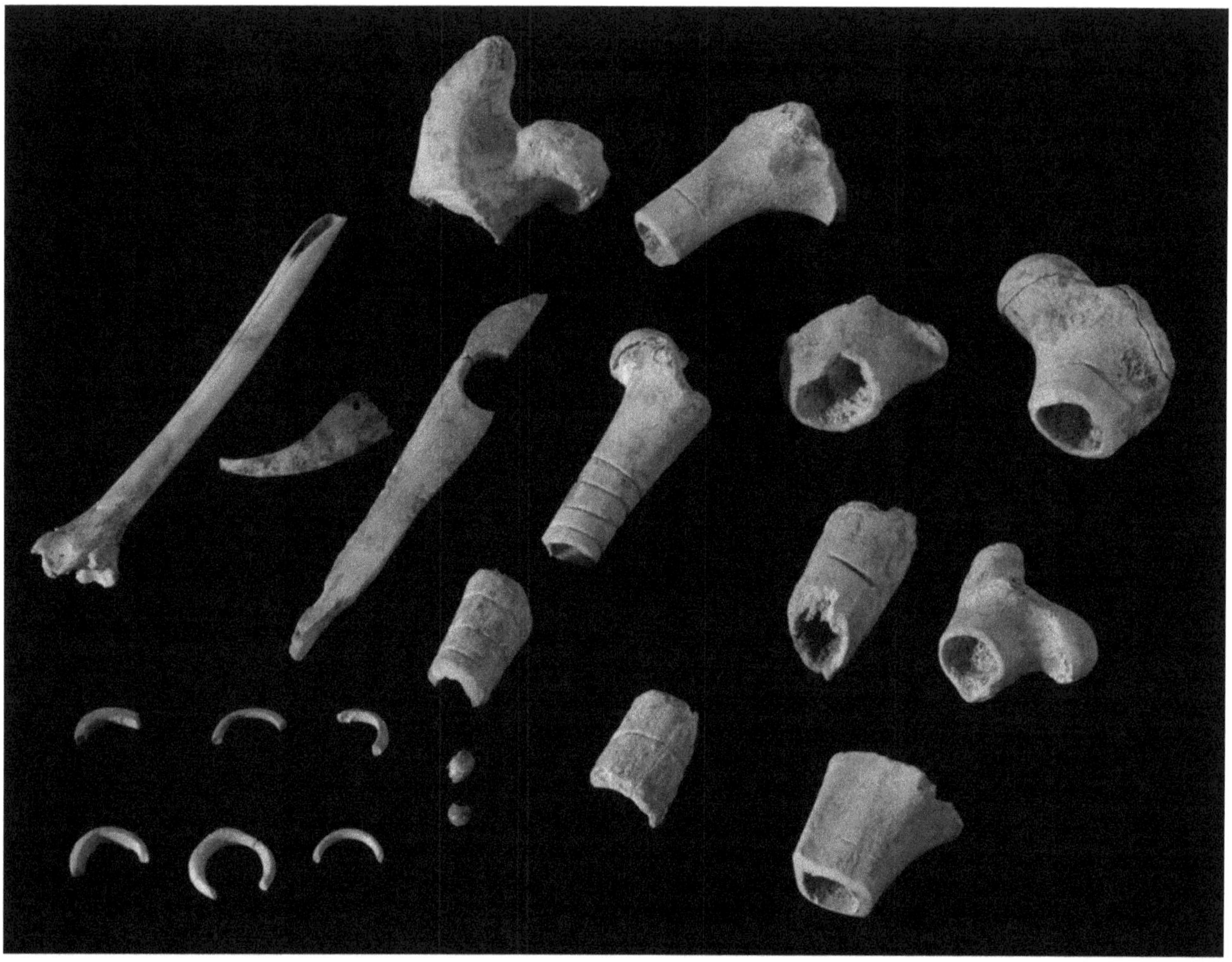

Figure 3. Bone industry recovered from different mine shafts of the Early Neolithic flint mine of Casa Montero (Madrid). So-called 'rings' on the lower left; bone shafts in various phases of production on the right.

coja in Spanish) is a large bush of the same genus, normally no more than 2 meters high with dense but superficial roots. The importance of these details relate to the debate concerning the use of long-lived organic remains for radiocarbon dating, and the frequent call to date short-life samples. While holm oak lives an average of 400 years, kermes oak rarely lives beyond 40. Unfortunately, both species are impossible to distinguish through a standard anthracological analysis. Nevertheless, some botanists suggest that soil conditions at Casa Montero would favor the presence of kermes oak, something that seems to be supported, among other things, by the fact that the excavated surface shows no evidence whatsoever of root alterations. As we will see below, it is not unlikely that most dated samples belong to the bushy short-lived species. However, short life samples dating to the same chronology are needed to reinforce this supposition.

Two individual fragments of *Quercus* Ilex/Coccifera from two different shafts were sent for radiocarbon dating (Figure 4; dates nº 1 and 2 published in Díaz-del-Río *et al.* 2006). Although we already expected an Early Neolithic chronology for most shafts, we did not anticipate such a date for both of them: at least two hundred years earlier than the earliest Neolithic date from the Madrid region. These dates are not unusual in other regions of Europe, such as Italy, where flint mines are one of the earliest Neolithic evidence. We did anticipate that both dates would be probably different since we assumed, based upon what we knew about other European flint mines, that the exploitation would have been prolonged in time. In any case, the resulting dates suggested that the two shafts were made at an interval of some centuries. Considering that the distance between these shafts was less than 25 meters, the expectations were that the more than 4000 shafts distributed throughout the four hectares under investigation would have been the result of a long term prehistoric exploitation. As a matter of fact, many of our colleagues openly expressed similar opinions, suggesting that the mine would have been in use until the third millennium BC. We knew, nevertheless, that the hypothesis of a long term use ran against other evidence, such as the lack of overlapping pits or the extremely consistent Early Neolithic chronology of all the scattered pottery fragments and other recovered artifacts.

Testing this alternative hypothesis required new contextualized and identified samples for dating. The rationale behind the second set of radiocarbon dates was precisely to confirm the complete time span of the mine and, if possible, observe the temporal growth of the mine field. In order to do so, we selected a collection of samples from eleven mine shafts covering 11 out of the 16 sampling units ('S.U.' in Figure 4). This essentially covered the entire extension of the site, north to south and east to west. The collection included 190 charred remains, 7 of which were not identified as to species, while the remaining 183 were assigned to only three taxa, in order of importance: *Quercus ilex/coccifera* (94%), *Juniperus Communis* (5%) and *Quercus* sp. (1%). We finally selected eight individual fragments of *Quercus ilex/coccifera* fragments, two of *Juniperus Communis* (common juniper) and a fragment of a bone shaft made out of a femur of *Sus* sp. for dating. Both the bone and the Juniper, another bushy species, where selected in order to evaluate the possible aging effect of old wood in the radiocarbon dates obtained from *Quercus* samples.

The complete set of radiocarbon dates are presented in Figure 4 and their spatial distribution in Figure 5. Unfortunately, the bone sample lacked enough collagen to be dated. The χ^2 test shows that all dates with the only exception of Beta-232890 are statistically identical. Beta-232890 dates the fourth stratigraphic unit filling pit 16.309, a mining pit that cuts through the filling of the only narrow and confined 'gallery' system at two depths documented at the site. Three alternative interpretations seem feasible. First, that the sample actually dates pit 16.309. It would then suggest the existence of an earlier phase of mining activity, previous to the main episode dated by the rest of the samples. Second, the date is slightly older than the rest because of an 'old wood effect'. Finally, it is a residual sample originally part of the infill of the gallery system, and would suggest that this system was the first technique to be applied at Casa Montero, a procedure quickly abandoned in favor of the more straightforward 'chimney' pit extraction method.

#	Lab Code	Sample #	S.U.	Shaft #	Strata #	Dated Material	Date BP	$^{13}C/^{12}C$	Cal BC 1 σ	Cal BC 2 σ
1	Beta-206512	CM/05/2384/2382	_	2384	2382	*Quercus ilex/ coccifera*	6400±40	-24.2 o/oo	5466-5325	5471-5318
2	Beta-206513	CM/05/2701/2229	_	2701	2229	*Quercus ilex/ coccifera*	6270±40	-26.2 o/oo	5300-5221	5324-5077
3	Beta-232884	CM/05/95/B1/7242/1	B1	7244	7242	*Quercus ilex/ coccifera*	6360±40	-25.4 o/oo	5462-5303	5469-5227
4	Beta-232885	CM/05/95/B2/7562/1	B2	7564	7562	*Quercus ilex/ coccifera*	6280±40	-24.9 o/oo	5303-5225	5359-5080
5	Beta-232886	CM/05/95/B3/7482/1	B3	7490	7482	*Quercus ilex/ coccifera*	6350±40	-25.6 o/oo	5452-5234	5466-5224
6	Beta-232887	CM/05/95/D1/7963/1	D1	7967	7963	*Juniperus Communis*	6290±40	-22.2 o/oo	5309-5225	5367-5085
7	Beta-232888	CM/05/95/D2/8142/1	D2	8147	8142	*Quercus ilex/ coccifera*	6240±40	*	5303-5081	5310-5066
8	Beta-232889	CM/05/95/ D3/15842/1	D3	15849	15842	*Juniperus Communis*	6290±40	-22.3 o/oo	5309-5225	5367-5085
9	Beta-232890	CM/05/95/ D4/16303/1	D4	16309	16303	*Quercus ilex/ coccifera*	6500±40	-25.6 o/oo	5512-5383	5534-5370
10	Beta-232891	CM/05/95/D5/8614/1	D5	8615	8614	*Quercus ilex/ coccifera*	6320±40	-26.2 o/oo	5338-5225	5460-5214
11	Beta-232892	CM/05/95/E3/9323/1	E3	9332	9323	*Quercus ilex/ coccifera*	6270±40	-26.2 o/oo	5300-5221	5324-5077
12	Beta-232893	CM/05/95/E4/9622/1	E4	9630	9622	*Quercus ilex/ coccifera*	6330±40	-25.6 o/oo	5363-5228	5463-5217
13	**	CM/05/95/ G3/16221/1	G3	16229	16221	*Sus sp.* Femur frag.	–	–	–	–

Figure 4. AMS Radiocarbon dates from Casa Montero (Madrid). Calibrations and χ^2 test performed with Calib 6.0.1. Samples number 1 and 2 were recovered during the first field season, and have been already published (Díaz-del-Río et al. 2006). S.U. Sampling Unit (*) "The sample was too small to do a separate $^{13}C/^{12}C$ ratio and AMS analysis. The only available $^{13}C/^{12}C$ ratio available to calculate a Conventional Radiocarbon Age was determined on a small aliquot of graphite. Although this ratio corrects to the appropriate Conventional Radiocarbon Age, it is not reported since it includes laboratory chemical and detector induced fractionation" (quoted from Beta Analytic report). (**) Not dated because of the lack of enough collagen in the sample.

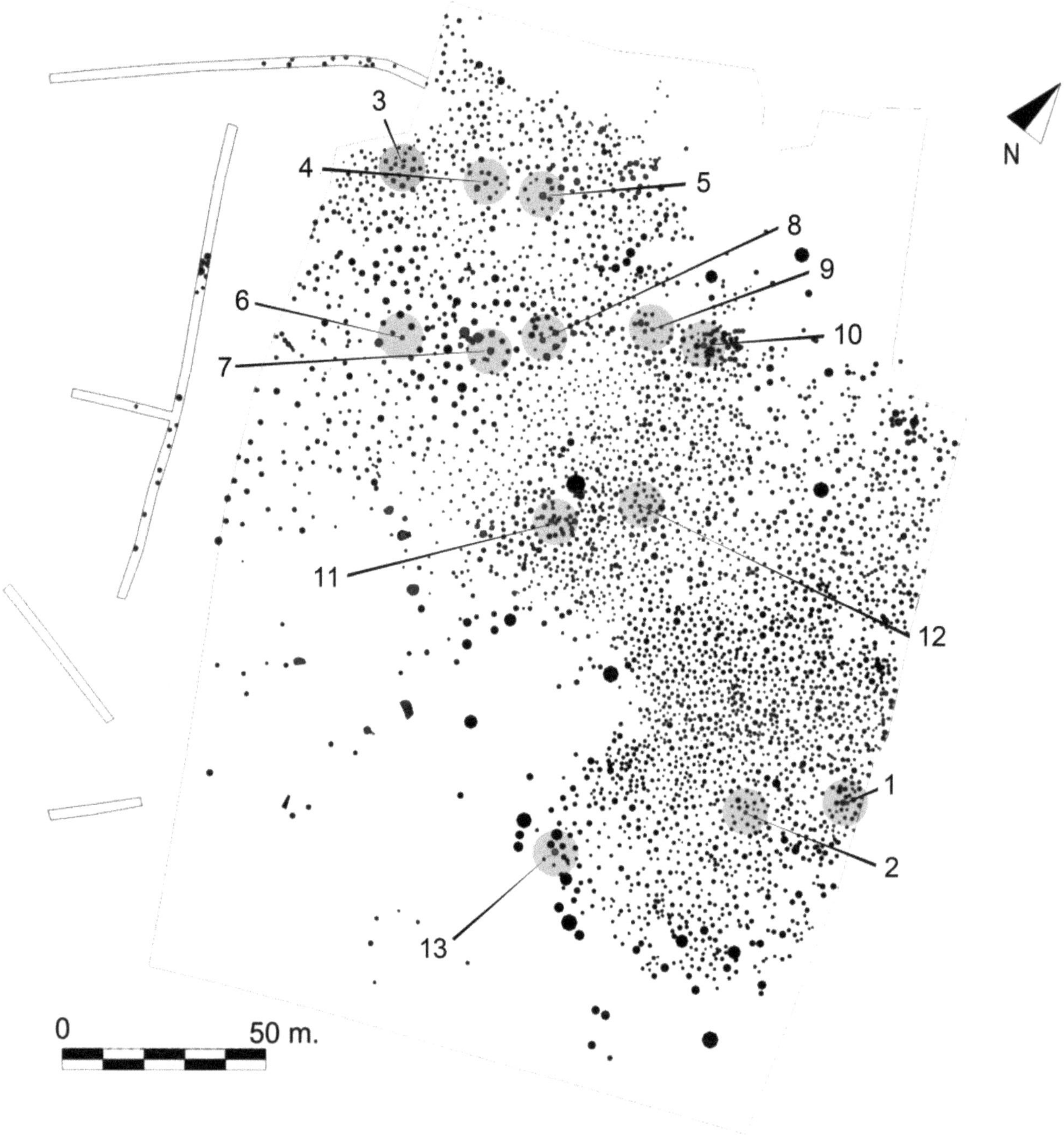

Figure 5. Distribution of the dated samples in the excavated area of Casa Montero flint mine. Numbers refer to figure 4.

Concerning the remaining eleven statistically identical dates (Figure 6), the calibrated average of all of them would suggest that there is a 95% probability that there was a single mining event at Casa Montero between 5317 and 5230 cal BC (2 σ). If we assume the more plausible hypothesis of more than one mining event, we could then consider the sum of probabilities of all these events occurring in a certain period of time: there is a 65% probability that all mining episodes occurred between 5337 and 5218 cal BC (1 σ), a time span of approximately a hundred years. In short, the radiocarbon dates do not indicate in which direction did the mine field expand over time. We are thus left to accept that the main episode of mining activity at Casa Montero lasted no more than a single century, four generations. This interpretation is not just *possible*, it is *probable*.

4. Combining dates with other evidence

Following Doelman's (2008) analysis of New South Wales quarries, it would seem reasonable to accept that close in time mining expeditions would create a fairly even and homogeneous refuse pattern, while longer time spans would increase the variability of the recovered sample. In addition, the high probability of a short time use for Casa

Montero, as shown by the radiocarbon dates, should be reinforced with a limited variability of the archaeological record, and not just in the refuse pattern.

Perhaps the most obvious evidence for a short time exploitation of the mining complex is the lack of intersection between shafts. The high density of mining shafts, with nearest neighbors frequently as little as a meter apart, indicates that Neolithic miners saw and avoided previous extractions. Both the surface accumulation of soil and the topography of mining areas would have been visible enough to allow the selection of new mining spots. This system of digging numerous narrow but deep pits offered the miners a certain security against potential wall collapse, and was labor efficient, reducing the investment required in removing high amounts of soil. Shafts do have a considerable depth variation (474±194cm mean), something reasonable considering the variation in the depth of flint seams throughout the mining area, although this is not the case for their width on surface (112±28cm mean). That is, there is a reduced variability in the way that each one of the mining pits where designed and executed. This consideration is reinforced by the fact that other soil stripping methods where feasible, such as wider pits or quarry fronts, as observed in the modern flint mining evidence located throughout the immediate surroundings of the Neolithic mine.

The infills of extraction pits are also homogeneous in both geological and sedimentological terms: the qualitative variation recorded during the excavation process by multiple observers can be reduced to five major groups. Unlike other contemporary mines such as Defensola (Galiberti ed. 2005), at Casa Montero the raw material was both procured and processed on-site. This resulted in a massive amount of discarded flint that was dumped back into the mine pits with the extracted soil. Thus, mine pits and their infill are the result of short time mining occurrences. The possibility of estimating the minimum amount of pits open at a time has been possible through the refitting of quartzite pebbles (see Capote, this volume). At Casa Montero, quartzite pebbles were brought from the close river terraces by miners, and used as hammers for percussion activities during the excavation and production process. Many broke while being used, and became part of the infill of open shafts. The results reveal that many shafts in each ten by ten meter sampling unit were actually refilled in one only mining event. That pottery fragments recovered from two shafts approximately 60 meters apart also could be refitted suggests that at least some mining events may have been bigger than the evidence obtained from sampling units.

The fact that flintknappers at Casa Montero followed the same pattern throughout the lifetime of the mine offers an additional support to our hypothesis. The recovered lithic remains amount to 65 tons, more than 1.5 million items, while the total stone processed at the site may have been more than 740 tons (>17 million items). The majority of these remains are knapping residues. The main goal of these knapping activities was the production of blades and in less quantity bladelets, although flakes were also occasionally produced, the most common final product being blades with mean dimensions of 5x2cm. The analysis of knapping waste has determined that the operative chains at different mine shafts was much the same, involving the systematic repetition of a single procedure: the methodical removal of the external opal parts of nodules in order to produce tools out of the quartz inner part (Castañeda *et al.* 2008).

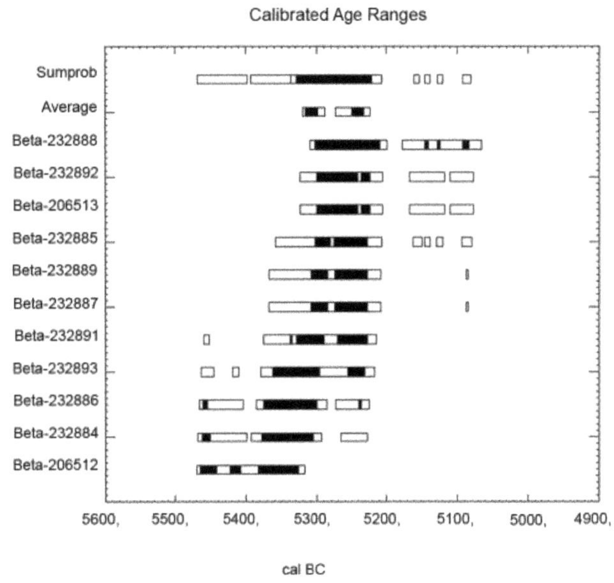

Figure 6. Plot of the eleven statistically identical dates from Casa Montero (Calib 6.0.1). The upper two represent the calibrated sum of probabilities and the calibrated average of all dates.

Other more circumstantial evidence for a short time exploitation of the mine comes from the faunal sample. Macrofaunal remains amount to 44 NR, more than 66% of which are bone industry (Figure 3). Out of this bone industry, more than 70% is represented by all the steps of ring production, from bone shafts to the final neatly polished rings. Remains of ring production were found throughout the excavated area. Bone shafts and their products are occasionally recovered from Early Neolithic sites in Iberia, although in small quantities except at a few Levantine sites (Yravedra *et al.* 2008). This kind of craft was somehow associated with the activities taking place during many mining events, and the homogeneity of its crafting followed the same pattern.

To sum up, most of the evidence recovered at the site suggests that the complete time span of Casa Montero may well have been a century. The timing of each one of these events could have been seasonal, perhaps spring, as suggested by Pete Topping for the English mines (Topping, this volume). At Casa Montero, timber or flint tools would have easily be blunted by muddy clay during the winter rainy season, while in summer drought-hardened soil would have increased the labor investment in the actual digging (as well as increasing the amount of water that had to be carried to the miners). In tactical terms, the best moment

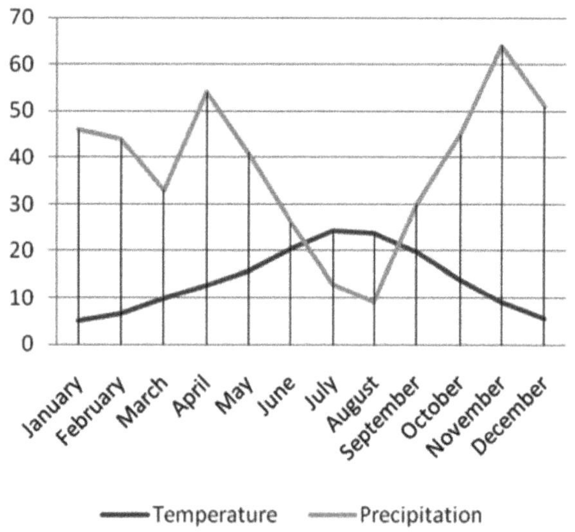

Figure 7. Mean rainfall (mm) and temperature (ºC) in Madrid. Source: Ministerio de Medio Ambiente. Instituto Nacional de Meteorología. Normal values of precipitation and temperature. 1960-1990.

to program such activity would either be spring (May to June), or fall (September to October), when the probability of rainfall is reduced and temperature has not reached its maximum (Figure 7). The recovery of a swallow in the lower stratigraphic unit of mine shaft nº 7209 would favor the former as a better hypothesis.

Additionally, the seasonal visiting of the mine would have avoided the 'who calls dilemma', that is how to concentrate enough labor force in a single spot in the context of a nonhierarchical decision making system. This is a reasonable scenario considering what we know about the amount and size of early Neolithic groups, and the possibly limited capacity of individuals to mobilize larger scale labor teams beyond the immediate domestic spheres. Early Neolithic evidence is scarce in the 8000 square kilometers Madrid region, an area where archaeological investigation is intense, although patchy. Only 13 locations are known to have Neolithic remains, six of which have been systematically excavated in recent times (only two have evidence of architectural remains, both unimpressive). Just one of these sites, has radiocarbon dates that may be contemporary to Casa Montero. Our project's survey of the sixty minute buffer zone surrounding Casa Montero has not increased the number of Neolithic sites. In fact, none have been recognized to date. While the mineralogical composition of pottery inclusions indicates the use of local clays, suggesting that those who mined at Casa Montero most likely dwelled in the region (Díaz-del-Río *et al.* 2011), the presence of cinnabar who's isotopic analysis refers to the Almadén district (200 km distance) points towards certain extra-regional connections. Early Neolithic groups were most probably very small and considerably mobile. If this was the case, the comparatively larger scale mining actions at Casa Montero would have necessarily required the mobilization of more than a few of these groups in a succession of collective actions presumably signaled through natural events (e.g., astronomical) that could allow the avoidance of the 'who calls dilemma'.

Casa Montero was abandoned sometime around 5200 BC. We have no other evidence of human activity for more than thirty centuries. In the second millennium BC some small group living nearby considered the previous mining area significant enough to use it as an occasional burial ground.

5. Coda

Throughout this paper we have argued that most mining events at Casa Montero took place during the lifetime of some few generations of Neolithic groups. Of course, the historical scenario in which flint mining occurred at Casa Montero does not necessarily parallel mining actions in other areas of Neolithic Europe. Many European flint mines were under exploitation for longer periods of time, or so it would seem from their radiocarbon dates. Nevertheless, one would have certain difficulties in deciding whether these radiocarbon dates actually represent thousands of years of small scale extractions or sets of short term highly active 'generational' mining episodes distanced in time. If so, flint mining may not have been a long term technical solution to a practical need, but an extraordinarily meaningful, timely and historically contingent social activity.

Acknowledgments

This paper has been possible because of the collective work developed in the context of the Casa Montero Project, especially by its core members Enrique Capdevila, Marta Capote, Cristina Casas, Nuria Castañeda, Cristina Criado and Aurora Nieto. Very special thanks to Mónica Ruiz-Alonso and Lydia Zapata for analyzing all charred remains on request; Manuel García-Heras, Fernando Agua and María Ángeles Villegas for their thorough analysis of pottery from Casa Montero; and Ignacio Montero and Mark Hunt for the isotopic analysis of cinnabar. We would also like to express our deep gratitude to Antonio Gilman for his sharp comments and thorough editing. Pete Topping and Dave Field offered us some useful comments and highlighted some important comparative issues. This paper has been written in the framework of the 'Proyecto de Investigación Arqueológica en el yacimiento de Casa Montero (Madrid). Producción y circulación de sílex en el Neolítico de la Meseta' sponsored by Autopista Madrid Sur C.E.S.A. through a collaboration agreement between the Consejería de Cultura y Deportes de la Comunidad de Madrid, the Consejo Superior de Investigaciones Científicas (CSIC) and Autopista Madrid Sur Concesionaria Española S.A for the research, preservation and scientific divulgation of the Casa Montero mine.

References

Bustillo, M.A., Castañeda, N., Capote, M., Consuegra, S., Criado, C., Díaz-Del-Río, P., Orozco, T., Pérez-Jiménez, J. L. and Terradas, X. 2009. Is the macroscopic classifi cation of Flint useful? A petroarchaeological analysis and characterization of Flint raw materials from the Iberian Neolithic mine of Casa Montero. *Archaeometry* 51 (1), 175-196.

Capote, M., Castañeda, N., Consuegra, S., Criado, C. and Díaz-Del-Río, P. 2008. Flint mining in Early Neolithic Iberia: a preliminary report on 'Casa Montero' (Madrid, Spain), in P. Allard, F. Bostyn, F. Giligny and J. Lech (eds.), *Flint mining in prehistoric Europe : interpreting the archaeological records*, 123-137. Oxford, BAR Publishing. British Archaeological Reports International Series 1891.

Castañeda, N., Capote, M., Criado, C., Consuegra, S., Díaz-Del-Río, P., Terradas, X. and Orozco, T. 2008. Aproximación a las cadenas operativas líticas de la mina de sílex de Casa Montero. In *Actas del IV Congreso de Neolítico Peninsular* (Alicante, 27-29 de noviembre de 2006). Volume II. Alicante, 231-234.

Díaz-Del-Río, P., Consuegra, S., Castañeda, N., Capo-te, M., Criado, C., Bustillo, M.A. and Pérez-Jiménez, J. L. 2006. The earliest fl int mine in Iberia. Antiquity 080, http://www.antiquity.ac.uk/projgall/diazdelrio307/

Díaz-Del-Río, P., Consuegra, S., Domínguez, R., Mar-tín-Bañón, A., Vírseda, L., Agua F., Villegas, M.A. and García-Heras, M. 2011. Identifi cación de una tradición tecnológica cerámica con desgrasante óseo en el Neolítico peninsular. Estudio arqueométrico de materiales cerámicos de Madrid (5300-3400 cal AC). *Trabajos de Prehisto-ria* 68 (1): 97-122.

Doelman, T. 2008. *Time to Quarry: The Archaeology of Stone Procurement in Northwestern New South Wales, Australia.* Oxford, BAR Publishing. British Archaeological Reports International Series 1801.

Galiberti, A. ed. 2005: *Defensola. Una miniera di selce di 7000 anni fa.* Siena, Protagon Editori Toscani.

Gero, J. M. 1991. Genderlithics: Women's roles in stone tool production, in J.M. Gero and M.W. Conkey (eds.), Engendering Archaeology. Women and Prehistory, 163-193. Cambridge, Blackwell Publishers.

Karimali, E. 2005. Lithic Technologies and Use, in E. Blake and A. B. Knapp (eds.), *The Archaeology of Mediterranean Prehistory*. 180-214. Blackwell Studies in Global Archaeology. Malden, Blackwell Publishing.

Longworth, I. H. and Varndell, G. 1996. *Mining in the deeper mines. Excavations at Grimes Graves, Norfolk,* 1972-6. London, British Museum Press.

Schild, R. 1995. Tomaszów, Radom Province. *Archaeologia Polona* 33, 455-465.

Yravedra, J., Maicas R.; Consuegra, S. and Díaz-Del-Río, P. 2008. Anillos para una mina. Industria ósea y fauna de la mina de sílex neolítica de Casa Montero. In *Actas del IV Congreso de Neolítico Peninsular* (Alicante, 27-29 de noviembre de 2006). Volume II. Alicante, 240-247.

Working in the flint mine: Percussion tools and labour organisation at Casa Montero (Spain)

Marta CAPOTE

Abstract
This article provides an analysis of labour organisation at the Spanish mine of Casa Montero, based on the study of percussion tools. The study examines both individual pieces and refits, and was designed to throw light on tasks performed at the mine, workforce sizes, specialisation and changes in these aspects over time. To place this study in context, comparisons are made with other European mines using the information available on the percussion tools used. The results show that, even though flint mining was a very widespread phenomenon in Neolithic Europe, it was not at all uniform, at least not in terms of the way work was socially organised. The percussion tools of Casa Montero were less elaborate and less intensely used than most of those recovered at other mines, while the workforce size required at some moments was larger than might have been expected. At Casa Montero, work was organised in a relatively simple way while mobilizing more people than the size of contemporaneous groups would have us believe. Finally, although working procedures appear to have remained fairly stable over the lifespan of the mine, the scale of labour changed from some mining events to others.

Keywords
Iberian Peninsula. Labour organisation. Percussion tools. Specialisation. Standardisation. Labour intensity. Workforce size.

1. Introduction: Mines and work organisation

Prehistoric mines and quarries are the product of collective efforts. The persons involved in these efforts were engaged not only in the procurement of raw materials, but in social relationships that were reproduced and reinterpreted in the organisation of their activities. Work organisation analysis can be used to examine the way in which communities and individuals distributed and played their roles. Key questions in such analyses revolve around the number of persons making up the groups involved, the factors that determined the right to take part in activities, whether participants were organised on the basis of egalitarianism or through the imposition of authority, whether learning was part of the process, and how different tasks were distributed among group members. It is difficult to answer all these questions based only on the archaeological data recovered from mining sites, but important insights can be obtained by examining the number of extraction structures open at each mining site, the number of tools used, the intensity of their use, tool standardisation, and any changes in organisational patterns over time.

2. Percussion tools and work organisation in European mines

Neolithic percussion tools can be classified into two groups depending on the kind of action they once performed: incisive tools such as picks, levers and wedges, and pounding tools such as hammers, hammerstones and pounders. Pounding tools are found in almost every mine. These may be either antler tools, such as those found at Harrow Hill, Spiennes, Jablines, Villemaur-sur-Vanne "Les Grand Bois Marot", Villemaur-sur-Vanne "Les Orlets", Krzemionki, Polany Kolonie II, Sümeg-Mogyorósdomb or Kranaselsky (Holgate 1995a, 152; Collet 2004; Bostyn and Lanchon 1992, 111-114; Labriffe *et al.* 1995a, 332; Labriffe *et al.* 1995b, 343; Borkowski 1995a, 513; Schild 1995, 484, 487; Bácskay 1995, 388-390; Charniausky 1995, 266) or stone tools.

Regardless of the different raw materials used, from the point of view of work organisation the most important distinction between them lies in their degree of elaboration.

Unmodified cobbles of different raw materials have been documented at several sites, e.g., flint hammerstones at Grimes Graves (Holgate 1995a, 152), limestone and quartzite hammerstones at Rijckholt-St. Geertruid (Felder *et al.* 1998, 43-47), quartzite cobbles at Löwenburg (Diethelm 1997, 64), Miskolc-Avas Hill (Simán 1995, 377), Sümeg-Mogyorósdomb (Bácskay 1995, 389-390) and in the Den of Boddam (Saville 2008, 7-8), and sandstone hammers at Tusimice (Lech and Mateiciucová 1995, 273-274). Other cobbles show some kind of conditioning for hafting, such as that used in a hammer from Feuerstein (Leitner 2008) and more than 1100 grooved hammers from Kleinkems (Diethelm 1997, 63). Finally, some pounding tools, such as the notched hammers from Rijckholt-St. Geertruid (Felder *et al.* 1998, 48) and Lousberg (Weiner 1995), were configured by knapping. At Defensola, knapped and sometimes partially or totally pecked hammers have been found, along with flint hammerstones sometimes regularised by pecking (Galiberti 2005).

Incisive tools include both antler and stone (mainly flint) tools. Antler picks are the tools most commonly found in Neolithic European mines; they have been documented at Harrow Hill, Grimes Graves, Jablines, Serbonnes, Villemaur, Lowenburg, Tusimice and Sümeg-Mogyorósdomb (Holgate 1995a, 1995b; Clutton-Brock 1984; Holgate 1995a; Bostyn and Lanchon 1992; Labriffe *et al.* 1995a, 1995b; Diethelm 1997; Lech and Mateiciucová 1995; Bacskay 1995).

At Defensola and Tomaszów, picks were exclusively made of flint (Galiberti 2005; Schild 1995) and at Rijckholt-St. Geertruid the number of antler picks is anecdotal compared to the more than 14,000 flint picks discovered (Felder *et al.* 1998, 47). At Kvarnby-S.Sallerup (Rudebeck 1987), Spiennes or Krzemionki both stone and antler incisive tools have been recovered. The use of antler or stone tools might be linked to the hardness of the deposits excavated or represent different moments in time (Collet *et al.* 2006, 69). At Krzemionki four distinctive types of structures have been recognised, each linked to particular geological conditions, chronologies and specific toolkits (Borkowski *et al.* 1991; Borkowski 1995a; 1995b).

Neolithic miners were clearly able to adapt to the physical conditions imposed by the geological settings they encountered, the fact that most mines were located in chalky settings at least partly explaining the predominance of antler tools. The shapes and types of raw material used in the production of percussion tools, however, were determined not only by their efficiency but by the social organisation of work. It should be remembered that, regardless of the tools and techniques used, the operational chain almost always involved the same steps: excavation, extraction, flint preparation and knapping, and finally waste disposal. However, the way these tasks were temporally and spatially distributed, and the way that groups were socially organised, varied. This organisation is the focus of the present work, with special attention paid to the type of specialisation, the workforce size, and the transformations of these organisational patterns over time.

Specialisation

In agreement with Costin and Hagstrum's analysis of craft production (1995), this work does not attempt to establish whether percussion tools were specialised, but examines the kind of specialisation that was the context of their use. Costin and Hagstrum distinguish among different kinds of specialisation using the following concepts: *context*, which refers to the existence, or not, of the control of labour by elites ('attached labour' refers to that mobilized and controlled by elites, whereas 'independent labour' is not thus governed); *concentration*, which refers to the separation, or not, of production areas from consumption areas; *constitution*, which deals with the size and composition of working units; and *intensity*, which deals with the amount of time invested relative to other economic activities (work can therefore be full-time or part-time). The specialisations defined by these four concepts are manifested in the technology used through particular *labour intensities*, *standardisations* and *skills*.

Neolithic flint mining is usually interpreted as a periodic communal activity, which in Costin and Hagtrum's terminology corresponds to a community specialisation characterized by an independent context, a nucleated concentration, household constitution, and part-time intensity. However, this should not be taken for granted; each case must be examined separately, and even if almost every instance of flint mining can be described using these four general concepts, differences in work organisation may still exist. For example, the expression "part-time" merely describes a non-continuous activity, but obviously the implications of a *periodic* activity (taking place seasonally or at regular intervals) and of a *spasmodic* activity (taking place only a few times within certain periods) are different (see Díaz-del-Río and Consuegra in this volume). With respect to context, labour mobilisation by powerful elites might be argued in some cases, but it is unlikely for other European Neolithic mines; however, an independent context may refer both to community-based decision making and to collaboration and agreement between communities. As for concentration, mining might seem to inevitably imply a nucleated strategy since it represents the intensive exploitation of a particular point in the landscape rather than a more opportunistic exploitation of different and dispersed resources. However, this might may be more justified in some cases than others, depending on the regional availability and quality of raw materials. Further, group-specific mobility patterns and sizes are always important factors determining the use of a nucleated or other strategy. Moreover, a region may have just one or many more mines. Finally, the size of the workforce and its composition would depend on the size and organisation of the groups engaged in mining as much as on the number of groups that might work together in a mine. Thus, no single interpretative template of work organisation can be applied to every mine. To do so would give a false impression of homogeneity that would not account for the different social contexts of mining in Neolithic Europe.

Labour intensity

Labour intensity may be examined via the analysis of percussion tools, focusing on two features: the amount of work devoted to their production, and the amount of work performed with them. Contrary to the assumption by some authors that the elaboration of tools was minimal or indeed non-existent (see, for example, Sidéra 1995, 123), the production of antler and bone tools was often a complex process involving several stages that took place outside the mine before mining itself could begin (Clutton-Brock 1984, 26, Boguszewski 1991, Bostyn and Lanchon 1992, 105-114). For example, the making of antler tools required the gathering of antlers, the softening of their hard outer layer or *compacta* before cutting them (performed by immersing them in water for two days [Boguszewski 1991, 46] or using localized heat in the areas that were to be cut [Clutton-Brock 1984, 26]), and removing some of the tines in order to produce L-shaped of single-hafted picks. The effort invested in the configuration of these tools was therefore notable, and often required several days. In addition, and despite differences between mines, flint tools all show some degree of configuration for their use as mining instruments. In the Kvarnby-S. Sallerup mine in Sweden, crudely knapped picks have been identified (Rudebeck 1987, 153). At Defensola, picks were not only knapped but also pecked, and while picks from Rijckholt-St. Geertruid, Spiennes and Krzemionki, along with hammers from Rijckholt-St. Geertruid, Valkenburg aan de Geul (Brounen 1995) and Lousberg are not as elaborate as these Defensola picks, they were clearly configured to comply with specific requirements.

Stone tools made from other raw materials, mostly pounders and hammerstones, show greater variation in their degree of elaboration. Unmodified cobbles used as hammerstones have been found at Grimes Graves, Spiennes, Rijckholt-St.Geertruid, Feuerstein, Löwenburg, Krzemionki, Tusimice, Sümeg-Mogyosrósdomb and Miskolc-Avas Hill (although in most of these mines antler and flint tools have been found that required greater investment in terms of manufacturing time). In contrast, the cobbles used at Kleinkems show at least some conditioning for hafting. Finally, some very elaborate stone percussion tools are known, such as the cigar-shaped pick-axes of Krzemionki.

Several factors need to be taken into account when assessing the intensity of the work carried out using these tools: the number of tools found, their signs of wear and reconditioning, and estimates of the number of tools that would be needed at a particular mining site. Not every publication provides the total number of tools found, although when figures are provided they are usually quite high. For example, more than 14,000 picks, apart from other types of tool, have been documented at Rijckholt-St. Geertruid, more than 1100 hammers have been cited for Kleinkems, and more than 500 antler tools and over 800 quartzite cobbles for Sümeg-Mogyorósdomb, despite the small size of the mine at just 1500 square meters and the fact that only a fifth of it has been excavated. In addition, 215 artefacts have been cited for 26 mining structures at Villemaur-sur-Vanne 'Le Grand Bois Marot', 284 tools for 128 mining structures at 'Les Orlets', and for Grimes Graves some 343 antlers, 171 antler fragments, an axe and 6 hammerstones have been catalogued among the material recovered from five mines or groups of shafts and galleries (Longworth and Varndell 1996, 96-98).

The extent of tool wear is difficult to establish, although apparent size reduction through use can provide some indication. At Rijckholt, unused picks have an average length of 161mm, while tools discarded after using both ends have an average length of 137.30mm (Felder *et al.* 1998, 48-49). The existence of tools used to recondition mining implements, such as the quartzite and sandstone whetstones used at Spiennes to sharpen antler tools, is also an indication that wear was heavy enough to render tools unusable without some degree of maintenance. At this site, refits also help our understanding of the progressive wear of tools (Collet *et al.* 2008). Clutton-Brock's (1984) comparison of similar antler tools from the mine of Grimes Graves and the contemporary enclosure of Durrington Walls showed the mining tools had a greater degree of wear.

It is also important to note that a great number of tools were probably used. Sieveking estimated that 400 antlers would have been needed at Grimes Graves every year, while Legge estimated that between 100 and 150 picks per shaft would have been necessary (Clutton-Brock 1984, 15-16).

In conclusion, the percussion tools found in European flint mines generally show signs of relatively large labour investments, both in terms of their manufacture and use.

Standardisation

Costin and Hagstrum distinguish between *intentional* and *mechanical* standardisation. Intentional standardisation is the result of the effort required to make products or tools comply with technical or other (e.g., stylistic) requirements. Mechanical standardisation is the unintentional result of the way work is socially organised. Indications of both types of standardisation in percussion tools can be seen in European Neolithic flint mines, i.e., tools were intentionally configured to furnish them specific technical features, but they were also used in a standardised manner.

Both incisive and pounding percussion tools present signs of intentional standardisation. The similarity of antler tools is partly the result of the fact that antlers are very similar to one another. However, deliberate selection was also involved, as suggested by the dimensions of the

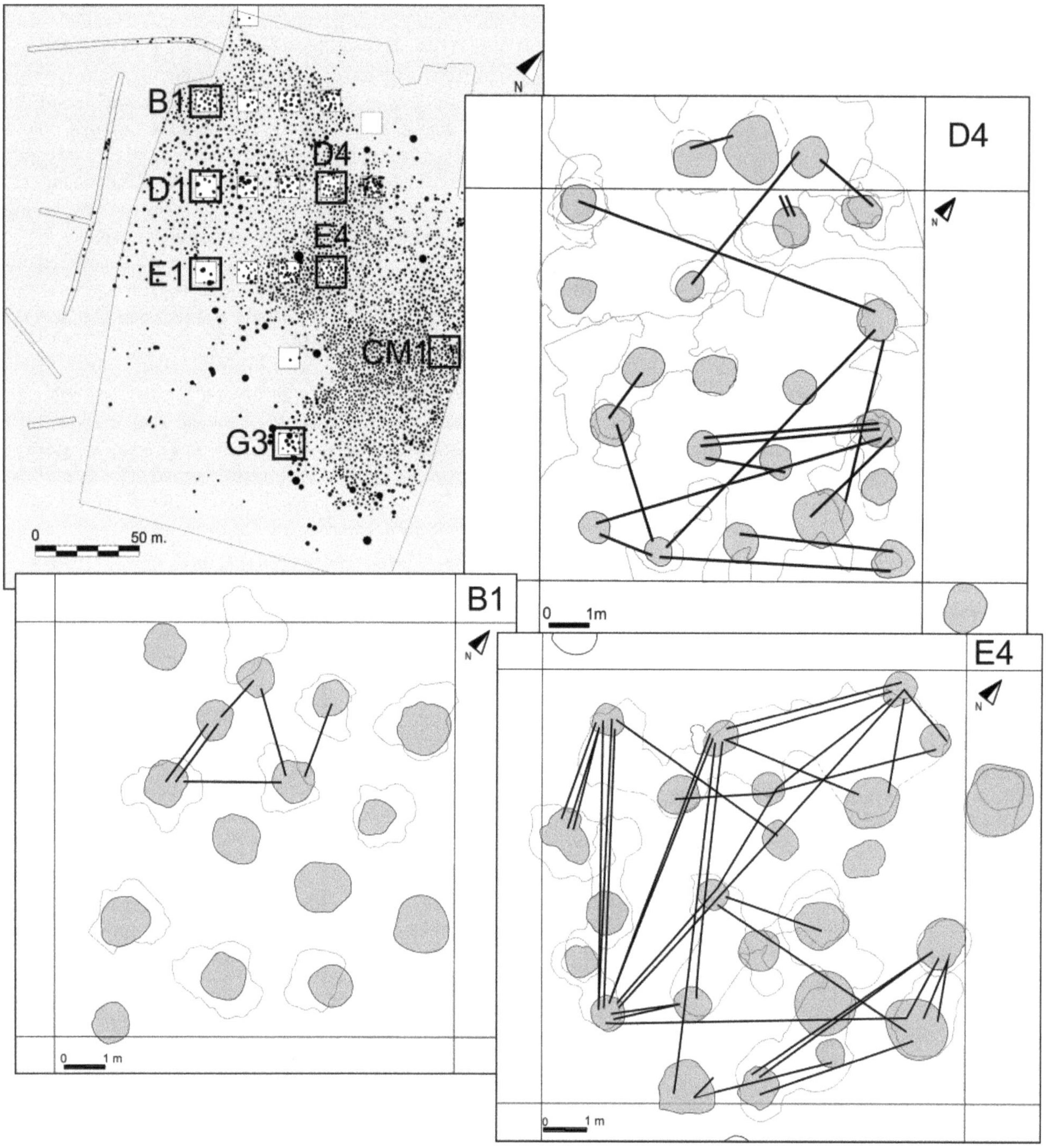

Figure 1: Localisation of selected sampling units where refits of material from different shafts have been attempted and detail of the three sampling units in which these refits have been achieved.

antler tools from Grimes Graves and Durrington Walls (Clutton Brock 1984, 25, 36). The antlers recovered from the former site show more uniformity in their length and are, on average, longer and heavier than at the latter. The deer in the area of Grimes Graves may simply have been bigger, but the fact that their dimensions also vary less suggests specific lengths were selected. In fact, Grimes Graves antlers are more often those of mature animals than those at Durrington Walls (Clutton-Brock 1984, 23). Sidéra (1995, 133) also indicates that parts of the antlers not used at contemporary settlements and burial sites were employed in the Villemaur-sur-Vanne 'Le Grand Bois Marot' mine.

After particular blanks had been selected, the process of tool configuration was performed with the aim of producing similar tools that conformed to technical requirements. Antler picks were made either by cutting off some of their points, leaving just the distal one in order to create L-shaped picks, or by hafting single points in wooden handles. Each standardised type of antler pick would be later used in a specific manner.

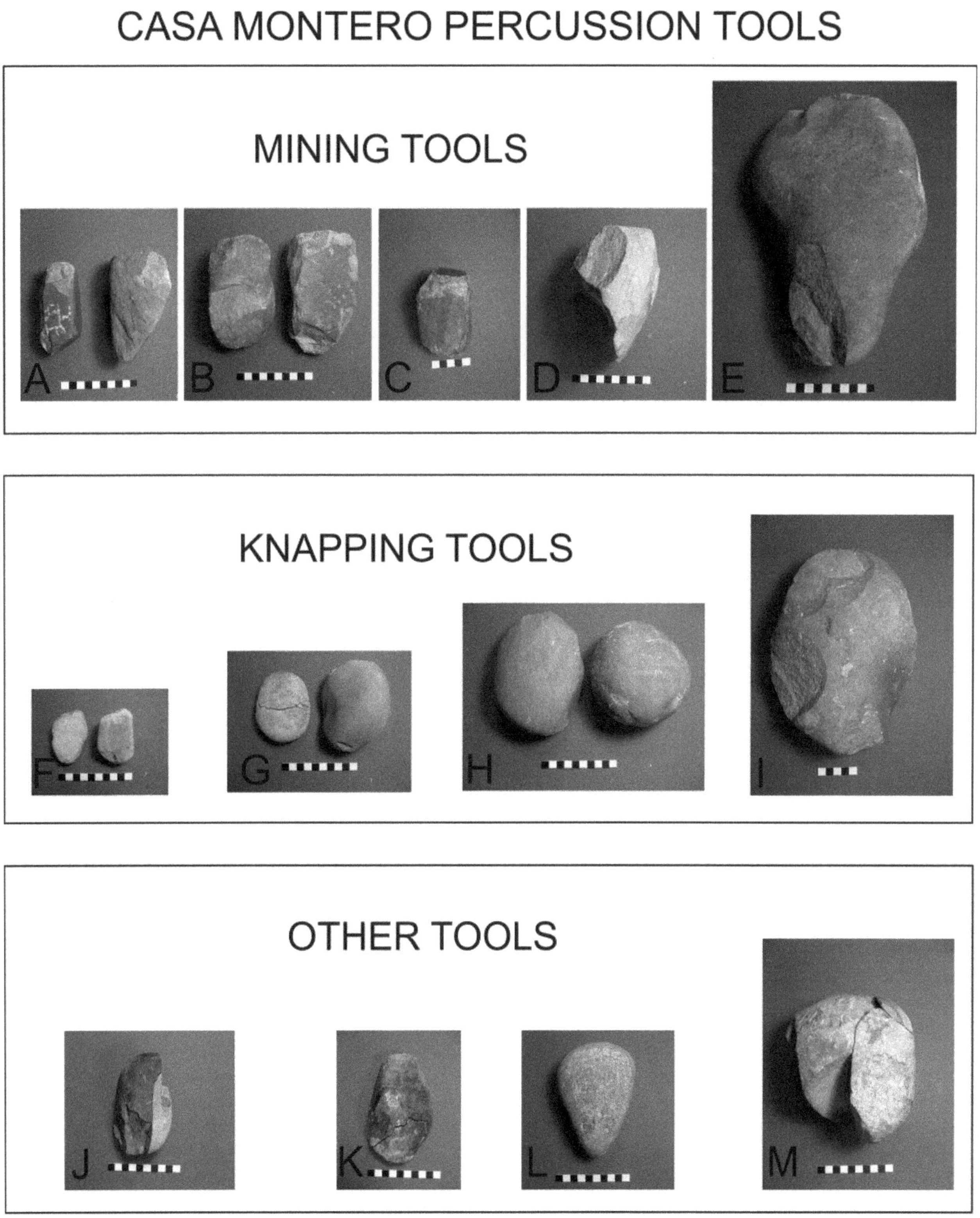

Figure 2: Percussion tools from Casa Montero. A-Bipolar A; B- Bipolar B; C-Bipolar C; D-Wedge; E- Big Hammerstone; F, G, H, I- Different sizes of knapping hammerstones; J, K-Thermally altered tools; L- Grinder; M- Anvil.

Flint tools were also manufactured with the goal of producing standardised tools. In the region of Rijckholt-St.Geertruid, several mines (Rijckholt-St.Geertruid, Valkenburg aan de Geul and Lousberg) have been found to contain the same type of notched hammer (Weiner 1995). At Rijckholt, flint picks are also fairly standard, and were repaired and sharpened on site to continue to meet technical requirements.

The picks and hammers from Defensola show considerable variation in terms of their degree of elaboration, size and weight (Galiberti 2005, 139), but nonetheless they have relatively uniform shapes and active parts.

Intentional standardisation is sometimes made manifest through the existence of uniform toolkits, or even particular toolkits linked to different types of extraction structure. Such is the case at Rijckholt, Spiennes, Valkenburg, Krzemionki, Sümeg and Kvarnby. At Krzemionki, four extraction strategies have been distinguished, each requiring its own, specific toolkit.

Mechanical standardisation is the result of recurrent patterns in the use and abandoning of tools. Mining percussion tools show specific traces of use and greater wear than similar tools used in other contexts (Clutton-Brock 1984, 38; Sidéra 1995, 123), once again reflecting particular ways of organising labour. At Rijckholt, for instance, three groups of picks with standardised lengths and degrees of wear have been distinguished: unused picks, picks with one used end, and picks showing wear at both ends (Felder *et al.* 1998, 48-49). If wear depended exclusively on tool function, all these picks would show a similar pattern, but the existence of these three groups suggests that more picks than eventually used were taken into the mine, and that among those that were used some were more curated than others despite their having similar characteristics. Other factors besides technical considerations must therefore be taken into account if we are to explain this different treatment of the same type of tool.

Skill

Of all the factors that Costin and Hagstrum define as manifestations of specialisation, skill is probably the most difficult to analyse. Given the time invested in the manufacture of percussion tools, and the difficulty involved, it is reasonable to think that at least moderate skill would be necessary in this task. However, little else can be said if we are to rely exclusively on their analysis. Some data indicate that, in mines, skills different to those required in other contemporary contexts were necessary. Sidéra (1995, 132), who compared tools from the Villemaur-sur-Vanne 'Le Grand Bois Marot' mine with those from other sites of similar chronology, reached the conclusion that the manufacture of mining tools required know-how that was different and complementary to that needed for making the tools found at settlements and burial sites. There is no doubt that the efficient use of mining tools required some degree of skill, but these tools have no specific attributes that can be used to compare the skills required in different mines.

Workforce size

The size of mining groups seems to have been variable across Europe, as indicated by the differences in the scale of mining activity detected. While at Rijckholt mining must have been relatively large-scale (see Díaz-del-Río and Consuegra, this volume), in other mines such as Jablines only two shafts were opened at any one time (Bostyn and Lanchon 1992, 217). Other data provide information on the size of workforces. Felder estimates that, at the mining structures of Rijckholt, up to six persons might have worked together at any one time, while at those of Grimes Graves up to 21 may have been involved (Felder 1979, 60-62). At the other end of the spectrum is the Arnhofen mine, where shafts are so narrow that only one or two persons could have worked them simultaneously (Rind 2003).

There is also some experimental evidence that the number of persons working in some mines was considerably small. For Jablines, Bostyn *et al.* (2005, 30-31) showed that three persons would be enough to dig a shaft 2.5 meters deep in nine days (55 hours of work).

Certainly, the analysis of percussion tools from different mines strongly suggests that workforce sizes varied. For example, over 14,000 picks have been recovered at Rijckholt, while just four tools have been recovered at Feuerstein.

The scale of mining sites, the number of persons working in each extraction structure, and the variety of activities carried out at a mine, therefore provide an indication of the workforce size.

Changes in labour organisation over time

Data from some mines show that labour organisation changed over time. For example, at Krzemionki, the four types of structures seen, along with their respective toolkits, are not only adapted to different geological conditions but are interpreted as belonging to different periods. Sidéra (1995) reports how, in the same area in the Paris Basin, two mines with different chronologies show different ways of structuring mining activities. At Serbonnes - Middle Neolithic - antler tools were manufactured in the mine using the same fragments and techniques as at contemporary settlements, enclosures and burial sites. Their manufacture did not require too much time and they were directly used without additional hafts, while at Villemaur - Final Neolithic - percussion tools were configured outside the mine using specific techniques and antler parts not employed in other contexts. These tools are more elaborate and more standardised than those of Serbonnes, and are also hafted. Differences also exist in the number of mining structures at each mine. While at Serbonnes there are only 300 shafts, at Villemaur there are thousands. Sidéra concluded that, in the Final Neolithic, mining was more intensive and that mining technology had become more specific. In the same way that mining technology evolved so that specific tools and techniques were reserved for mining, it is possible that specific persons became devoted to this activity.

At other mines, such as Grimes Graves, different working strategies have been detected (Longworth and Varndell 1996, 89), but it has not been possible to establish how mining evolved and which strategies belong to which periods.

3. The materials studied: percussion tools from the Spanish mine of Casa Montero

The percussion tools analysed from the Casa Montero mine are mainly unmodified cobbles and fragments involved in different percussion tasks. The vast majority of these remains are quartzite cobbles. A few quartz items and a very small number of igneous rocks (such as granite) have been found. All of these raw materials can be found on the banks of the Jarama and Henares rivers which run quite close to the site.

A very small group of flint percussion tools - unretouched discarded pieces resulting from different phases of the operational chain and that show macroscopic traces of possible use – has also been analysed. They show no conditioning and appear to have been chosen in an opportunistic manner for short percussion tasks. Ten flint picks had been already identified by Nuria Castañeda and Cristina Criado. Most of the 24 flint tools subsequently analysed have been interpreted as wedges or slashing tools.

As part of the analysis of the cobble percussion tools, refits have been attempted with materials from selected sampling units (Figure 1) (including pieces from different shafts and different units). Refits between pieces recovered from different shafts have been obtained from three sampling units: B1, D4 and E4. Thus, these shafts were filled with residues from ongoing percussion activities. Since shafts were refilled shortly after they had been excavated, these refits suggest that related shafts were opened and refilled in the same mining event.

A total of 513 complete or almost complete tools (both individual pieces and refits) have been analysed; the rest are fragments. These 513 tools have been interpreted according to their shape, size, weight and the number, location, use traces and other characteristics of their working surfaces. Almost all the working surfaces documented show macroscopic percussion marks, the most common being pits, accidental extractions and battered ridges. Alterations have also been recorded. From every sampling unit a small number of pieces showing thermal alterations has been found. The main alteration affects their colour, the pieces taking on a reddish appearance. Some of these pieces are also fire-cracked.

The *Used Area Index* analyses the intensity of the use of cobble tools. This index is obtained by dividing the total working area by the estimate of the total surface area of the cobbles.

The *Shape Ratio* (Grace 1989) studies the relationship between the size of the working surfaces and the area left for handling or hafting the tools. This ratio is more variable in some tools than in others, although fairly regular for some.

The arrangement of working surfaces also provides information on the way tools were handled. Most of the tools show bipolar use, but interesting differences between tool types have been recorded (see below).

Tasks performed at the mine

Most of the tools examined seem to have been used in very heavy percussion tasks, where the contact materials would have been very hard (probably rock in most cases). These heavy percussion activities would have been mainly linked to shaft excavation, flint extraction and flint-knapping. More than 50% of the percussion tools found are related to flint-knapping. Hammerstones of different sizes have been documented that might have been used in all the steps of the operational chain (Figure 2).

Large hammerstones, *Bipolar B* tools, *Bipolar C* and incisive tools such as wedges and *Bipolar A* tools could have been used in the process of breaking up and extracting nodules (Figure 2). Anvils and (exceptionally) grinders have also been identified. It is not clear what materials were processed using them since no residues have been recovered from their working surfaces. In any event, these tools were certainly not involved in extraction or knapping.

Finally, some cobbles were recycled for use in association with hearths after their use in percussion activities (Figure 2).

Specialisation

- Labour intensity

At Casa Montero, neither the cobble tools nor flint percussion tools show any sign of conditioning prior to their use in percussion tasks. With respect to the cobble tools, some time would have had to be spent in the procurement of cobbles from river deposits and transportation to the mine, although this would have been minimal given the deposits are located one km from Casa Montero. Indeed, cobbles may simply have been gathered on the way to the mine. No time was invested in their modification and only in very few cases may some time have been spent in their hafting.

The mean value for the *Used Area Index* is 0.04, with most values between 0 and 0.1 (Figure 3). These low figures are partly due to the fact that only some areas of the cobbles were used, the rest of the surface being in contact with the hand or the haft. But even so, in most cases less than 10% of the surface was used, suggesting that tools were discarded when they still had usable areas.

- Standardisation

The cobble percussion tools at Casa Montero show little intentional standardisation; they were not modified but simply chosen for their size, weight and shape. However, some degree of mechanical standardisation resulted through repeated working procedures. The *Shape Ratio* (Figure 4) is fairly regular for the mine's hammerstones, and very little

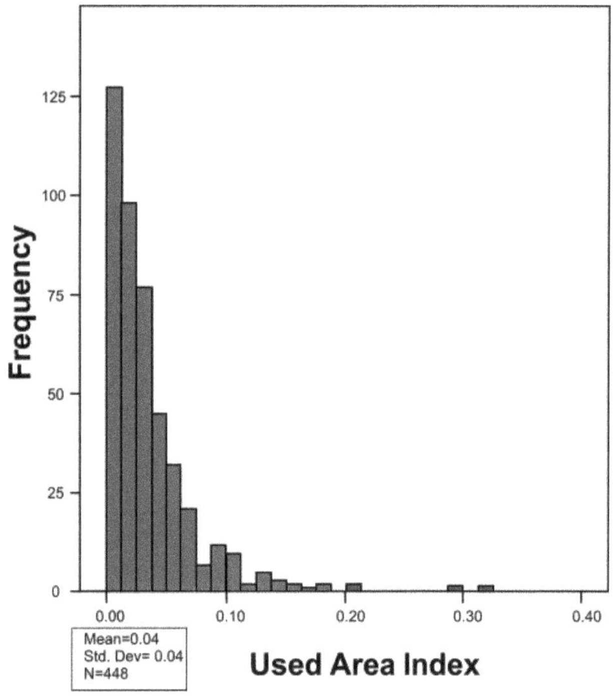

Figure 3: Histogram that shows the distribution of the values of the Used Area Index of complete tools.

variation is seen for its *Bipolar A* tools, *Bipolar C* tools or grinders. Thus, each of these types of tools, regardless of differences in size and sometimes shape among cobbles within it, was manipulated in a regular manner.

The arrangement of the working surfaces is also fairly uniform for these same types of tool. Most of the tools show bipolar use, almost exclusively so for *Bipolar A* tools and hammers. The grinders normally show such use, with the obverse or reverse faces - but almost never any lateral faces-involved. The hammerstones predominantly reflect bipolar use and sometimes lateral use, but no obverse or reverse use. The lack of any statistically significant difference among the sampling units in terms of the type of tools found in each further supports the idea that working procedures were repeated time and again.

- *Skill*

Although no specific attributes of these tools can be clearly linked with specific skills, their repeated use in a standardised fashion must have required some degree of know-how. Other mining activities would also require some experience.

Thus, these tools seem to be associated with low labour intensity. They reflect almost no intentional standardisation and only moderate mechanical standardisation. No skill was necessary in their manufacture, but at least moderate skill would have been necessary for their proper use in flint-knapping. These characteristics agree with what Costin and Hagstrum call *community specialisation*, i.e., an independent context, nucleated concentration, household constitution (household-based productive units aggregated in this case for a communal activity) and part-time intensity.

Workforce size

Based on the refits from sampling units B1, D4 and E4 (Figure 1), the minimum number of shafts open in each mining event can be estimated. Differences are seen between different areas of the mine. With respect to sampling unit B1, just 5 shafts are connected by refits, while for sampling units D4 and E4, 18 and 17 shafts are connected respectively. Both in D4 and in E4, the few shafts with no refits with other shafts were also refilled during the same mining episode, as shown by the horizontal excavations connecting them with other shafts that do have refits and which have the same infill. Hence, at the centre of the mine, at least 21 shafts were excavated during each mining event, while at the margins there is no evidence for more than five being opened in a single event. Consequently the workforce would have been relatively large during some events and smaller during others. The possibility that instead of more persons more time was involved is considered unlikely because we would have to explain why a single group would spend in the mine much more time than was necessary in order to meet its own raw material needs (given that we have no reason to believe that they would be producing a surplus for exchange).

Labour intensity and differences over time

A substantial difference exists between the central area of the mine and the margins, not only in terms of the total weight of percussion tools recovered but also in terms of the mean weight of percussion tool remains per shaft (Figure 5). In some areas more shafts were excavated than in others, but also, particularly in some areas at the centre of the mine (sampling units D2, D3, E2 and E4), more percussion tools per shaft were used. The intensity of the use of these tools is similar throughout the mine, with no significant difference in the *Used Area Index* of tools between sampling units. Thus, regardless of the number of shafts excavated, the intensity of percussion tasks per shaft was greater in some areas than in others. If these different areas were exploited at different times, some mining events must have involved more work than others, independent of the use of different extraction methods. For example, in sampling unit D4, where flint seams were more intensively exploited by horizontal excavations, the weight of percussion tools per shaft is in fact smaller than that recorded for other sampling units.

4. Conclusions

The attributes of the percussion tools from Casa Montero point to a specialisation in which independent individuals from one or more groups gathered periodically for projects

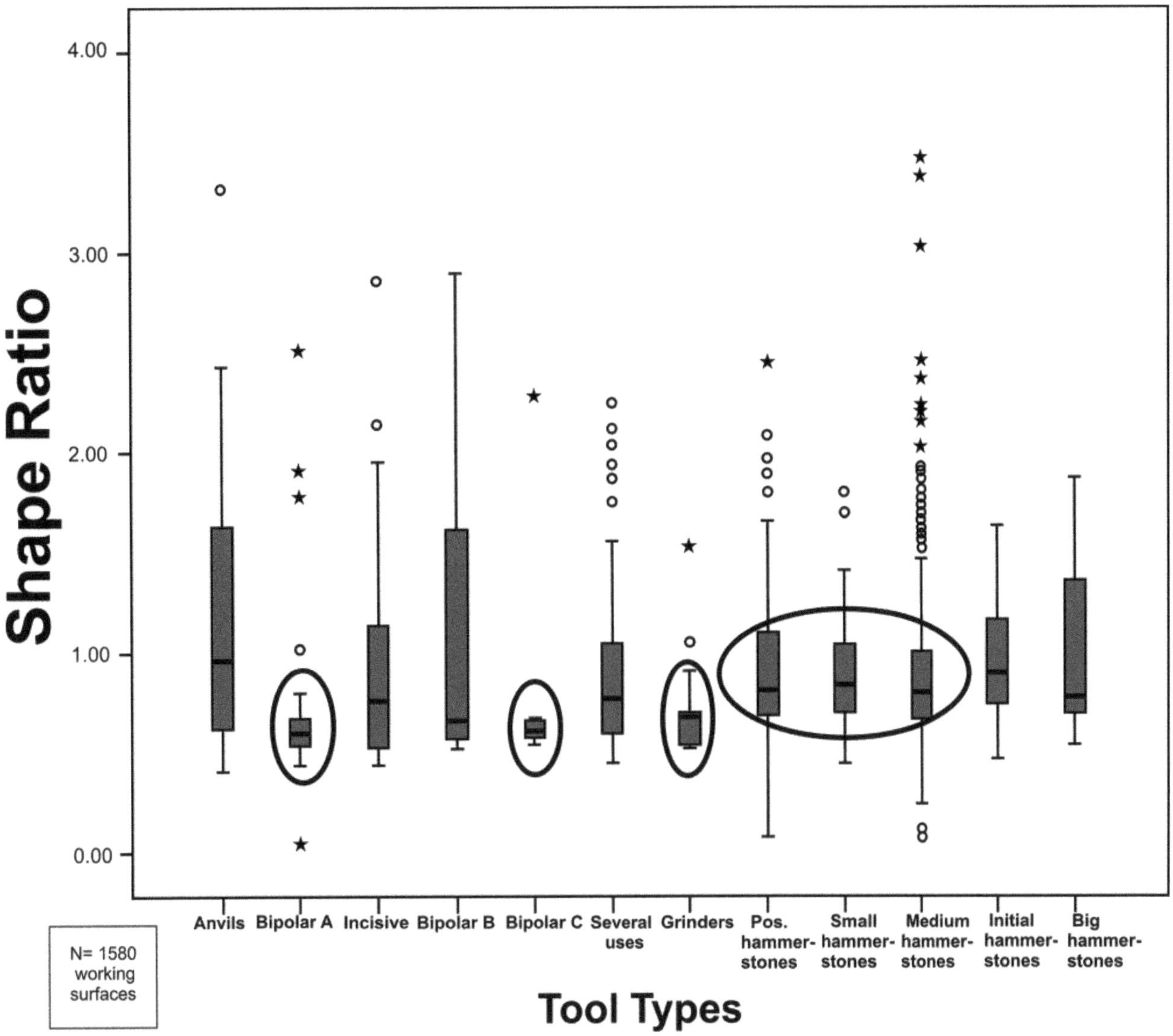

Figure 4: Shape Ratio for each different type of tool. Some types of tools present remarkably low variability, suggesting they have been handled in a regular manner despite differences in size and shape.

that entailed a low labour intensity, low intentional standardisation, moderate mechanical standardisation, and at least a moderate degree of skill.

Cobble tools were systematically used in the same way, in a standardised procedure that was repeated with little variation. The flint percussion tools, on the other hand, are exceptional and the consequence of opportunistic behaviour. Along with cobble tools, other elements such as wooden wedges or digging tools must have formed part of the relatively standardised mining toolkit.

Percussion tools were involved in shaft digging, raw material extraction, flint-knapping, grinding activities and hearth-related use. In the relatively short period of use of the mine, working techniques did not change significantly, with tasks being performed using the same tools in almost the same way every time. This may be linked to mechanisms of knowledge acquisition and transmission that ensured procedures were perpetuated. What did change was the scale of collective labour. This could merely be a result of a changing demand for raw material, but also of a changing capacity or need for the mobilisation of shared labour by communities. Givenz the large scale of some mining events in a regional and chronological context of very small groups, an unsolved question is how and why such a workforce was mobilised.

From a methodological point of view, the efficiency of analysing a minority element such as the cobble tools of Casa Montero should be noted; this strategy allowed us to tackle the question of the number of shafts opened in a mining single event with a relatively small investment in time.

Neolithic flint mining in Europe was not a uniform phenomenon, at least not in terms of labour organisation. There was considerable variability in labour intensity, the degree of elaboration of percussion tools, tool standardisation, the scale of mining events and the size of the workforce involved.

Sampling Unit	Number of pieces	Total weight of cobbles and fragments (Kg)	Number of shafts	Cobble weight per shaft (Kg)
A2	4	2.75	2	1.38
B1	308	138.71	15	9.25
B2	166	89.55	8	11.19
B3	429	133.39	14	9.53
D1	184	58	5	11.6
D2	223	150.92	9	16.77
D3	284	233.41	10	23.34
D4	684	325.86	22	14.81
E1	30	30.46	3	10.15
E2	638	240.95	14	17.21
E3	476	257.72	19	13.56
E4	1074	355.65	21	16.94
G3	86	32.36	5	6.47
CMI	36	11.5	10	1.15

Figure 5: Total weight of cobbles and fragments for each sampling unit and average weight per shaft in each case.

Even given the important differences among European mines, Casa Montero seems to be a special case. From the point of view of labour intensity, the percussion tools found in most European Neolithic mines reflect considerable time investments in their making, that large quantities of them were needed, and that they were intensely used. At Casa Montero, however, these tools were obtained close to the mine, used without further modification, and discarded when they might still have been useful, as suggested by their low mean *Used Area Index* value.

Percussion tools from other Neolithic mines show signs of intentional standardisation, while those from Casa Montero were not selected or transformed in order to comply with technical requirements. Indeed, those of Casa Montero seem rather more opportunistic in character. Mining activities at this site seem to have been less structured and developed than in other parts of Europe, maybe due to the short life-span of the mine.

The differences documented here are not merely adaptations to different geological settings, nor are they strictly technical. Since this work examines variables related to labour organisation, these differences must reflect variations in the way that mining work was socially organised.

Acknowledgements

The author wishes to thank all the members of the Casa Montero project for their inestimable help and support and to stress that without their efforts in the excavation of the site and in the ulterior study of the material and documentation this work would not have been possible.

This paper was carried out as part of the project "Proyecto de Investigación Arqueológica en el yacimiento de Casa Montero (Madrid). Producción y circulación de sílex en el neolítico de la Meseta", funded by Autopista Madrid Sur C.E.S.A. in the framework of the Collaboration Agreement between the Council for Culture and Sports of the Community of Madrid, the Spanish National Research Council (CSIC) and Autopista Madrid Sur Concesionaria Española S.A. for the research, conservation and diffusion of the archaeological site of Casa Montero (Madrid, Spain).

References

Bácskay, E. 1995. The flint mine of Sümeg-Mogyorósdomb. In J. Lech (ed.), *Archaeologia Polona* 33. Special Theme: Flint Mining. Dedicated to the Seventh International Flint Symposium, 388-390. Warsaw, Institute of Archaeology and Ethnology, Polish Academy of Sciences.

Boguszewski, A. 1991. Experimental use of antler tools in flint mines in Krzemionki, Poland. In *Archeologie experimentale. Actes du Colloque International "Expérimentation en archéologie: Bilan et Perspectives.* Tome 2: La terre. L'os et la pierre, la maison et les champs, 46-48. Paris, Éditions Errance.

Borkowski, W., Migal, W., Salacinski, S. and Zalewski, M. 1991. Possibilities of investigating Neolithic flint economies, as exemplified by the banded flint economy. *Antiquity* 65(248), 602-27.

Borkowski, W. 1995a. Prehistoric flint mines complex in Krzemionki (Kielce Province). J. Lech (ed.), *Archaeologia Polona* 33. Special Theme: Flint Mining. Dedicated to the Seventh International Flint Symposium, 506-524. Warsaw, Institute of Archaeology and Ethnology, Polish Academy of Sciences.

Borkowsi, W. 1995b. Élements d'anayse du système d'aménagement d'une minière: l'exemple de Krzemonki, in J. Pelegrin and A. Richard (eds.), *Les mines de silex au Néolithique en Europe. Avancées récentes*, 67-72. Vesoul, Comité des Travaux Historiques et Scientifiques.

Bostyn, F. and Lanchon, Y. 1992. *Jablines. Le Haut Château (Seine-et-Marne). Ne minière de silex au Néolithique.* Paris, Editions de la Maison des Sciences de l'Homme.

Bostyn, F., Cayol, N., Gologny, F., Lo Carmine, A. and Maigrot, Y. 2005. *Creusement experimental d'un puits d'extraction de silex sur la minière de Flins-sur-Seine (Yvelines).* Les Mureaux, Mémoires et Travaux du Paléoscope 1, Association L'Homme Retrouvé.

Brounen, F. T. S. 1995. Neolithic flint extraction at Valkenburg aan de Geul (Limburg, Netherlands). In J. Lech (ed.), *Archaeologia Polona* 33. Special Theme: Flint Mining. Dedicated to the Seventh International Flint Symposium, 445-453. Warsaw, Institute of Archaeology and Ethnology, Polish Academy of Sciences.

Charniausky, M. M. 1995. Ancient flint mines in Belarus. In J. Lech (ed.), *Archaeologia Polona* 33. Special Theme: Flint Mining. Dedicated to the Seventh International Flint Symposium, 263-269. Warsaw, Institute of Archaeology and Ethnology, Polish Academy of Sciences.

Collet, H. 2004. Les mines Néolithiques de Spiennes. État des connaissances et perspectives de recherche. In *Actes du XIVéme Congrès de l'UISPP.* Oxford, British Archaeological Reports International Series 1303. http://minesdespiennes.org/textes/spiennes.etatdesconnaissances.hcollet.pdf

Collet, H., Hubert, F. and Joris, J. P. 2006. The flint mines of Petit-Spiennes (province of Hainaut, Belgium): an update, in G.Körlin and G. Weisberger (eds.), *Stone Age- Mining Age*, 67-72. Bochum, Deutsches Bergbau-Museum.

Collet, H., Hauzeur, A. and Lech, J. 2008. The prehistoric flint mining complex at Spiennes on the occasion of its discovery 140 years ago, in P. Allard, F. Bostyn, F. Giligny and J. Lech (eds.), *Flint Mining in Prehistoric Europe. Interpreting the archaeological records*, 41-77. Oxford, British Archaeological Reports International Series 1891.

Costin, C. L. and Hagstrum, M. B. 1995. Standardization, labor investment, skill, and the organization of ceramic production in Late Prehispanic Highland Peru. *American Antiquity* 60(4), 619-639.

Clutton-Brock, J. 1984. *Neolithic antler picks from Grimes Graves, Norfolk, and Durrington Walls, Wiltshire: a biometrical analysis.* Excavations at Grimes Graves, Norfolk 1972-1976, Fascicule I. London, British Museum Publications Limited.

Diethelm, I. 1997. Neolithic flint mining in the Three-Country Corner (Basel Region), Kleinkems (Germany) and Löwenburg (Switzerland), in memoriam Prof. Elisabeth Schmid, in R. Schild and Z. Sulgostowska (eds.), *Man and Flint. Proceedings of the VIIth International Flint Symposium*, 63-64. Warsaw, Institute of Archaeology and Ethnology, Polish Academy of Sciences.

Felder, P. J. 1979. Prehistoric Flint Mining at Rijckholt-St. Geertruid (Netherlands) and Grimes Graves (England). *Staringia* 6, 57-62.

Felder, P.J., Rademakers, P., Cor M. and de Grooth, M. E. Th. 1998. *Excavations of Prehistoric Flint Mines at Rijckholt-St- Geertruid (Limburg, The Netherlands) by the "Prehistoric Flint Mines Working Group" of the Dutch Geological Society, Limburg Section.* Bonn, Deutsche Gesellschaft für Ur- und Frühgeschichte.

Galiberti, A. 2005. Utensili per l'attività estrattiva e per lavori di supporto ad essa ralativi. In A. Galiberti (ed.), *Defensola. Una miniera di selce di 7000 anni fa,* 127-140. Siena, Protagon Editori Toscani.

Grace, R. 1989. *Interpreting the function of stone tools.* Hypertext version of Interpreting the Function of Stone Tools: The quantification and computerisation of microwear analysis. British Archaeological Reports International Series 474.

http://www.hf.uio.no/iakh/forskning/sarc/iakh/lithic/bar/bar1.html#anchor253048

Holgate, R. 1995a. Neolithic Flint mining in Britain. In J. Lech (ed.), *Archaeologia Polona* 33. Special Theme: Flint Mining. Dedicated to the Seventh International Flint Symposium, 133-161. Warsaw, Institute of Archaeology and Ethnology, Polish Academy of Sciences.

Holgate, R. 1995b. Harrow Hill near Findon, West Sussex. In J. Lech (ed.), *Archaeologia Polona* 33. Special Theme: Flint Mining. Dedicated to the Seventh International Flint Symposium, 347-350. Warsaw, Institute of Archaeology and Ethnology, Polish Academy of Sciences.

Labriffe, P. A. de, Augereau, A. and Sidéra, I. 1995a. Villemaur-sur-Vanne "Le Grand Bois Marot", Aube district. In J. Lech (ed.), *Archaeologia Polona 33*. Special Theme: Flint Mining. Dedicated to the Seventh International Flint Symposium, 322-335, Warsaw, Institute of Archaeology and Ethnology, Polish Academy of Sciences.

Labriffe, P. A. de, Augereau, A. and Sidéra, I. 1995b. Villemaur-sur-Vanne, "Les Orlets", Aube district. In J. Lech (ed.), *Archaeologia Polona* 33. Special Theme: Flint Mining. Dedicated to the Seventh International Flint Symposium, 335-345, Warsaw, Institute of Archaeology and Ethnology, Polish Academy of Sciences.

Lech, J.and Mateiciucová, I. 1995. In J. Lech (ed.), *Archaeologia Polona* 33. Special Theme: Flint Mining. Dedicated to the Seventh International Flint Symposium, 271-276. Warsaw, Institute of Archaeology and Ethnology, Polish Academy of Sciences.

Leitner, W. 2008. Steinzeitlicher Silexabbau im Kleinwalsertal. *Archäologie in Deutschland* 4, 28-29.

Longworth, I. and Varndell, G. 1996. *Mining in the deeper mines.* Excavations at Grimes Graves, Norfolk 1972-1976, Fascicule 5, London, British Museum Publications Limited.

Rind, M. M. 2003. *Wer Anderen eine Grube gräbt...*, Bücherbach, Archäologie in Landkreis Kelheim.

Rudebeck, E. 1987. Flintmining in Sweden during the Neolithic period: new evidence from the Kvarnby-S. Sallerup area, in G. de G. Sieveking and M. H. Newcomer (eds.), *The human uses of flint and chert*, 151-157. Cambridge, Cambridge University Press.

Schild, R. 1995. Polany Kolonie II, Radom Province. In J. Lech (ed.), *Archaeologia Polona* 33. Special Theme: Flint Mining. Dedicated to the Seventh International Flint Symposium, 480-488. Warsaw, Institute of Archaeology and Ethnology, Polish Academy of Sciences.

Saville, A. 2008. Flint extraction and processing from secondary flint deposits in the north-east of Scotland in the Neolithic period, in P. Allard, F. Bostyn, F. Giligny and J. Lech (eds.), *Flint Mining in Prehistoric Europe. Interpreting the archaeological records*, 1-10. Oxford, British Archaeological Reports International Series 1891.

Sidéra, I. 1995. Relations minières/hábitat: un problème de méthode. Le potentiel des artefacts osseoux, in J. Pelegrin and A. Richard (eds.), *Les mines de silex au Néolithique en Europe. Avancées récentes*, 115-134. Vesoul, Comité des Travaux Historiques et Scientifiques.

Simán, K. 1995. Prehistoric mine on the Avas Hill at Miskolc. In J. Lech (ed.), *Archaeologia Polona* 33. Special Theme: Flint Mining. Dedicated to the Seventh International Flint Symposium, 371-382. Warsaw, Institute of Archaeology and Ethnology, Polish Academy of Sciences.

Weiner, J. 1995. Les outils d'extraction à encoches en silex et pierre de la mine Néolithique Final du Lousberg, Aachen (Rhénaqnie septentrionale- Westphalie, Allemagne), in J. Pelegrin and A. Richard (eds.), *Les mines de silex au Néolithique en Europe. Avancées récentes*, 93-106. Vesoul, Comité des Travaux Historiques et Scientifiques.

Mining tools use in a mining context or how can the expected become unexpected

Xavier TERRADAS, Ignacio CLEMENTE and Juan F. GIBAJA

Abstract

In archaeological research it is common practice to infer the function of prehistoric tools from their typology and the context in which they were recovered. However, analysis of function based on the inspection of macro- and microscopic use-wear evidence is the only way to provide credible support for such inferences. The present work reports data gathered during the use-wear analysis of a sample of antler and stone artefacts found in the Neolithic mine of Casa Montero. These data certify that the antler tool recovered were not mining tools, while the knapped stone tools provide new information regarding activities undertaken in this flint mining context, as well as the taphonomic processes occurring at the site. The proposed uses of these tools indicate human habitation near the site.

Keywords

Mining tools. Antler picks. Lithic production context. Use-wear analysis. Neolithic. Casa Montero.

1. Introduction

In certain archaeological contexts, affirmations are all too often made without having gone through the necessary steps to justify them. Such affirmations are often made with regard to the production contexts of ancient societies. In mining complexes – places specialised in the acquisition of lithic raw materials and the first steps in their transformation into products – the tools found are commonly associated with the specialised tasks of these production processes. However, detailed use-wear analysis of these tools' actual use shows that such associations should not be lightly made. The aims of the present work are to:

- Present the data gathered during the analysis of function of tools recovered from the Neolithic flint mine of Casa Montero. This analysis involved the examination of the macro- and microscopic evidence of the use made of flint and antler tools (i.e. that visible on the surfaces and edges of these instruments). Comparisons were made with experimental reference tools with the goal of obtaining diagnostic data regarding their use, the material worked, and the taphonomic process at work at the site. There have been very few such studies on materials recovered in mining contexts.

- To study the supposed flint and antler mining tools collected following a sampling strategy (which, in the case of the flint tools involved waste materials, blanks and retouched pieces). The results of this work provide evidence to help determine whether the recovered artefacts were mining tools, whether they were used in complementary activities (maintenance and repair), or whether they were used for other purposes.

2. The Casa Montero flint mine

The Casa Montero mining complex is found in Vicálvaro -southeast of the Madrid region-, in the centre of Iberia. It lies on a small plateau at an altitude of some 650m, near the confluence of the Rivers Jarama and Henares. The site was discovered as a result of the archaeological impact assessment of Madrid's M-50 highway belt and has been subject to different phases of excavation between 2003 and 2006 (Consuegra *et al.* 2004; Díaz-del-Río *et al.* 2006, 2008; Capote *et al.* 2008), during which some 4000 vertical shafts were recorded over an area of some 4ha. To date, 324 of these shafts, which are between 1 and 10m deep and nearly 1m wide, have been excavated. The excavation work focused exclusively on the material filling the shafts since no associated work or dwelling areas have been identified. These filling materials include a very large amount (about 65t) of lithic production waste.

The site lies on a small anticline of alternate green clays (illite and smectites) and carbonates (dolomitic marls and

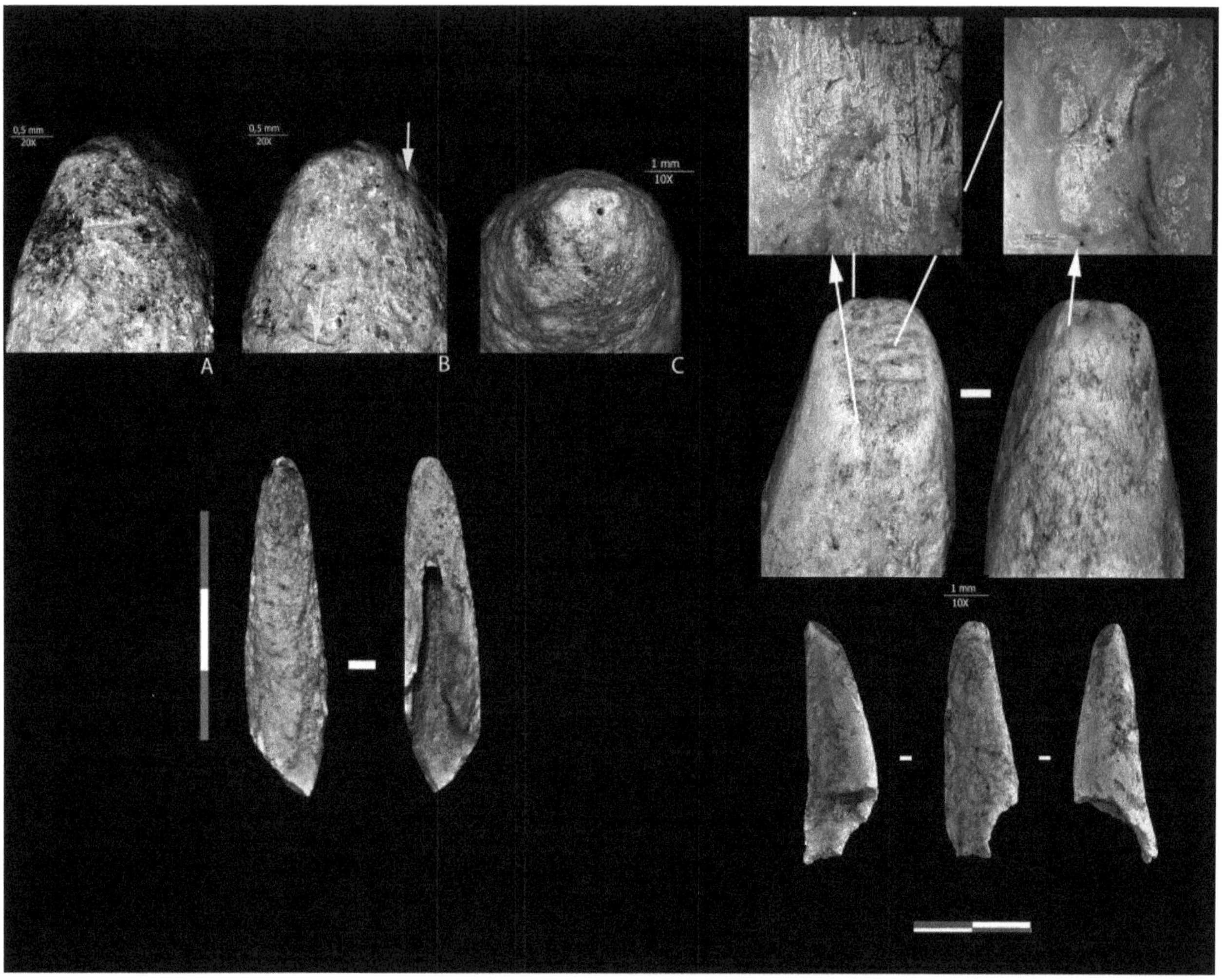

Figure 1. Pieces nº 9562 (left) and 7142 (right) elaborated on antler. The images on the top have been obtained by means of a metallographic microscope (200X).

dolostones), attributable to an intermediate Unit of the Miocene (Bustillo and Pérez-Jiménez 2005). Four flint layers can be seen over this sequence, although the exploitation of this material focused preferentially on obtaining the opaline cherts in the upper layers (Castañeda *et al.* 2008).

Although other stages of exploitation are known, the greater part of the site's mining activity was undertaken over a relatively short period of the early Neolithic (5400–5200 cal BC) (see Díaz-del-Río and Consuegra in this volume); this period is discussed in the present work. Exploitation focused on the digging of mine shafts for the extraction of blocks of flint, the transformation of which was mainly directed towards obtaining cores for the production of blades. Neither the distribution of these raw materials nor of the products made from them is currently known; this is largely the consequence of our not knowing of other, contemporaneous Neolithic sites in the area.

Dating studies (see Díaz-del-Río and Consuegra in this volume) suggest that the Casa de Montero mine was exploited for just a few hundred years during the early Neolithic. The many shafts dug, the fact that they do not overlap, and the lack of any dwelling structures in the area suggest that the mine grew over a series of short, repeated visits linked to obtaining siliceous raw materials and the first stages of its transformation.

3. Antler tools

The material recovered from the Casa Montero mine includes only three fragments of deer antler. Given the finds made in analogous mining contexts, in which different types of antler mining tool were made (Gurina 1976; Boguszewski 1995; Lech 1997; Russell 2000; Bostyn *et al*,. 2005), one might think that those from Casa Montero might also correspond to tools used in the excavation of the mining shafts. However, such a conclusion was in doubt even during the first observations of these materials; indeed, they were finally linked to purposes other than mining (Capote *et al.* 2008; Yravedra *et al.* 2008). Consuegra *et al.* (2004) and Capote *et al.* (2008) indicated that only a study of the evidence of the use of these artefacts could provide valid information regarding their true function.

Reference experimental studies

The majority of experimental investigations into the use of antlers for the digging of mine shafts have been prospective, i.e., preliminary work into the questions posed, the elements involved and their properties (Clemente 1997; Terradas and Clemente 2001; Gibaja 2003). The control of variables has often been scant, but the results obtained, although not definitive, have been useful for laying the ground for future experimental work and determining what variables might be of importance. Two main types of experimental work have been undertaken:

- To determine the efficiency of antler picks and to make estimates regarding the time that would be required to dig out a certain volume of earth. Such was the pioneering work of Lane Fox in 1875 (cited in Russell 2000), which centred on the Neolithic mining galleries of Cissbury (England), and the work undertaken by French researchers on the Neolithic quarries of Plancher-les-Mines (Jeudy *et al.* 1995).

- To determine the working efficiency of antlers on different rocks and sediments. For example, Semenov and Korobkova (cited in Gurina 1976), who worked on the Byelorussian mines of Krasnovo Sela and Karpovtsev, reported antler picks to be more efficient than stone picks for digging galleries in chalk.

Controlled experiments in this context, which have been much fewer in number, have focused on the study and recording of the traces of use left on tools. Such experiments require the design of an experimental program that focuses on specific objectives and hypotheses, a methodology to be followed that identifies the variables to be controlled and the techniques to be used, and the rigorous and systematic recording of the results obtained (Clemente 1997; Terradas and Clemente 2001; Gibaja 2003).

Work of this kind has been undertaken by Maigrot for the Flins-sur-Seine site (Bostyn *et al.* 2005), which followed on from earlier work on the mines of Jablines (Bostyn and Lanchon 1992). Boguszewski (1995) suggested that antlers were not used as picks but as wedges and/or levers, and attributed the splintering of antler material at the point of the cutting tine to such use. Most experimental works use as their reference the picks recovered at Grimes Graves (Russell 2000); these were configured from thick antlers, the crown and the lateral tines being removed, leaving only the basal tine which formed the active digging point.

Bostyn *et al.* (2005) showed that the evidences of use on antler tools fell into two main areas:

- On the trunk of the antler where the tool is held. Here, the wear on the surface can be very extensive and invasive, sometimes occupying the entire gripping area. The location of these evidences is completely different to these documented on the antlers of free-living deer, which is usually seen on the tips of the tines. In fact, the wear seen in experimental work tends to regularise the natural ridges of the antler, and to increase its whiteness. Under the microscope, the surface is seen to acquire a uniform sheen and granular texture. The invasive polishing caused by use is associated with numerous, relatively fine crisscrossed striations, which are organised in a transversal way to the longitudinal axis of the antler.

- On the basal tine, that acts as a pick point. The apex becomes squashed and cracked, causing antler material to splinter and be lost. These removals can become large, reaching 3-5cm in length and 1-1.5cm in width. The experimental use of an antler pick over two hours on limestone led to the loss of 2cm (7g) of material at the active point (Teno and Delgado 2002-2003).

- On the external curvature of the basal tine. Under the microscope, this wear appears as numerous lines orientated along the longitudinal axis of the tine. These striations can be several centimetres long and over 1mm wide, and are encrusted by other, finer striations. Microscopic observation of these striations reveals them to be linear depressions with irregular but clear outlines. In contrast, the striations seen on the antlers of free-living deer are much wider and shorter, and lie obliquely to the longitudinal axis of the tine.

The antler tools of Casa Montero

The post-depositional changes that have occurred at Casa Montero have provoked the degradation of its antler artefacts to the extent that the microscopic analysis of their surfaces is sometimes impossible. Such is the case of the largest antler artefact (nº 5580) which, from its shape, might correspond to a basal tine. An additional problem is the degree of fracturing of some pieces (nº 7142 and 9562), which impedes an understanding of their original size and hinders the reconstruction of the correct use of these tools. However, despite these limitations, the preliminary diagnoses of Capote *et al.* (2008) and Yravedra *et al.* (2008), which indicate these pieces were not picks, is here confirmed.

Given the state of preservation of the antler items from the Casa Montero mine, only two could be examined. Object nº 9562 (Figure 1) was originally described as a pointed object made from antler that showed evidences of hollowing and abrasion along the perimeter of its fractured area (Yravedra *et al.* 2008). Some small scars can be seen along with areas of abrasion parallel to the long axis of this piece of tine. The latter were possibly caused by pressure at the tine point. The artefact looks as though it may have been rubbed down to conform to a circular, pointed shape. The internal spongy bone tissue has been removed, perhaps related to the way the tool was held during use. Although the tool was suggested to be a pressure flaker (Capote *et al.* 2008), it shows none of the characteristic evidences of such instruments. Pressure flakers show numerous traces along their sides, while the contact area appears as a bevelled edge. In addition, they show deep, wide striations running backwards from the point (Maigrot 2003).

Piece n° 7142 (Figure 1), made from an antler tine, was originally interpreted as a small spatula-chisel, the surface of which was crossed by traces transversal to the long axis of the object, and with slight scarring on the back side (Yravedra *et al.* 2008). Indeed, on the bevelled edge several deep and wide grooves can be seen that invest the piece with a certain roughness and which are clearly functional in nature. This is the active area that made contact with the material worked. The pressure exerted on the tool during its use led to the small scars on its back side. On the edge running along the platform, and in raised areas between the grooves, areas of bright, closed/compact-patterned micropolishing and of rough appearance can be seen. Over this micropolished surface there are a number of striations of different morphology that run perpendicular to the distal border of the bevelled edge. The location and characteristics of all these traces indicate this tool came into contact with a material that was rigid and rather hard. The instrument was used such that the bevelled edge would lift a fine layer of the material being worked, and the raised areas would wear down the surface of the cut material like a file. The optical characteristics of the micropolished surface suggest the tool may have been used on a woody material.

The entire tool is impregnated with a red pigment. This is characteristic of pieces representing the antler industry at Casa Montero (Yravedra *et al.* 2008), although its function remains unknown.

4. Knapped stone tools

Sampling strategy and conservation of collected lithic material

The study of the function of knapped stone tools in the context of the Casa Montero mine presents challenges for two reasons. Firstly, the number of items recovered is huge (more than 65t have been collected), and secondly, the context of the site is one of production, in which the great majority of items found are lithic wastes with no use at all. Clearly, the microscopic analysis of the surfaces and edges of these artefacts obliges a reduced number of samples be studied. Thus, in the present work, representative sampling of the material was necessary, both from a qualitative and quantitative point of view. A total of 221 pieces were finally examined. The selection criteria outlined below were followed:

- Shafts located in grid squares B4, D3, D5 and G3 (Capote *et al.* 2008; Díaz-del-Río *et al.* 2008) were selected for general sampling; this allowed for a good spatial representation of the excavated area.

- All the retouched materials gathered in these grid squares were examined to determine whether the transformation of their edges might be related to their use.

- Unretouched flakes and blades with sharp edges from shafts located in grid squares B1, B3, B4, D1, D2 and E4 were examined; such objects are sometimes associated with specific activities such as the cutting of meat, skins or vegetables.

- Special attention was paid to certain stone tools, such as a number of picks showing modifications at their apices, and blades with shiny edges perhaps used as sickles.

- No remains with surface patinas nor which showed scaring along their perimeters were examined.

- Finally, all objects under 3cm in length were excluded from analysis.

The selected pieces were subjected to a preliminary examination with the aim of detecting organic and inorganic residues adhered to their surfaces. The pieces with no such residues were then cleaned with a soapy solution. All the remains were examined under a binocular microscope (10-90X) and metalographic microscope (50-400X).

The number of pieces showing evidence of use was relatively small (n=40), perhaps due to:

- The lithic production context of the samples (Terradas 2002); it is to be expected that the majority of items should be wastes produced by the mining and initial transformation of the raw material.

- The abundance of raw material. This might have led to the use of tools in an expedient and opportunistic manner; the necessary conditions for leaving evidences of use may not often have arisen.

- The very large number of altered remains; many of the collected pieces could not be adequately examined.

The macro- and microscopic analysis of the recovered lithic material showed it to have suffered different natural and man-made alterations, hindering its use-wear analysis. For example:

- Many of the recovered items had a white patina on their surfaces making it difficult to identify the micropolishing produced by the working of any material.

- All the recovered items showed strong soil sheen, both in the raised and lower parts of their microtopograhy (Figure 2). This prevented the inspection of some micropolishings, especially those produced via contact with soft materials (meat, fresh skin and fish) as well as those still in their initial stages but produced via contact with harder materials.

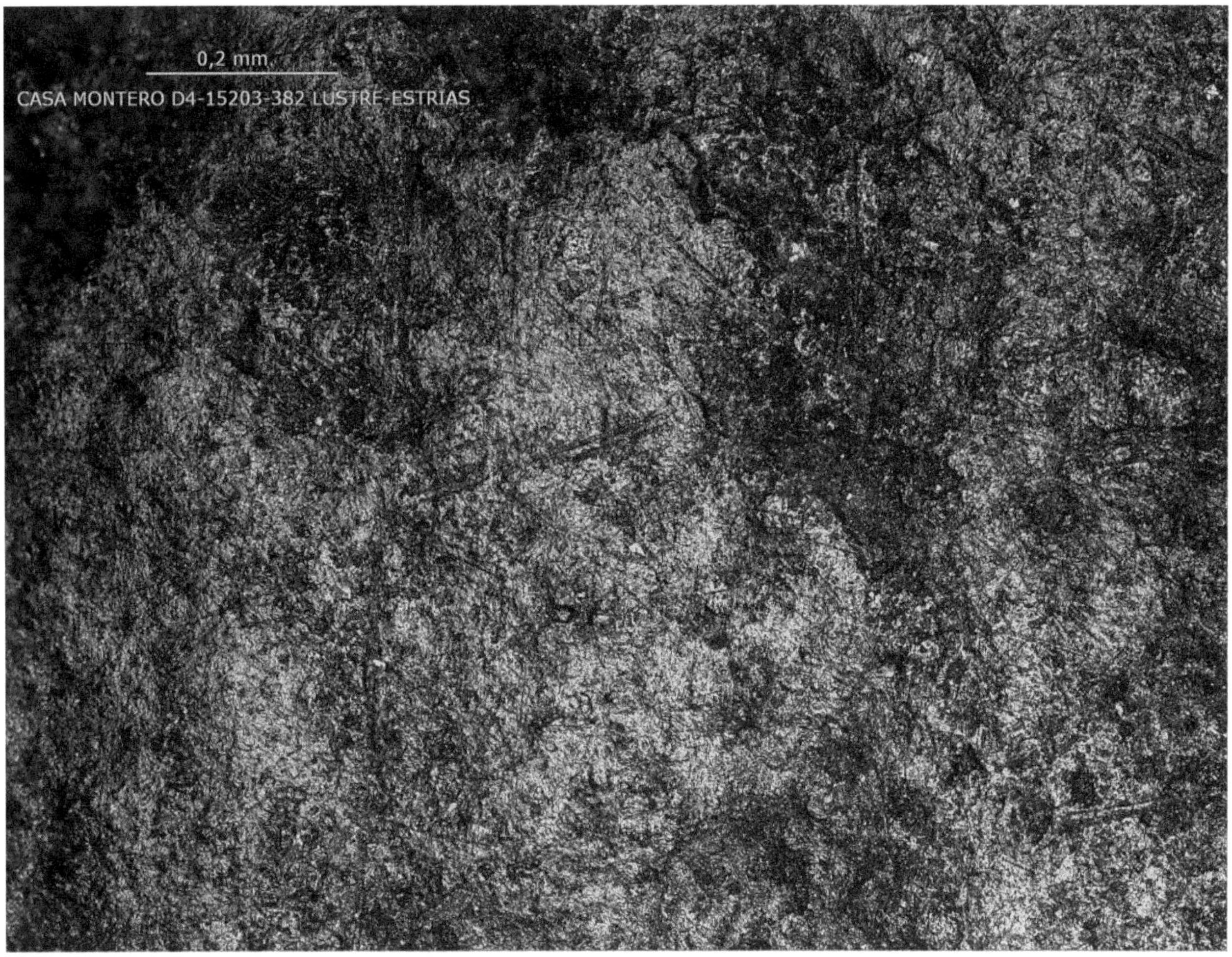

Figure 2. Microscopic evidences of soil sheen and striations chaotically arranged on the surface of a flint flake.

- Some pieces showed characteristic evidences of thermal alterations (cracking, scales, gloss,); however, in no case could this be associated with any intentional heating.

- A considerable number of items showed evidence of mechanical alteration, the consequence of the taphonomic processes to which they were subjected during their knapping, accumulation and eventual discarding in the shafts. Some pieces showed scars of variable size along their edges, abundant striations with no apparent organisation, or compact, flat polished plaques (known as 'G-polished plaques'), characteristic of rubbing between stone objects.

Another problem was the scant development of any trace of use on the majority of the tools examined, a consequence of their being used for little time. This is coherent within the context of Casa Montero where the abundant availability of raw materials made it unnecessary to use tools until they were no longer usable (Russell 2000; Terradas 2002). This poor evidence of traces made it difficult to identity the materials worked with the tools.

Together, these problems explain the large number of pieces that could not be analysed, the scarcity of used tools, and the impossibility of identifying the materials worked by many used tools.

Use-wear analysis

Of the 221 pieces selected for analysis, only 40 (18%) showed possible evidences of use. Forty two (19%) were unused and 139 (63%) were un-analysable owing to the alterations they had suffered. Twenty (50%) of the used tools provided insufficient evidence for the materials they worked to be clearly identified. However, the types of activity performed with all 40 of these instruments were discernable.

- The tools used on soft material were employed in cutting actions. These pieces were blades over 50mm in length with sharp, unretouched edges (20-30º). These morphological features are similar to those of tools used in butchering.

- The tools used with harder materials were for scraping. The majority of these tools were retouched flakes (scrapers and side-scrapers) with obtuse edges (40-80º), and were of considerable size (60-130mm). These features confer great effectiveness for the transformation of hard materials.

The type of material worked and the mode of use was discernible for the 20 best lithic remains. Special attention is paid to types of instrument, picks and sickles, owing to their singularity and clear traces of use preserved. It is difficult, however, to reach firm interpretations or establish proposals based on such a small number of items; this paper therefore dares provide only more general contributions regarding the processing of the materials worked.

The processing of animal materials

Seven tools showed evidences of the processing of different animal materials; four with butchering, and three with the scraping of hard animal parts.

The butchering tools were a flake and three blades with very sharp, unretouched edges (20-50°) – very effective for this type of work. Their function was discerned from the presence of small scars associated with open-patterned polishings and compact micropolished spots, possibly arising through contact with bone while butchering.

The scraping of hard materials –bone or antler– was performed with a blade (88x29x15mm), the distal right side of which showed semi-abrupt retouching associated with very compact micropolished spots. Given the length of the used area (some 50mm) this tool may have been used for very little time in some activity related to the sharpening or repair of another bone/antler tool.

Two flakes were used to scrape a semi-hard material, probably bone/antler but perhaps wood. These pieces, whose use-wear analysis suggests they were used for only a short period of time, are very different in size (74x63x30 and 35x25x8mm).

The processing of plant materials

Seven tools were found that were used to work wood, as well as another for cutting plants, along with two sickles or blades used for harvesting cereals. None of these pieces showed well developed evidences of use; they were probably used for only a short period after their manufacture. Of these seven tools, five were used to scrape, one to cut and one for cutting and scraping. The tools used to scrape were all retouched flakes (three side-scrapers, one scraper and one denticulate flake) and were probably very effective given their edge angle (70-90°), the length of the active area (over 20mm but reaching 70mm in one case), and their large size. The cutting tool was a blade 79mm in length, with an unretouched active area. With its 40° cutting edge angle the tool would be ideal for this function. The blade used to simultaneously cut and scrape was fragmented and had been slightly modified by abrupt retouching, thus adapting it to the scraping activity it was to perform.

It is likely that these types of instrument were used in the repair and maintenance of mining tools made of wood. Other authors have made similar proposals when examining side-scrapers or denticulate or notched tools from mining contexts in England and Italy. However, their hypotheses were not confirmed by use-wear analysis (Russell 2000; Galiberti *et al.* 2001).

Sickles

During the preliminary examination of the collected material, two blades showing a harvesting gloss similar to that generated by the cutting of cereals were noticed. This function was confirmed by use-wear analysis.

- One of these sickles (70x25x5mm) showed a modification to its distal curvature via abrupt retouching. One of its edges showed a well developed harvest micropolishing with striations, probably caused by abrasive soil particles. This suggests that cutting was performed close to the ground in order to harvest the grain and the stalk together. Over time the edge became dull and was re-sharpened. The distribution of the polishing shows that the blade was inserted diagonally into the handle.

- The extremes of the other sickle (42x10x3mm) were modified by abrupt retouching to produce a rather straight blank probably designed to facilitate the tool's setting in a handle. This sickle was also used over a long period and close to the ground. The distribution of the polishing on the blade surface indicates that it was inserted diagonally into its handle. The presence of remains of ochre indicates the possible use of this mineral in the resin that fixed the tool to the handle (Figure 3).

The finding of these sickles suggests that the group that exploited the mine practiced agriculture, despite the lack of evidence of any dwelling areas. The use of local flint for the manufacture of the sickles, the abundance of raw material (which would obviate the need to bring tools from elsewhere), and the fact that these sickles seem not to have been used in other activities, suggests there were cultivated fields in the vicinity of the mine.

As part of a wider research project (Ibáñez *et al.* 2008), we have compared the sickles of Casa Montero with those from other contemporaneous contexts in the Iberian Peninsula and southern France. The present sickles are intermediate between those from Andalusia and the Levant (made from blade fragments or flakes set diagonally and in-line into the handle to form a toothed instrument) and those of the Peninsular northeast and southern France (whole blades or fragments of blades arranged parallel to the handle).

The Casa Montero sickles are similar to those from La Draga (Banyoles, Girona) and Revilla del Campo and La Lámpara sites (Ambrona, Soria) (Gibaja 2008), all of

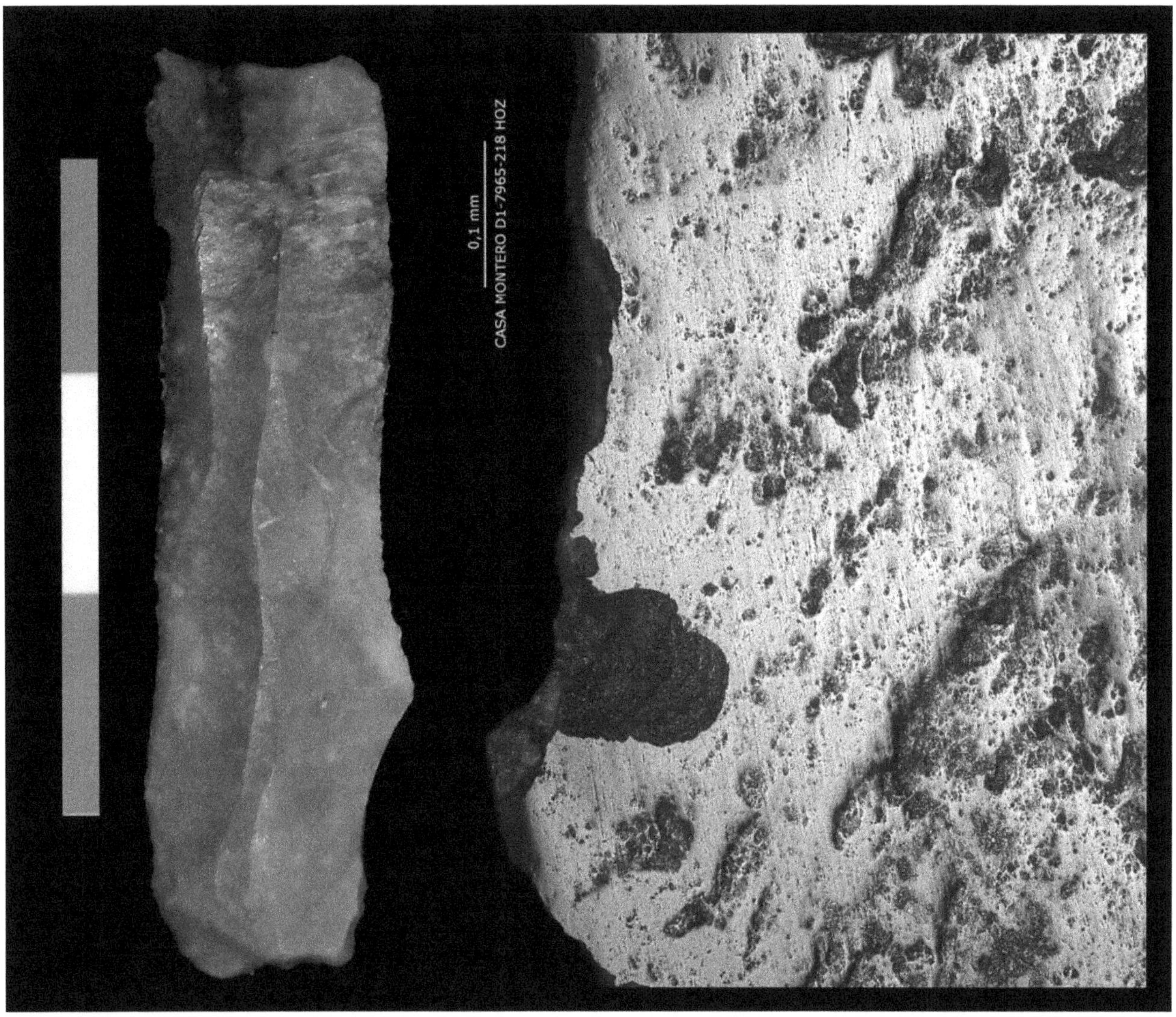

Figure 3. Blade used as a sickle where remains of ochre can be observed. Detail (100X) of polishing developed from its use for harvesting cereal.

which are early Neolithic and made from a single flint blade set diagonally into the handle.

Mining picks

Five pieces were examined that showed evidences of having been used as mining picks. These were pointed flakes about 150mm long, obtained during the first stages of the exploitation of flint cores. Use-wear analysis confirmed three had been used in mining activities.

- Item UE134-135-431 showed numerous longitudinal striations at its distal end, as well as compact-patterned polished plaques. These evidences of use gradually disappear away from the active area, and are practically absent at the centre of the tool. However, some areas the edges of the central and proximal areas also showed striations, chaotically arranged, and perhaps linked to the tool's hafting (Figure 4).

- Item UE113-249 is very similar to the above piece, showing modifications in the distal area caused by use. These include compact, smooth polished plaques with numerous longitudinal striations. These polished plaques might be related to contact with other stone elements.

- Item UE9623-E4-694 is a crested blade of notable smaller size (90x31x16mm). The traces of use it shows are less developed but are of the same type as those above: very compact polished plaques associated with striations and hinged or abruptly ended scars that suggest the apical area was subject to percussion.

Given the light, relatively fragile nature of these tools, which show no clear evidence of having been set in handles, and the absence of the fractured, rounded and abraded areas typical of violent percussion, they may have not have been used as picks in the strict sense. Rather, they may have been tools (employed for only a short period) for use with clayey sediments, but which from time to time came into non-violent contact with harder rock. Given the diversity of lithologies recorded at the Casa Montero context (Bustillo and Pérez-Jiménez 2005; Bustillo *et al.* 2009) and the diversity of mining tools recovered (Capote *et al.*

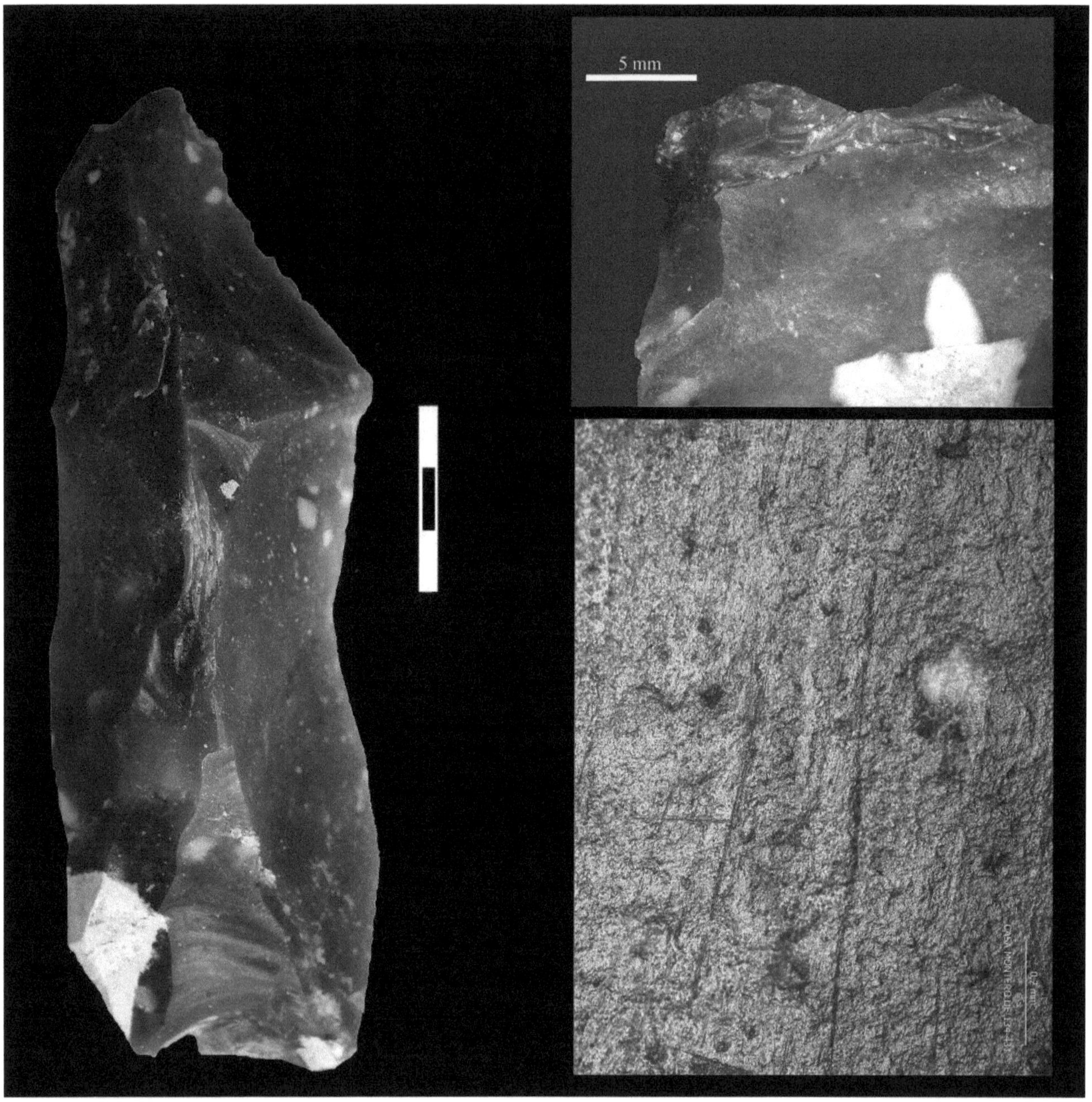

Figure 4. Pointed flake used as a mining tool with distal hinged or abruptly ended scars related to percussion, and detail of a compact-patterned polished plaque with longitudinal striations (100X).

2008; see also Capote in this volume), they may have been used in the extraction of flint blocks (to unseat them) for later fracturing with hammers and mallets, or for freeing them from their clay matrix following their fracturing.

5. Conclusion

The function of the antler and stone tools from Casa Montero is very difficult to analyse. This, plus the mining context of the site and the limited number of tools recovered, limits the making of any interpretations. Paradoxically, the majority of the supposed mining tools recovered show no evidence at all of having been used in the extraction of flint. Such is the case of the antler tools, which cannot be interpreted as picks, unlike those found in other mining contexts in France, the United Kingdom or Poland where such tools were profusely used. This might be related to the geology of the Casa Montero mine, where flint layers are found between others of clay and carbonates, quite unlike that of other European mining sites which usually have a limestone setting. In further support of the present results, at the Italian mine of Defensola, which has a calcareous setting, no antler mining tools have been found either (Galiberti 2005).

The antler tools recovered at Casa Montero cannot even be linked to mining-associated activities such as knapping; they were used neither as pressure flakers nor retouching tools. The situation is similar for the stone tools recovered; only the flint picks had an undoubted mining function. However, these tools cannot really be considered picks or wedges; without handles, they appear to have been used in the extraction of the raw flint material from its background matrix.

Some tools seem to have been used in activities complementary to mining, such as the working of wood and the repair and maintenance of wooden tools (stakes, levers, handles etc.).

Clearly, the mining tools used at Casa Montero show great diversity (Consuegra *et al.* 2004; Capote *et al.* 2008), including mallets and hammers made mainly from quartzite (see Capote in this volume), the flint tools described in the present work, and wooden stakes (detected via their imprints left on the clay walls of the mine shafts). The association between the diversity of lithologies and mining tools in Casa Montero suggests the specialisation of mining tools according to the nature of the matrix to be excavated (clays, carbonates, flint). This hypothesis, however, needs to be rigorously tested by means of experimental works and use-wear analyses. In this respect, future work should include the examination of the imprints conserved in the walls of the mine shafts. Silicon casts have been made of some of these imprints, which should allow virtual microscopic examinations of the supposed wooden stakes.

Finally, the analysis of function of some items showed that subsistence activities suggestive of habitation were undertaken at the site, such as the harvesting of cereals or butchering. Bone rings have also been found in the past (Yravedra *et al.* 2008). Although no evidence of habitation has been found in the immediate vicinity of Casa Montero, these data show there must have been a settlement close by, or that small groups travelled to the area and stayed for short periods while they extracted flint in order to produce cores for their ulterior transformation into blades.

Acknowledgements

The authors thank Susana Consuegra, Pedro Díaz-del-Río and the remaining researchers of the Casa Montero project for having entrusted us with the materials discussed in this work. We also thank Yolaine Maigrot for providing us with the pictures showing the traces of use made on antler picks during experimental work. Finally, thanks are due to Antoni Palomo for allowing us to examine a set of antler pressure flakers and retouching tools used in experimental stone knapping.

References

Boguszewski, A. 1995. Les outils en bois de cerf de la mine néolithique de silex à Sèvres (Haut-de-Seine), in J. Pélegrin and A. Richard (eds.), *Les mines de silex au Néolithique en Europe*, 107-113. Paris, C.T.H.S.

Bostyn, F. and Lanchon, Y. (eds.) 1992. *Jablines, Le Haut Château (Seine-et-Marne) : une minière de silex au Neolithique.* Paris, Documents d'Archéologie Française 35, Maison des Sciences de l'Homme.

Bostyn, F., Cayol, N., Giligny, F., Lo Carmine, A. and Maigrot, Y. 2005. *Creusement expérimental d'un puits d'extraction de silex sur la minière de Flins-sur-Seine (Yvelines).* Les Mureaux, Mémoires et Travaux du Paléoscope 1, Association L'Homme Retrouvé.

Bustillo, M. A., Castañeda, N., Capote, M., Consuegra, S., Criado, C., Díaz-del-Río, P., Orozco, T., Pérez-Jiménez, J. L. and Terradas, X. 2009. Is the macroscopic classification of fl int useful? A petroarchaeological analysis and characterization of fl int raw materials from the iberian Neolithic mine of Casa Montero. *Archaeometry* 51 (2), 175–196.

Bustillo, M. A. and Pérez-Jiménez, J. L. 2005. Características diferenciales y génesis de los niveles silíceos explotados en el yacimiento arqueológico de Casa Montero (Vicálvaro, Madrid). *Geogaceta* 38, 243-246.

Capote, M., Castañeda, N., Consuegra, S., Criado, C. and Díaz-del-Río, P. 2008. Flint Mining in Early Neolithic Iberia: a Preliminary Report on 'Casa Montero' (Madrid, Spain), in P. Allard, F. Boston, F. Giligny and J. Lech (eds.), *Flint Mining in Prehistoric Europe Interpreting the archaeological records*, 123-137. Oxford, BAR, British Archaeological Reports International Series 1891.

Castañeda, N., Capote, M., Criado, C., Consuegra, S., Díaz-del-Río, P., Terradas, X. and Orozco, T. 2008. Las cadenas operativas líticas de la mina de sílex de Casa Montero (Madrid), in M. S. Hernández Pérez, J. A. Soler Díaz and J. A. López Padilla (eds.), *IV Congreso del Neolítico en la Península Ibérica*, vol. 2, 231-234. Alicante, Museo Arqueológico de Alicante.

Clemente, I. 1997. *Los instrumentos líticos de Túnel VII: una aproximación etnoarqueológica.* Madrid, Treballs d'Etnoarqueologia 2, CSIC.

Consuegra, S., Gallego, M. M. and Castañeda, N. 2004. Minería neolítica de sílex de Casa Montero (Vicálvaro, Madrid). *Trabajos de Prehistoria* 61 (2), 127-140.

Díaz-del-Río, P., Consuegra, S., Castañeda, N., Capote, M., Criado, C., Bustillo, M. A. and Pérez-Jiménez, J. L. 2006. The earliest fl int mine in Iberia. *Antiquity* 80 (307) (http://www.antiquity.ac.uk/projgall/diazdelrio307/).

Díaz-del-Río, P., Consuegra, S., Capote, M., Castañeda, N., Criado, C., Vicent, J. M., Orozco, T. and Terradas, X. 2008. Estructura, contexto y cronología de la mina de sílex de Casa Montero (Madrid), in M. S. Hernández Pérez, J.

A. Soler Díaz and J. A. López Padilla (eds.), *IV Congreso del Neolítico en la Península Ibérica*, vol. 1, 200-207. Ali-cante, Museo Arqueológico de Alicante.

Galiberti, A. 2005. *Defensola. Una miniera de selce di 7000 anni fa*. Siena, Protagon Editori Toscani.

Galiberti, A., Sivilli, S. and Tarantini, M. 2001. La mineria neolitica della Defensola (Vieste-Foffi a): Lo stato delle ri-cherche. *Origini* XXIII, 85-100.

Gibaja, J. F. 2003. *Comunidades Neolíticas del Noreste de la Península Ibérica. Una aproximación socio-económica a partir del estudio de la función de los útiles líticos*. Oxford, BAR Publishing, British Archaeological Reports In-ternational Series 1140.

Gibaja, J. F. 2008. La función del utillaje lítico documentado en los yacimientos neolíticos de Revilla del Campo y La Lámpara (Ambrona, Soria), in M. Rojo, M. R. Garrido and I. García Martínez de Lagrán (eds.), *Paisaje de la memoria: Asentamientos del neolítico antiguo en el Valle de Ambrona (Soria, España)*, 451-493. Valladolid, Arte y Arqueología 23, Universidad de Valladolid.

Gurina, A. 1976. Гурина, А. 1976. *Древние кремнедобывающие шахты на территории СССР [Minas de sílex prehistóricas en el territorio de la URSS]*. Leningrado, Nauka.

Ibáñez, J. J., Clemente, I., Gassin, B., Gibaja, J. F., González, J. E., Márquez, B., Philibert, S. and Rodríguez, A. 2008. Harvesting technology during the Neolithic in South-West Europe, in L. Longo and N. Skakun (eds.), *'Prehistoric Technology' 40 Years Later: Functional Studies and the Russian Legacy*, 183-196. Oxford, BAR Publishing, British Archaeological Reports International Series 1783).

Jeudy, F., Jeunesse, C., Monnier, J. L., Pelegrin, J., Pétre-quin, A. M., Pétrequin, P. and Praud, I. 1995. Les carrières néolithiques de Plancher-les-Mines (Haute-Saône). Exem-ples d'une approche intégrée, in J. Pélegrin and A. Richard (eds.), *Les mines de silex au Néolithique en Europe*, 241-280. Paris, C.T.H.S.

Lech, J. 1997. Remarks on Prehistoric Flint Mining and Flint Supply in European Archaeology, in A. Ramos-Millán and M. A. Bustillo (eds.), *Siliceous Rocks and Culture*, 611-637. Granada, Monográfi ca Arte y Arqueología, Uni-versidad de Granada.

Maigrot, Y. 2003. *Etude technologique et fonctionnelle de l'outillage en matières dures animales. La station 4 de Chalain (Néolithique fi nal, Jura, France)*. Unpublished PhD thesis, Université de Paris I.

Russell, M. 2000. *Flint Mines in Neolithic Britain*. Gloucestershire, Tempus Publishing Ltd.

Teno, D. and Delgado, M. 2002-2003. Uso y efectividad de "picos mineros" de asta. *Boletín de Arqueología Experimental* 5, 20-25.

Terradas, X. 2002. Los contextos de producción lítica y las actividades extractivas de materias primas minerals en sociedades cazadoras-recolectoras prehistóricas, in J. M. Mata-Perelló and J. R. González (eds.), *Primer simposio sobre la minería y la metalurgia antigua en el Sudoeste europeo*, 51-60. Manresa, Universitat Politècnica de Catalunya.

Terradas, X. and Clemente, I. 2001. La experimentación como método de investigación científica: aplicación a la tecnología lítica, in. L. Bourguignon, I. Ortega and M. C. Frère-Sautot (eds.), *Préhistoire et approche expérimentale*, 89-94. Montagnac, Editions Monique Mergoil.

Yravedra, J., Maicas, R., Díaz-del-Río, P. and Consuegra, S. 2008. Anillos para una mina. Industria ósea y fauna de la mina de sílex neolítica de Casa Montero (Madrid), in M. S. Hernández Pérez, J. A. Soler Díaz and J. A. López Padilla (eds.), *IV Congreso del Neolítico en la Península Ibérica*, vol. 2, 240-247. Alicante, Museo Arqueológico de Alicante.

Prehistoric flint mines of the gargano: an overview

Massimo TARANTINI, Attilio GALIBERTI and Fabrizio MAZZAROCCHI

Abstract

For years, flint mining archaeology in the Gargano focused on the ancient Neolithic mine of Defensola, a consequence of its dimensions, antiquity and state of preservation. However, non-systematic surveys led to the discovery of a network of mines, particularly involving the northeastern part of the Gargano. Research is now concentrated on the entire range of Gargano prehistoric flint mines. To date, 18 mining sites have been identified. Extraction activities began in the early 6th millennium BC, coinciding with the Neolithisation process, and concluded at the end of the Copper Age, about half way through the 3rd millennium BC. Two mining systems can be distinguished: 1) sub-horizontal mining that exploited compact formations – an activity that required technological skill; such mining was characterized by the involvement of sometimes extremely large underground areas, 2) and vertical shaft mining; this was usually practised in formations affected by tectonic activity or which were crumbly, and involved access shafts leading to single, small, subterranean chambers. The mining system chosen appears to be closely linked to geo-morphological factors. However evidence also exists that social and cultural factors influenced the choice made, with different extraction systems being preferred at different times.

Keywords

Gargano. Neolithic. Copper Age. Sub-horizontal mines. Shaft. Mining systems. Geophysical survey.

1. The Gargano Prehistoric Flint Mines Project: a second phase of research

The Gargano Prehistoric Flint Mines Project originated in 1981 following the discovery of the Defensola flint mine as a consequence of earthmoving for construction purposes. However, systematic investigations of the Gargano mines only began in 1986, coordinated by Attilio Galiberti.

Flint mines had been known to exist in the Gargano since the 1930s (Rellini *et al.* 1934), but they did not become the subject of systematic study. Since all of these were Copper Age mines, it was believed that flint mining activities in the Gargano had taken place only during this period. However, the discovery of impressed wares at Defensola showed flint mining to have begun in the ancient Neolithic. Radiocarbon dating certified Defensola to be the earliest Neolithic flint mine in Europe, predating the beginning of mining in the rest of Europe by almost a thousand years; until then, the earliest known mine in Europe was Tomaszòw I, in Poland, dated to 6260±60 BP (Schild 1995). This dating closely linked the flint mines of the Gargano area with the Neolithisation process.

European prehistoric mining complexes generally consist of a series of (sometimes spectacular) shafts (e.g., see the Grimes Graves, Spiennes, Krzemionki and Casa Montero sites). A peculiarity of the Defensola mine is that excavation started from entrances opened on the slope of a hill, to develop in sub-horizontal manner.

It quickly became clear that the Defensola mine was extremely well preserved. It is still partially accessible without the need for archaeological excavation; Neolithic materials have been found inside in their functional positions, e.g., a vessel with traces of food remains left in a resting area far from the extraction points (Figure 3.6), or a stone lamp on the floor near an extraction point that was still active when the mine was abandoned. This abandonment may have been rapid, the consequence of a seismic event - an earthquake - that caused the collapse of the entrance areas. The time of abandonment is now dated to 5380-5200 BC (Figure 2).

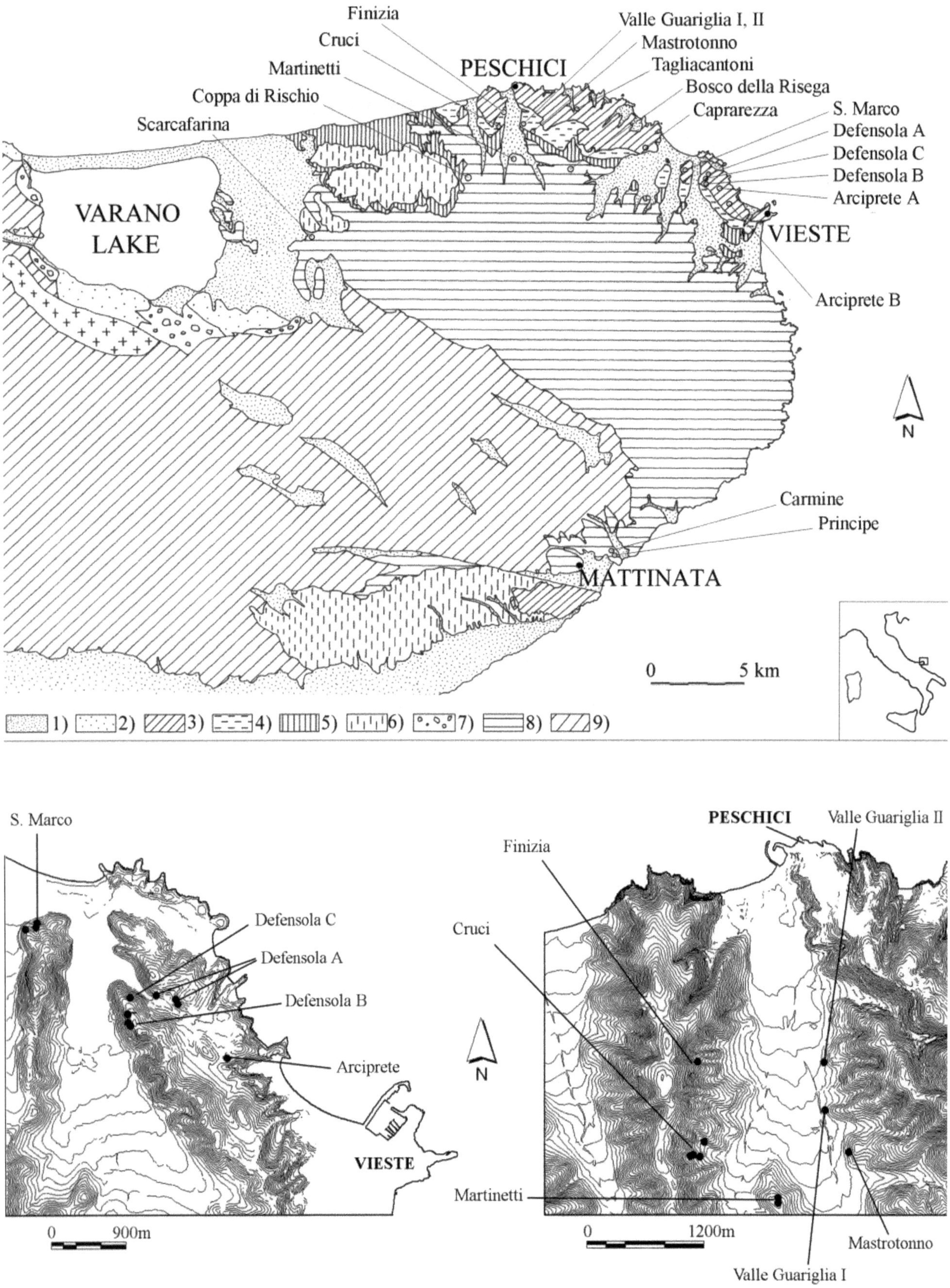

Figure 1. 1) Geological map of the Gargano with location of the flint mines. 1 Quaternary; 2 Mio-Pliocene; 3 M. Saraceno sequence; 4 Scaglia; 5 Scisti a Fucoidi; 6 M. Acuto e M. S. Angelo formations; 7 Breccia di Cagnano; 8 Maiolica; 9 M. Sacro sequence (geological data from Bosellini et al. 1999; Morsilli et al. 2004). 2) Mine locations in the Vieste district. 3) Mine locations in the Peschici district (drawing by M. Tarantini; topographic elaborations by G. Corrente).

For years, researched focused on the study of the Defensola mine as a single entity, but in 2005 a monograph was published (Galiberti 2005) marking the end of this phase. A second phase of research is now in progress, which, along with continued research at the Defensola mine, started by surveying the area for new mining sites. The results of those preliminary surveys were published in the mid 1990s (Di Lernia *et al.* 1997; Basili *et al.* 1995). However, only since 2001, starting with the PhD thesis of one of the present authors (Tarantini 2005), has any systematic investigation of the identified mines been undertaken. All the accessible underground parts of these mines have now been explored and surveyed, and more attentive surface surveying has identified many new entrances. The diachronic perspective adopted has revealed important changes in the extraction methods used over time (the Gargano affords the opportunity of analysing mining dynamics over a time span of some 3500 years, from the beginning of 6th to half way through the 3rd millennium BC) (Tarantini 2007, 2008).

Since 2006, the Gargano Prehistoric Flint Mines Project has been organized along different research lines thanks to funding from the University and Research Ministry, which recognises the project to be of national interest. Field and laboratory investigations (some in their initial phases) are now concentrated on the following areas:

1. The study of the geological and morphological context, age, and the mining techniques used at each mining complex.

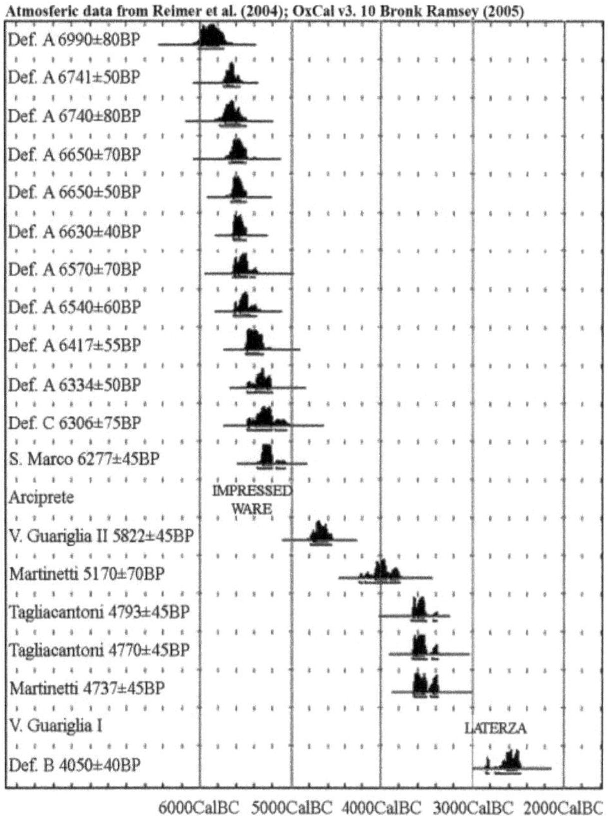

Figure 2. Chronological or cultural data available for the Gargano prehistoric flint mines.

2. Geophysical surveys (closely tied to research area 1) orientated towards determining the internal layout of each sub-horizontal mining structure for which only the entrance is known, or to verify the presence of new mining structures when only a few shafts are known. These geophysical surveys are performed in different ways, including the use of ground penetrating radar, electrical tomography and electromagnetic induction. The objective is to integrate different results while testing each technique. The final aim is to be able to understand large mining areas using non-destructive analytical systems (Tarantini *et al.* 2010).

3. Research on mining tools and other lithic materials in the mining context. These mining tools - picks and mallets - are made of coarse flint. With the exception of Defensola A, they are generally surface collected. Their number varies notably from mine to mine, from very few to hundreds of units. The picks and mallets of the Neolithic and Copper Age contexts show marked differences. Lithic studies are limited to analysing material coming from two workshop areas strictly associated mining activities: one in the interior of Defensola A, the other at one of the entrances of Defensola B. The aim is to identify the purpose of production following a technological approach coupled with use-wear analysis.

4. The study of the regional context of mining activities. These mines, of course, are part of a broader archaeological landscape, and are a very important component of the general organisation of the territory in the Neolithic and Copper Age. This is one of the most critical lines of research since, although a great tradition of surveying exists in the Gargano, it has never been systematic. In addition, only a few excavations have been undertaken, and all quite some time ago. The aim is to improve our scattered data by the use of a geographical information system.

5. To track the movement of flint away from the Gargano. The mined flint had a local use value, but surely also an exchange value. To trace the circulation of flint out of the Gargano, an approach based on its geochemical characterisation has been adopted. The choice of this method was determined by the difficulty of visually characterising Gargano flint, a consequence of its variability. All the Gargano flint-bearing formations have been sampled, and about a hundred have been analysed for a broad spectrum of chemical elements using inductively coupled plasma mass spectroscopy (ICP-MS), inductively coupled plasma optical emission spectroscopy (ICP-OES), and X-ray fluorescence spectroscopy. The identification of specific geochemical markers (total concentrations of trace elements, distribution models of elements coherent from a geochemical point of view, isotope ratios) will allow the characterisation of Gargano flints and help determine the provenance of flint samples from Neolithic and Copper Age settlements in central and southern Italy.

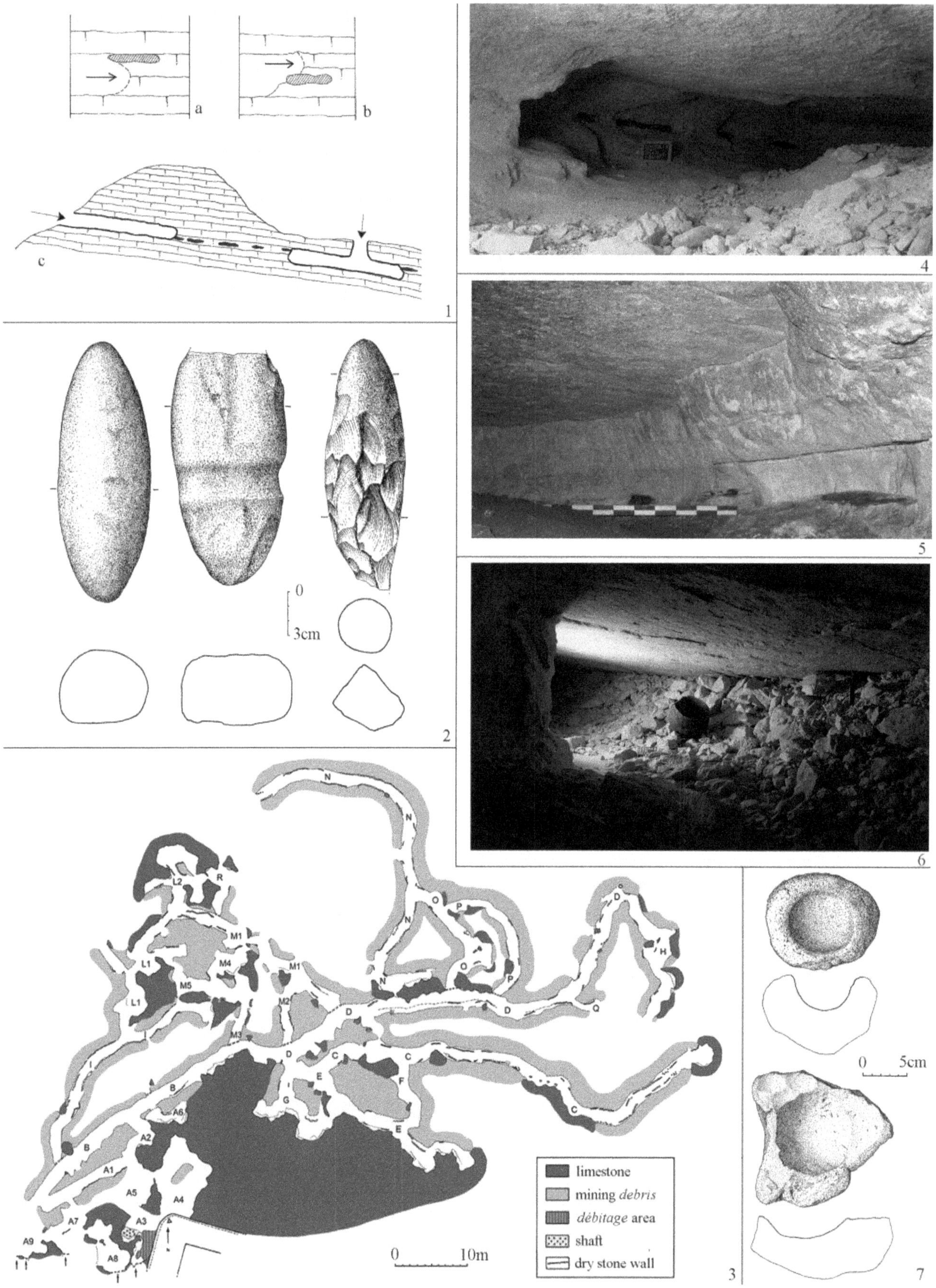

Figure 3. Sub-horizontal mining system. 1) Typology of extraction structures in compact formations (c) and extraction techniques for flint nodules in compact formations: (a) 'by collapse' and (b) 'in steps'. 2) Picks from Defensola A. 3) Defensola A plan. 4) An extraction step in Defensola A. 5) The interior of the S. Marco mine. 6) A vessel left in a resting area in Defensola A. 7) Limestone lamps from Defensola A (drawings and photographs by A. Galiberti and M. Tarantini).

2. Distribution, geological context and chronology of the mines

Eighteen mining sites are currently known. Some were discovered along with their original entrances, such as the sub-horizontal mines of Defensola B (original entrances are commonly almost completely covered by slope sediments). Sometimes - for example at S. Marco and at Defensola C, near Vieste - the original entrances were enlarged during modern times to create stables or storehouses; these cut across the mines, showing interesting sections of the ancient structures. Other mines were discovered as a consequence of earthmoving for construction, such as Defensola A and some shafts at Carmine, near Mattinata. Others were found during earthmoving for road works, such as Martinetti, Coppa di Rischio, Valle Guariglia I and II, Bosco della Risega and others at Principe. Research on these complexes has so far been limited to surveying those areas that can still be accessed directly, as well as the sections brought to light by earthmoving work. These sections sometimes extend for some hundreds of metres.

Although it is sometimes difficult to gain an overview, all of these sites can be described as mining complexes since they consist of a number of mines in close proximity. In other words, there is never an isolated mine, even though there are not hundreds or thousands of adjacent shafts present as there are in other European contexts.

The mines in the Gargano are scattered over an extremely large area, but are especially concentrated in the northeastern part of the Gargano Promontory (Figure 1.1). This is because the flint bearing formations are concentrated to the North of the Varano-Mattinata line. From a geological point of view (Bosellini *et al.* 1999; Morsilli *et al,.* 2004) the supporting structure of the Gargano Promontory is a Jurassic carbonate platform (the so-called *Piattaforma carbonatica apula*), which acted as a reef complex. Since the beginning of the Cretaceous, the clinostratified surface of the platform has been covered by a basin unit of white micritic limestone – the Maiolica formation - that formed the spectacular reef between Mattinata and Vieste and which shows different levels of tectonic alteration. The Maiolica is abundant in tabular and nodular flint and a number of mines were opened here.

An interruption in the sedimentation process linked to the platform is represented by the marl formation of Scisti a Fucoidi. Platform sedimentation started again at the beginning of the upper Cretaceous and went on for almost all the Palaeocene. This period saw the formation of the Scaglia, a formation of white and powdery micritic limestone with tabular and nodular flint. Its thickness and width are limited but some mines were opened here.

Finally, during the Eocene, the platform margin collapsed and a Nummulite platform later formed. The Nummulite limestone is limited to the area between Vieste and Peschici; it is rich in nodular flint of excellent quality. Numerous mines were opened in this compact formation. The contact between the Scaglia formation and Nummulite Limestone is clearly visible on the cliff at Vieste.

A series of isolated mining complexes and two major mining districts have been identified: 1) the district of Vieste, which covers an area of about 5 square kilometers (Figure 1.2), with mines cutting into the Nummulite limestone formation and that consist of a series of structures found on hill slopes (often fairly steep, as at Defensola A and B); and 2) the district of Peschici, which covers an area of 4 square kilometers (Figure 1.3), with mines cutting into the Scaglia and Maiolica formations; here a series of mining complexes are found in the Ulso Valley.

Some mines still present problems of dating and/or cultural attribution. However, 11 complexes have now been attributed to specific contexts by the radiocarbon dating of charcoal samples collected from the debris inside the mines and/or on the basis of pottery found at the surface (Figure 2). The beginning of mining activities has been dated to the early 6th millennium BC (Defensola A: 6990±80 BP, cal 2σ 6010-5720 BC), coinciding with the Neolithisation of southeastern Italy. Defensola A was active for about six centuries till 5380-5200 BC; other mines in the district of Vieste also date to this period (S. Marco and Defensola C). In addition, the Arciprete mine has been dated to the early Neolithic, thanks to the finding of impressed wares. Only one mine is dated to the 5th millennium BC (Valle Guariglia II). This suggests a decline in the intensity of mining activity, probably due to local population dynamics that are clearly reflected in the simultaneous depopulation of the nearby Tavoliere area, where the cycle of large Neolithic ditched villages (e.g., Masseria Candelaro village) comes to an end around this time (Cassano and Manfredini 2004).

Intense mining activity began again in the late Neolithic-early Copper Age, with the spreading of the Macchia a Mare facies (Martinetti, Cruci). It would seem to have ended in the final phase of the Copper Age, in the first half of the 3rd millennium cal BC (Defensola B: 4050±40 BP, cal 2σ 2850-2820, 2670-2470 BC).

3. Mining systems

In general terms, two types of mining technique can be distinguished in the Gargano: sub-horizontal mining and vertical shaft mining.

3.1. Sub-horizontal mining

The sub-horizontal mining system (Figure 3) exploits the geomorphology of the natural hilly environment, where flint-bearing limestone formations can be seen along the slopes as a sequence of compact, sub-horizontal or slightly sloping seams. The layer containing flint nodules is therefore easy to identify and can be mined in depth starting from different access points along the slopes.

In some cases, mining in this hilly environment leads to the formation of a characteristic cone of mine debris on the slope outside. These cones are sometimes entirely preserved, revealing a prehistoric mining landscape.

Inside the hill, the flint-bearing seams are mined following the joins between the layers of rock; this creates structures with a characteristic flat ceiling. Layers of limestone were removed in sufficient number to obtain a floor-ceiling height large enough to allow mining to proceed. In any event, the tunnels are rarely more than 60cm in height and therefore present considerable logistical difficulties. These difficulties increase for the archaeologist when the mines are partially filled by slope sediments, as at Defensola B and S. Marco. It should be remembered that all activities inside these mines required artificial lighting; this was provided by limestone lamps probably fed by animal fat (Figure 3.7).

The extraction techniques employed in these sub-horizontal mines entailed the almost unvarying use of a single model (i.e., the mining of individual seams starting from above, followed by an extraction step to remove the flint nodule exposed) (Figure 3.1b; 3.4). The only exception is that adopted in some specific contexts, where the nodules were extracted by 'collapse' (Figure 3.1a).

The stone tools used to cut through the rock were picks and mallets made of coarse flint. These show a remarkable degree of refinement, with part or all of their knapped surface having been hammered to obtain rounded shapes, which are more resistant to blows. Even the unknapped pieces were carefully worked and shaped. The picks are tapered and those found at Defensola A are 168mm long and weigh 860g on average; over half of these pieces have a narrowing in the middle, made by knapping or hammering, where they were joined to a handle (Figure 3.2).

The compact nature of the limestone formation is an essential safety feature of this mining system. A conservative estimate suggests that one of the two floors of Defensola A covered a surface area of at least 6500 square meters and was active for several centuries. Much of the limestone seam was removed to create large chambers supported by a few columns of rock and filled with large heaps of debris reaching up to the ceiling. This solution allowed for maximum exploitation of the limestone seam without compromising the stability of the roof.

The Defensola A mine consists of a network of tunnels of varying length that cut through previously accumulated debris heaped up to the ceiling. The main tunnel is 110m long and led to the extraction faces that were active at the time when the mine was abandoned. It takes about half an hour to cover this distance (Figure 3.3).

In the sub-horizontal system, the depth of excavation towards the interior of the hill ranges from about 20m in partially explored mines such as S. Marco, to over 100m in Defensola A (the only mine studied in depth). Recent geophysical surveys suggest that other mines are almost equally deep (Figure 4).

At Arciprete, another early Neolithic mine near Vieste, only one small entrance is known. To determine the size of the mine a ground penetrating radar survey was undertaken using a grid with 25m sides. This clearly showed an underground area corresponding to a sub-horizontal mine. One of the authors (F.M.) has later undertaken an electrical tomographic study of a 128m-long area to determine differences in electrical resistivity. Two anomalies were found, one about 33m long corresponding to an empty space (although it is unknown if this represents a mining space or a tectonic crack), and another about 55m long corresponding to a space full of conductive material such as *terre rosse*. This surely corresponds to a mine since it starts at the known mine entrance.

The Principe shaft, near Mattinata, has still not been dated nor has any cultural attribution been made. This is the only case in which a sub-horizontal excavation starts from a shaft. A ground penetrating radar survey on a grid of 25m sides has been performed and the data reveal a network of chambers and tunnels. A later electrical tomographic study involving a 128m long section was also undertaken, starting from a small and doubtful entrance on the hill slope (a fox den opened in mining debris). The anomalies corresponding to an empty space suggest a mine stretching for about 100m, with an entrance on the hill slope and at least another shaft along its internal extension (Tarantini *et al.* 2010).

If the geophysical survey results are reliable, the size of the Defensola mine is not exceptional; thus, in addition to the standardization of mining techniques, the size of mines is often comparable.

The long-running research undertaken at Defensola A allows a general hypothesis on the working of these mines to be proposed. Mining work was managed and planned for long-term exploitation, and systematically aimed at deep areas. This is known because large galleries were dug through the mining debris that had previously accumulated; these galleries provided more rapid access to the active extraction faces in the innermost parts of the mine. The manpower needed to construct these galleries should not be underestimated. The fact that some are bounded for most of their length by dry stone walls on both sides (in one case cemented with limestone mud) confirms the theory of the planned management of the mine. The evidence also suggests that extraction took place not in simple 'visits' to the extraction face, but involved relatively long stays (a day?) inside the mine. Certainly, there is evidence that meals were consumed inside the mine, and there appear to have been rest areas at more salubrious points away from the mine face; a number of intact pot-

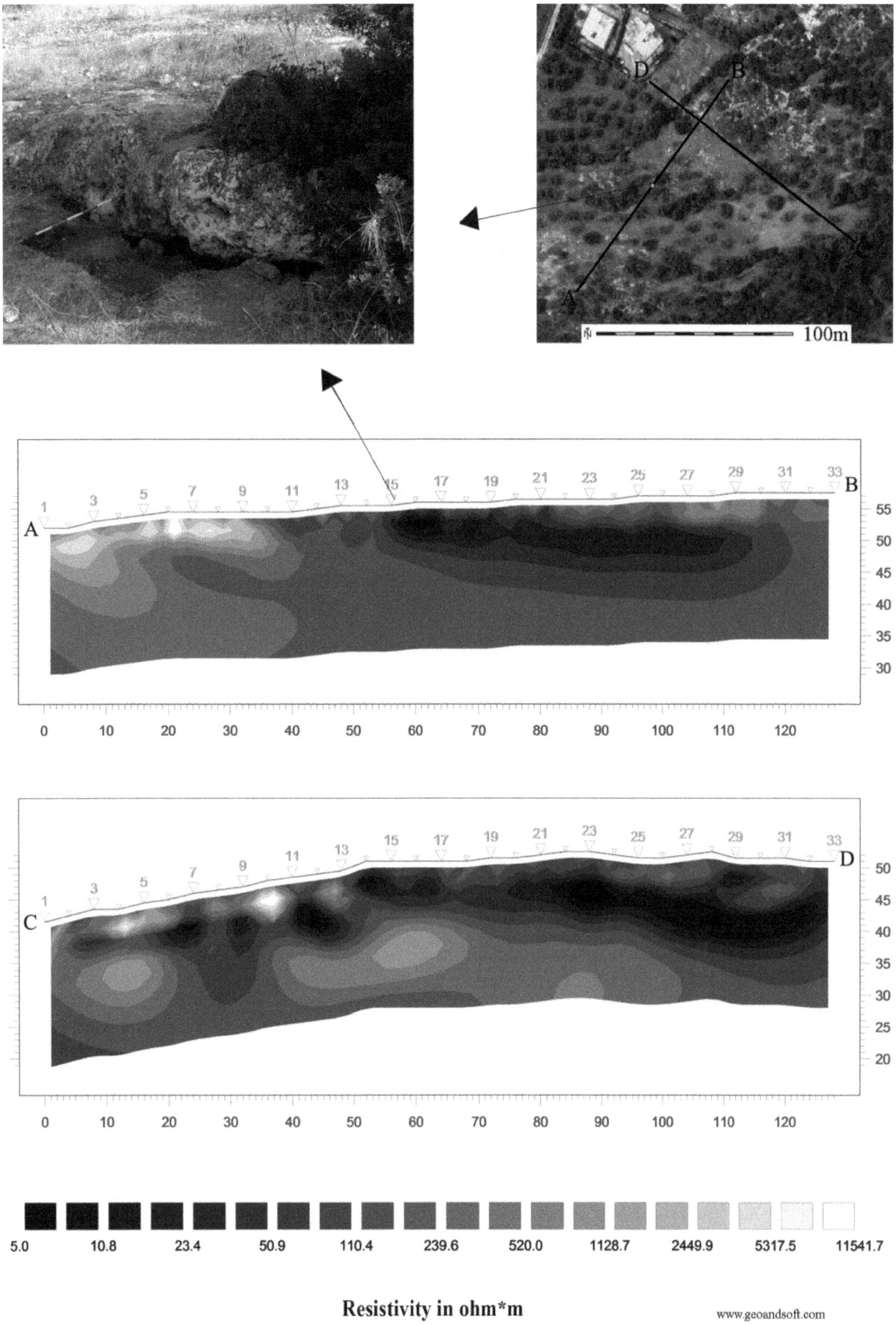

Figure 4. Electrical tomographies for the Arciprete mine.

tery vessels have been found in functional position in such places. These pieces of pottery were probably used to carry food, as indicated by the macroscopic and microscopic traces of organic residues on them (Di Lernia *et al.* 1993). Defensola A also has an area for flint knapping and the manufacture of tools used for activities secondary to mining proper, such as the construction or repair of pick handles, or the making of containers to transport debris and flint. These were likely made of wood, as suggested by use-wear analysis. Finally, this large mine was difficult to move around in and was confusing in terms of self-orientation; miners must therefore have undergone some form of 'initiation'.

The sub-horizontal mining system was used exclusively in the mines of the Vieste area, where it was used to exploit the Nummulite limestone formation, and in the Principe mine near Mattinata. It appears to be characteristic of the ancient and early-middle Neolithic, i.e., approximately from 6000 to 5200 cal BC. An extraction complex where sub-horizontal excavation was used is also attested to in the first half of the 3rd millennium BC (Defensola B), but here the standardised mining techniques of the ancient Neolithic were not employed; significantly, the geological formation involved is not compact limestone but a detrital and tectonized formation.

3.2. Vertical shaft mining

The second type of mining system used in the Gargano was vertical shaft mining (Figure 5). Vertical access to flint-bearing formations is the only option in flat terrain (which extends over large areas of central Europe). Extraction using vertical shafts is also the only option in formations that are heavily tectonized or crumbly; deep sub-horizontal mining would be impossible to carry out safely.

In the Gargano, vertical mining techniques were employed both in compact formations and, above all, in tectonized formations belonging both to the Scaglia and Maiolica formations. This type of mining system was also used in hilly environments. However, less information about this extraction system is available since in-depth research began only very recently. Some preliminary, general observations can, however, be made. First, it should be noted that the Gargano shafts are not very deep; the deepest is only three and a half metres (although it should be remembered that none has been excavated). Moreover, most of them are 'bell shafts' and generally the 'access' shaft is known but not the underground chamber. An example is a shaft at Carmine (Mattinata), which was discovered during earthmoving for construction (Figure 5.2); another, still intact and empty, can be found at Defensola B (Figure 5.1). The diameter of the shafts seems to be fairly standardized at around one metre. In a number of sites a single underground structure can be accessed both from a shaft and a horizontal tunnel dug into the slope, e.g., at Finizia (Figure 5.5) and Cruci. With the exception of the Principe site, where a shaft leads to a sub-horizontal mine dug into a context where the rock is compact, 'access' shafts lead to single subterranean chambers. These chambers are small and of variable height, though they are always greater in height than those of sub-horizontal mines. Unfortunately, however, their morphological layout can rarely be determined. In some mines there are traces of a specific excavation model with several shafts leading to a single underground chamber. However, some shafts are too narrow for people to pass through, and it is possible that these were used to allow light and/or air to enter. For example, the plan of the Bosco della Risega mine shows two shafts about a metre wide and another only 40cm wide (Figure 5.4).

Some of the debris from these mines was used to fill earlier shafts or mining structures, as in the case of Valle Guariglia I, which was found to be totally filled when it was discovered during earthmoving work. The rest of the debris was distributed over the land surface, and is made visible when cut through by roads. It is interesting to note, however, that the mining structures are not always fully filled: sometimes only the shaft is filled; such is the case of the Bosco della Risega mine. Generally speaking, the management of debris in this mining system required much less effort than that required in sub-horizontal mining, where debris had to be transported over long horizontal distances.

Vertical shaft mines also differ significantly from the sub-horizontal mines of the Neolithic in terms of the tools used to cut through the limestone. In contexts dating to after the 5th millennium, picks were more coarsely knapped, tools less carefully shaped, and the narrowing in the middle was created by broad grooves cut into longitudinal ridges. Frequently, these picks were obtained from thick blades, the ventral faces of which are often found intact (Figure 5.3).

No overall planimetric view of any mining contexts of this type is yet possible, although this will be obtained via the systematic geophysical surveys that have now begun. In some cases, however, the size of complexes can be estimated where roads have cut through them, leaving series of structures in close proximity visible. For example, at Martinetti, a road cuts through 12 mining structures over a distance of 80m, at Coppa di Rischio a road section about 180m long cuts through roughly 17 mining structures opened in a very powdery limestone, and at Principe-Carmine, near Mattinata, numerous structures can be seen along a stretch of road about 400m long, suggesting a mining area no smaller than 3ha.

According to the information currently available, the mining system with vertical access to flint-bearing formations is chronologically restricted to the Copper Age (*c.* 4000-2500 BC), with only one isolated structure (Valle Guariglia II) dating to the Middle Neolithic (LTL2715A: 5822±45 BP; cal 2σ: 4790-4550 BC).

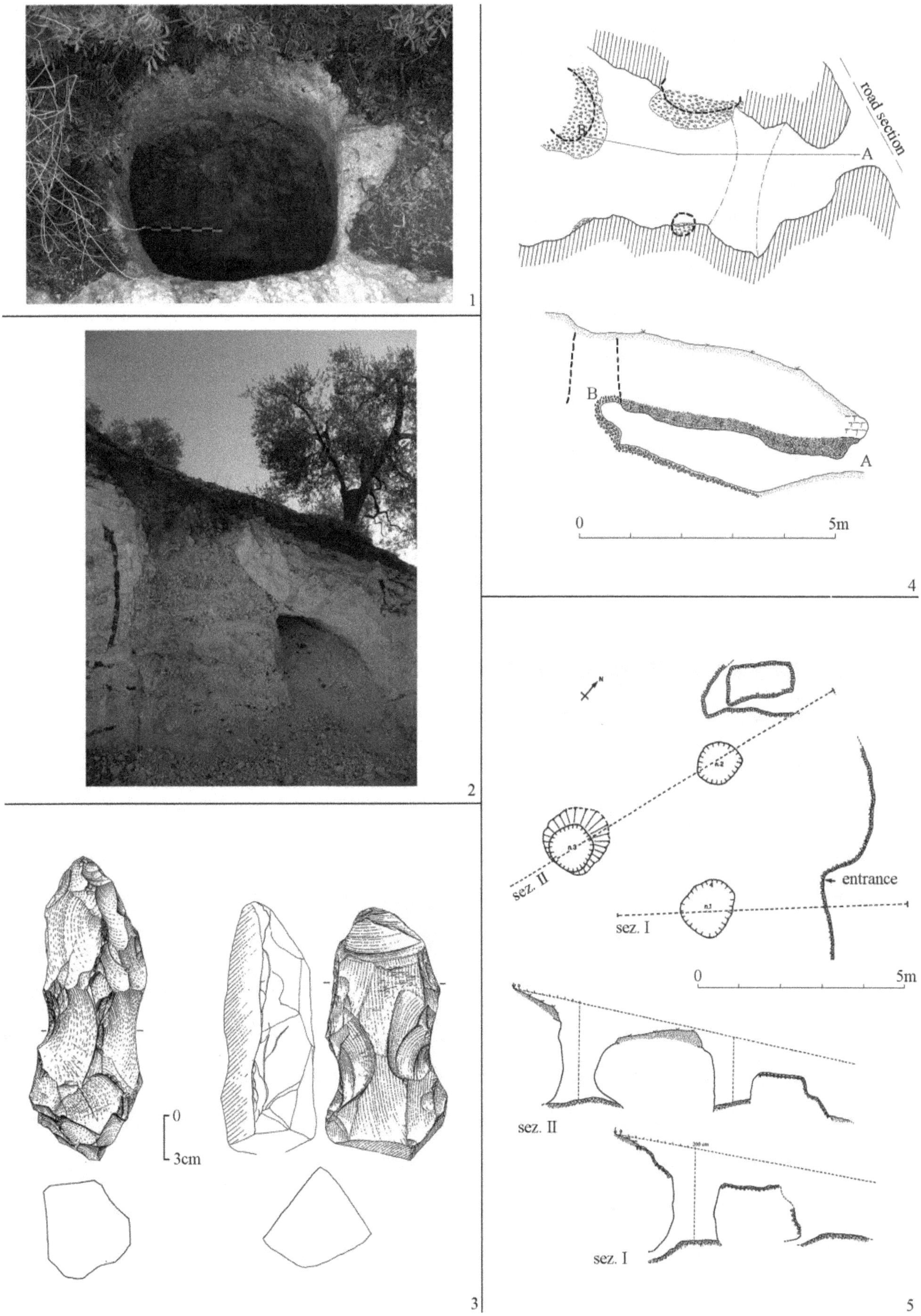

Figure 5. Vertical shaft mining system. 1) Defensola B, 2) one of the shafts at Carmine, 3) Picks from Valle Guariglia I and Cruci, 4) Coppa di Rischio mine, 5) Finizia shafts (drawings by A. Galiberti, O. Filippi and M. Tarantini).

4. Technical changes and the organization of mining work

The different mining systems adopted in the Gargano appear to be closely linked to geo-morphological factors, particularly the compactness of the rock formations present. However, evidence also exists that social and cultural factors influenced the mining system used, leading to different extraction systems being preferred at different times. For example, for the 6th millennium BC, the ancient and early-middle Neolithic, only sub-horizontal mines in compact formations are found. In contrast vertical shafts and mines dug mostly, but not exclusively, into formations affected by tectonic activity, began to appear from the 5th millennium. In the middle-final Neolithic and during the Copper Age (4th and 3rd Millennium BC) this mining system was employed almost exclusively.

For over half a millennium, during the ancient and middle-initial Neolithic, the constant use of a highly standardized mining model can be observed. In the 5th millennium, a simplified system became established, adapted in an opportunistic way to formations of varying compactness. Unlike Neolithic mines, with individual structures such as Defensola A showing continuous use over several centuries, Copper Age mines were smaller and their stability often precarious: this suggests that each mine represents a single extraction event.

This new extraction system required much less technical know-how than that needed during the previous period, and probably went hand-in hand with the way of organizing mining work. In the ancient and early-middle Neolithic, at least judging from Defensola A, mining work required a high degree of specialization and a specific knowledge of the mine environment (given the size of mines). In any event, on the basis of archaeological data from contemporary contexts in central and southern Italy, the existence of full-time specialists can be ruled out. Rather, the miners might be described as *part-time specialists*. In other words, mining activities were carried out by men who can be described as specialists because of the technological skills they possessed, but who worked on an occasional and irregular basis, possibly in accordance with the cyclical organisation of production activities characteristic of farming societies. It is important to stress that all known ethnographical data on New Guinean and Australian societies with a comparable level of technology and socio-economic organisation suggest that mining activities were never continuous in nature (De Grooth 1991, 154; Petrequin and Petrequin 2002, 361). The observations made at the Defensola mine also suggest a parallel between the working methods used at the Neolithic mining structures of the Gargano and one of the two trends identified by ethno-archaeological research in Irian Jaya. Here the skills required for undertaking extraction activities (and the corresponding rituals) were complex and were passed on within groups of specialists (Petrequin and Petrequin 2002, 370).

In the Copper Age, with the establishment of vertical shaft mines generally of modest depth and size, the technology required was significantly more simple. This probably made flint extraction work less specialized, which might indicate a different organisation of mining work. No longer, was it the job of group of people with highly specialized skills as in the Neolithic, but rather – once again with reference to the trends identified for the Petrequins in Irian Jaya – anyone could have been a potential flint extractor, with work organized in a manner reminiscent of the collective activities of a group of farmers. In the same period, the Gargano saw a general typological and structural modification of lithic production. Compared to the Neolithic a greater typological variety is seen along with a higher overall degree of laminarity, the emergence of flaked artefacts (which later became established as one of the most important typologies), and a general attention to the profiling of fairly commonplace objects such as *tranchets* (Palma Di Cesnola 1987). Therefore, in the Copper Age, the extraction of flint seems to have been open to all, at least in terms of the technical skills required, with specialist skills shifting towards those of *débitage*.

References

Basili, R., Di Lernia, S., Fiorentino, G. and Galiberti, A. 1995. Review of prehistoric flint mines in the "Gargano" Promontory (Apulia, Southern Italy). *Archeologia Polona* 33, 413-434.

Bosellini, A., Morsilli, M. and Neri, C. 1999. Long-term event stratigraphy of the Apulia Platform Margin (Upper Jurassic to Eocene, Gargano, Southern Italy). *Journal of Sedimentary Research*, 69(6), 1241-1252.

Cassano, S. M. and Manfredini, A. 2004. *Masseria Candelaro. Vita quotidiana e mondo ideologico in una comunità neolitica del Tavoliere*. Foggia, Claudio Grenzi editore.

De Grooth, M. E. T. 1991. Socio-economic aspects of Neolithic flint mining: a preliminary study. *Helinium* 31(2), 153-189.

Di Lernia, S., Franchi, R. and Pallecchi, P. 1993. Manufacture characteristics, provenance problems and content residues: an archaeometric approach to the Neolithic pottery of the Defensola mine (Vieste, FG - Italy). *Quaternaria Nova* 3, 151-75.

Di Lernia, S., Fiorentino, G., Galiberti, A. and Basili R. 1997. Topography of Gargano mining sites between geological context and quarrying techniques: a preliminary investigation, in A. Ramos-Millán and M. A. Bustillo (eds.), *Siliceous rocks and culture. VI International Flint Symposium (Madrid 1991)*, 195-209. Granada, Universidad de Granada.

Galiberti A. (ed.) 2005. *Defensola. Una miniera di selce di 7000 anni fa.* Siena, Protagon.

Morsilli, M., Rusciadelli, G. and Bosellini, A. 2004. *The Apulia carbonate platform margin and slope, Late Jurassic to Eocene of the Maiella and Gargano Promontory: physical stratigraphy and architecture.* Field Trip Guide Book – P18 – 32nd International Geological Congress, Florence – Italy (August 20-28 2004). APAT, Rome.

Palma Di Cesnola, A. 1987. Studio sistematico del primo Eneolitico del Gargano. 1. Studi e considerazioni sulla facies di Macchia a Mare. *Atti 5° Convegno di Studi sulla Preistoria e Storia della Daunia* (San Severo, 1983), 85-113.

Petrequin, P. and Petrequin, A. M. 2002. *Ecologie d'un outil: la hache de pierre en Irian Jaya (Indonésie).* Paris, CNRS édition.

Rellini, U., Baumgaertel, E. and Leopold, H. M. R. 1934. Secondo rapporto preliminare sulle ricerche preistoriche condotte sul promontorio del Gargano (1932-33). *Bullettino di Paletnologia Italiana* 54, 1-64.

Schild, R. 1995. PL2, Tomaszòw, Radom Province. *Archeologia Polona* 33, 454-465.

Tarantini, M. 2005. *Miniere di selce e tecniche minerarie sul Gargano tra VI e III millennio a.C.* Unpublished PhD thesis, University of Siena.

Tarantini, M. 2007. Le miniere neolitiche ed eneolitiche del Gargano. Tecniche estrattive e dinamiche diacroniche. *Atti XXXIX r.s. Istituto Italiano di Preistoria e Protostoria* (Firenze 2004), vol. I, 343-353.

Tarantini, M. 2008. Changements techniques au IVe millénaire. Les mines de silex au Gargano (Italie) dans le contexte italienne, in M. H. Dias-Meirinho, V. Léa, K. Gernigon, P. Fouéré, F. Briois and M. Bailly (eds.), *Les industries lithiques taillées des IVe et IIIe millénaires en Europe occidentale*, 331-346. Oxford, BAR Publishing. British Archaeological Reports International Series 1884.

Tarantini, M., Mazzarocchi, F., Mondet, M., Rossi, C., Salvini, R. and Tessaro, C. 2010. Geophysical surveys on Gargano prehistoric flint mines. *Origini* 32, 161-187.

Two Flint Caches from a Lower-Middle Paleolithic Flint Extraction and Workshop Complex at Mount Pua, Israel

Ran BARKAI and Avi GOPHER

Absract

Two lithic caches were discovered during excavation of one of the tailing piles at the Lower-Middle Paleolithic extraction and workshop complex of Mt. Pua, Israel. These caches, both of which consisted of 13 items, including a Levallois core in each cache and a handaxe in the first one, were intentionally placed on top of the exhausted extraction front and covered by a massive cap stone. In this paper we provide a detailed description of the archaeological context of the caches, discuss the interpretations of lithic caches in the archaeological literature and conclude by arguing that they had a symbolic connection to the quarrying activity, the successful exploitation of an exhausted extraction front and the initiation of a new quarrying locality.

Keywords

Caches. Flint extraction. Mt. Pua. Lower-Middle Paleolithic.

1. Introduction

The survey of the summit of Mt. Pua in Northern Israel conducted in 1997-2000 revealed a Paleolithic surface-quarrying complex and hundreds of stone heaps strewn with knapped flint items (Barkai *et al.* 2002, 2006). The finds of the survey identified the site as belonging to the Late Acheulian (Lower Paleolithic) and/or early Mousterian (Middle Paleolithic) cultural complexes (Barkai *et al,*. 2002, 2006). The tailings (quarry debris heaps) are covered with flint nodules and Paleolithic artifacts such as tested nodules, cores, roughouts, blanks, knapped lithic waste material and shaped items ('tools'). Preliminary mapping of the site identified approximately 1500 tailing heaps (Figure 1), varying in size from <1 to >15 meters in diameter and from <0.3 to >3 meters in height. Most, if not all, of the extraction debris heaps lie adjacent to limestone outcrops containing flint nodules. Numerous flint nodules have eroded from the outcrop due to natural weathering processes. However, specific breakage patterns and impact marks observed on the outcrops, as well as massive hammerstones bearing impact marks, indicate human exploitation of the flint nodules using a method of extraction called 'surface quarrying' (e.g. Claris and Quartermaine 1989). Our preliminary reconstruction of the extraction techniques demonstrates that Paleolithic hominins took advantage of master joints in the limestone outcrops, expended these joints using massive hammerstones, smashed the limestone blocks, extracted the flint nodules and piled the extraction waste in proximity of the extraction front. Test pits excavated at two different heaps indicate that the tailings are placed on top of exhausted flint sources, covering exploited extraction fronts. Our interpretation relates this behavior to the organization of flint procurement and exploitation strategies practiced at the site. More specifically, we suggest that expended flint sources were intentionally covered to be marked as potential sites of future manipulation (Barkai *et al.* 2002, 2006, 2009).

This paper deals with the finds excavated at Pua Workshop heap No.3 (henceafter PW3, Figure 2) and focuses on two cache deposits recovered in a deep test pit. We thoroughly examine the archaeological context of these caches and discuss their significance in late Lower-Middle Paleolithic quarrying/production activities.

2. Fieldwork at the Mount Pua Quarrying complex

During the fieldwork at the Mount Pua Quarrying complex one large linear stone pile was excavated partially (PW3, Figure 2-3) and one small circular stone pile was excavated completely (Pua Workshop pile no. 100). The objective of

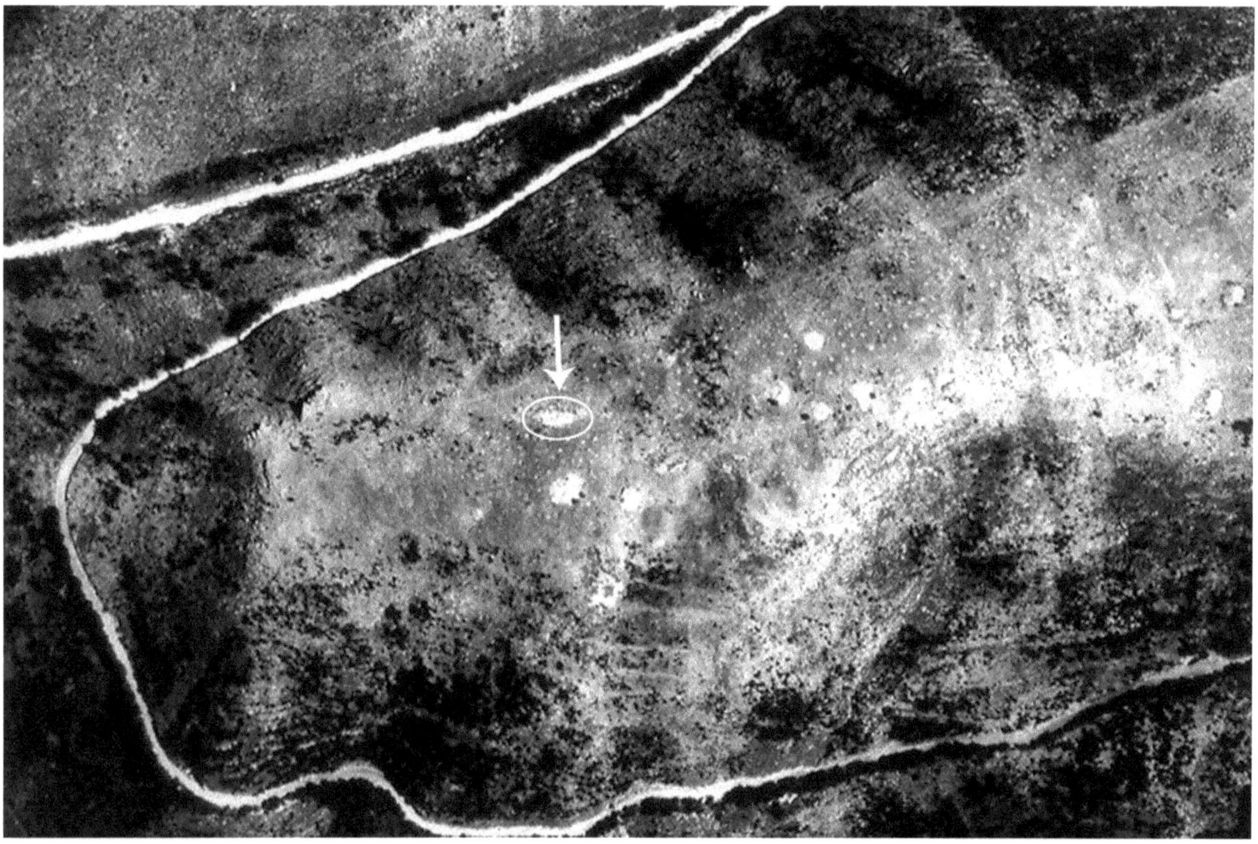

Figure 1. Mt. Pua flint extraction and workshop complex. White dots are extraction and reduction localities. Pua workshop pile No. 3. is marked by a circle and an arrow.

these initial excavations was to elucidate the formation and content of these waste piles, and to compare the characteristics of large and small tailings piles. This paper, however, deals with the test excavation of the large pile only.

The large linear tailings pile (PW3) is 30m long and 12m wide and is located in the northeastern part of the Mt. Pua extraction complex (Figure 1). It covers the area of some 350 square meters (Figure 2). A 2x2m grid was set on this pile, with one 4 square meters unit, G-24 chosen at random for excavation (Figure 3). Unit G-24 is located at the center of the northern third of PW3 and looks much like other parts of the pile. The excavation of the pile consisted in controlled removal of broken limestone blocks and the collection of all flint items from the limestone quarry debris, down to 90cm, at which point an exhausted flint extraction front was reached (Figure 3). After the removal of a massive stone block, two flint caches were discovered at a depth of ca. 70-90cm below the surface level of Unit G-24, topping the exhausted extraction front. Each of the stone caches included 13 large flint artifacts stacked one on top of the other. Each of the caches also contained a Levallois core and one cache contained a hand axe (probably a rejected bifacial roughout). The two caches also included cores, cortical flakes and large flakes. The archaeological context of these two lithic concentrations allowed them to be interpreted as caches purposefully placed on top of the exhausted quarry surface.

Figure 2. A close-up view at Pua workshop pile No. 3. Note a person (A.G.) as a scale.

PW3 as a whole was first systematically surface collected in 2x2m squares covering 120 square meters of this large-scale heap (squares D-I/20-24). These squares cover all the different parts of the heap (upslope, midslope and downslope). All flint items were collected including those that were not knapped. The study concentrated on knapped items—those bearing at least two scars in the case of nodules or cores and dorsal and ventral faces in the case of flakes/blanks. The unworked flint collected from these squares weig-

Figure 3. Excavation square G-24 at Pua workshop pile No. 3 at the close of the excavation.

hed 811kg. The kanpped flint assemblage from the surface collection was made up of 2699 items including cores (n=348, 13% of the assemblage), core trimming elements (n=86, 3%), cortical flakes (n=670, 25%), flakes (n=531, 20%), blades (n=42, 1.5%), naturally backed knifes (n=40, 1.5%), shaped items ('tools', n=385, 14%) and unclassifiable chunks of knapped flint (n=597, 22%).

The second stage of fieldwork included the excavation of the 2x2m sq. G-24 in the central part of heap PW3 (Figure 3), in order to examine the depth of the deposit and possibly reach the 'virgin soil'. Limestone blocks and waste material were removed in an attempt to 'peel' the heap from top to bottom – generally in horizontal spits. This, however, was difficult to accomplish due to eastward inclination of the heap. All unflaked (natural) flint was weighed (94.4kg for the whole excavated volume) while the flaked flint assemblage was studied and classified (Figure 4). At the top of the heap, limestone blocks of various sizes could be easily removed and flint was abundant. Some 25cm below surface flint quantities decreased. In the next 40cm the excavated volume had little flint, which the exception of the two caches treated in this paper (Figures 5-7). Below the caches the quantity of flint decreased sharply and ceased some 10-20cm lower, on a surface constituting of large bedrock surface covering 2/3 of the square's base area. Remnants of flint nodules are still attached to the limestone bedrock karrens, and apparently most of the flint nodules had been extracted from this part of the natural slope before the heap was formed.

The deposition of the heap seems to have occurred in stages and reached over 70-90cm in thickness in the area excavated, depending on the inclination. The sediment fill between stones in the lower part of the section is composed of red loam (*terra rossa*) – possibly washed in from up slope by water and/or, formed *in situ* from weathering of the karrens.

The finds from sq. G-24 are similar in nature to the surface finds described above but flint preservation is better. Another difference between the two assemblages is the presence of small waste artifacts (items smaller than 4cm) in the excavation which were completely absent in the surface collection of PW3. Such small items are absent from the two caches as well. The excavated lithics are presented in Figure 4 (including the two caches and the surface collection).

3. The archaeological context of the caches

The first cache (No.1) was found under a large block (ca. one meter long) in the northeastern part of Sq. G-24. It included a concentration of 13 flaked items piled one on

G 24	Primary Elements	Flake	Blade	Core Trimming Elements	Core	Natural Backed Knife	Chunk	Shaped items	Varia	Total
Surface Collection	34	33	5	4	12	4	40	17		149
Excavation (large items)	118	84	14	12	44	2	242	81		597
Excavation (items smaller than 4 cm)	45	83	6	3			221			358
Cache No. 1	5	3		1	2		1	1		13
Near cahce No. 1					1		1			2
Cache No. 2	3	3			3	1	1	1	1	13
Level 3 (around and below caches)	1	1			4		7	1		14
Total	*203*	*203*	*25*	*20*	*66*	*7*	*516*	*105*	*1*	*1146*
	17.71%	*17.71%*	*2.18%*	*1.75%*	*5.76%*	*0.61%*	*45.03%*	*9.16%*	*0.09%*	*100.00%*

Figure 4. Sq. G-24 Lithic assemblage (including surface collection and excavation).

top of the other in an area of less than ½ square meters. All items are large and the rest of the sediment around was sterile. Two additional artifacts were found in close proximity to the cache but are not necessarily part of this concentration. Since the discovery of this cache was unexpected, no pictures of this cache are available. Due to space limitations, in this paper we focus on the context of the caches and provide neither a detailed description nor illustrations of the items deposited in the caches.

Cache No. 1 includes five cortical flakes (120-404gram in weight, 76-111mm in length), three flakes (93-248g); one Levallois core made on a nodule (542g), one handaxe, most probably a roughout, made on a nodule (336g, 99mm in length, 86mm in width, 45mm in thickness); one flake core (507g); one core trimming element and one unclassifiable chunk of knapped flint.

The second cache (No. 2) was found 20cm lower, but not directly underneath cache No. 1, located slightly to the east of the first cache, in the northeastern corner of sq. G-24 on a some 30x50cm-large rock bench (Figure 5-7). Since the natural slope of the bedrock below the heap inclines from west to east, it appears that, although cache No. 1 is higher than the second one, they might belong to the same stratigraphic horizon, above the bedrock and below the stone heap. 13 items were found here as well, while the sediment around was sterile. The 13 items were piled one on top of the other representing a specific concentration (Figure 7). Cache No. 2 includes three cortical flakes (88-420g in weight), three flakes (90-280g), two large flake cores (780 and 1420g); one tested nodule (532g), one roughout (568g), one Levallois core (240g); one Naturally Backed Knife and one unclassifiable chunk of knapped flint. Very few flint items were found as the excavation proceeded to bedrock after the removal of the caches. Theses finds appear as Level 3 in Figure 4.

Both caches are very similar to each other and do not seem to represent a concentration resulting from knapping that took place at the spot since they were composed of only large items. It appears that the artifacts included in the caches were mostly selected according to their size and other specific properties such as production techniques or special significance (in the case of Levallois cores and the handaxe). It is clear that the artifacts in the two caches do not represent a single reduction sequence since all the waste material and by-products involved in their produc-

Figure 5. Cache No. 2 in Sq. G-24. The large flake at the top of the cache as apeared at the beginning of the exposure of this cache. The flake is marked by circle and an arrow.

tion are absent from the caches. The specific location of the caches rules out the possibility of a post-depositional contribution to the formation of the two caches. The large items could not have penetrated the dense heap and reach its bottom after the heap was formed. The possibility that the artifacts were present on the exhausted extraction front prior to the formation of the heap and remained undamaged under the pile during the formation of the heap in an unintentional manner seems rather unlikely. Thus we suggest that the two concentrations of flint artifacts are an intentional deposit of carefully selected items. In both cases the artifacts were piled one on top of the other and no other flint items, including very small fragments, were found in close proximity.

The caches are composed of relatively large flint items (Figures 6-7) and the first cache was completely covered by a massive limestone block. The second cache was deposited on top of the exhausted limestone outcrop (Figure 8) and thus both caches are directly related to massive limestone blocks, either from bottom or top. Both caches were sealed between the limestone bedrock underlying cache No. 2 and the limestone block covering cache No. 1. We suggest that these two caches were intentionally placed on top of the exhausted extraction front prior to the formation of the heap using waste material of the extraction process and products of the flint knapping process. The caches were covered by the large stone block and thus protected during the subsequent formation process of stone heap PW3 above them.

In summary, we present the sequence of events that led, in our opinion, to the formation of this special archaeological context:

Stage 1: An extraction front for flint quarrying was established at the specific location labeled as square G24 in our excavation grid. The extraction front was most probably much larger than that seen in the 4 square meters unit excavated by us at random.

Stage 2: Flint nodules had been extracted from this extraction front until it became exhausted. The flint nodules were most probably not reduced on top of the extraction front since no flaking waste material was left at that place.

Stage 3: The extracted flint nodules were reduced elsewhere and specific flint items were taken from the knapping location and brought to the exhausted extraction front. It is, of course, impossible to indicate whether the large items placed on top of the exhausted extraction front were actually produced from nodules previously extracted from this specific front or from nodules originating in other localities. The question whether knapped flint items were

Figure 6. Cache No. 2 in Sq. G-24. The pile of large items located at the bottom of the excavation.

brought back to the specific place where the raw material was extracted remains open.

Stage 4: The two caches, each containing 13 large flint items, were placed on top of the exhausted extraction front (Figure 8) and a large limestone block was placed on top of the upper cache. It is of course difficult to determine if the two caches were deposited simultaneously or whether the lower cache was placed earlier than the upper one. In any case, both caches were placed, the artifacts were piled one on top of the other and a cap stone covered the area of deposition.

Stage 5: Extraction limestone debri from flint quarrying conducted elsewhere (most probably in close proximity) was piled on top of the sealed caching locality. Knapped flint items and flint nodules were added as well. At the end of this process, that might have been multi-stages and of unknown duration, the two caches were covered by a large mass of stones up to a thickness of 90cm. Despite of the heavy covering mass, the caches were not damaged or displaced due to the protection by the cap stone.

Stage 6: The large heap was created covering a large area, embracing the caches at its bottom. We cannot tell if additional flint caches were placed in other parts of the PW3 front or rather we have been exceptionally lucky in encountering the two deposits placed below this huge heap. This, again, remains a question to be answered by future fieldwork.

4. Discussion and conclusions

Caching behavior provides an opportunity to study episodes of intentional human activity within a specific archaeological context. In this discussion we limit ourselves to caches comprised of lithic artifacts and do not consider caching behavior of other items or materials such as human remains, cultic objects or waste material/garbage. A more comprehensive study of caches is of course needed, but it is beyond the scope of this study. We prefer to focus on the rare discovery of Paleolithic stone caches deposited within a quarrying and workshop complex and discuss the possible significance of this context. We do hope that this will promote research of caching behavior in prehistory and its importance in understanding human behavior, decision making and cultural perception.

Caches of flint artifacts or flint raw material seem to be more abundant in post-Paleolithic archaeological contexts in the Levant and Europe (Neolithic and later, e.g. Astruc *et al.* 2003; Barzilai and Goring-Morris 2007; Bertola *et al.* 1997; Bradley 1987; Hamon and Quilliec 2008) than in earlier Acheulian or Mousterian sites. Whether this pattern indicates a diachronic increase in caching from the Paleolithic to the Neolithic or rather reflects research intensity, scale of excavation or just random discoveries requires further fieldwork and analyses.

The only other case known to us of a flint cache from a Lower Paleolithic context in Israel was found during Garrod's excavation at Tabun Cave some 80 years ago. In layer E at Tabun, 29 handaxes were cached near the cave entrance by the cave's wall (Garrod and Bate 1937). Unfortunately, neither detailed description nor photographs or drawings are available.

We are not familiar with any published studies of Lower Paleolithic stone caches in Europe, although the recent discovery from Sima de los Huesos in Spain might be relevant to our discussion. A finely flaked red quartzite handaxe was found in association with hominin remains, and the researchers suggest that both the handaxe and the human accumulation have symbolic significance (Carbonell and Mosquera 2006). Notwithstanding the fact that this is an isolated artifact and not a concentration, its special context might indicate a special-purpose deposition. Considering that a single handaxe was included in one of the caches described in this paper, this case of caching behavior that cannot be ignored when discussing Lower Paleolithic caching behavior from Mt. Pua.

As for the Early Stone Age of Africa, a 'stone cache strategy' has been suggested by Isaac (e.g. 1978) and Potts (1984; 1988). They claim that early Hominins employed strategic planning in their technological organization, anticipating future need of stone tools for carcass processing and transported raw materials or tools to specific locations for future use. While such behavior is indeed possible and the claim that artifacts were moved from place to place is not disputed, as far as we have understood, stone caches have not yet been discovered and the 'stone cache strategy' is not backed by archaeological data.

In this short review of Lower Paleolithic stone caches one cannot avoid mentioning the embarrassing incident of the Japanese site of Kamitakamori. The 'site' has received much attention due to the discovery of caches containing colorful handaxes claimed to be half a million years old, later to be exposed as a fraud planted by a archaeologist (for a comprehensive review see Kaner 2002; Normile 2002; Kobayashi 2004). This unfortunate case, however, should not cast a shadow over genuine lithic caches but rather reinforce the need for a careful and detailed description of such exceptional archaeological contexts.

Two major interpretations of the function and meaning of caching behavior prevail in the archaeological and anthropological literature. The dominant interpretation usually foregrounds functional and practical aspects, while the other suggests that caching had a ritual and/or symbolic purpose. The functional interpretation defines a cache as an act by artisans who had the intention of coming back and use the items but for some reasons did not. Potts (1994) for example suggested that already in the very early stages of tool making early hominids used caching as a strategy of secondary raw material storage in areas poor in raw material but important in their routes as hunter-gatherers. This means that raw material and/or tools storages are expected to be found at sites where certain scheduled activities such as seasonal hunting or movement of game herds took place. Caching in extraction sites or next to them is interpreted through the functional prism as the caching of a surplus to be collected and used in the future. Ethnographic studies further support this idea. Alyawara of Australia left extracted stone in the extraction site for future use when too much raw material was extracted, or the amount to be carried in one trip was excessive (Binford and O'Connel 1984). Another study of Australian aborigines demonstrated that after extraction and reduction, the unused blades were bundled together and buried at the quarry in caches to be recovered at a later date, but usually they were simply left on the surface at the place of production (Patton 1994). Most studies of Neolithic and later stone caches from Europe and North America mostly follow this line of argument, focusing on caching for future use as a reaction to unexpected danger, storage of surplus items or a way to retain the freshness of

Figure 7. Cache No. 2 in Sq. G-24. A close-up view of the concentration of large items.

the artifacts (e.g. Bertola *et al.* 1997; Bradley 1987, 1990; Hurst 2007; Lintz and Dockall 2002).

Another line of research highlights the special nature of caches and their context. Caching is seen as a ritual act reflecting world views – a way to define social and historical ties of a community or individuals to specific loci (Edmonds 1998). In the case of flint caches discovered within raw material extraction contexts, such as Mt. Pua, some scholars have argued that the act of nodule extraction could be seen as a transformation, a changing relationship between man and his environment (like marriage, house building, travel etc.) that had to be ritualized (Rudebeck 1998). It has been suggested that caching flint at quarrying and extraction sites might have had a symbolic objective of insuring land fertility and continued appearance of flint nodules and/or protecting the quarry men from dangers of their job and assuring their success. Caching might have taken place at the beginning or end of a quarrying operation (Cooney 1998). Ethnography supports the symbolic/ritual interpretation of human behavior in stone extraction sites, demonstrating world-wide examples of rites and beliefs associated with extraction of stone from the ground (e.g. Burton 1984; Jones and White 1988; Taçon 1991).

As for the interpretation of the two caches found in the extraction and workshop complex of Mt. Pua, we would like to begin by emphasizing the claim that 'distinction should be made between deposits which could be rather easily recovered, as hoards buried near or inside the settlements, and contexts where recovery of hoards would have been either difficult or impossible' (Bradley 1987, 351).

It is clear that the Mt. Pua caches belong to the second category of hoards whose recovery from the ground after deposition would be rather difficult or impossible, which prompts us to discard the functional interpretation. The description provided above of the two caches and circumstances of their deposition point, in our view, to an intentional deposition of well selected artifacts conducted in the course of an operation of flint quarrying and stone-tool production. The caches had been placed on top of an exhausted extraction front, most probably at the end of the process of extracting flint nodules from this specific location and just before the stage of backfilling that spot by quarrying debri from another, recently opened extraction front. We suggest that the caches mark the end of one, most probably successful extraction stage, and the initiation of a new flint quarrying stage.

At the moment we have no interpretation of the particular selection of large items for the caches and the deposition of 13 items in each. The fact that a Levallois core was included in each of the caches and the handaxe deposited in cache No. 1 deserves special attention. Handaxe production and Levallois technology are the most prominent techno-

Figure 8. The location of the two caches on top of the exhausted extraction front (marked by circles) at the bottom of Sq. G-24.

logical procedures conducted by Lower and Middle Paleolithic flint knappers in the region and some scholars have suggested that, on top of their functional properties, such significant objects must have had special social meaning (e.g. Kohn amd Mithen 1999; White and Ashton 2003). The two Levallois cores deposited in the Mt. Pua caches were shaped and reduced elsewhere. The Levallois blanks produced from these cores were not cached and the two cores are the only clear manifestation of this specific technology in the caches. So in the case of the Levallois cores, only the unusable remains of the Levallois technology, the exploited cores were deposited in the caches. It is hard to say whether the two Levallois cores were used to exhaustion. It is however clear that the continued production of Levallois blanks from these cores would have required an investment in reshaping the cores according to the Levallois concept. One might be bold enough to suggest that the two exhausted Levallois cores were deposited on top of the exhausted extraction front from which the raw material used for their production was extracted. As for the handaxe found in cache No. 1, it appears indeed to be a roughout discarded in a very early stage of production. This biface was shaped on a nodule using few bifacial blows, so in terms of their place within the lithic production sequence, the Levallois cores and the handaxe present two extremes – the beginning and the end of the knapping process. A detailed description of the rest of the items found in the caches is beyond the scope of this paper, since there is not enough space for illustrating each item. We hope to provide a description of all artifacts in the caches elsewhere. By way of generalizing, we would say that the rest of the components of the two caches are not different than the rest of the finds collected on the surface and in the excavation of heap PW3. The only clearly distinguishing feature of the items in the caches is their size, but a more detailed study of the artifacts might reveal other significant characteristics.

In conclusion, we would like to make it clear that the two caches found within the specific context of flint extraction and reduction is intentional. Ruling out the functional explanations of this caching behavior, we argue that these caches had a symbolic role connected to the quarrying activity--the successful exploitation of an exhausted extraction front and the initiation of a continued, new quarrying locality.

This case represents one of the earliest manifestations of caching behavior conducted for symbolic or ritual purposes. Since such behaviors are not commonly reflected in the archaeological record of the Lower and Middle Paleolithic periods, this is a rare opportunity to study aspects of behavior, decision making and world views of such early hominins.

References

Astruc, L., Abbes, F., Ibanez Estevez, J.J. and Gonzalez Onzalez, J. 2003. "Dépôts", "réserves" et "caches" de matériel lithique taillé au Néolithique précéramique au Proche-Orient : quelle gestion de l'outillage? *Paléorient* 29, 59-78.

Barkai, R., Gopher, A. and La Porta, P. C. 2002. Paleolithic landscape of extraction: flint surface quarries and workshops at Mt. Pua, Israel. *Antiquty* 76, 672-680.

Barkai, R., Gopher, A. and La Porta, P. C. 2006. Middle Pleistocene Landscape of Extraction: Quarry and Workshop Complexes in Northern Israel, in N. Goren-Inbar and G. Sharon (eds.), *Axe Age: Acheulian Toolmaking - from Quarry to Discard*, 7-44. Oxford, Equonox Publishers.

Barkai, R. and Gopher, A. 2009. Changing the face of the earth: Human behavior at Sede Ilan, an extensive Lower-Middle Paleolithic quarry site in Israel, in B. Adams and B. Blades (eds.), *Lithic Materials and Paleolithic Societies*, 174-185. Oxford, Blackwell.

Barzilai, O. and A.N. Goring-Morris. 2007. Blade caches in the southern Levant, In L. Astruc, D. Binder, and F. Briois (eds.), *La Diversité des Systèmes Techniques des Communautés du Néolithique Pré-céramique vers la Caractérisation des Comportements Sociaux,* 277–294. Proceedings of the 5th International Neo-Lithics workshop. Antibes, Éditions APDCA.

Bertola, S., Di Anastasio, G. and Peresani, M. 1997. Hoarding unworked flint within humid microenvironments. New evidence from the Mesolithic of the Southern Alps. *Prehistoire Europeenne* 10, 173-185.

Binford, L. and O'Connel, J. 1984. An Alyawara day: The stone Quarry. *Journal of Anthropological Research* 40, 406-432.

Bradley, R. 1987. Stages in the chronological development of hoards and votive deposits. *Proceedings of the Prehistoric Society* 53, 351-362.

Bradley, R. 1990. *The Passage of Arms: An Archaeological Analysis of Prehistoric Hoards and VotiveDdeposits.* Cambridge, Cambridge University Press.

Burton, J. 1984. Quarrying in tribal societies. W*orld Archaeology* 16, 234-247.

Carbonell, E. and Mosquera, M. 2006. The emergence of a symbolic behaviour: the sepulchral pit of Sima de los Huesos, Sierra de Atapuerca, Burgos, Spain. *Comptes Rendus Palevol* 5, 155-160.

Claris, P. and Quartermaine, J. 1989 The Neolithic quarries and axe factory sites of Great Langdale and Scafell Pike: A new field survey. *Proceedings of the Prehistoric Society* 55, 1-25.

Cooney, G. 1998. Breaking stone, making places: The social landscape of axe production sites, in A. Gibson and D.

Simpson (eds.), *Prehistoric Ritual and Religion*, 108-118. Phoenix Mill, Stroud, Gloucestershire, Sutton Publishing.

Edmonds, M. 1998. Sermons in stone: Identity, value and stone tools in later Neolithic Britain, in M. Edmonds and C. Richards (eds.), *Understanding the Neolithic of North-Western Europe*, 248-276. Glasgow, Cruithne Press.

Garrod, D.A.E. and Bate, D.M.A. 1937. *The stone Age of Mount Carmel.* Oxford, Clarendon Press.

Hamon, C. and Quilliec, B. 2008. *Hoards from the Neolithic to the Metal Ages*. Oxford, BAR Publishing, British Archaeological Reports Intrnational Series 1758.

Hurst, S. 2007. An analysis of caching behavior. *Lithic Technology* 31, 101-126.

Isaac, G. 1978. The food-sharing behavior of protohuman hominids. *ScientificAmerican* 238 (4), 90-108.

Jones, R. and White, N. 1988. Point blank: Stone tool manufacture at the Ngilipitji quarry, Arnhem Land, 1981, in B. Meehan and R. Jones (eds.), *Archaeology with Ethnography: An Australian Perspective*, 51-87. Dept. of Prehistory, Research School of Pacific Studies, The Australian National University.

Kaner, S. 2002. Trouble in the Japanese lower and middle palaeolithic. *Before Farming* 2/4, 1-19.

Kobayashi, T. (ed.) 2004. *Recent Paleolithic Studies in Japan. Proceedings for Tainted Evidence and Restoration of Confidence in the Pleistocene Archaeology of the Japanese Archipelago.* Tokyo, The Japan Archaeological Association.

Kohn, M. and Mithen, S. 1999. Handaxes: products of sexual selection? *Antiquity* 73, 518-526.

Lintz, C. and Dockall, J. 2002. The Spreen cache: A study of a prehistoric curated collection of broken tools from 41RN108, Runnels County, Texas. *Lithic Technology* 27, 13-37.

Normile, D. 2002. Japanese fraud highlights media-driven research ethic. *Science* 292, 34.

Patton, R. 1994. Speaking through atones: A study from northern Australia. *World Archaeology* 26, 172-184.

Potts, R. 1984. Home bases and early hominids. *American Scientist* 72, 338-347.

Potts, R. 1988. *Early Hominid Activities at Olduvai.* New York, Aldine de Gruyter.

Potts, R. 1994. Variables versus models of early Pleistocene hominid land use. *Journal of Human Evolution* 27, 7-24.

Rudebeck, E. 1998. Flint extraction, axe offering and the value of cortex, in M. Edmonds and C. Richards (eds.), *Understanding the Neolithic of North-Western Europe,* 312-327. Glasgow, Cruithne Press.

Taçon, P. 1991. The Power of Stone: Symbolic Aspects of Stone Use and Tool Development in Western Arnhem Land, Australia. *Antiquity* 65, 192-207.

White, M. and Ashton, N. 2003. Lower Palaeolithic core technology and the origins of the Levallois Method in North-Western Europe. *Current Anthropology* 44, 598-609.

A new Neolithic quarry complex at Har Gevim, Israel: An introduction

Avi GOPHER and Ran BARKAI

Abstract

In this paper we briefly present the flint quarry of Har Gevim in the Negev, Israel, assigned to Pre-Pottery Neolithic B period based on the lithics found on it. Drawing on the observations made during a short survey conducted in 2009, we sketch out the site's location and geology and describe its other features. In the last section of the paper we situate Har the Gevim quarry complex in the regional context and relate it to other Neolithic quarry sites.

Keywords

Flint quarrying. Southern Levant. Pre-Pottery Neolithic B.

1. Introduction

The flint quarry complex of Har (Hebrew for Mt.) Gevim was first discovered in the course of a desert excursion to the Arava area, south of the Dead Sea. The presence of hundreds of round stone feature within the massive stony cover of the mountain gave it an appearance of a strange ruin of some sort. Dr. Zeev Meshel of Tel Aviv University who was familiar with some of our previous work on several flint extraction complexes such as Mt. Pua, Sede Ilan and Ramat Tamar (Barkai *et al.* 2002, 2006, 2007, 2009; Gopher and Barkai 2006), decided to bring it to our attention when on a visit to Har Gevim he noticed that the surface of the site was strewn with flint nodules and other flint features.

During a short visit to the site in 2007 we concluded that it was a flint quarry complex most probably dating to the Pre-Pottery Neolithic period. The scale of the site, the abundance of natural flint sources on and around it and the many rounded-amorphous features we observed throughout most of its area attracted our interest and we decided to survey the site.

In early autumn of 2009, we carried out the first (several-days long) survey of the site's features and the surrounding area. Our preliminary observations are presented in this short paper, whose objective is to introduce this Neolithic flint quarry complex and its characteristics.

2. Location and geology

The quarry complex of Har Gevim is located some 60km south of the Dead Sea, east of Kibbutz Ein Yahav in the eastern Negev, on a series of three eroded small ridges cut by deep gullies and rather steep cliffs in the south and east (230-220masl). In terms of major view lines, it is open to the east and overlooks an impressive landscape of the Arava (the rift) valley (-20 up to 0mbsl) and the Transjodanian Mountains (plateau) (at elevation ca. 1500 masl) to the east of the rift (Figure 1).

Figure 1. A view from Har Gevim to the east.

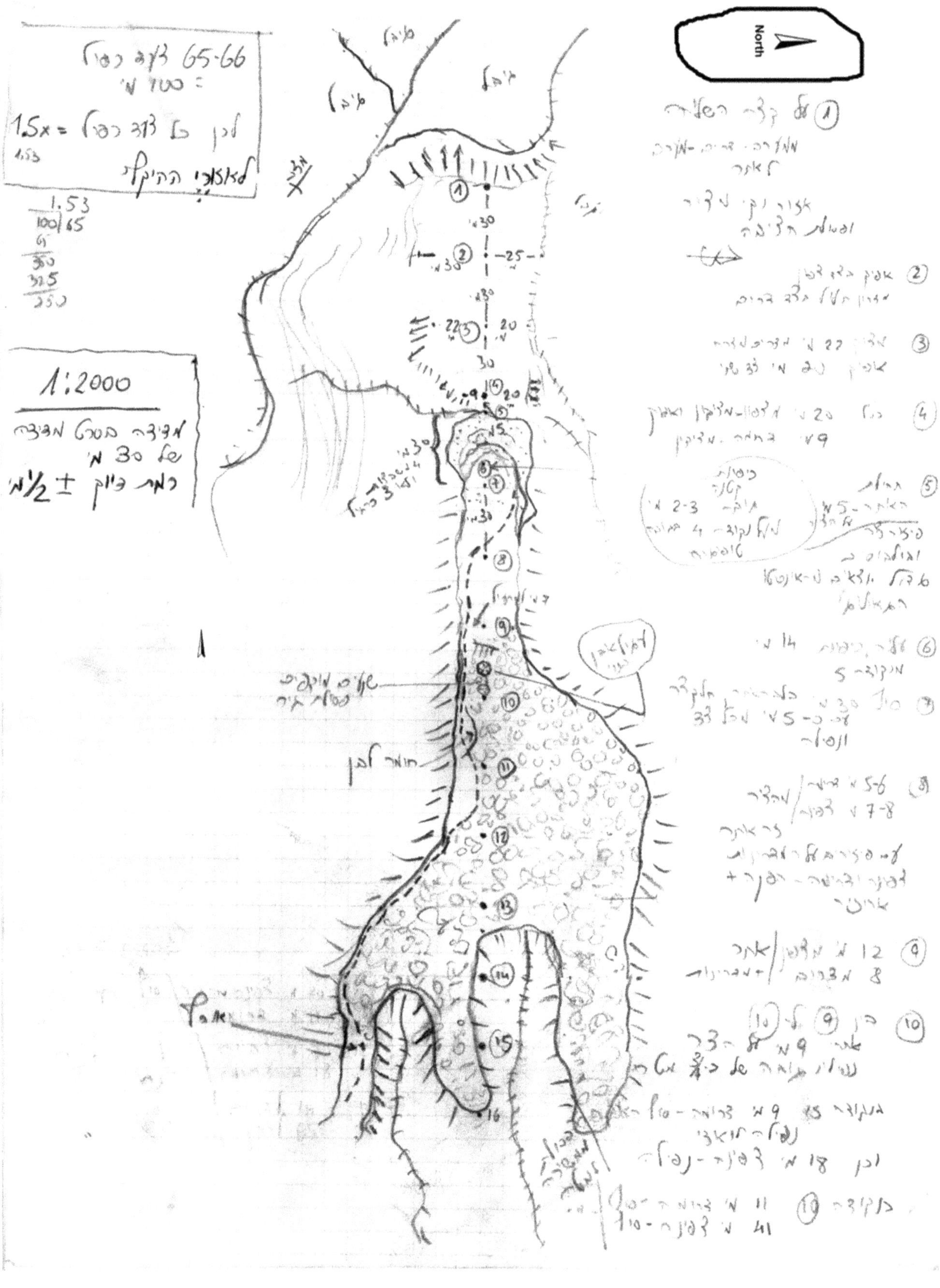

Figure 2. A schematic plan of the Har Gevim Quarry site.

The site appears as a narrow strip of a ridge in the west with mainly natural flint nodules and flint debitage. It widens eastwards, splitting into three separate, smaller ridges, and becomes densely covered by limestone waste mostly in the form of rounded-amorphous features with flint nodules and debitage.

The site covers some 12 dunams (12,000 square meters, some 3 acres, or 1.2 hectares) and, according to our estimates, has some 500-600 features (Figure 2). It appears that the whole ridge or most of its area was 'pealed' of its rocky cover to extract the underlying flint and the features are basically the waste products of this operation.

Generally the geological strata of the whole area are inclined eastwards – towards the rift valley. They all belong to the Ovdat formation group of the tertiary Eocene era and consist of alternating beds of medium-hard limestone, soft chalky limestone and flint horizons.

The number of flint horizons visible in the area is large including a black-cracked (a low quality) top flint layer eroding out from the summits and the upper part of the slopes. The section shows at least a dozen more horizons of flint nodules of varying quality below it. The flint horizons relevant to the quarrying are covered by 2-5m of alternating soft, medium and hard limestone. It seems that two horizons of flint were chosen for extraction. One horizon was situated above a medium-hard limestone layer within the abovementioned cover and another below this very layer. The nodules and small plates of flint are of very high quality. While in most of the area this medium-hard limestone layer ranges in thickness between 70-120cm and is sometimes covered by a couple of meters of soft limestone and softer sediments, the medium-hard limestone on-site, where Neolithic people decided to quarry was shallow consisting of only 25-40cm thickness and jointed both horizontally and vertically. The choice of this specific part of the formation for extraction still needs to be studied in detail since it is not clear whether this shallow cover is due to erosion, nature of slope, structural reasons or else. It is, however, a clearly efficient choice.

3. Site description

None of the described features have been excavated. We have only conducted a survey collecting no finds. Some representative flint items have been collected, photographed and returned to their original place.

The dense part of the site is fully covered by stone waste and shows generally rounded to amorphous features with some open space in the center (Figure 3, and see also Figures 5, 6). At the bottom of some of the features bedrock can be seen but we didn't find flint *in situ*. The density of features indicates a well organized, long term operation involving initial preplanning of the extraction from the whole complex for achieving greater efficiency. It looks as if the area were stripped off its rock, flint nodules were extracted from it and the limestone waste discarded and piled around the extraction front.

A detailed look at the selected features in the Har Gevim complex may shed some light on the quarrying techniques. Two major types of features appear:

- Relatively short bedrock quarry fronts. The rock was possibly split at the joints to reach the flint horizon below it (Figure 4) and the limestone blocks piled around it. We cannot say whether they are continuous before we excavate some of the features.

- Rounded to amorphous features with piled limestone around them and a central area, generally free of stone waste (Figure 5, 6). Here too it is impossible to reconstruct extraction without excavating some of these features.

Figure 3. Two general views on Har Gevim from the east.

Figure 4. Two flint horizons near by the site of Har Gevim.

4. The lithic finds and dating

The only kind of recovered finds at the site are lithic. The amount of flint is relatively small, at least as far as one can see on the surface and in the generally free-of-stones parts in the center of the features. However, this is an impression previous to excavation.

Beyond the low density of flint, the lithics include mainly cores and core trimming elements and some other blanks, mainly blades and some flakes. The most indicative finds are Naviform blade cores typical of the PPNB (Quintero and Wilke 1995; Khalaily 2006 and Figure 7). From a first glance it appears that the typical 'elegant' PPNB blades are missing, probably taken away from the quarry while some of the less successful products of the Naviform blade industry were left on site. Products of other lithic production trajectories such as flakes were also found in small numbers. The number of shaped tools observed on the surface is low, and the absence of arrowheads in particular is remarkable. A single broken Neolithic bifacial tool, possibly an axe, and some retouched pieces constitute major finds, but none of them are particularly noteworthy.

We thus presume that some of the extracted nodules were shaped on-site as naviform cores while others were taken away from the quarry for further use, mainly for blade production. The few cores we found are bidirectional blade cores, and therefore, following southern Levant accepted assignments, the industry is dated to the PPNB – 10600-9000 cal BP (Kuijt and Goring-Morris 2002).

5. Har Gevim in context: a broader view

Neolithic flint mining in the Near East is not very well-known. Although the researchers began to look into the question of flint trade in the region a long time ago (e.g. Kozlowski 1999 and references therein) and flint knapping, especially that of the PPNB blades, was recognized as a specialized craft (Quintero and Wilke 1995), we are far from approaching the wealth of data and interpretations of Neolithic flint mining in Europe.

Considering the scarcity of data available at the moment, we can only briefly suggest two major Neolithic flint economy models related to quarry sites:

I. Quarry sites located near or directly related to consumers' sites. The PPNB Wadi Huaijer, located near the PPNB mega site of Ain Ghazal in Jordan (Quintero 1994; 1996) and Mt. Pua where Neolithic mines found near the site of Ain Miri (Gopher and Barkai 2006) are the most illustrative representatives of this model. It also applies to the large PPNB site of Yiftha'el in the lower Galilee with thousands of blade cores on-site, located at a distance of 2-3km from the quarry (Khalaily 2006).

II. Quarry sites situated far away from known Neolithic sites, located on conspicuous landscape features and with

Figure 5. Rounded/amorphous features with an open space at the centre.

Figure 6. Top: plan and section of a feature. Bottom: plan of another feature.

Figure 7. Blade cores and blades from Har Gevim.

views open to large distances. This could be the case for Ramat Tamar in the southern Negev (Taute 1994; Schyle 2007). However, in this case, beside the quarry and small workshops, a 'miners' site or a knapping site, is found nearby, in the valley below the quarry. They produced bifacial tools, mostly axes that were probably exported to the north and maybe to the northeast, since Neolithic sites in this southern desert region do not have bifacial tools at all.

Isolated, located away from Neolithic sites and overlooking a major rift valley landscape, Har Gevim seems to belong to the second model (see Figure 1). But, in contrast to the case we mention above, no knapping site has yet been found in its vicinity. Considering the fact that this is *terra incognita* and we have not yet surveyed the area, this may still change. We hope to carry out such a survey and look for knapping sites since knapping in the quarry site itself does not seem to be very intensive compared to, for example, the finds in the Neolithic quarry site of Ramat Tamar.

Har Gevim appears to be a large-scale site of production of blades, possibly catering to the needs of large PPNB communities, located to its north and east, on the Transjordanian plateau. At the moment we cannot say exactly what kinds of products were transported from the Har Gevim quarry site: whether they were nodules, cores or blades. However, the density and the composition of lithics found on site lead us to believe that mostly nodules were extracted and transported from the site. This suggestion however requires further study, which we look forward to conducting in the coming few years.

Acknowledgments

We thank Dr. Z. Meshel for telling us about the site at Har Gevim and two members of Kibutz Ein Yahav who helped us doing our work at the site – H. Weizer and S. Shtookelman.

References

Barkai, R., Gopher, A. and La Porta, P. C. 2002. Paleolithic landscape of extraction: flint surface quarries and workshops at Mt. Pua, Israel. *Antiquity* 76, 672-680.

Barkai, R., Gopher, A. and La Porta, P. C. 2006. Middle Pleistocene Landscape of Extraction: Quarry and Workshop Complexes in Northern Israel, in N. Goren-Inbar and G. Sharon (eds.), *Axe Age: Acheulian Toolmaking - from Quarry to Discard*, 7-44. Oxford, Equinox Publishers.

Barkai, R., Gopher, A. and Weiner, J. 2007. Quarrying Flint at Neolithic Ramat Tamar: An Experiment, in L. Astruc, D. Binder and F. Briois (eds.) *Systemes Techniques et communates du Neolithique Preceramique au Proche-Orient*. Actes du 5 Colloque International Frejus, 2004, 25-32. Antibes.

Barkai, R. and Gopher, A. 2009. Changing the face of the earth: Human behavior at Sede Ilan, an extensive Lower-Middle Paleolithic quarry site in Israel, in B. Adams and B. Blades (eds.), *Lithic Materials and Paleolithic Societies*, 174-185. Oxford, Blackwell.

Gopher, A. and Barkai, R. 2006. Flint extraction sites and workshops in prehistoric Galilee, Israel, in G. Korlin and G. Weisgerber, (eds.), *Stone Age Sites – Der Anschnitt* 19, 91-98, Bochum: Deutsches Bergbau Museum Bochum.

Khalaily, H. 2006. *Lithic traditions during the late Pre-Pottery Neolithic B and the questionof the Pre-Pottery Neolithic C in the southern Levant*. Unpublished Ph.D dissertation. Beer Sheva, Ben Gurion University of the Negev.

Kozlowski, S. K. 1999. *The Eastern Wing of the Fertile Crescent*. Oxford, British Archeological Reports International Series 760.

Kuijt, I. and Goring-Morris, N. 2002. Foraging, farming and social complexity in the Pre-Pottery Neolithic of the southern Levant: A review and synthesis. *Journal of World Prehistory* 16/4, 361-440.

Quintero, L. A. 1994. Neolithic flint mining in Jordan. *Neo-Lithics* 2(94), 2.

Quintero, L. A. 1996. Flint mining in the Pre-Pottery Neolithic: Preliminary report on the exploitation of flint at Neolithic Ain Gahzal in highland Jordan, in S. K. Kozlowski and H-G. Gebel (eds.), *Neolithic Chipped Stone Industries of the Fertile Crescent and their contemporaries in adjacent regions*. Studies in early Near Eastern Production, Subsistence and Environment 3, 233-242. Berlin, Ex Oriente.

Quintero, L. and Wilke, P. 1995. Evolution and economic significance of naviform core and-blade technology in the southern Levant. *Paleorient* 21(1), 17-33.

Taute, W. 1994. Pre-Pottery Neolithic Flint Mining and Flint Workshop Activities Southwest of the Dead Sea, Israel (Ramat Tamar and Mesad Mazzal), in H. G. K. Gebel and S. K. Kozlowski (eds.), *Neolithic Chipped Stone Industries of the Fertile Crescent*, 495-509. Berlin, Ex Oriente.

Schyle, D. 2007. *Ramat Tamar and Mezad Mazal, the early Neolithic economy of mining and production of bifacials southwest of the Dead Sea.* Bibliotheca neolithica Asiae meridionalis et occidentalis. Berlin, Ex Oriente.

Reassessment of a putative chert quarry in Oman

Gerhard TRNKA

Abstract

In a 1997 article, 'A new type of exploitation in Oman (Arabia),' Gerd Weisgerber described a flint extraction site at Jebel Qatar near the little village of al-Hajir, close to Quriat on the Omani coast. The eocene limestone banks are described as having been quarried for 'flint' nodules by manually breaking the limestone ('pounding work') using quarzite and dolomite pebbles, as indicated by working and knapping traces. I examined this site in 2006. The local raw material is bright greyish through to reddish eocene chert and is localized in the eastern part of the hill (Figure 4d-f), an area with abundant fossils (Figure 6). At the proposed mining site (Figure 3) no chert could be found. According to my observations and assessment, the fragmentation of the limestone bedding has been caused by natural cracking and other erosive influences (Figure 4a-c), coupled with relatively recent exploitation of stone by the local population. The proposed quarzite hammer stones with traces of use wear are actually cores for flakes (Figure 5). The site remains important because of its prehistoric shell middens and further investigation is therefore warranted.

Keywords

Arabian peninsula. Oman. Reevaluation of chert quarrying.

The Arabian peninsula with its great desert and semi-desert terrain is one of the most impressive landscapes on earth. In terms of geology, Oman, in particular, is a sort of 'open book' providing fascinating glimpses into the earlier phases of our planetary history.

In 2006, in the course of pursuing lithic raw material studies for the Vienna-Lithothek (VLI), I undertook field prospection both for stone age sites and for natural geological outcrops of flint in both the south and east of the country.

An important background consideration in this fieldwork was the report from Gert Weisgerber entitled 'A new type of flint extraction in Oman (Arabia)' ('Eine neue Form des Silexabbaues im Oman [Arabien]', Weisgerber 1997). In addition to providing an overview of the presence of flint in Oman, and a chronological scheme (Weisgerber 1997, 155-157), a prehistoric settlement with associated 'mine' was identified to the north of Quarayat (or Qurayyāt; following the western convention, also 'Quriat') in the Jebel Qatar range on the Arabian coast, southeast of Muscat in northeastern Oman (Figure 1a-d).

The hill lies to the north of the modern settlement of al-Hajir near Quriat and was included in a prospection survey in 1983. Adjacent to a prospective quarry for flint nodules, extensive shell remains of 'shell midden' type were documented in association with artefacts including ceramics. In summary, the 'mine' of al-Hajir is described as a 'tiered quarry works' where tightly packed limestone boulders up to 60cm in size were manually pounded with pebbles of different sizes and weight. The material ranges from brownish quartzite to grey dolomite (Weisgerber 1997, Figure 6), washed out from the gravels of the nearby Wadi Satari. The chalk embedded flint nodules are of a poor quality due to intensive tectonic and erosive processes. Inside, they display a dark, sometimes red-brown colouration, with an outside that has a white weathering layer that is extremely dehydrated; nevertheless, there is also a grey-brown variant, with a yellow or white outer surface. In the western part of the hill a number of quarries ('mines') were identified by Weisgerber (1997, Figure 1) and described by him as mine workings of a kind where local conditions have resulted in the presence of step like boulders of limestone and resulting sharp edged limestone flakes, chert debitage and chert pebbles (Weisgerber 1997, Figure 2-5).

As, according to the Geological Map of Oman (1993 - Signatur EM_D) the neogene limestone is of late Eocene to early Miocene date (the Dhofar Group, Schelf-Facies (Fi-

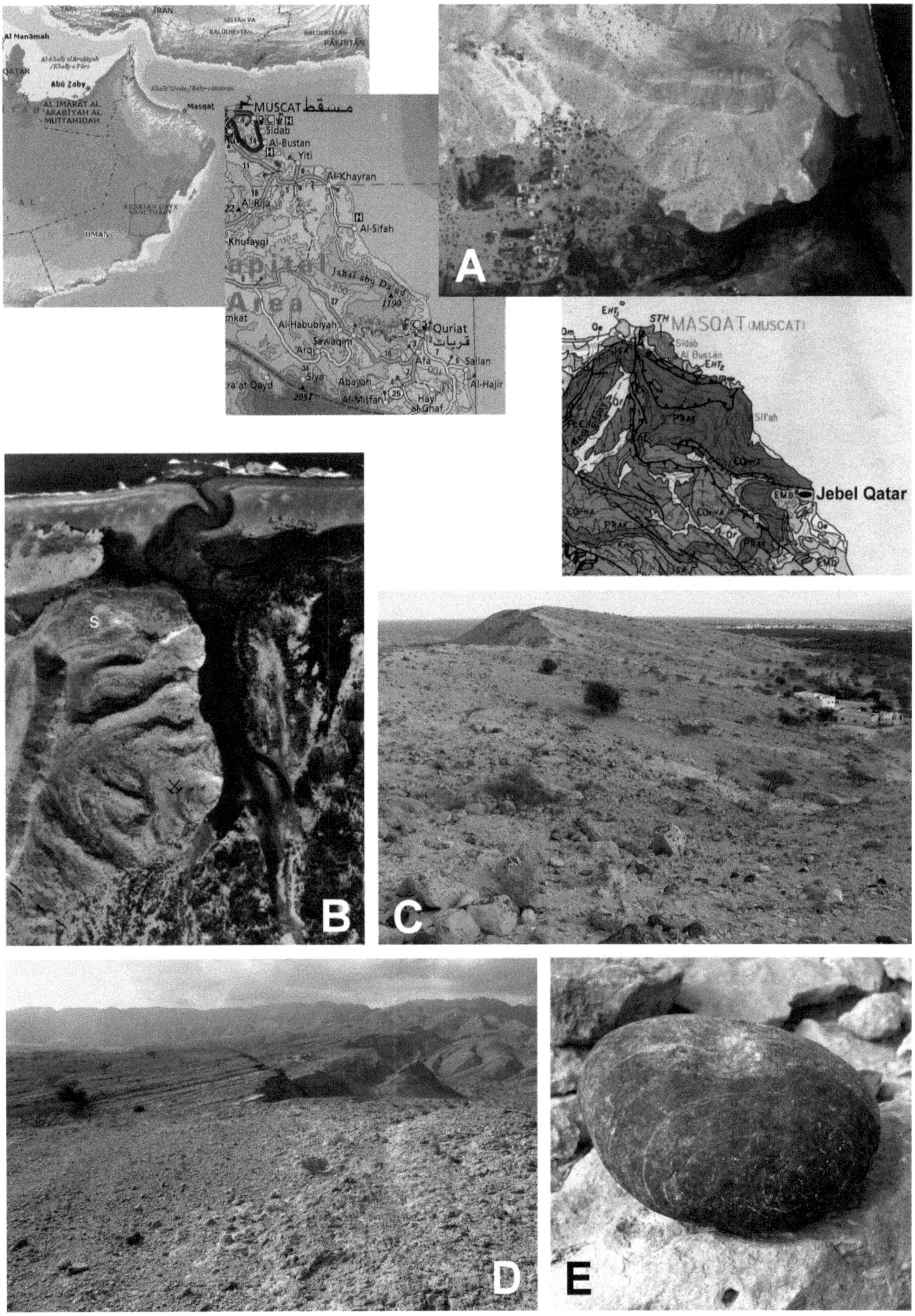

Figure 1. Quriat–al–Hajir, Jebel Qatar (Northeast Oman). Maps – Encarta, Freytag & Berndt, Google Earth, Geological Map of Oman 1993; B – after Weisgerber 1997, fig. 1; C–D – Jebel Qatar, the views from west and east (28.11.2006). E – Pebble with percussion scars.

gure 1), Gerd Weisgerber proposed that the sedimentary rock was a late Eocene lithoclastic limestone with presence of marly limestone (Weisgerber 1997, 153).

In this context, it is germane to note what occurred on the 28th to 29th November 2006 when an attempt was made to locate this findspot. Jebel Qatar is a high, east-west oriented ridge of limestone massif, falling from north to south. The eastern boundary is formed by the sea coast, while to the west the ground rises towards the mountains of the interior.

The southern boundary is defined by the modern settlement and a lagoon with palm groves (Figures 2a, 3a,c,d).

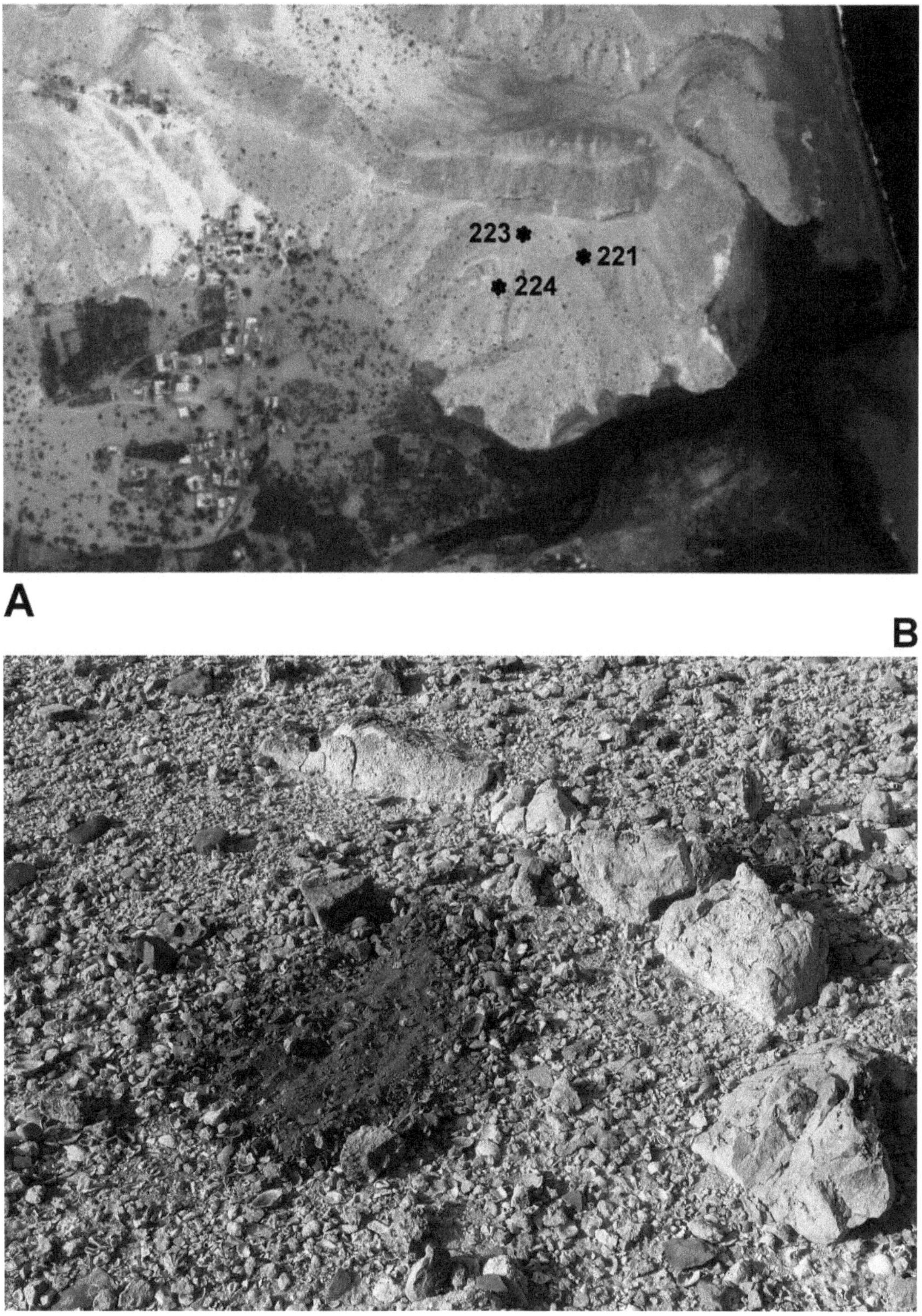

Figure 2. Quriat-al-Hajir, the hill of Jebel Qatar (Northeast Oman). A – Google Earth (GPS 221, 223–224), B – 'shell middens' GPS 223 (28.11.2006).

Figure 3 Quriat – al-Hajir, Jebel Qatar (Northeast Oman). A–B – after Weisgerber 1997, figs 3 and 5; C-F – GPS 224 (28.11.2006); G – limestone debris.

Towards the north, the limestone elements of the hill break up forming a wide dry valley (Figure 2a). The north-south erosion of the hill is a continuous, on-going process where the limestone bedrock breaks up in a block-like or, alternatively, net-like fashion, and breaks away to be deposited down-slope (Figure 4a-c). In this way, the eastern part of the hill is divided away from the section to the south by a channel-like depression that becomes wider and deeper, and where the generally strongly weathered sedimentary rock is made up of light grey chert (samples: GPS 221, Figure 4d-f; 28.11.2006: 23,27975°, E 58,91580°, height 55.9m) (Figure 6). The two chert samples are from the strongly weathered parent limestone. They may be characterised as translucent light grey, sometimes trending towards reddish, chert with fossil inclusions, crazing and a weathered cortex. Exposed surfaces sometimes display a deep milky-white patination. On the crest of the hill (Figure 1d), as well as on its southern flanks, 'shell-middens' have been found, or at least, the remains of quite thick horizons characterised by marine mollusc shells and mollusc remains intermixed with ash above the bedrock (Figure 2b). There is frequent representation of quartzite tools and manufacture in the form of hammer stones, cores (Figure 5a, right), debitage (Figure 5b, 5-8) and working debris, most of it with a green colour, but some of it also displaying brown, light grey and even some darker colouration (Figure 5a); less frequent are chert fragments (Figure 5a and 5b, 1-3) and chert cores (Figure 5b, 4) relating to the local raw material. Flat oval pebbles with percussion scars on both faces, and sometimes controlled knapping (Figure 1e), could well have served as anvil stones for the breaking open of shellfish. These waste deposits of exploited marine molluscs have been dated to the Bronze Age (second millennium BC) and also to the 'pre-Islamic' period (Weisgerber 1997, 153), but, in my opinion, due to the lack of well-characterised archaeological material or excavation, cannot really be precisely dated.

Figure 4. Quriat-al-Hajir, Jebel Qatar (Northeast Oman). A-C – Erosion processes on the limestone outcrop; D-F – chert sample GPS 221 (28.11.2006).

Figure 5. Quriat-al-Hajir, Jebel Qatar (Northeast Oman). Quarzite- and chert-spectrum – GPS 223 (29.11.2006).

Figure 6. Quriat-al-Hajir, Jebel Qatar (Northeast Oman) – chert sample GPS 221 (28.11.2006).

GPS 223 (28.11.2006): N 23,28042°, E 58,91483°, height 62.6m. Area of 'shell-middens' with associated quartzite tools (Figures 1d and 2b).

Of particular interest here was the putative grounds for visiting the Jebel Qatar, viz. the mining and extraction of flint nodules. The locale was unambiguously locatable thanks to the previously available photographs (Weisgerber 1997, Figures 2-3 and 5; GPS location 224, Figure 3), and it was also clear that the expected blocky weathering of the upper limestone deposit was unambiguously visible, but within that formation, no chert nodules were actually present. Rather, in one or two places, it was possible to detect small chert fragments of poor quality, not approaching that which would be considered mineable.

Much more also becomes apparent, namely that in places 'freshly' modified, fragmented limestone debris, with sharply angled faces and percussion facets, are to be found lying on the surface (Figure 3f). This residue must have been recently created, presumably during the extraction of limestone for the adjacent settlement, material that is today no longer in demand. It follows that the 'mines' are in reality the naturally eroded remains of recent quarrying in the limestone massif. In terms of extraction activities or working floors, nothing can be determined with certainty, not even the presence of locations where breaking took place or production debitage accumulated. The so-called hammer stones employed for the extraction and breaking up of limestone blocks (Weisgerber 1997, Figure 6) are too small, and in a general sense inappropriate, for any robust mechanical applications. These shaped and, to a degree, also broken quartzite pieces (Figure 5a, right) were cores (compare also Wesgerber 1997, Figure 6) that can be associated with the rich detritus that should principally associate with the 'shell-middens' (Figure 5b, 5-8). My assessment of the heavily weathered limestone outcrops is that they have not been artificially modified. The overall situation observable on the hill is that the limestone has broken up vertically in long lateral sections, creating a banded or netlike pattern of erosion (Figure 4, a-c) that then deposits itself down-slope.

GPS 224 (28.11.2006): N 23,27957°, E 58,91412°, height 52.3 m (Figure 3). Recent weathering of the limestone massif with visible clints and grikes. Sample of fragmented piece (Figure 3g). REM-EDX analysis by Michael Götzinger (Institut für Mineralogie und Kristallographie, Universität Wien) has determined that the limestone contains silicic acid.

Acknowledgements

For help with the organisation of the Oman trip and interpreting, I would like to thank Frau Mag. Petra Unterlechner, Senftenberg. English translation: Dr Timothy Taylor.

References

Weisgerber, G. 1997. A new type of exploitation in Oman (Arabia), in R. Schild and Z. Sulgostowska (eds.), *Man and Flint.* Proceedings of the VIIth International Flint Symposium Warszawa - Ostrowiec Świętokrzyski September *1995*, 153-157.

Flint procurement strategies of the early hunter-gatherers of eastern Uruguay

José LÓPEZ MAZZ, Andrés GASCUE and Gustavo PIÑEIRO

Absract

The southeastern area of the American continent was probably first populated because of the high productivity of its coastal wetlands and lagoons. These small groups had a high mobility pattern, covering considerably long distances over the course of the year, and developed a wide-spectrum hunting technology. Essential to this technological system was the supply of quality raw material for flaking. This social and economic mobility is key in order to understand these prehistoric production systems. In this paper we present petrographic information of lithic remains produced by the ancient settlers of the the 'Rincón de Los Indios' archaeological site (9300-7600 cal BP). We discuss organizational patterns of lithic production and strategies of flint procurement.

Keywords

South America. Prehistoric quarrying. Petrography. Hunter-gatherer mobility.

1. Introduction

Evidence of early human occupation (Pleistocene-Holocene transition) of South America comes from different archaeological regions, including the Andean ranges, the Caribbean and Pacific coasts, Patagonia and the Pampa (Dillehay 2000; Politis *et al.* 2004). Over recent years, an increasing amount of data has been collected in Uruguay (30°-35°S to 58°-54°W) from archaeological sites located near important watercourses (e.g., the Negro, Uruguay and Tacuarembó rivers and their tributaries), and from hills and lagoons near the Atlantic coast (Suárez and López Mazz 2003). These early Americans lived in small groups and covered long distances over the course of the year (Dillehay 2000). The early settlements of southeastern Uruguay are located on a Quaternary coastal plain home to landscape features such as coastal lagoons, lowlands, mid-altitude plains and the seashore itself, the origin of which lies in changes in the sea level during the Pleistocene-Early Holocene (Bracco 1995; Martin and Suguio 1989).

This paper discusses the petrographic information provided by lithic material produced by ancient settlers (*c.* 9300-7600 years BP cal) and recovered from the Rincón de Los Indios archaeological site (López Mazz 2001), and assesses organisational models of lithic production and strategies of flint procurement. Models of lithic technology require data regarding distances to quarries; the present work provides information in this respect.

Knowledge of the area's lithic economy enables discussion regarding the type of societies to which its early hunter-gatherers belonged. They appear to have made use of a wide range of animal resources, including fishes, rodents and deer, and made many products from the native palm tree *Butia capitata* (López Mazz 2001). The most important feature of the lithic technology of these groups was the production of projectile points, bifacial tools, unifacial tools, flakes and grinding stones (Gascue *et al.* 2009a; López Mazz *et al.* 2009).

2. Early South American hunter-gatherers

The first Americans to arrive in the Far South did so before the raising of the La Plata River to its present level (*c.* 14,000-9000 years BP), possibly attracted by the high productivity of the coastal wetlands and lagoons of the ancient Palaeo-Paraná delta (now underwater) (Dillehay 2000; López and Gascue 2007; Politis *et al.* 2004). These peoples developed a wide-spectrum hunting technology focused on the fauna of open environments. The variability of projectile points and *bolas* (polished stones for use in a *bolas* sling) show that large, medium and small animals

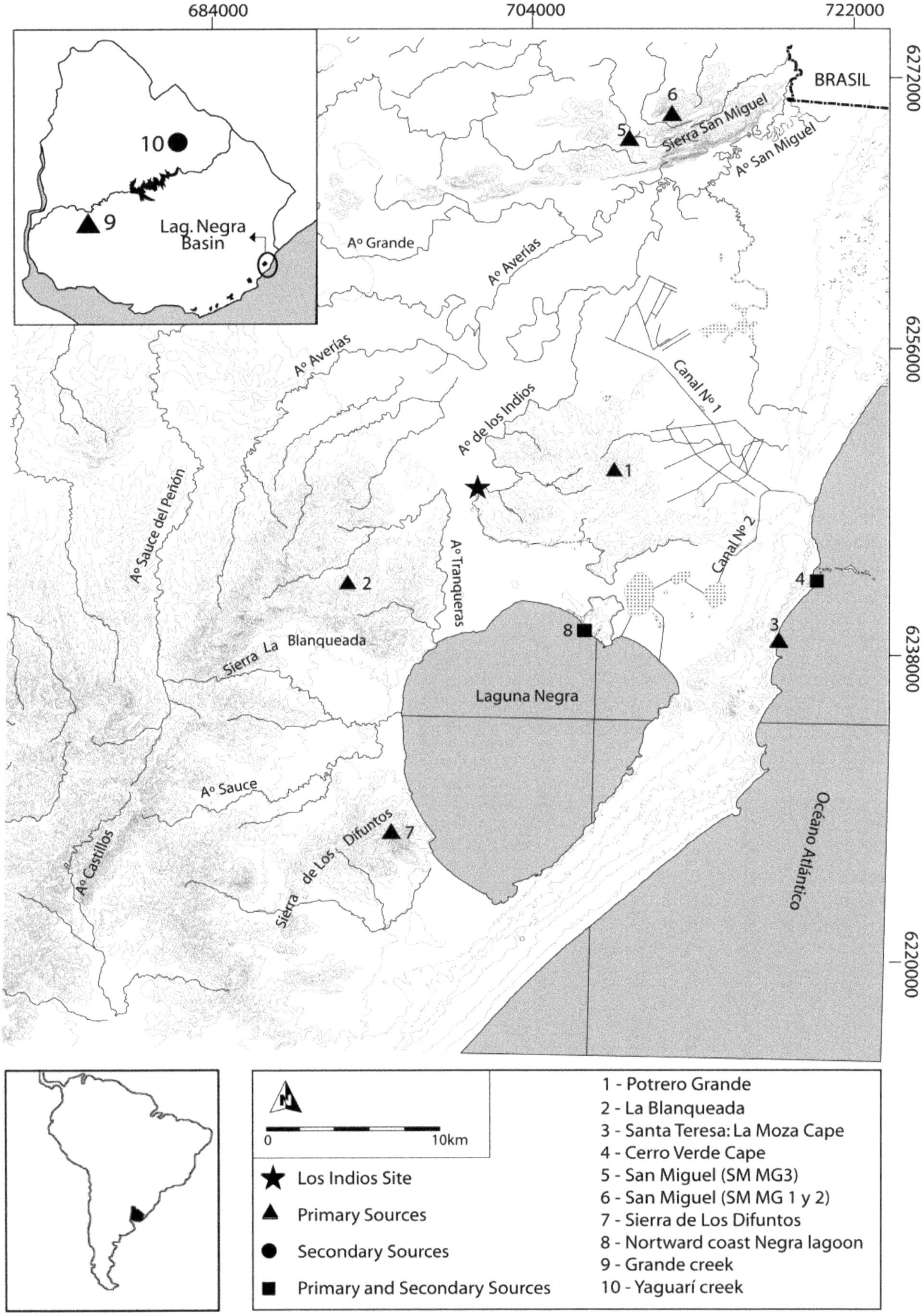

Figure 1. Site location and sources of the analysed raw materials.

were hunted. Essential to this technological system was the supply of good quality raw material for flaking (López Mazz *et al.* 2009; Gascue *et al.* 2009a).

3. The Rincón de Los Indios archaeological site

The Rincón de Los Indios archaeological site is a multi-mound, multi-component site located on a peninsula-shaped area surrounded by the Las Maravillas swamp. The occupation process dates between 9300 years BP to the period of contact with European conquerors. The site was excavated between 1995 and 2010; current evidence suggests that it was an area of logistic use in the lowlands close to the sea (Figure 1).

The oldest level of the site shows that it was occupied between 9300 and 7700 BP. The site is located at a strategic point for hunting and fishing activities as well as for intra-regional communication. The formation processes of the site are related to intensive human occupation (*longue durée*) and the changing lowland environments of the Early Holocene (López Mazz 2001). Lithic material recovered in excavations shows the use of different technologies with each

Resources	%
Quartz	52.4
Quartzite	20.4
Chalcedony	14.6
Rhyolite	2.8
Microgranite	2
Granite	1.8
Slates	1.2
Silicified limestone	0.8
Other quartz-rich rocks	0.6
Other suitable for polishing rocks	0.5
Ochre	0.1
N/I	2.7

Figure 2. Macroscopic classification of raw materials from the Rincón de Los Indios site.

ID	Source	Location	Origin
LN MG2	Secondary	Northern coast, Negra Lagoon	Local
LN MG10	Primary	Northern coast, Negra Lagoon	Local
PG MG1	Primary	Potrero Grande	Local
LN MG11	Primary	de los Difuntos Hill	Regional
Blanq MG1	Primary	de los Difuntos Hill	Regional
Moza MG9	Primary	Santa Teresa Headland, Atlantic coast	Regional
Moza MG7	Primary	Santa Teresa Headland, Atlantic coast	Regional
CV MG8	Primary	Cerro Verde Headland, Atlantic coast	Regional
SM MG1	Primary	San Miguel Hill	Regional
SM MG2	Primary	San Miguel Hill	Regional
SM MG3	Primary	San Miguel Hill	Regional
AGDE MG1	Primary	Mouth of the Río Grande (Paso de Lugo)	Extra-regional
Yaguarí D'	Secondary	Río Yaguarí (Paso Casildo)	Extra-regional

Figure 3. Sources the geological samples and their classification.

ID	Macro classification	Minerals	Structures
Moza MG 9	Brown opaque chalcedony	Porphyroclasts and porphyroblasts of quartz. 50% matrix of fine-grained quartz and Fe oxide. Chalcedony in voids.	Microcracks and transgranular fractures Voids cemented with quartz.
PG MG1	Quarzite	Quartz deformed and fractured – Chalcedony layers and quartz in voids.	Microcracks and transgranular fractures
Blanq MG1	Quarzite	Quartz + K feldspar+ chlorite+ crystal pseudomorphs. Chalcedony layers in voids.	Microcracks and transgranular fractures
LN MG 11	Quarzite	Qz + opaque minerals + Fe-oxides	Stockwork
LN MG 2	White quartz	Qz with undulose extinction + K feldspar+ garnet	Microfaults/ mineral cracks
SM MG1	Rhyolite	Tabular sanidine, quartz, isometric magnetite, apatite	Massive
SM MG2	Basalt	Pg + Px + opaque minerals	Massive
SM MG3	Opal	Qz with undulose extinction + brown microcrystalline minerals	Faults, cracks, voids
CV MG 8	Smokey quartz	Quartz + albite + K feldspar + epidote	Perthites / displaced fractures filled with quartz
Moza MG7	Smokey quartz	Quartz + albite + turmaline + biotite + epidote	Perthites /fractures
LN MG 10	Micro-crystalline quartz	Two grain size generations of microcrystalline quartz	Ribbon quartz / strained quartz veins/fractures
Yaguarí MG D	Chalcedony layers (agate)	Chalcedony fibres. Lengthfast.	Layers of twisted-untwisted quartz fibres
A GDE MG 1	Red Silicified Limestone	Chalcedony + calcite + opaques + Qz + Fe-oxides	Gastropods/ charophytes

Figure 4. Petrography of sampled outcrops and potential quarries.

Texture-fabric	Matrix	Secondary minerals	Lithological classification
Oriented. Mosaic quartz	Fine	Fe oxide. Chalcedony rims	PROTOMILONITE
Cataclastic / Fluidal	Fine to medium	Fe oxide. Chalcedony bands and rims	QUARTZITE CATACLASITE
Cataclastic / fluidal	Fine to medium	Disseminated Fe oxide. Chalcedony rims.	QUARTZITE CATACLASITE
Cataclastic and grano-blastic	Fine	Fe oxide. Feldspar silicification	QUARTZITE CATACLASITE
Granoblastic to cata-clastic	Very coarse	Feldspar alteration (kaolinite?).	QUARTZITE CATACLASITE
Micro porphyritic	Very Fine	Sanidine pseudomorphs. Fe oxide	RHYOLITE
Intergranular	Very Fine	Clay transformation. Fe oxide.	THOLEITIC BASALT
Microcrystalline	Very Fine	Brown microcrystalline minerals	QUARTZITE BRECCIA
Pegmatitic	Medium	Epidote in feldspars. Mn oxides in cleava-ge filosilicates.	CATACLASTIC PEGMATITE
Pegmatitic	Coarse	Epidote in feldspars. Mn oxides in cleava-ge filosilicates	CATACLASTIC PEGMATITE
Granoblastic and mortar	Very Fine	Fe oxide in fractures	QUARTZ VEIN
Fibrous sequence layers	Very Fine	Fresh	CHALCEDONY LAYERS
Microcrystalline	Very Fine	Fresh	SILCRETE

ID	Macro classification	Minerals	Structures
iV_7-38	Brown opaque chalcedony	Porphyroclasts and porphyroblasts of quartz. Fe oxides. Chalcedony in voids.	Microcracks and transgranular fractures.
iV_6-9	Pale brown quartzite	Porphyroclasts and porphyroblasts of quartz + epidote. Voids filled with undulose quartz.	Ribbon quartz. Faults, microcracks.
iV 7-17	Pale brown quartzite	Porphyroclasts and matrix of undulose quartz. Chalcedony.	Stockwork
iV_6-4	Sedimentary quartzite	Quartz porphyroclast. Matrix: quartz + epidote + opaques.	Layers of ribbon and fragmented quartz.
iV_6-8	Brown opal	Quartz and chalcedony; Fe oxide rims in voids	Chalcedony in cracks
iV_6-5	Red rhyolite	Tabular feldspar + porphyrocrystals of quartz. Matrix: quartz + amphibol	Massive
iV_6-29	Reddish-brown rhyolite	Tabular sanidine. Matrix; microcrystalline quartz + K feldspar + opaques	Massive
iV_6-35	Pale brown quartzite	Holocrystalline quartz	Massive
iV_6-31	Translucid quartz	Quartz (one crystal)	Microcracks
iV_7-42	Quartz	Quartz sacharoidal	Faults, cracks
iV_7-14	Red quartzite	Sacharoidal quartz vein in a garnet + epido-te + amphibol granite.	Quartz veins in granite
iV_7-16	Chalcedony layers (agate)	Cryptocrystalline and fibrous chalcedony	Twisted layers of quartz fibres. Cracks.
iV_7-37	Translucid chalcedony	Chalcedony + opaques	Bands, not always with orientated chalcedony
iV_7-33	Chalcedony layers	Criptocrystalline chalcedony	Layers / microcracks
iV_6-44	Red silicified limestone	Chalcedony + Ca-sparite + opaques + quartz	Gastropods / charophytes – Voids with chalcedony
iV_6-32	Orange silicified limestone	Chalcedony + Ca-sparite + opaques + quartz	Veins and voids partially filled with chalcedony
iV_6-30	Gneiss	Quartz + feldspars + amphibol	Bedding (So) and foliation (S1)
iV_7-12	Quartzite	Quartz + zoisite + opaques + albite + green - brown biotite + epidote + apatite + zircon	Faults
iV_7-18	Quartzite	Quartz + K-Feldspar	Massive

Figure 5. Petrography of archaeological samples from the earliest layer of the Rincón de Los Indios site.

Texture-fabric	Matrix	Secondary minerals	Lithological classification
Oriented.	Fine	Fe oxides. Chalcedony rims	PROTOMILONITE
Cataclastic.	Fine to medium	Some Fe oxides.	QUARTZITE-CATA-CLASITE
Cataclastic and granoblastic zones	Very fine l to medium	N/I	QUARTZITE-CATA-CLASITE
Oriented. Granoblastic.	Fine	Epidote	QUARTZITE-PROTO-MILONITE
Unequigranular xenomorphic	Very Fine l	Disseminated Fe oxides. Chalcedony rims.	HORNFELS
equigranular to microporphyrytic	Very Fine	Feldspars pseudomorphs.	TRACHITE
Porphyrytic	Very Fine	Fe oxide	RHYOLITE
Granoblastic	Very Fine	Some Fe oxide	METASEDIMENTARY QUARTZITE
Pegmatitic	Very coarse	N/I	PEGMATITE
Pegmatitic	Very coarse	N/I	PEGMATITE
Granoblastic	coarse	Fresh / some Fe oxide over cracks	QUARTZ VEIN
Layered. Cracks.	Very Fine	Fe oxides rims	CHALCEDONY LAYERS
Layered	Very Fine	Fresh	CHALCEDONY LAYERS
Oriented fibres.	Very Fine	Fe oxide dendrites. Opalescence of chalcedony	CHALCEDONY LAYERS
Microcrystalline.	Very Fine	Silicification, Fe oxide. Quartz rims.	SILCRETE
Microcrystalline	Very Fine	Silicification, Fe oxide. Quartz rims.	SILCRETE
Grano-nematoblastic	Very Fine	Defective section	SLATE
Oriented.	Very Fine	Epidotized quartz	MILONITE
granoblastic	Medium	Fresh	ALCALI GRANITE

Figure 6. Photomicrograph of pegmatite quartz: C° Verde MG8 sample (left) and iV 7-42 (right). Scale x 100, cross-polarised light.

kind of raw material; Figure 2 shows a macroscopic classification of these materials and their frequency of recovery.

Flint-knapped products of the early occupants of the Rincón de Los Indios site include flaked instruments (unifacial and bifacial), cores, flakes and debris, and projectile points made by uni- and bifacial reduction (Gascue *et al.* 2009a; López Mazz *et al.* 2009). The most important economic activities identified at the site by the latter authors include the:

- gross transport to the site of different varieties of quartz, quartzite, rhyolite and microgranite.

- reduction of cores of the above raw materials by unipolar and bipolar knapping (mainly quartz), oriented towards the production of blanks for instrument-making (*sensu* Inizan *et al.* 1995).

- use of flakes without modification by retouching.

- production of tools by unifacial and bifacial knapping of good quality raw material.

- use of tools.

- production of projectile points from good quality raw materials.

- reactivation of projectile points.

- recycling of projectile points damaged by use.

- discarding of projectile points with irreparable damage or that required much repair.

- production of *bolas* from granite and rhyolites.

The aims of the present work were to a) verify the classical macroscopic classification of the lithic raw materials used by the prehistoric populations of eastern Uruguay, b) to explore the petrographic characteristics that influenced the selection of raw materials, and c) to evaluate, using lithological data, the origin of the lithic resources used by the early inhabitants of the Rincón de Los Indios site.

4. Methods and Constraints

Moving to produce, is often an important strategy of the modern-day ethnographic societies of America (Politis 1996; López Mazz *et al.* 2009). Movement is part of human life, but normally leaves no material trace. Movements made in the past are therefore very difficult to investigate (Close 2000). Nevertheless archaeology must attempt to measure the economic distances travelled if it is to explain social processes. Spatial analysis provides information on the economic organisation of prehistoric peoples, while the comparative petrographic study of lithic raw materials from outcrops and lithic artifacts provides a means of reconstructing their territoriality and circulation patterns.

This paper reports information generated by local, regional and extra-regional geological surveys and technological lithic analyses of the artifacts from the Rincón de Los Indios site. The ethnographic definitions of Politis (1996) and López Mazz and Bracco (2010) used in this work, including definitions of scales of distance, are listed below:

- **Local** sources refers to quarries located at a distance of under **10km**.

- **Regional** sources refers to quarries located between **10 and 100km** away.

- **Extra-regional** sources refers to quarries located more than **100km** away.

- Primary sources correspond to geological outcrops.

Figure 7. Photomicrograph of quartzite cataclasite: PG MG1 sample (left) and iV 6-9 (right). Scale x 100, cross-polarised light.

- Secondary sources correspond to sedimentary deposits.

- Quarrying strategies (*sensu lato*) imply the work of selection, exploitation and transportation of the geological material to the hunter-gatherer system point of production.

Materials examined

Petrological thin sections of 19 archaeological materials from the earliest level of the Rincón de Los Indios site, and 13 geological materials from quarries and potential quarries (here referred to as the geological sample) were examined using a Leitz Wetzlar Laborlux 12 pol S model petrological microscope at 100-300X magnification. The mineral identification criteria of Kerr (1965) and Mason's classification of metamorphic rocks (1990) were adhered to.

Samples of local flint raw material were collected from quartz veins and rhyolithic rocks in outcrops in the nearby hills of La Blanqueada, Potrero Grande, De Los Difuntos and San Miguel, from siliceous rocks on coastal headlands, and from coastal gravels (northern coast of the Negra lagoon, the Atlantic coast, and the La Moza and Cerro Verde outcrops). One sample taken from an extra-regional archaeological quarry – a Palaeocene silcrete here referred to as silicified limestone – was collected from the Rió Grande in western Uruguay. Another extra-regional sample (banded chalcedony) was collected from north-central Uruguay (Río Yaguarí gravel deposits); this material originated in vesicles within tholeitic basalts (see Figure 1 and Figure 3 for location).

The archaeological samples were selected from among the most common (near 90%) flint-type lithic materials found at the Rincón de Los Indios site, i.e., quartzite, quartz and chalcedony. However, some mistakes in the classification of the flint-like material of the archaeological elements are to be expected given the lack of important structural and contextual geological information. The selection of the geological samples may have involved some subjective bias, and the natural variation of both the geological and archaeological materials may have had some influence on the results obtained. However, in almost all cases, a number of well known textural features (e.g., chalcedony bedding, saccaroidal and porphyritic texture, volcanic voids) easily linked the geological and archaeological materials.

5. Results

The petrographic features observed in some of the materials making up the regional and extra-regional offer of potential lithic resources (Figure 4) are described analytically and graphically. Others materials seen to have been used by the early inhabitants of the Rincón de Los Indios site are also described (Figure 5). The analysed geological sample was composed of 31.1% variable-genesis quartzite, 18.8% quartz and pegmatite dikes, 12.5% banded chalcedony, 9.4% silcretes (silicified limestone), 9.4% mylonites or protomylonites, 12.5% volcanic rocks and 6.3% other materials (microgranite and metapelite).

Petrography of the geological samples

Siliceous metamorphic rocks such as quartz-breccia, quartzites-cataclasites and protomylonites made up the largest group of materials. Subvolcanic (dikes and pegmatites) and volcanic rocks of local and regional origin, and silicified limestone and banded chalcedony of extra-regional origin, were also identified.

Figure 4 shows that the geological samples near the archaeological site are affected by dynamic metamorphism, even samples previously classified macroscopically as quartz dikes. This is reflected by the presence of quartz veins (ribbon quartz-type), faults and cracks, strained crystals and the frequent dispersion of the extinction angle of the quartz crystals. The dominant dynamic metamorphism type seems to be characteristic of shear zones from eastern

Uruguay (Sánchez-Bettucci *et al.* 2001). The rhyolites, trachytes and acidic subvolcanic rocks are characteristic of the Cretaceous extensive margins of the Merin Lagoon and St. Lucia basins.

The silicified limestone sample from western Uruguay is characterized by the presence of fossils of pulmonate gastropods and charophytes (according to the description of Martínez *et al.* [1997] and Tófalo and Pazos [2009]). The reworked sample of banded chalcedony originated within the basalt vesicles and cavities of the centre-north of the country, and is composed of monomineral banded chalcedony layers associated with quartz, chlorite and calcite minerals.

Petrography of the archaeological samples

The archaeological samples are composed of several flint-like lithologies (micro-crystalline quartz, hydrothermal quartz, banded chalcedony and silicified limestone) followed in smaller proportion by microgranites and acidic volcanic rocks (see Figure 5).

Protomylonitic textures are the most frequent and, except for the pegmatites and quartz veins, the crystal size is fine to very fine (<0.1mm), particularly for the banded chalcedony, silicified limestone material and some of the volcanic and metamorphic materials.

6. Discussion

The acidic composition of the geological and archaeological samples analysed, a consequence of the dominance of varieties of quartz and alkaline feldspars, dictated their aptitude (i.e., concoidal fracturing and hardness) for knapping. The grain size of the archaeological samples was not as fine as expected, suggesting that coarse-grained quartz (i.e., pegmatite quartz) and single crystal were commonly used for knapping. Some authors (Crabtree 1975; Gil Sevillano 1997; Ratto 1988) confirm wear and tear resistance of some varieties of coarse-grained quartz to be desirable for tools to be subjected to rigorous use.

The distinctive textures of dynamic metamorphism of the region (cataclastic to mylonitic) do not appear to have compromised the use of the above materials, since they are well represented in the archaeological sample. However, the use of such resources is generally limited to the manufacture of quickly made, non-standard instruments (Gascue *et al.* 2009a).

The macroscopic description of the samples made during the technological analysis sometimes did not fit the observed petrography (Figure 5). Petrographic analysis provided important information for establishing that the percentage of massive chalcedony (14%) was much more than suggested by macroscopic analysis (see Figure 5, sample iV_7-38). Previous authors have warned that the strong presence of chalcedony at the site reflects an extra-regional resource (Gascue *et al.* 2009a).

The data obtained showed that some archaeological lithic remains identified as chalcedony are in fact other very homogeneous lithologies of excellent quality for knapping, and which are available at a regional scale (e.g., protomylonites). However, the presence of chalcedony in voids in the geological sample of protomylonite from the Atlantic coast (Moza MG9, Figure 4) reinforces the hypothesis that this chalcedony comes from a small regional resource. A resource of excellent quality for knapping, present in the San Miguel Hills and traditionally identified as 'San Miguel Opal' (Curbelo and Martinez 1992), is actually a very fine-grained brecciated quartzite.

The results show that the present analytical method is useful for improving the identification of exploited lithic resources, and suggests that the first occupants of the Rincón de Los Indios site made use of a greater diversity of raw materials than originally identified at the macroscopic level (Gascue *et al.* 2009a). This diversity covers trachytes, pegmatites (igneous) and metamorphic rocks such as mylonites and metapelites.

The results of the petrographic analysis tend to confirm some of the hypotheses expressed earlier regarding the location of sources of some varieties of quartz and quartzite, banded chalcedony and silicified limestone present in the Rincón de Los Indios archaeological record. The geological origin of the ochre component of the lithic materials from the site is not discussed in this paper. However, sesquioxide enrichments can be recognised around stockwork-type outcrops with ferro-siliceous near-extensional plains, and in some residual weathering deposits.

Local and regional mining activities

The great petrographic similarity between samples Moza MG9/iV 7-38, Moza MG7/iV 6-31 and C° Verde MG 8/iV 7-42 (Figures 2, 4 and 5) confirms that the early occupants of the Rincón de Los Indios site undertook displacements to the Atlantic coast to acquire pegmatite quartz and protomylonites.

In turn, the similarity of some geological and archaeological samples of quartzite (PG MG1/iV 6-9 and LN 11/iV MG 7-17; Figure 7) supports the assertion of both local and regional exploitation of this resource in the nearby hills of la Blanqueada, de Los Difuntos and Potrero Grande. Finally, it is confirmed that the rhyolite of San Miguel was truly a resource exploited by the peoples in question (see samples iV 6-29 and SM MG1, Figures 4 and 5).

Extra-regional mining activities

The great petrographic similarity of the sample of silicified limestone from the Río Grande with the entire archaeological sample confirms the extra-regional origin of this resource (Figure 8). However, given the spatial dispersion

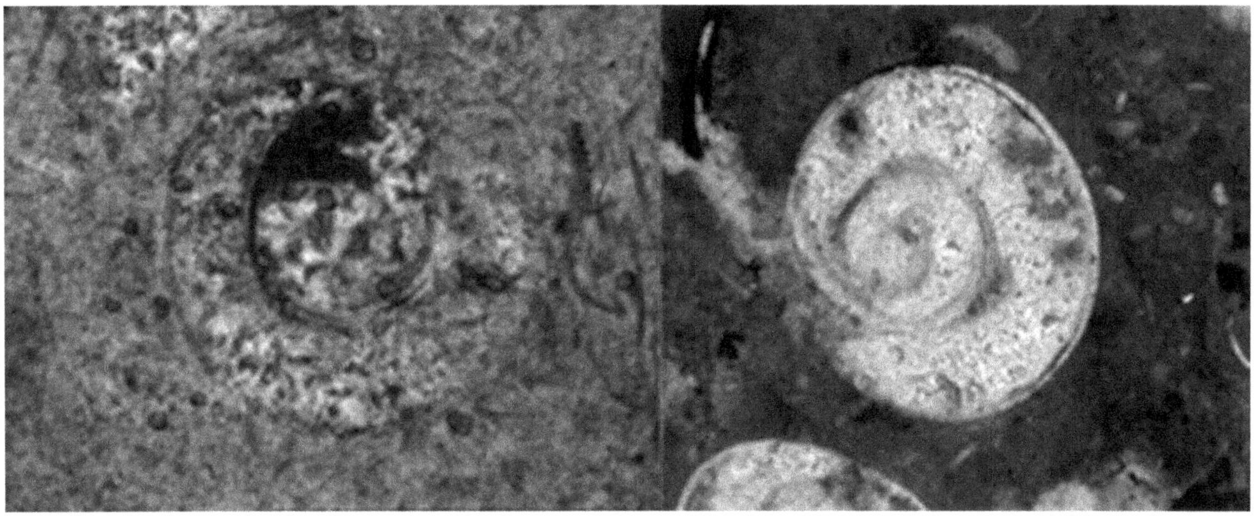

Figure 8. Photomicrograph of silicified reddish limestone showing a section of a pulmonate gastropod shell replaced by chalcedony: A GDE MG 1 sample (left) and iV 6-44 (right). Scale x 100, cross-polarised light.

of this resource (Calizas del Queguay formation) the catchment area may have been the coastal region of the Río Uruguay, the north and south of the Río Negro (Bossi *et al.* 1998; Florines 2004; Gascue 2009; Tófalo and Pazos 2001), and even the Santa Lucia river basin (Martínez *et al.* 1997; Beovide 2004). It is difficult to determine the exact origin for this raw material since the elements considered diagnostic (fossils and mineral composition) are shared by the brecciated facies of the silcretes in the mentioned areas (Martínez et al. 1997; Tófalo and Pazos 2009). Similarly, comparison of the geological sample of banded chalcedony with the three samples recovered from the Rincón de Los Indios site (MG Yaguarí D '/ iV iV 7-16 7-37 7-33 iV) confirms that this raw material comes from northern Uruguay. The nearest source is located in the pebble alluvial beds of the Río Negro (Gascue *et al.* 2009b), although an abundant primary source is located over 400 km away in northern Uruguay and southern Brazil (Arapey Formation; Bossi *et al.* 1998).

7. Conclusions

The lithic material from the oldest level of the Rincón de Los Indios archaeological site represents the dominant pattern of early South American technology and raw material use (Suarez and López Mazz 2003). Early South American knappers may have obtained the good quality raw material they used via networks of exchange. In any event, flint procurement strategies would have had to be combined with residential and economic mobility.

The technological organization of the early inhabitants of eastern Uruguay involved the acquisition of a variety of local, regional and extra-regional mineral raw materials from primary (geological outcrops) and secondary (fluvial and coastal gravels) quarries. These groups undertook significant movements across the region to acquire good quality rocks, showing them to be hunter-gatherers with high residential mobility able to exploit a diversity of environments over the year. They developed a conservative technology (*sensu* Binford 1998) using good quality extra-regional raw materials. The extra-regional silicified limestone and chalcedony brought from the west and north of Uruguay was exclusively used for the production of projectile points and underwent different recycling processes. Currently it remains unknown how these raw materials were circulated. The *bolas* recovered were made from local and regional materials (granite and rhyolite). However, some *bolas* made from raw materials from northern outcrops have been found in early settlements in southern Uruguay.

References

Beovide, L. 2004. Recursos y Organización del espacio prehistórico costero, Uruguay, in L. Beovide, I. Barreto y C. Curbelo (eds.), *X Congreso Uruguayo de Arqueología: La Arqueología Uruguaya Ante Los Desafíos Del Nuevo Siglo. Montevideo, Uruguay.* Montevideo, CD-ROOM edition.

Binford, L. 1998. *En busca del pasado.* Barcelona. Crítica. 3ª ed.

Bossi, J., Ferrando, L., Montaña, J., Campal, N., Morales, H., Gancio, F., Schipilov, A., Piñeiro, G. and Sprechman, P. 1998. *Geocarta. Carta geológica del Uruguay - 1:500.000.* Versión 1.01. Montevideo, Geoeditores S.R.L.

Bracco, R. 1995. Cronología de la Laguna de Castillos. In. J. Bossi, J. Montaña and, R. Bracco, *Causas geológicas del paisaje rochense*, 14-17. Montevideo, Uruguay.

Close, A. 2000. Reconstructing Movement in Prehistory. *Journal of Archaeological Method and Theory* 7 (1), 49-74.

Crabtree, D. E. 1975. Comments on lithic technology and experimental archaeology. In E. Swanson (ed.), *Lithic Technology: making and using stone tools,* 105-114. Chicago, Mouston Publishers, The Hauge.

Curbelo, C. and Martínez, E. 1992. Aprovechamiento de Materias Primas Líticas para un Área Arqueológica Relacionada con la Sierra de San Miguel, Depto. Rocha. *Ediciones del Quinto Centenario* 1, 121-139.

Dillehay, T. 2000. *The settlement of the Americas.* New York, Basics Books.

Florines, A. 2004. Relevamiento Arqueológico de la Localidad Rupestre del Arroyo Chamangá, Flores, in L. Beovide, I. Barreto y C. Curbelo (eds.), *La Arqueología Uruguaya Ante los Desafíos del Nuevo Siglo.* AUA, CD-ROOM Edition, Montevideo.

Gascue, A. 2009. Tecnología Lítica y Patrones de Asentamiento en la Cuenca de Arroyo Grande (Soriano), in J. M. López Mazz y A. Gascue (eds.), *Arqueología Prehistórica Uruguaya en el Siglo XXI*, 133-150. Montevideo, Biblioteca Nacional and Facultad de Humanidades y Ciencias de la Educación.

Gascue, A., López Mazz, J. M., Villarmarzo, E., De León, V., Sotelo, M. and Alzugaray, S. 2009a. La organización de la tecnología lítica de los pobladores tempranos del este de Uruguay. *Intersecciones en Antropología* 10, 63-73.

Gascue, A., Baeza, J. and Bortolotto, N. 2009b. Arqueología del Valle del Río Negro. In *II Jornadas de Investigación*, Montevideo, Facultad de Humanidades y Ciencias de la Educación.

Gil Sevillano, J. 1997. Lithic tool making by Amazonian palaeoindians: a case-study on materials selection. *Journal of materials science letters* 16, 465–468.

Kerr, R. 1965. *Mineralogía Óptica.* Madrid, Ediciones Castilla.

López Mazz, J. 2001. Las estructuras monticulares (cerritos) del litoral atlántico uruguayo. *Latin American Archaeology* 12, 231-255.

López Mazz, J. and Bracco, D. 2010. *Minuanos. Apuntes y notas para la Historia y la arqueología del territorio genoa-minuan (indígenas de Uruguay, Argentina y Brasil).* Montevideo, Linardi y Risso.

López Mazz, J. and Gascue, A. 2007. El valle del arroyo Balizas: Estructuras monticulares y sitios superficiales del litoral atlántico uruguayo. *Cazadores –Recolectores del cono Sur* 2: 89-103.

López Mazz, J., Nadal, O., Suárez, X., De León, V. and Salvo, X. 2009. La gestión regional de los recursos minerales en las tierras bajas del Este: el espacio como variable de la producción lítica, in J. M. López Mazz y A. Gascue (eds.), *Arqueología Prehistórica Uruguaya en el Siglo XXI*, 85-100. Montevideo, Biblioteca Nacional - Facultad de Humanidades y Ciencias de la Educación.

Martin, P. and Suguio, K. 1989. *International Symposium of Global Changes in South America during the Quaternary.* Special Publications 2, 8-12. Sao Paulo.

Martínez, S., Veroslavsky, G. and Verde, M. 1997. Primer registro del paleoceno en el Uruguay: paleosuelos calcáreos fosilíferos en la cuenca de Santa Lucía. *Revista Brasileira de Geociências* 27 (3), 295-302.

Mason, R. 1990. *Petrology of the Metamorphic Rocks.* 2nd ed. New York, Routledge.

Politis, G. 1996. Moving to produce. Nukak mobitity and settlement patterns in Amazonia. *World Archaeology* 27, 492-511.

Politis, G., Messineo, P. and Kaufmann, C. 2004. El poblamiento temprano de las llanuras pampeanas de Argentina y Uruguay. *Complutum* 15, 207-224.

Ratto, N. 1988. Eficacia funcional y técnicas de caza. *Shincal* 1, 32-43.

Sanchez-Bettucci, L., Cosarinsky, M. and Ramos, V. A. 2001. Tectonic Setting of the Late Proterozoic Lavalleja Group (Dom Feliciano Belt), Uruguay. *Gondwana Research* 4 (3), 395-407.

Suárez, R. and López Mazz, J. 2003. Archaeology in the Pleistocene Holocene transition n Uruguay: an overview. *Quaternary International* 109-110, 65-76.

Tófalo, O. R. and Pazos, P. J. 2009. Paleoclimatic implications (Late Cretaceous-Paleogene) from micromorphology of calcretes, palustrine limestones and silcretes, southern Paraná Basin, Uruguay. *Journal of South American Earth Sciences* 29 (3), 665-675.

www.ingramcontent.com/pod-product-compliance
Lightning Source LLC
LaVergne TN
LVHW072121210926
848482LV00028B/724